GIS Approaches for Remote Sensing and Photogrammetry

GIS Approaches for Remote Sensing and Photogrammetry

Editor: Matt Weilberg

R CALLISTO REFERENCE

www.callistoreference.com

Callisto Reference,
118-35 Queens Blvd., Suite 400,
Forest Hills, NY 11375, USA

Visit us on the World Wide Web at:
www.callistoreference.com

ISBN: 978-1-63239-930-4 (Hardback)

Cataloging-in-Publication Data

GIS approaches for remote sensing and photogrammetry / edited by Matt Weilberg.
    p. cm.
Includes bibliographical references and index.
ISBN 978-1-63239-930-4
1. Geographic information systems. 2. Remote sensing. 3. Photogrammetry. I. Weilberg, Matt.
G70.212 .G57 2018
910.285--dc23

Table of Contents

Preface...IX

Chapter 1   Co-Registration of Terrestrial and UAV-Based Images – Experimental Results.......................1
             M. Gerke, F. Nex, P. Jende

Chapter 2   Automatic Orientation and Mosaicking of Archived Aerial Photography using
             Structure from Motion..9
             J. A. Gonçalves

Chapter 3   Exterior Orientation of Hyperspectral Frame Images Collected with UAV for
             Forest Applications...13
             A. Berveglieri, A. M. G. Tommaselli

Chapter 4   Precision Analysis of Point-And-Scale Photogrammetric Measurements for
             Corridor Mapping: Preliminary Results...19
             P. Molina, M. Blázquez, J. Sastre, I. Colomina

Chapter 5   Need for Reliable Sensor Calibration from the Perspective of a National
             Mapping Agency..25
             S. Baltrusch

Chapter 6   Odometry and Low-Cost Sensor Fusion in TMM Dataset..29
             A. M. Manzino, C. Taglioretti

Chapter 7   Spatial Data Quality and a Workflow Tool..37
             M.Meijer, L.A.E. Vullings, J.D. Bulens, F.I. Rip, M. Boss, G. Hazeu, M.Storm

Chapter 8   Towards Efficiency of Oblique Images Orientation...42
             W. Ostrowski, K. Bakuła

Chapter 9   Change Detection and Land Use / Land Cover Database Updating using Image
             Segmentation, GIS Analysis and Visual Interpretation...48
             Jean-François Mas, Rafael González

Chapter 10   Trajectory Adjustment of Mobile Laser Scan Data in GPS Denied Environments...................53
             P. Schaer, J. Vallet

Chapter 11   Low-Level Tie Feature Extraction of Mobile Mapping Data (MLS/Images) and
             Aerial Imagery..57
             P. Jende, Z. Hussnain, M. Peter, S. Oude Elberink, M. Gerke, G. Vosselman

Chapter 12   Calibrating Cellular Automata of Land Use/Cover Change Models using a
             Genetic Algorithm...65
             J.F. Mas, B. Soares-Filho, H. Rodrigues

Chapter 13   **Study of Lever-Arm Effect using Embedded Photogrammetry and On-Board GPS Receiver on UAV for Metrological Mapping Purpose and Proposal of a free Ground Measurements Calibration Procedure**..69
M. Daakir, M. Pierrot-Deseilligny, P. Bosser, F. Pichard, C. Thom, Y. Rabot

Chapter 14   **Sub-Camera Calibration of a Penta-Camera**...75
K. Jacobsen, M. Gerke

Chapter 15   **Autonomous Navigation of Small UAVs based on Vehicle Dynamic Model**.............81
M. Khaghani, J. Skaloud

Chapter 16   **Role of Tie-Points Distribution in Aerial Photography**...87
S. Kerner, I. Kaufman, Y. Raizman

Chapter 17   **Comparing National Differences in what People Perceive to be *There*: Mapping Variations in Crowd Sourced Land Cover**...91
A. Comber, P. Mooney, R. S. Purves, D. Rocchini, A. Walz

Chapter 18   **The EnMAP Contest: Developing and Comparing Classification Approaches for the Environmental Mapping and Analysis Programme – Dataset and First Results**.............96
A. Ch. Braun, M. Weinmann, S. Keller, R. Müller, P. Reinartz, S. Hinz

Chapter 19   **A Semi-Automatic Procedure for Texturing of Laser Scanning Point Clouds with Google Streetview Images**..103
J. F. Lichtenauer, B. Sirmacek

Chapter 20   **Automatic Extraction and Topology Reconstruction of Urban Viaducts from Lidar Data**..109
Yan Wang, Xiangyun Hu

Chapter 21   **Semantic Interpretation of InSAR Estimates using Optical Images with Application to Urban Infrastructure Monitoring**...114
Yuanyuan Wang , Xiao Xiang Zhu

Chapter 22   **Integration of Remote Sensing Data and basic Geodata at different Scale Levels for Improved Land use Analyses**...122
G. Waldhoff, S. Eichfuss, G. Bareth

Chapter 23   **A Task-Driven Disaster Data Link Approach**..127
L.Y. Qiu, Q. Zhu, J. Y. Gu, Z.Q. Du

Chapter 24   **3D GIS based Evaluation of the Available Sight Distance to Assess Safety of Urban Roads**..135
M. Bassani, N. Grasso, M. Piras

Chapter 25   **Integrated Estimation of Seismic Physical Vulnerability of Tehran using Rule Based Granular Computing**..142
H. Sheikhian, M. R. Delavar, A. Stein

Chapter 26   **A Graph based Model for the Detection of Tidal Channels using Marked Point Processes**..149
A. Schmidt, F. Rottensteiner, U. Soergel, C. Heipke

Chapter 27   **Assessing Modifiable Areal Unit Problem in the Analysis of Deforestation Drivers using Remote Sensing and Census Data**..156
J.F. Mas, A. Pérez Vega, A. Andablo Reyes, M.A. Castillo Santiago,
A. Flamenco Sandoval

Chapter 28   **Development and Testing of Geo-Processing Models for the Automatic Generation of Remediation Plan and Navigation Data to use in Industrial Disaster Remediation**..160
G. Lucas, Cs. Lénárt, J. Solymosi

Chapter 29   **Change Detection based on Persistent Scatterer Interferometry - Case Study of Monitoring an Urban Area**..167
C.H. Yang, U. Soergel

Chapter 30   **Generalisation and Data Quality**..175
N. Regnauld

Chapter 31   **Image based Recognition of Dynamic Traffic Situations by Evaluating the Exterior Surrounding and Interior Space of Vehicles**..179
A. Hanela, H. Klöden, L. Hoegner, U. Stilla

Chapter 32   **Simple Approaches to Improve the Automatic Inventory of Zebra Crossing from MLS Data**..187
P. Arias, B. Riveiro, M. Soilán, L. Díaz-Vilariño, J. Martínez-Sánchez

Chapter 33   **Development of a Cartographic Strategy and Geospatial Services for Disaster Early Warning and Mitigation in the ECOWAS Subregion**..193
L. A. Gueye, M. S. Keita, J. O. Akinyede, O. Kufoniyi, G. Erin

Chapter 34   **Installed base Registration of Decentralised Solar Panels with Applications in Crisis Management**..200
Rosann Aarsen, Milo Janssen, Myron Ramkisoen, Filip Biljecki, Wilko Quak,
Edward Verbree

Chapter 35   **Location-Based Infrastructure Inspection for Sabo Facilities**..205
M. Nakagawa, T. Yamamoto, S. Tanaka, Y. Noda, K. Hashimoto, M. Ito,
M. Miyo

Permissions

List of Contributors

Index

Preface

It is often said that books are a boon to mankind. They document every progress and pass on the knowledge from one generation to the other. They play a crucial role in our lives. Thus I was both excited and nervous while editing this book. I was pleased by the thought of being able to make a mark but I was also nervous to do it right because the future of students depends upon it. Hence, I took a few months to research further into the discipline, revise my knowledge and also explore some more aspects. Post this process, I begun with the editing of this book.

Geographic Information Systems (GIS) use satellites to gather relevant data and information for accurate representations of the Earth and its geographic features. It provides a simple solution so as to sort, classify and monitor the various environmental and geographical (natural and human) processes that are occurring in the world. GIS often rely on maps of various scales to interpret new information. Precise GIS models have diverse industrial applications. This book on GIS approaches for remote sensing and photogrammetry will help new researchers by foregrounding their knowledge in this branch. It aims to serve as a resource guide for students and experts alike and contribute to the growth of the discipline.

I thank my publisher with all my heart for considering me worthy of this unparalleled opportunity and for showing unwavering faith in my skills. I would also like to thank the editorial team who worked closely with me at every step and contributed immensely towards the successful completion of this book. Last but not the least, I wish to thank my friends and colleagues for their support.

Editor

1

CO-REGISTRATION OF TERRESTRIAL AND UAV-BASED IMAGES – EXPERIMENTAL RESULTS

M. Gerke, F. Nex, P. Jende

University of Twente, Faculty of Geo-Information Science and Earth Observation (ITC),
Department of Earth Observation Science, The Netherlands {m.gerke,f.nex, p.l.h.jende}@utwente.nl

KEY WORDS: Photogrammetry, Orientation, Data, Integration, Multisensor, Accuracy, Estimation

ABSTRACT:
For many applications within urban environments the combined use of images taken from the ground and from unmanned aerial platforms seems interesting: while from the airborne perspective the upper parts of objects including roofs can be observed, the ground images can complement the data from lateral views to retrieve a complete visualisation or 3D reconstruction of interesting areas. The automatic co-registration of air- and ground-based images is still a challenge and cannot be considered solved. The main obstacle is originating from the fact that objects are photographed from quite different angles, and hence state-of-the-art tie point measurement approaches cannot cope with the induced perspective transformation. One first important step towards a solution is to use airborne images taken under slant directions. Those oblique views not only help to connect vertical images and horizontal views but also provide image information from 3D-structures not visible from the other two directions. According to our experience, however, still a good planning and many images taken under different viewing angles are needed to support an automatic matching across all images and complete bundle block adjustment. Nevertheless, the entire process is still quite sensible – the removal of a single image might lead to a completely different or wrong solution, or separation of image blocks.
In this paper we analyse the impact different parameters and strategies have on the solution. Those are a) the used tie point matcher, b) the used software for bundle adjustment. Using the data provided in the context of the ISPRS benchmark on multi-platform photogrammetry, we systematically address the mentioned influences. Concerning the tie-point matching we test the standard SIFT point extractor and descriptor, but also the SURF and ASIFT-approaches, the ORB technique, as well as (A)KAZE, which are based on a nonlinear scale space. In terms of pre-processing we analyse the Wallis-filter. Results show that in more challenging situations, in this case for data captured from different platforms at different days most approaches do not perform well. Wallis-filtering emerged to be most helpful especially for the SIFT approach. The commercial software pix4dmapper succeeds in overall bundle adjustment only for some configurations, and especially not for the entire image block provided.

1. INTRODUCTION

Multiplatform image data is very interesting for many applications. Unmanned Aerial Vehicles (UAV) are getting more mature and fully automatic processing workflows are in place which help turning the image set into point clouds or more advanced products. At the same time and due to the availability of easy-to-use end-user software also hand-held cameras are used by researchers from a variety of disciplines to model objects. Examples are as-is-modelling of buildings, archaeology/cultural heritage, cadastre/city modelling. In order to model the outer faces of buildings entirely, with great detail, and with a minimum amount of occlusions, the object should be photographed from many different viewpoints. Those should be at different heights and enclosing a variety of angles with the object. In case of complex architectures such as intrusions, extrusions (like balconies), a UAV can offer favourable viewpoints to avoid or minimize occlusions. In addition the roof should be captured from conventional nadir-looking views.
The processing pipelines proposed and implemented in research and commercial products work well, especially when the following conditions are met:

- Approximate position and viewing direction of cameras are known: to more reliably find matching mates for each image and exclude unlikely matches.
- Sequence of image acquisition resembles overlapping configuration (adjacent images also have similar time stamp)
- Viewing direction change between overlapping images is small (i.e. perspective distortion is small): because most key point descriptors, or image matchers, are not invariant with respect to large perspective distortions.

- Lighting conditions for overlapping images are similar: again, some key point descriptors are sensitive to global grey value distribution changes
- Object does not show repetitive patterns: similarity between areas in the object will lead to wrong matches. Especially in buildings with symmetrical façade-object arrangements (windows/doors) this is a problem.

In practice, however, an image block configuration might not be perfect: GPS which is used to estimate approximate camera location might compute largely wrong positions because of signal obstruction and multipath effects in urban canyons, or close to buildings in general. In addition, when several image blocks, taken from different platforms are merged, the valuable adjacency information from the image capture time gets lost, at least for matches between the blocks.
When object planes enclose large angles – like right-angled building facades – special attention has to be taken when images are captured to ensure that the image matcher still finds enough valid tie points in adjacent images. The lighting might become problematic, as well. First, when we do not have diffuse light, parts of the building may be covered by strong shadow, e.g. casted by extrusions of the building itself. Second, depending on solar elevation angle, building height and camera height, glare might lead to underexposure of building details, and third, when images from different sensors are matched. Repetitive patterns which are standard in many architectural designs are a problem especially when images at very high resolution are used: single shots then only cover parts of the façade and the context might get lost completely: when a single window is photographed on two images it might not be possible to decide whether this is actually the same or just another window of the same type.

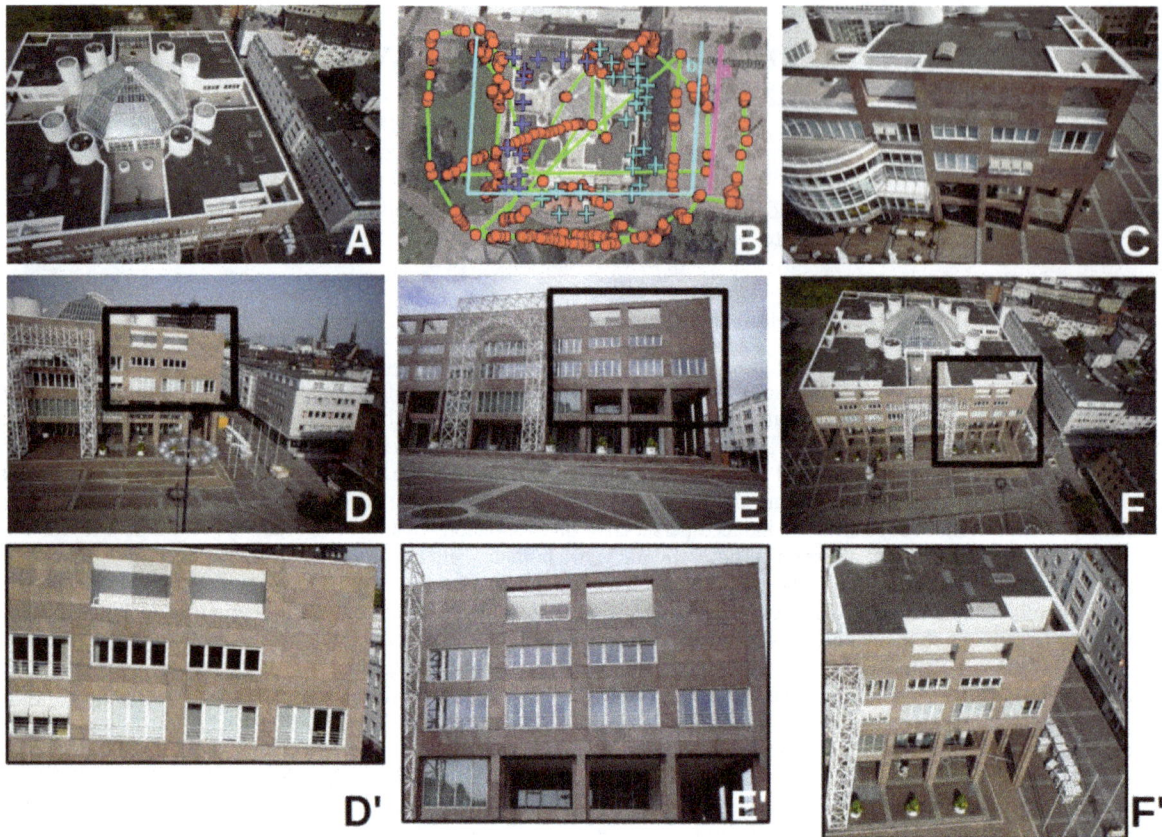

Figure 1: Overview on benchmark dataset. A,C,D,F: UAV images from different locations and under different angles, B: GPS-based image position approximations overlaid on google earth view (red dots), GCP/CPs (crosses), magenta and turquoise lines: used facades for software test, section 4, subset a and b, respectively, E: terrestrial image. D', E', F': cut outs of the respective image.

A challenging dataset, composed of terrestrial and UAV-based images is provided to the research community in the framework of the multi-platform photogrammetry benchmark (Nex et al., 2015), which is supported by the ISPRS scientific initiative 2014-2015[1].

One of the released datasets is a combined UAV/terrestrial image block, and the task for participants is to co-register the images and fine adjust the bundle block using ground control points provided. In Fig. 1 some overview is given including sample images. Fig. 1 B) shows the approximate positions of images around the centre building (municipality hall in Dortmund, Germany). In the northern part, where images were captured in a narrow street, the positions are largely off the right location. Images A,D,E,F show the same part of the building from different perspectives. Although the same camera (Sony Nex 7) was used for both – terrestrial image captures and on board the UAV – the colour and grey value distribution is quite different, in particular between terrestrial shots (E) and the UAV-based images. In addition, the sky/clouds are reflected in the windows in the terrestrial shots. During the terrestrial acquisition the weather was very bad – the sky was cloudy and the campaign was interrupted frequently by heavy rain. During that day it was not possible to operate the UAV. Therefore, the acquisition has been conducted about 4 weeks later. Then the weather was very good, led to no diffuse light and clear sky (compare D and E). As far as the building architecture is concerned the window and other elements do not show large variations, but large repetitions. Note also that image C) shows

another façade of the building, but actually the window elements and their arrangement are similar to the façade shown in A,D,E,F.

In this paper we present experiments and their results focussing on the issues: how do current state-of-the-art tie point matching algorithms perform on this dataset and which influence does image pre-processing have? To this end, several image combinations (same platform/across platform) are matched. We perform outlier removal based on a RANSAC approach exploiting the epipolar constraint in stereo matches, using the essential matrix. The Wallis filter (Wallis, 1976) was applied in a separate experiment. Apart from testing different stereo matching techniques we analyse the performance of one software package. The entire structure-from-motion workflow including image matching and bundle adjustment is tested. The solution offered by pix4d (pix4dmapper) is used for those tests.

2. RELATED WORK

In literature we find a multitude of approaches to tie point extraction, description and matching. The aim of this paragraph is not to give a comprehensive overview, the interested reader might want to refer to (Dahl et al., 2011, Levi and Hassner, 2015) for some general overview. Urban and Weinmann (2015) tested state-of-the-art key point extractors and detectors in the context of co-registration of terrestrial laser scans. To this end the authors compared the keypoint extractors SIFT, SURF, ORB and A-KAZE on depth-images derived from the single scans. Used approaches are of a different type in the sense that ORB is based on corner detection, and the others extract blobs in the scale space. As far as keypoint descriptors are concerned

[1]http://www2.isprs.org/commissions/comm1/icwg15b/benchma rk/data-description-Image-orientation.html

we can distinguish between gradient-based descriptors, finally encoded in floating numbers, and descriptors computed from intensity differences and encoded as binary strings. The latter one is supposed to be more computational efficient, but many of those are known to be more sensitive to noise. In addition to the aforementioned key point extractors/detectors we add the SIFT-based ASIFT and the KAZE approaches when we evaluate the performance of the methods for the benchmark dataset. One pre-requisite for all used descriptors is that they are both, scale and rotation invariant.

2.1 SIFT/ASIFT

The Scale Invariant Feature Transform SIFT (Lowe, 2004) became a standard in computer vision and photogrammetry. It works in scale space which is derived by image convolution with a Gaussian kernel. Extrema in the DoG (Difference-of-Gaussian) constitute keypoint candidates. After removing candidates along edges or in low contrast regions, a higher order function is fitted to derive sub-pixel accuracy. Scale invariance is achieved implicitly through the realisation of a scale pyramid. The keypoint descriptor is derived by computing gradient histograms in all directions. The area around the point is subdivided into 4x4 regions and in each region the orientation histogram is computed in 8 angular bins. Those 4x4x8 bins are concatenated and stored as a 128-dimensional descriptor, along with the dominant scale. Since the main gradient direction is derived as well and rotations are normalised accordingly, the descriptor is rotation invariant.
Morel and Yu (2009) extended the SIFT approach in order to achieve invariance under affine image transformations (ASIFT-affine SIFT). To this end, the images are stepwise rotated around both axis. For each of such re-projections SIFT points are computed and described.

2.2 SURF

In contrast to SIFT, the Speeded-Up Robust Features SURF detector (Bay et al., 2008) does not work with the DoG, but with the Hessian matrix in the scale space pyramid. Local maxima of the matrix determinant in image- and scale space constitute candidates for keypoints. Similar to SIFT, uncertain hits are removed and accuracy is increased through sub-pixel interpolation. Rotation invariance of the descriptor is also derived by first computing the dominant gradient direction. Image subdivision in 4x4 regions is also done similar to SIFT, but in this case Haar wavelets are computed to describe the local gradients in the frequency domain. Four descriptors per sub region are computed, leading to 4x4x4=64 entries per point.

2.3 ORB

The full name of the acronym ORB (Rublee et al, 2011), namely Oriented FAST and rotated BRIEF, already tells that this is a combination of an improved version of the FAST feature detector (Rosten and Drummond, 2005) and the rotational invariant BRIEF (Calonder et al., 2010) descriptor.
The FAST (Features from Accelerated Segment Test) descriptor finds keypoints basically by comparing intensity differences around each pixel in the image of interest, where the pattern of tests has a circular shape. The two main extensions within ORB are that those detections are done in scale space, i.e. adding scale invariance, and that rotation angle information is added.
As far as keypoint description is concerned, the authors of ORB added rotation invariance and unsupervised learning to select an ideal pairing of pixel samples to the BRIEF descriptor.

2.4 KAZE/A-KAZE

One basic idea behind the KAZE (Alcantarilla et al., 2012) point extractor is to use a non-linear scale space to enable scale invariance. Compared to the Gaussian scale space derivation as done by other approaches, a non-linear scale space, in this case realized by nonlinear diffusion filtering, preserves edges while reducing noise at the same time. In terms of computational time, however, it is reported that the employed additive operator splitting (AOS) is quite inefficient. Therefore, in Alcantarilla et al. (2013) the authors propose A-KAZE, where the A stands for accelerated. Here, the fast explicit diffusion (FED) is used to compute the scale representation.
Similar to SURF, in both cases, the Hessian matrix is employed to find salient points. For KAZE, point description is undertaken by a modified variant of SURF (M-SURF, Agrawal et al., 2008) adapted to the non-linear scale space. AKAZE, however, utilizes a binary description based on a modified version of the Local Difference Binary method proposed by Yang and Cheng (2012).

2.5 Matching and inlier filtering

In order to match keypoints for each candidate in an image the closest mate in terms of descriptor-space distance is searched for. This descriptor-based matching is then followed by the so-called ratio test. The distance in descriptor space between the best and second best match is computed, and the ratio should not exceed a certain threshold. In this way outliers can be removed since the assumption is that also in the descriptor space a valid match should not be isolated.
In order to enhance filtering inliers, a two-step approach is pursued, exploiting the perspective camera model: since the camera models are known for the benchmark description, a RANSAC-based filtering is done using an estimation of the essential matrix E. To this end, the image coordinates are first normalized employing the camera calibration matrix K (Hartley and Zisserman, 2008). The essential matrix can be computed using the 5-point algorithm by Nistér (2004) or Li and Hartley (2006). We used the MATLAB implementation provided by the latter. In a second step the filtered points are again processed by a RANSAC-based filtering using the fundamental (F)-matrix estimation. We found out that using only F-matrix-based filtering in cases with a large number of outliers might not converge. In those cases the geometric constraints imposed by the essential matrix help. However, since the camera calibration might not be known too well, we also observed that still a significant number of outliers are present afterwards. Therefore, an F-matrix-based filtering applied to the remaining matches reduces the number of outliers considerably.

2.6 Image enhancing prior to point extraction

As far as image pre-processing is concerned Jazayeri and Fraser (2008) reported that an image enhancement with the Wallis filter (Wallis, 1976) helped to significantly improve corner point detection. Therefore, after testing the tie point matching with the original images we will perform the same with Wallis-filtered images on selected pairs. This filter is an adaptive contrast filter, working in local windows. In contrast to many other global filters, image details will remain and contrast and brightness is balanced over the entire image.

3. TIE POINT MATCHING

In order to test the 6 point extractor/descriptor combinations for the benchmark test data, a setup has been defined as follows.

We selected 10 different image combinations which reflect all the challenges we are facing in this dataset as mentioned in section 1. In Figures 2 and 3, the four selected terrestrial and UAV images, respectively, are shown.

Figure 2: Terrestrial images used for the tests

3.1 Stereo pair matching in original images@25%

The images are resampled to 25% of the original size. This is simply done for practical reasons: the computation time we need to perform all the experiments can be reduced. We may assume that the number of matches is smaller compared to a higher resolution, so the visualization of matches is better legible. Anyhow, we believe that this reduction of resolution is valid since we are only interested in the relative performance between the approaches, given several typical stereo image combinations.

However, in a later step we will show experiments with full resolution images for selected stereo pairs. We also add results from matching of pre-processed images, in particular after Wallis filtering.

All experiments have been conducted with the Matlab/OpenCV[2] implementation of the approaches used. An exception is ASIFT for which we use the C-implementation provided by the authors (Morel and Yu, 2009).

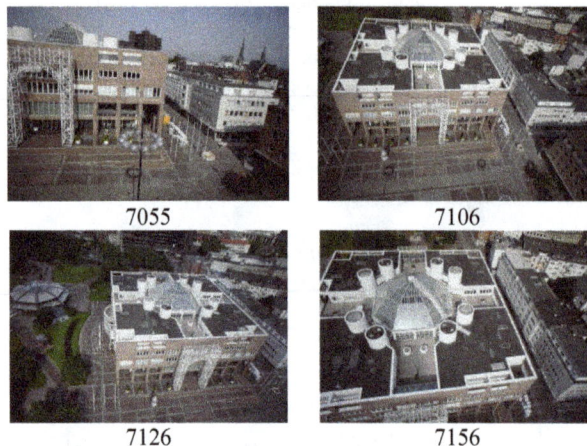

Figure 3: UAV images used for the tests

Pair 1 – within terr-1: 2315- 2342 (standard)

In Fig 4., the upper part the matches of SURF are displayed while in the lower part for each approach the number of inlier

matches after E-/and F-based filtering (blue) is shown and compared to the real number of inliers (from visual inspection). Compared to all other pairs, this is the simplest case for matchers: a similar scale, similar viewing direction, just small baseline, same time of capture, i.e. no illumination changes. In terms of the number of reliable matches, the SURF approach outperforms all the others; however, all methods would give enough reliable matches for practical applications.

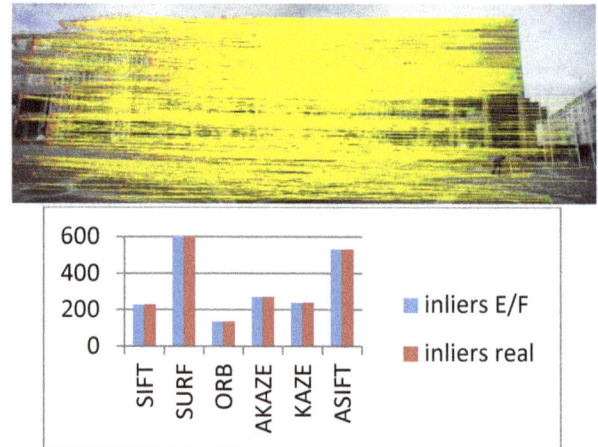

Figure 4: pair-1 results. Upper: SURF matches, lower: for each number of matches after E-/and F-based filtering (blue) and real inliers (red)[3]

Pair 2 – within terr-2: 2392-2342 (perspective/rotation)

This pair is a bit more challenging than the first one in two respects: the viewpoint and image rotation changes in a way to include the 2nd facade. The rotation is quite small, though. In addition the orientation of one camera changes from landscape to portrait mode. This is done in close range projects for two reasons. Sometimes the opening angle of the camera in landscape mode does not allow for acquiring the façade in the entire vertical direction. Additionally, by rotating the camera by 90° the self-calibration of interior parameters is supported, in particular the estimation of the principal point.

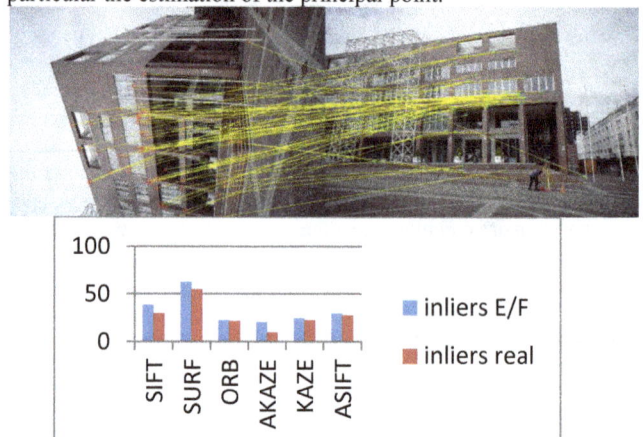

Figure 5: pair-2 results. Upper: SURF matches, lower: for each number of matches after E-/and F-based filtering (blue) and real inliers (red)

The result (see Figure 5) show that all matchers yield much fewer valid tie point connections compared to pair 1. Again

---

[2] OpenCV 3.0, including the API MEXOpenCV to use OpcnCV functions in Matlab: https://github.com/kyamagu/mexopencv

[3] A visualization of matches from all six approaches for all 10 matching pairs is available as additional material on research gate, see http://www.researchgate.net/profile/Markus_Gerke

SURF delivers most inliers, but especially the A-KAZE inliers might be too few in a practical setup.

Pair 4 – within UAV-1: 7106 -7126 (standard)
In this case, we test a pair similar to pair 1 in the sense that it resembles a simple situation: Similar viewing direction of images, with another viewpoint shifted in façade direction. Similar to pair 1, SURF and ASIFT provide the most matches, but SURF only achieved a comparable result to SIFT (Fig. 6).

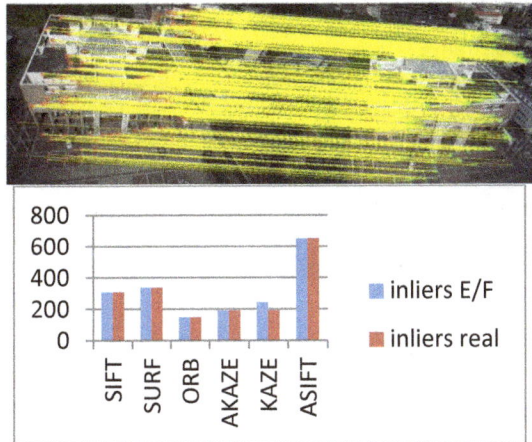

Figure 6: pair-4 results. Upper: ASIFT matches, lower: for each number of matches after E-/and F-based filtering (blue) and real inliers (red)

All other matchers provide a similar number of valid matches as in the example of pair 1. Another remarkable observation is that the ORB and A-KAZE matches are not as well distributed as the matches from the other methods

Pair 5 – within UAV-2: 7106-7055 (oblique- horizontal)
This pair is typical for UAV image blocks in the context of 3D building modelling: the camera is tilted to include different angles with the nadir direction. In this way, horizontal views (for façades) can be connected to vertical views (for the roof). In addition more complex object structures like in- or extrusions can be modelled since occlusions get minimized.

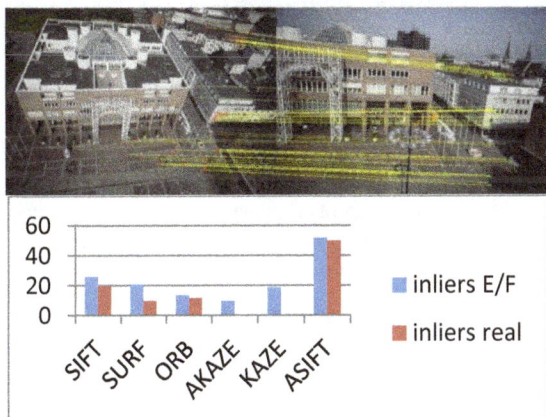

Figure 7: pair-5 results. Upper: ASIFT matches, lower: for each number of matches after E-/and F-based filtering (blue) and real inliers (red)

Again, the ASIFT result is significantly better than the others (Fig. 7). While standard SIFT still yields around 50% of the number of ASIFT matches, the number of real inliers from the others is much less. Especially (A)KAZE do not deliver valid matches at all.

Pair 6 – within UAV-3: 7106-7156 (oblique-vertical)
Again, this dataset is typical for 3D building modelling, like pair 5, but this time a nadir view is connected to the slanted one.

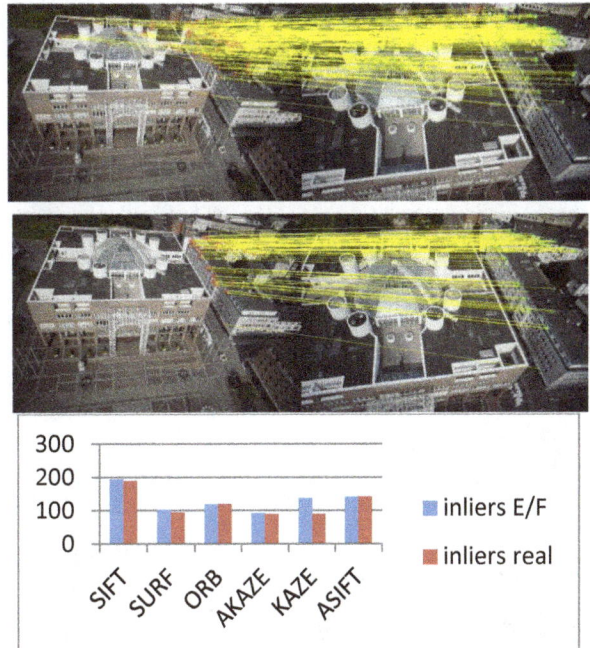

Figure 8: pair-6 results. Upper: SIFT, middle: ORB, lower: for each number of matches after E-/and F-based filtering (blue) and real inliers (red)

Although the nature of perspective transformation imposed by the different camera nick is similar as in the previous case, here many more matches in general can be found – all methods yield around 100 valid matches, around double compared to SIFT. Remarkably SIFT delivers more matches than ASIFT. However, it is also visible from the match visualization (Fig. 8) that the distribution of matches from (A)SIFT is much better than that one from the other approaches – ORB, AKAZE and KAZE just return matches in a very small area of the scene.

Pairs 3 and 7 – within terr-3: 2315 -2793 and within UAV-4: 7055-7336 (wrong pair)
The object arrangement is quite similar to pair 2 and pair 5, respectively. However, the images of the pair show different parts of the building, so *each match is a false* match. Therefore in the bar chart (Fig. 9) the "real" inliers are omitted since there cannot be any real inliers.

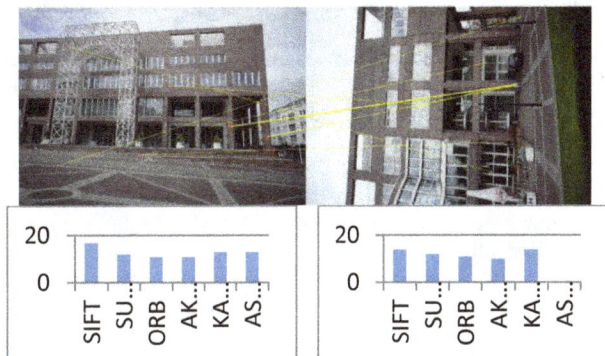

Figure 9: pair-3 results. Upper: SIFT matches for terrestrial wrong matches, lower: for each number of matches after E-/and F-based filtering (blue) (left: terrestrial, pair-3, right: UAV, pair-7)

Although all extractor/descriptor combinations deliver only a few inlier matches after E-/F-filtering, it is remarkable that for instance AKAZE does not produce fewer matches compared to pair-2 where a valid image combination was used. When the matches are visualized (Fig. 6), one can see that most of the matches are at window frame corners, and columns. For the airborne case, however, ASIFT-based matching did not result in inliers.

Pair 8 – across terr/UAV-1: 2315-7055 (similar viewing direction)
From a geometrical point of view this pair is standard and thus comparable to pair 1. The building façade is photographed with the same camera, from the same direction, only the UAV took a higher altitude as the terrestrial shot. However, this pair is special in the sense that the illumination of the scene is quite different, see the description of the acquisition campaign in section 1.

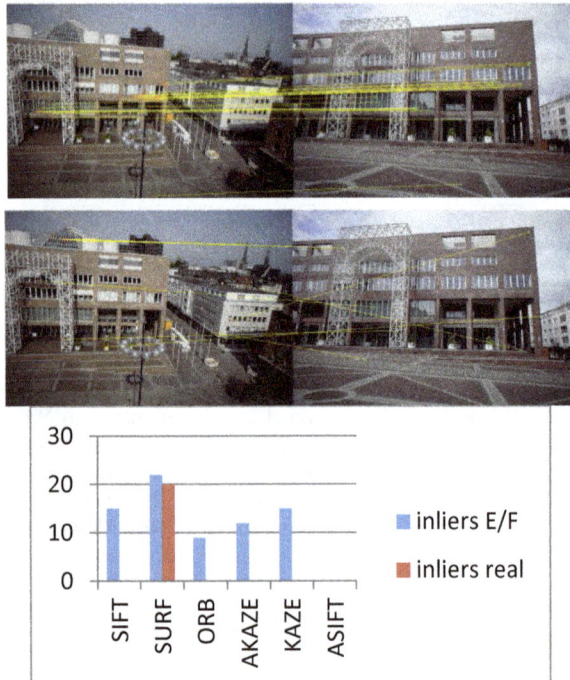

Figure 10: pair-8 results. Upper: SURF, middle: SIFT, lower: for each number of matches after E-/and F-based filtering (blue) and real inliers (red)

Achieved results are poor, cf. Fig. 10: although in all cases inliers remain after RANSAC-based filtering, all of them are invalid, except for the SURF results which are still acceptable. However, with about only 20 inliers, this result is worse compared to pair 1 (more than 600).

Pair 9 – across terr/UAV-2: 2315-7106 (horizontal- oblique)
This combination is similar to pair 5 (within UAV-2): a horizontal view is combined with a slanted view from the air. Here, basically all matches fail. Almost all remaining inliers after RANSAC filtering (10 to 15) are actually wrong matches. When comparing to the previous pair, this result is reasonable since the geometric transformation imposed by the camera tilt makes the task not easier, and in addition the images show the same unfavourable radiometric differences as in pair 8.

Pair 10 – across terr/UAV-3: 2315-7126 (horizontal-oblique)
The idea for this combination is the same as for pair 9: a classical terrestrial view is combined with a slanted oblique view. This pair, however, is even more challenging than the previous one since the common area is more towards the background of the UAV image. Hence, the scale difference between both images is quite large. As one could expect, no approach produced usable results.

3.2 Alternative setups

In order to experimentally analyze the impact of image rescaling and Wallis filtering, we performed the same experiments with a selected pair (pair 8) again. We selected pair 8 because on the one hand it seems to be challenging as the two images were captured by the two different platforms at different days, but on the other hand - from a geometrical point of view - it should not be too difficult (see description above).

Full resolution matching
In practical projects it is important to reach the highest geometric accuracy, therefore it is advised to match full resolution imagery. In our experiments, however, we found out that – at least in the used setups and implementation – there is no significant improvement in terms of real inliers. Although the number of keypoints grows almost linearly with the image resolution, the absolute number of inliers, i.e. not the inlier ratio, remains somewhat stable for all approaches.

Wallis filtering
After Wallis filtering, the matching result for pair 8 improved significantly, but only for SIFT matching. While SIFT does not produce a single real inlier in the original images (cf. Fig. 10, middle), it results in more than 100 real inliers in the Wallis-filtered images, see Fig. 11.

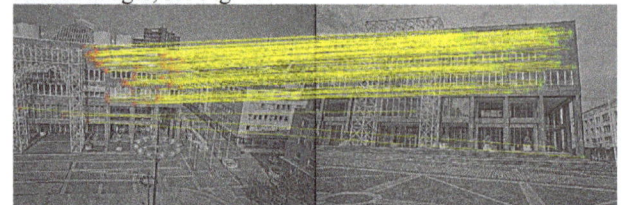

Figure 11: pair-8 results: SIFT matching in Wallis-filtered images.

3.3 Summary of tie point matching results

The results give some interesting insights to the performance of the selected tie point extractors and matchers. Since in terms of outlier removal and image scale we used the same settings for all approaches, we can at least compare the results relatively to each other. For the standard case, namely to match images taken from the same sensor and platform, and showing similar illumination conditions, all matchers perform satisfactorily, see pairs 1 and 4. Under certain image transformations other than similarity, most approaches show difficulties. In those cases ASIFT and SURF performed best, but also SIFT works well, especially when nadir UAV images are combined with slanted UAV views. When it comes to image pairs across platforms, and in this case being taken under quite different illumination conditions, the number of inliers drops drastically. Only in the simplest combination, when the geometry of image acquisition is similar (pair 8), a fairly decent result was obtained from

SURF only. We might suspect that the employed Hessian matrix is less sensitive to large illumination changes.

After Wallis-filtering, only the SIFT result improved significantly – from no real inliers in the original image to around 100 in the Wallis-case. To use the full resolution of the original image, in turn, does not lead to an improvement of descriptor-based matching.

4. STRUCTURE-FROM-MOTION TESTS

When end-users are working with software packages they often have no influence on the entire workflow. Especially the keypoint extraction and description is normally hard coded and the algorithm behind is not disclosed. We tested the initial performance of such software, in particular with the challenging data of this benchmark. To this end, the pix4dmapper by Pix4d[4] has been analysed. For all experiments we did not use the ground control point information provided in the benchmark dataset in order to be independent from external tie information.

We undertook the following experiments with this software. First, we defined three sub-datasets, considering only terrestrial or UAV images as well as both datasets:
a) all images showing only one façade of the building (East façade, the same side as used for the tie point matching experiments);
b) three façades: all terrestrial images except for North side where the largest problem with initial GPS location is observed. This set, however, includes already all UAV images;
c) entire dataset.
The location of subsets for a) and b) are also indicated in Figure 1, B).

By analyzing subset a), we can focus only on the across-platform matching and bundle adjustment performance since the approximate GPS location is reasonable for all images. Also for subset b), the GPS locations are good, but the geometry of buildings is more challenging since multiple oriented façades plus roof and ground planes are involved. The full dataset (subset c)), finally, is the most challenging since the poor GPS locations from the terrestrial images showing the North façade are included. For all subsets we performed several software runs: only terrestrial, only UAV, combination terrestrial and UAV. This was done twice: on original and Wallis-filtered images, and on the first pyramid (half image resolution) only.

Configuration	original	Wallis
UAV only		
UAV_set a) – East Facade	☑	☑
UAV_set c) – all	☑	☑
Terrestrial only		
Terrestrial_set a) – East Facade	☑	☑
Terrestrial_set b) – East, South, West	☑	☒
Terrestrial_set c) – all	☒	☒
Combinations		
UAV_set a) and Terrestrial_set a)	☒	(☑)
UAV_set c) and Terrestrial_set b)	☒	☒
UAV_set c) and Terrestrial_set c)	☒	☒

Table 1: Experimental results with pix4dmapper on several image block subsets.

[4] http://www.pix4d.com

Table 1 gives an overview on the results obtained. The tick mark indicates that all images are adjusted in one connected bundle block. The cross mark is used when the block got divided into sub blocks. There is one case (first combination with Wallis-filtered images), where pix4dmapper resulted in a stable block, but 5 UAV images got excluded from the solution.

The software had no problems when only the UAV images were used. The quality report reveals that images are well connected, see Figure 12. The results shown are from set c (all UAV images), using the original images. Using the Wallis-filtered images yields a similar result.

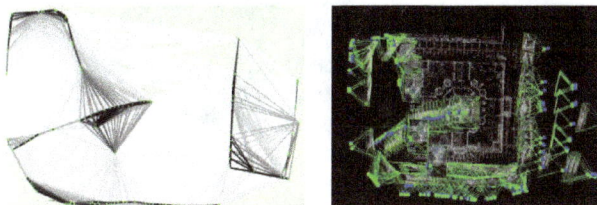

Figure 12: Image connectivity (left), tie point cloud and camera locations (right), all UAV images

The terrestrial image set was solved without bigger problems as expected in the one-façade-only case (set a). The next challenging set b was only solved as one image block when the original images got employed. The entire terrestrial image block, however, (set c) was not adjusted successfully at all. This observation might support the assumption that good GPS location observations are necessary to support the entire matching and adjustment process.

A typical block configuration is shown in Figure 13: the terrestrial block got separated; in particular the images from the North façade are not connected to the others.

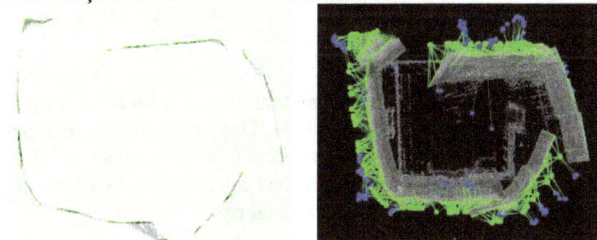

Figure 13: non-connected terrestrial image-block, Wallis-filtered images

For the combined UAV- and terrestrial image configurations, only the most simple one (set a) got solved, but only for the Wallis-filtered images and even in that case some cameras are not included in the block, cf. Fig. 14. This observation somehow backs up our observation from section 3, namely that the matching across the multi-platform dataset, especially the fact that the illumination is quite different, challenges commercial state-of-the-art software, as well.

5. CONCLUSIONS

It turns out that the ISPRS benchmark dataset is a challenging, but at the same time also realistic example for close range/UAV image blocks. State-of-the-art tie point matching approaches show good results in some published work. However, in this case still the traditional SIFT, in combination with Wallis filtering outperforms all other approaches for a mixed-platform and illumination image pair. The selected images are all from the released dataset, i.e. all the experiments can also be conducted by interested researchers.

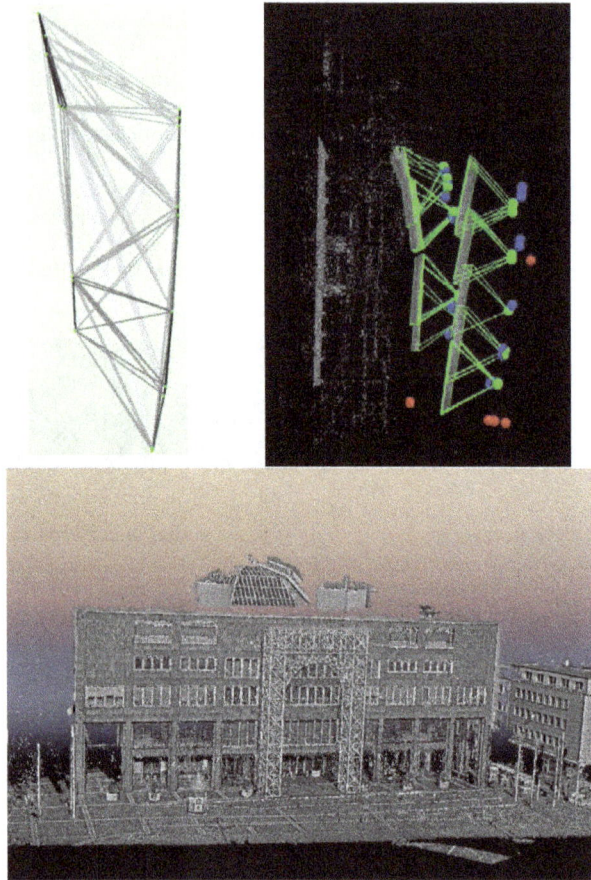

Figure 14: UAV and terrestrial, set a, Wallis-filtered image. Upper left: image connections, upper right: tie point cloud and camera locations (red: not adjusted), lower: densified point cloud.

Using pix4dmapper we found out that the entire block of images gets split-up into several sub-blocks. Our assumption is that one reason for this failure is a combination of a low number of inter-platform image matches, with a bad approximate geo-location for images, especially in the terrestrial part.

ACKNOWLEDGEMENTS
Data acquisition and pre-processing was made feasible through the funds provided by ISPRS (Scientific Initiative) and EuroSDR. We thank Pix4D for providing us a research license of pix4dmapper.

REFERENCES

Agrawal, M., Konolige, K. and Blas, M. R., 2008. CenSurE: center surround extremas for realtime feature detection and matching. *Proceedings of the European Conference on Computer Vision*, Vol. IV, pp. 102–115.

Alcantarilla, P. F., Bartoli, A. and Davison, A. J., 2012. KAZE features. *Proceedings of the European Conference on Computer Vision*, Vol. VI, pp. 214–227.

Alcantarilla, P. F., Nuevo, J. and Bartoli, A., 2013. Fast explicit diffusion for accelerated features in nonlinear scale spaces. *Proceedings of the British Machine Vision Conference*, pp. 13.1–13.11.

Bay, H., Ess, A., Tuytelaars, T. and Van Gool, L., 2008. Speeded-up robust features (SURF). *Computer Vision and Image Understanding* 110(3), pp. 346–359.

Calonder, M., Lepetit, V., Strecha, C. and Fua, P., 2010. BRIEF: binary robust independent elementary features. *Proceedings of the European Conference on Computer Vision*, Vol. IV, pp. 778–792.

Dahl, A. L., Aanæs, H. and Pedersen, K. S., 2011. Finding the best feature detector-descriptor combination. *Proceedings of the International Conference on 3D Imaging, Modeling, Processing, Visualization and Transmission*, pp. 318–325.

Hartley, R. and Zisserman, A., 2008. Multiple view geometry in computer vision. University Press, Cambridge, UK.

Jazayeri, I. and Fraser, C.S., 2008. Interest operators in close-range object reconstruction, *ISPRS Archives of the Photogrammetry, Remote Sensing and Spatial Information Sciences, Vol. XXXVII-B5*, pp. 69-74.

Levi, G. and Hassner, T., 2015. LATCH: Learned Arrangements of Three Patch Codes, *arXiv preprint* arXiv:1501.03719.

Li, H and Hartley, R., 2006. Five-Point Motion Estimation Made Easy. *Proceedings of the 18th International Conference on Pattern Recognition, ICPR*, pp.630-633.

Lowe, D. G., 2004. Distinctive image features from scale-invariant keypoints. *International Journal of Computer Vision* 60(2), pp. 91–110.

Morel, J-M. and Yu, G., 2009. ASIFT: A New Framework for Fully Affine Invariant Image Comparison. *SIAM Journal on Imaging Sciences* 2(2): 438–469.

Nex, F., Gerke, M, Remondino, F., Przybilla, H.-J., Bäumker M. and Zurhorst, A., 2015. ISPRS Benchmark for multi-platform photogrammetry. *ISPRS Annals of the Photogrammetry, Remote Sensing and Spatial Information Sciences, Vol. II-3/W4*, pp. 135-142.

Nistér, D., 2004. An efficient solution to the five-point relative pose problem. *IEEE Transactions on Pattern Analysis and Machine Intelligence* 26(6), pp. 756–770.

Rosten, E. and Drummond, T., 2005. Fusing points and lines for high performance tracking. *Proceedings of the International Conference on Computer Vision*, Vol. 2, pp. 1508–1515.

Rublee, E., Rabaud, V., Konolige, K. and Bradski, G., 2011. ORB: an efficient alternative to SIFT or SURF. *Proceedings of the IEEE International Conference on Computer Vision*, pp. 2564–2571.

Urban, S. and Weinmann, M., 2015. Finding a good feature detector-descriptor combination for the 2D keypoint-based registration of TLS point clouds. *ISPRS Annals of the Photogrammetry, Remote Sensing and Spatial Information Sciences, Vol. II-3/W5*, pp. 121–128.

Wallis, K.F. (1976) Seasonal adjustment and relations between variables *Journal of the American Statistical Association,* 69(345) pp. 18-31.

Yang, X. and Cheng. K-T., 2012. LDB: An ultra-fast feature for scalable augmented reality on mobile devices. *IEEE Int. Symposium on Mixed and Augmented Reality*, pp. 49-57.

2

AUTOMATIC ORIENTATION AND MOSAICKING OF ARCHIVED AERIAL PHOTOGRAPHY USING STRUCTURE FROM MOTION

J. A. Gonçalves

Faculdade Ciências - Universidade Porto,
Rua Campo Alegre 4169-007 Porto, Portugal - jagoncal@fc.up.pt

KEY WORDS: Archived aerial photos, Structure from Motion, Bundle adjustment, Auto-calibration, DSM, Orthomosaics

ABSTRACT:

Aerial photography has been acquired regularly for topographic mapping since the decade of 1930. In Portugal there are several archives of aerial photos in national mapping institutes, as well as in local authorities, containing a total of nearly one hundred thousand photographs, mainly from the 1940s, 1950s and some from 1930s. These data sets provide important information about the evolution of the territory, for environment and agricultural studies, land planning, and many other examples. There is an interest in making these aerial coverages available in the form of orthorectified mosaics for integration in a GIS.

The orthorectification of old photographs may pose several difficulties. Required data about the camera and lens system used, such as the focal distance, fiducial marks coordinates or distortion parameters may not be available, making it difficult to process these data in conventional photogrammetric software.

This paper describes an essentially automatic methodology for orientation, orthorectification and mosaic composition of blocks of old aerial photographs, using Agisoft Photoscan structure from motion software. The operation sequence is similar to the processing of UAV imagery. The method was applied to photographs from 1947 and 1958, provided by the Portuguese Army Geographic Institute. The orientation was done with GCPs collected from recent orthophototos and topographic maps. This may be a difficult task, especially in urban areas that went through many changes. Residuals were in general below 1 meter. The agreement of the orthomosaics with recent orthophotos and GIS vector data was in general very good. The process is relatively fast and automatic, and can be considered in the processing of full coverages of old aerial photographs.

1. INTRODUCTION

In Portugal there are several archives of aerial photos in national mapping institutes, as well as in local authorities, containing more than one hundred thousand photographs, from the decades 1930, 1940 and 1950s. These data sets provide important information about the evolution of the territory, for environment and agricultural studies, land planning, legal boundaries disputes, and many other examples. There is an interest in making these aerial coverages available in the form of orthorectified mosaics for integration in a GIS.

Aerial photography has been acquired regularly for topographic mapping since the decade of 1930. The Portuguese Army started the production of topographic maps at scale 1:25,000, by stereoscopic restitution, in this period. A local company – Sociedade Portuguesa de Levantamentos Aéreos Ltd. (SPLAL) – made the aerial coverages, resulting in 40,000 photos that are now property of the Army Geographic Institute (IGeoE - Instituto Geográfico do Exército). The camera used was a Zeiss RMK S1818, with film size of 18 cm and a 204 mm lens (Redweik et al., 2010).

Other full coverages were also made. After the second world war, in 1947, the Royal Air Force made a uniform coverage of all Portugal, in scale 1:30,000, in a total of 10,000 photos, also property of IGeoE. In 1958 the United Sates Air Force made a full coverage of Portugal and Spain, resulting in some 12,000 photos for all Portugal. These two coverages (RAF47 and USAF58) were made using cameras with film size 23 by 23 cm and 152 mm lens. Figure 1 shows photos of SPLAL and RAF47 flights.

These collections were analysed with great detail by Redweik et al. (2010). It became possible to know some characteristics of the cameras used, which are important for a photogrammetric processing of those images. Anyway, conventional photogrammetric treatment of these photographs, in order to produce orthophotos would be time consuming and relatively expensive. At least, up to this moment no significant works of orthorectifying old aerial photos have been done.

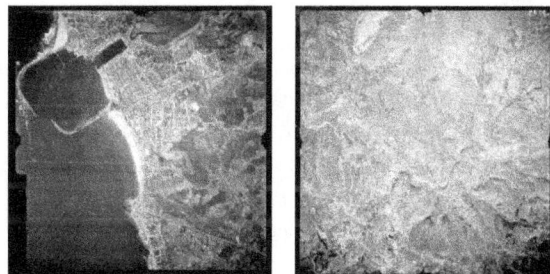

Figure 1 – Examples of photos from SPLAL and RAF flights, both from 1947.

The more common situation is that calibration certificates, or even basic knowledge of camera characteristics, are missing, making the use of those images even more difficult. In cases of flat terrain, simple bi-dimensional image georeferencing with polynomials may be used, as is the case of Alhaddad, et al. (2009) who describe a study of urban sprawl analysis using archived photographs. An alternative method, applicable in generic cases, is the Direct Linear Transformation (DLT) orientation method (Abdel-Aziz and Karara, 1972), which does not require camera information, but has the disadvantage of needing many ground control points. Ma and Buchwald (2012), for example, describe its use in small blocks of archived aerial photographs.

More efficient and largely automatic methods to orthorectify archived aerial photos are needed. Important recent developments in digital photogrammetry and computer vision algorithms can provide solutions for this task. This paper deals with a methodology based on those approaches, to orient and build orthomosaics of archived aerial photographs, similar to what is done with imagery acquired by unmanned aerial vehicles (UAV).

Many of these archived photos are already digitised by IGeoE, in a photogrammetric scanner, with a resolution of 0.021 mm. Some of the SPLAL and USAF photos were tested in this work.

2. PHOTOGRAMMETRIC PROCESSING OF UAV IMAGERY

UAVs are now commonly used to acquire coverages of hundreds of aerial images, with compact cameras. Processing methodologies commonly used are largely automated, for example by commercial software packages such as Pix4D, Agisoft Photoscan and Trimble UAS Master, or open source, such as MicMac (Deseilligny and Clery, 2011). These methodologies, in part originated in computer vision, are designated as Structure from Motion (SfM), (Abdel-Wahab et al., 2012). This method inspired the solution proposed for archived aerial photos.

2.1 UAV image processing with SfM

UAV images are acquired by compact cameras for which only approximate parameters are known (e.g. nominal focal length) and can have radial distortions that are not calibrated. This situation is similar to what happens with old photos, for which calibration certificates are not available. UAVs are supported by navigation GPS receivers, which can provide approximate position of the camera projection centres. The processing steps are the following:

1. Apply algorithms similar to SIFT (Scale Invariant Feature Transform), (Lowe, 2004), which give many conjugate points between overlapping images.

2. Do a free bundle adjustment in order to obtain a relative orientation of all the images. This operation is called in the SfM approach, "Image alignment". With the navigation GPS positions an approximate georeferencing can be done.

3. If more accurate georeferencing is needed, ground control points (GCPs) can be provided, in order to do an aerial triangulation, incorporating an auto-calibration.

4. Dense matching algorithms are used to generate a dense point cloud, by multi view stereo matching. A regular grid digital surface model (DSM) is obtained from the point cloud.

5. Images are orthorectified (true ortho if the DSM is used) and a continuous mosaic is built.

This workflow is similar in several programs. The one used in this work to process the old photos was Agisoft Photoscan.

2.2 Camera calibration

The distortion model used is the Brown distortion model (Agisoft, 2015), which considers the following parameters to describe a camera:

f focal length,

c_x, c_y principal point coordinates,

k_1, k_2, k_3 radial distortion polynomial coefficients,

p_1, p_2 tangential distortion coefficients.

There are also parameters to account for skew effect and for x and y different scales, which are treated as 2 focal distances.

The bundle adjustment considers an auto-calibration, in which the user chooses which of these parameters he wants to adjust. This process requires many and well distributed conjugate points, as well as control points.

3. PROPOSED METHODOLOGY

3.1 Image preparation

This software expects digital images, and the old photos are digitised, film based photos. In a standard photogrammetric treatment of digitised photos, an interior orientation, with fiducial marks, should be done. In this case it will be replaced by a resampling of all photos, such that all images have the same size and the same position with respect to the photographic coordinate system. That is done by an image registration, through an affine transformation.

Since in a general case, coordinates of fiducial marks will not be available, the chosen approach will be to take one of the images as a reference image and register all the others to it, using the fiducial marks. Figure 2 shows examples of fiducial marks of photos of the three flights referred before.

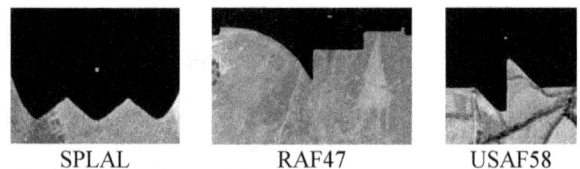

SPLAL   RAF47   USAF58

Figure 2 – Examples of fiducial marks of photos from the SPLAL, RAF47 and USAF58 flights.

In a case where the fiducial marks are not clearly identified, reference points in the image frame can be used, provided that they are stable from image to image.

In the manual registration done in all tests, the residuals of the affine transformation with 4 fiducial marks were normally under 1 pixel. In future works involving the processing of large numbers of photos, some automation can be easily done, in a similar way to what is done in an automatic interior orientation, due to the repetitive pattern of the marks.

3.2 Image orientation

The following step is the image alignment, in which conjugate points are obtained automatically. Figure 3 shows conjugate points obtained between two overlapping images of the SPLAL flight. That is followed by the "image alignment". GCPs can be provided at this stage in order to orientate all the block. Figure 4 shows the result of image orientation, including the sparse point cloud and the GCPs.

The image orientation is optimized in a bundle adjustment with camera auto-calibration, where the user chooses the parameters to be considered. Statistics of the residuals found on the GCPs and the conjugate points are provided, and once they are small, the result can be accepted. Accuracy assessment can be done more easily over the final products.

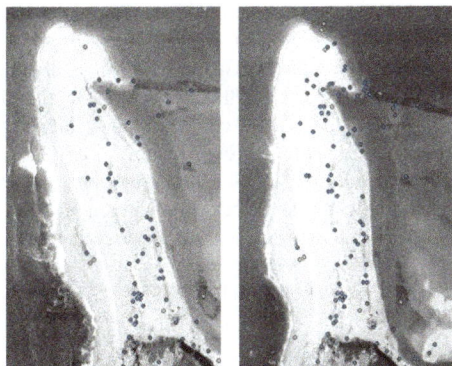

Figure 3 – Examples of conjugate points obtained between two consecutive images of the SPLAL flight.

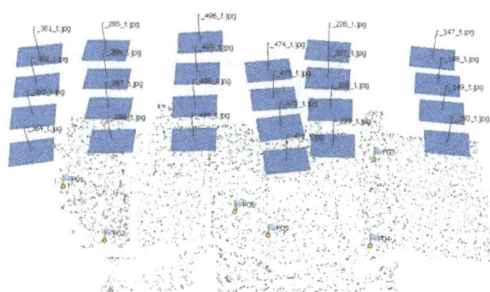

Figure 4 – Graphic representation of the oriented images, the sparse point cloud and the GCPs.

3.3 Accuracy assessment over the DSM and orthomosaic

The following step is, as in the UAV workflow, the generation of a DSM and an orthoimage mosaic. Over these two products we can analyze the global accuracy of the exterior orientation obtained. The orthoimage can be overlaid with other georeferenced images, from present time, or vector data, in order to check the planimetric accuracy. Vertical accuracy can be checked by comparing heights of the DSM and check points.

3.4 Improvement of the GCP search

The easiest way to collect GCPs for the image orientation is from orthoimages of the present time. Due to the significant changes that can occur, especially in areas around cities, finding GCPs in old photos may be very difficult, especially if the user does not know the area. A strategy was followed which facilitates this task.

For most of the available photos an approximate position of the image centre is known, in a similar way to what happens with UAV images. This allows for the creation of an orthoimage with an approximate georeferencing, without any GCPs. Although it may have large errors (100 meters or so) it can be overlaid with present time georeferenced imagery, in order to start the identification of common points to be used as GCPs. The orientation process is then iterated, now with the collected GCPs.

4. APPLICATION TO SPLAL AND USAF PHOTOS

The methodology was applied to several blocks of the available images, some with up to 40 photos. Results for two cases are described below, one with photos of the city of Porto from 1947 and another from a rural mountainous area, with USAF photos from 1958.

4.1 Images from the city of Porto (SPLAL, 1947)

A total of 24 photos were available from the SPLAL flights of 1947, covering all the city of Porto. The image alignment was done with a total of 6581 conjugate points. GCPs were collected from present time orthophotos (planimetric coordinates) and 1:1000 scale topographic maps (heights), provided by the local authority. Heights within the area range from sea level to 150 meters. Although the area is familiar to the author, collection of good quality GCPs, at ground level, was very difficult, due to changes occurred in more than 60 years. For a total of 8 points, a root mean square errors smaller than 1 meter was obtained in .

The auto-calibration gave a new value for the focal distance of 200.9 mm, instead of the nominal 204.4 marked on the photos. Tests were made considering radial or other distortion parameters but the results did not improve in terms of residuals. The resulting orthoimages (with or without distortion parameters) did not suffer significant changes, so it was concluded that no significant distortions existed.

Figure 5 shows the orthomosaic composed with the 24 photos. In order to assess the planimetric accuracy, check points should be used, but that was not done due to the difficulty in finding trustable points for this purpose. It was preferred to do a qualitative analysis by overlaying vector data, such as street axis, as shown in figure 6. The agreement of the vector data with the existing streets of the time was always very good throughout the image, letting us conclude that the results of the orientation and camera calibration were acceptable.

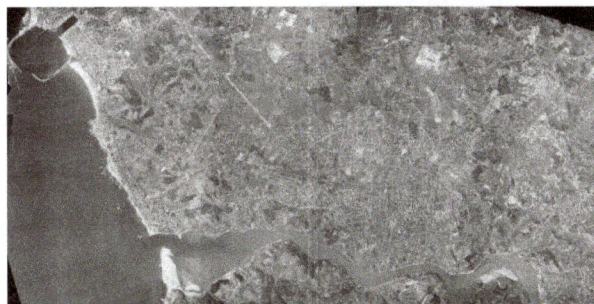

Figure 5 – Mosaic of 24 orthorectified photos of the city of Porto (SPLAL flight of 1947).

Figure 6 – Vector data overlaid on the orthophoto. Extension of the area; 750 m by 400 m.

4.2 Photos from rural mountainous area

Another set of 8 images of the USAF58 flight, in the region of Montalegre, a mointainous rural area in north Portugal, were also processed. Since the area was not familiar to the author, initial identification of GCPs in present time orthos was very difficult. The strategy of building an initial mosaic, with approximate georeferencing coming from the projection centers was very useful. After that, a total of 14 points were easily obtained, mainly in boundaries between fields, which frequently keep unchanged. Heights were obtained on topographic maps.

Figure 7 shows a sample of the orthomosaic (900 m by 500 m). The Open Street Maps (OSM) layer was overlaid for quality assessment of the mosaic, which was in general good. Focal distance of the camera was changed from nominal value 152.42 mm to 152.70 mm. The effects over the resulting orthophoto of considering or not the radial distortion parameters was negligible.

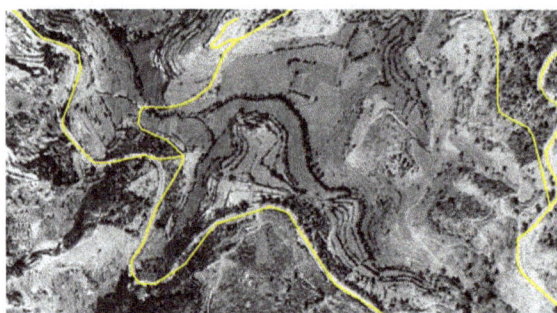

Figure 7 – Portion (900 m by 500 m) of the USAF58 orthophoto in a mountainous area, with OSM vector layer.

Figure 8 shows a sample of the extracted DSM in the mountainous area. In order to assess its vertical accuracy, spot heights from topographic maps of scale 1:10,000 of the area were used as check points. Height errors were calculated for a total of 1183 points. Table 1 contains the statistics of these errors.

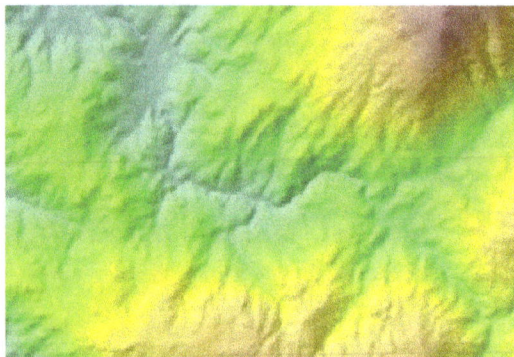

Figure 8 – Sample of the extracted DSM (5 km by 3 km).

Table 1 – Statistics of the height errors found in 1183 altimetric check points

	ΔH (m)
Minimum	-25.0
Maximum	22.7
Mean	1.2
RMSE	4.7

5. CONCLUSIONS

The method proposed aims at providing orthomosaics of archived aerial photographs. It comprises the determination of exterior image orientation and camera auto-calibration, in an essentially automatic manner. Ground control is obtained from present time data. The method overcomes the lack of information of detailed camera characteristics, such as rigorous focal distance, or fiducial mark coordinates.

Tests were made with images of old photographs of Portugal. It proved to be applicable to process very large archives of aerial photos, in acceptable time and with reduced costs.

REFERENCES

Abdel-Aziz Y. I, Karara H.M. (1971). Direct Linear Transformation from Comparator Coordinates into Object Space Coordinates. Proceedings of the ASP Symposium on Close-Range Photogrammtery, Falls Church, VA, pp. 1-18., 1971.

Abdel-Wahab, M., Wenzel, K., & Fritsch, D. (2012) Efficient Reconstruction of Large Unordered Image Datasets for High Accuracy Photogrammetric Applications, ISPRS Ann. Photogramm. Remote Sens. Spatial Inf. Sci., I-3, 1-6,

Agisoft, 2015. Agisoft documentation: http://www.agisoft.com /pdf/photoscan-pro_1_2_en.pdf (Visited in January 2014).

Alhaddad, B., Cladera, J.R., Burns, M.C. (2009). Monitoring Urban Sprawl from historical aerial photographs and satellite imagery using Texture analysis and mathematical morphology approaches. Proceedings of "Engineering of Reconfigurable Systems and Algorithms - ERSA'09", July 13-16, 2009. Nevada, USA.

Deseilligny, M.P., Clery, I. 2011. Apero, an open source bundle adjustment software for automatic calibration and orientation of set of images. In: Proceedings of ISPRS International Workshop on 3D Virtual Reconstruction and Visualisation of Complex Architectures, Trento, Italy, 2–4 March, 2011, Trento, Italy, pp. 269–276.

Lowe, D. 2004: Distinctive Image Features from Scale-Invariant Keypoints. International Journal of Computer Vision. Vol. 60, No. 2, pp. 91-110.

Ma, R., Buchwald, A. (2012). Orthorectify Historical Aerial Photographs Using DLT. Proceedings of the ASPRS 2012 Annual Conference, Sacramento, California. March 19-23, 2012.

Redweik, P., Roque, D., Marques, A., Matildes, R., Marques, F. (2010). Photogrammetric Engineering & Remote Sensing, Vol. 76, No. 9, September 2010, pp. 1007–1018.

ACKNOWLEDGEMENTS

Images were provided by the "Instituto Geográfico do Exército" (IGeoE).

Data for the city of Porto was provided by the geographic information division of the Porto City Hall.

Global orthoimages of Portugal are provided by "Direcção Geral do Território".

3

EXTERIOR ORIENTATION OF HYPERSPECTRAL FRAME IMAGES COLLECTED WITH UAV FOR FOREST APPLICATIONS

A. Berveglieri [a] and A. M. G. Tommaselli [b]

[a, b] Univ Estadual Paulista – UNESP, Faculty of Science and Technology, Presidente Prudente, Brazil,
[a] Graduate Program in Cartographic Sciences, [b] Department of Cartography
[a] adilsonberveg@gmail.com, [b] tomaseli@fct.unesp.br

KEY WORDS: Forest, Hyperspectral Image, Orientation, Sparse Control, UAV

ABSTRACT:

This paper describes a preliminary study on the image orientation acquired by a hyperspectral frame camera for applications in small tropical forest areas with dense vegetation. Since access to the interior of forests is complicated and Ground Control Points (GCPs) are not available, this study conducts an assessment of the altimetry accuracy provided by control targets installed on one border of an image block, simulating it outside a forest. A lightweight Unmanned Aerial Vehicle (UAV) was equipped with a hyperspectral camera and a dual-frequency GNSS receiver to collect images at two flying strips covering a vegetation area. The assessment experiments were based on Bundle Block Adjustment (BBA) with images of two spectral bands (from two sensors) using several weighted constraints in the camera position. Trials with GCPs (presignalized targets) positioned only on one side of the image block were compared with trials using GCPs in the corners. Analyses were performed on altimetry discrepancies obtained from altimetry checkpoints. The results showed a discrepancy in Z coordinate of approximately 40 cm using the proposed technique, which is sufficient for applications in forests.

1. INTRODUCTION

The monitoring of recovered and native forests is a widely recognized global need which requires updated geospatial information. Aerial and orbital imagery can be used for large areas but the spatial and temporal resolutions are limited and these techniques are not cost-effective for small areas.

In remaining forests, for example, the use of images collected by Unmanned Aerial Vehicles (UAVs) is feasible due to the lower costs and the possibility of images acquisition with suitable spatial and temporal frequency. In recent years, UAVs have been increasingly adopted as platforms for applications of photogrammetry and remote sensing, as discussed by Colomina and Molina (2014). Remondino et al. (2011) reported some UAV systems for photogrammetric applications as well as Eisenbeiss (2011) who described the potential of UAVs for mapping tasks.

New types of hyperspectral sensors have been introduced to UAV applications, such as the Rikola camera with a Fabri-Perot interferometer (FPI), which acquires sequence 2D images in frame format. A review presented by Aasen et al. (2015) reported important studies and the potentiality of hyperspectral sensors to derive information, e.g., about vegetation, plant diseases, environmental conditions, and forest. Honkavaara et al. (2013) performed experiments using UAV with a FPI-based hyperspectral camera to collect hyperspectral and structural information and to estimate plant height and biomass. The trials demonstrated great potential for precision agriculture and indicated feasibility for other research topics.

In spite of the instability of UAV platforms, accurate data can be provided depending on the requirements, as commented by Eisenbeiss (2011), which enables the use of UAVs in inaccessible areas, for example, forests. However, ground control is required to indirectly orient the images, although this task could be complicated in areas with dense vegetation such as tropical forests.

Considering these needs this study performed a preliminary assessment on the image orientation in vegetation area using images acquired with a hyperspectral frame camera (Rikola) on board a lightweight UAV with a dual-frequency GNSS receiver for image georeferencing. The main purpose is to assess the results of hyperspectral image orientation using sparse ground control only at the beginning of flying strips, i.e., simulating a configuration outside forest areas.

The experiments were conducted using presignalized targets arranged along the flight trajectory over a vegetation area. The altimetry accuracy obtained with Bundle Block Adjustment (BBA) was analysed based on the discrepancies at altimetry checkpoints, which demonstrated the feasibility of the proposed technique.

2. HYPERSPECTRAL FRAME CAMERA

The Rikola company has made a 2D hyperspectral frame camera based on FPI with a RGB-NIR sensor, as shown in Figure 1(a). This FPI into the lens system is used to collimate the light, and then the spectral bands are a function of the interferometer air gap. Changes in the air gap enable to acquire a set of wavelengths for each image (Figure 1(b)). The range of air gap in FPI can provide wavelengths from 500 nm to 900 nm using two CMOS image sensors.

After acquiring images, the data is converted from analog to digital in 16-bit. Time information, collected by GPS receiver, is recorded for each image trigger and can be synchronized by a microcontroller to work in time interval based self-triggering mode. Both GPS position and irradiance data are recorded.

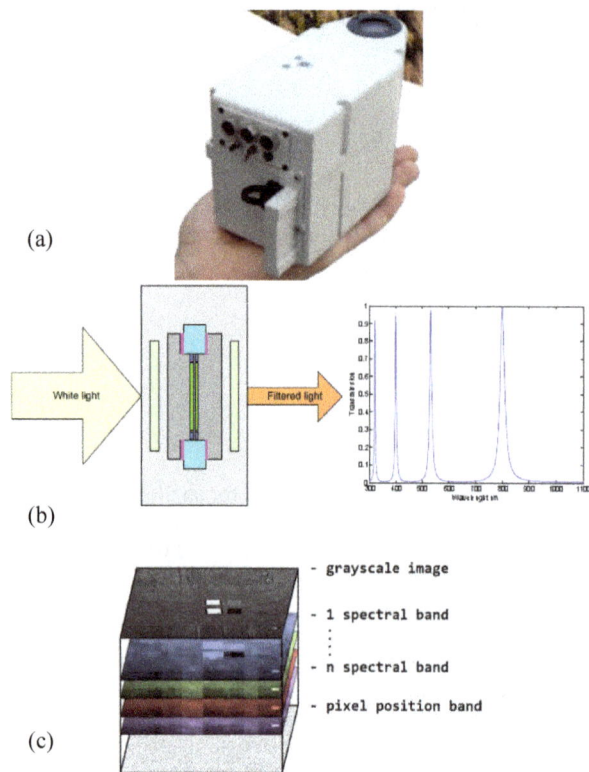

(a)

(b)

(c)

Figure 1: (a) Hyperspectral frame camera. (b) Fabry-Perot principle. (c) Image data cube. Source: adapted from (b) Rikola Ltd. and (c) Aasen et al. (2015).

The hyperspectral camera acquires the images as binary data cube with a number of bands using a predefined sequence of air gap values to reconstruct the spectrum of each pixel in the image, as described by Honkavaara et al. (2013) and depicted in Figure 1(c). Each data cube (or image) has information on the wavelengths of bands, radiometry, and pixel position.

According to Mäkeläinen et al. (2013), hyperspectral frame sensors are suitable for lightweight UAV imaging, since the sensor has approximately 600 g. In addition, the images in frame format allow conventional aerial triangulation for exterior orientation determination, which is essential for photogrammetric application that requires accurate results.

In general, photogrammetric projects with that type of sensor are planned with UAVs at heights of 100-200 m, resulting in a GSD of 5-15 cm, which is sufficient for UAV application as agriculture and forest canopy monitoring.

3.   METHODOLOGY

The methodology is based on the assessment of minimum configuration of ground control for small forest areas, refining the camera Perspective Centre (PC) positions collected during the flying survey . The following sections describe the required steps.

3.1   Camera calibration

The first step is the camera calibration, since accurate data are needed. Typically focal length (f), principal point (x_0, y_0) and

lens distortion coefficients (K_1, K_2, K_3, P_1, P_2) are the parameters to be determined, which define the Interior Orientation Parameters (IOPs). The mathematical model is based on the collinearity equations (Kraus, 2007) with addition of parameters of the Conrady-Brown model (Fryer and Brown, 1986).

The estimation of the IOPs can be done by a self-calibrating bundle adjustment considering constraints imposed to the ground coordinates, as proposed by Kenefick et al (1972).

3.2   Positioning of control targets and GNSS surveying

Ground control points are used to indirectly estimate EOPs or to refine the observed values when using GNSS to determine the coordinates of the exposure stations. Usually such control is planned to be distributed in the block borders (Kraus, 2007).

In tropical forests, GCPs are not available or suitable areas to set them up can be inaccessible due to dense vegetation, which complicates the acquisition of ground coordinates and the visibility from aerial images. Thus, a few presignalized targets or natural points have to be used as ground control outside the forest or in clearings. Depending of the vegetation features natural points are scares on unsuitable to be used. Considering that UAV operation is always on site it is straightforward to install signalised target around the take-off and landing area. These targets are installed before the flying survey and will be used to improve the PC coordinates determined with GNSS. A GNSS base station is also defined close to the GCPs to improve the GNSS post-processed solution since real time GNSS network are not dense in Brazil.

3.3   Image data acquisition

The aerial survey is planned to acquire hyperspectral images over a forest area. The camera PC positions are determined by a GNSS receiver during the flight using a reference band of the image cube, which is defined by the camera manufacturer. Since the attitude angles are not directly determined, they are later estimated in the image orientation by BBA using initial values from the flight plan.

3.4   Image orientation

The image cube is formed by a set of spectral bands that can be oriented separately. For image orientation, the image frames are connected with tie points to enable the BBA from a minimum of GCPs and weighted constraints in the camera PC positions. The dual-frequency GNSS receiver installed on board UAV allows the acquisition of accurate PC positions during the aerial survey. The attitude angles were considered as unknowns. The control targets are used as GCPs considering the accuracy resulting from GNSS surveying.

4.   EXPERIMENTS AND RESULTS

This section describes the methodology to perform an aerial survey to collect hyperspectral frame images over a vegetation area. The experiments were performed to assess the resulting accuracy of the image orientation using GCPs only in one side of the image block, which is likely to occur in UAV flights over forest areas. The results were compared with those achieved when using GCPs in the border corners.

4.1 Performing terrestrial camera self-calibration

The camera self-calibration procedure was performed in a 3D terrestrial calibration field composed of coded targets with Aruco format (Garrido-Jurado et al., 2014). These targets were automatically located in the images using a software adapted by Silva et al. (2014) that identifies each target image coordinate with its respective ground coordinate.

A total of twelve images was captured for camera calibration using the hyperspectral camera specified in Table 1. Previously the camera was configured to acquire data cubes with twenty five spectral bands, more details in Tommaselli et al. (2015).

Camera model	Rikola FPI2014
Nominal focal length	9 mm
Pixel size	5.5 µm
Image dimension	1023 × 648

Table 1. Technical details.

The images were collected from several camera stations in the calibration field using different positions and rotations to avoid correlation between IOPs and EOPs. Figure 2 exemplifies two images with Aruco target identified, which were used in the camera calibration procedure.

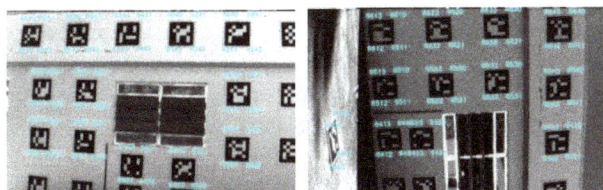

Figure 2. Hyperspectral images with Aruco targets identified. Both images were used in the camera calibration procedure.

Two spectral bands (one of each sensor, 605.64 nm and 689.56 nm) were selected for camera self-calibration. The procedure was performed with seven constraints in the object space. The coordinates of two ground points were fixed $(X_1, Y_1, Z_1, X_2, Y_2, Z_2)$ and a Z coordinate of a third point (Z_3) orthogonal to these two previous points. IOPs and EOPs were considered as unknowns in the BBA procedure, being calculated by the calibration multi-camera (CMC) software (developed in-house by Ruy et al. (2009)). The estimated values of the IOPs and the *a posteriori sigma* are presented in Table 2.

Parameter	Sensor 1 (689.56 nm)		Sensor 2 (605.64 nm)	
	Estimated value	Standard deviation	Estimated value	Standard deviation
f (mm)	8.6905	0.0118	8.6488	0. 0321
x_0(mm)	0.4165	0.0050	0.3665	0.0138
y_0(mm)	0.3997	0.0039	0.4365	0.0107
K_1(mm^{-2})	-4.74×10^{-3}	1.06×10^{-4}	-4.13×10^{-3}	3.04×10^{-4}
K_2(mm^{-4})	-1.78×10^{-6}	2.21×10^{-5}	-6.34×10^{-5}	6.54×10^{-5}
K_3(mm^{-6})	-6.71×10^{-7}	1.43×10^{-6}	4.75×10^{-6}	4.37×10^{-6}
P_1 (mm^{-1})	-2.86×10^{-5}	8.85×10^{-6}	-1.73×10^{-4}	2.48×10^{-5}
P_2 (mm^{-1})	-1.15×10^{-4}	1.17×10^{-5}	-3.06×10^{-4}	3.19×10^{-5}
a posteriori sigma (*a priori* = 1)	0.06		0.43	

Table 2. IOPs determined by self-calibration.

4.2 Aerial survey

Before carrying out the aerial survey, five presignalized control targets, as shown in Figure 3(a), were geometrically positioned along the border of the test area. Figure 4(b) displays the block geometry and Figure 3(b) shows an example of the control target appearing in the image.

Sensormap Company performed the aerial survey using a lightweight UAV (octopter) platform equipped with the Rikola hyperspectral camera and a dual-frequency GNSS receiver, as shown in Figure 4(a).

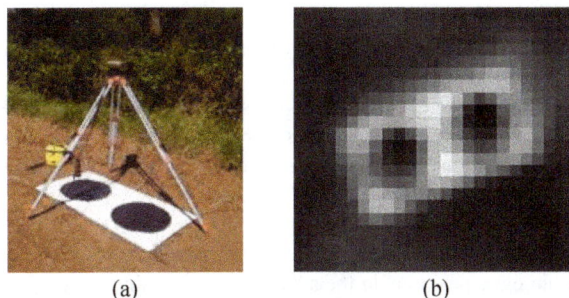

(a)   (b)

Figure 3. (a) GNSS surveying using a presignalized target. (b) Control target appearing in the aerial hyperspectral image.

The hyperspectral camera was configured to acquire spectral image cubes at a flying height of 160 m with flying speed of 4 m/s. Two flying strips were collected covering a range of 800 m over a vegetation area (composed of forest and sugarcane, as shown in Figure 4(b)), which resulted in images with a GSD of 11 cm. The flight planning was performed with longitudinal and lateral overlap of 80% and 30%, respectively.

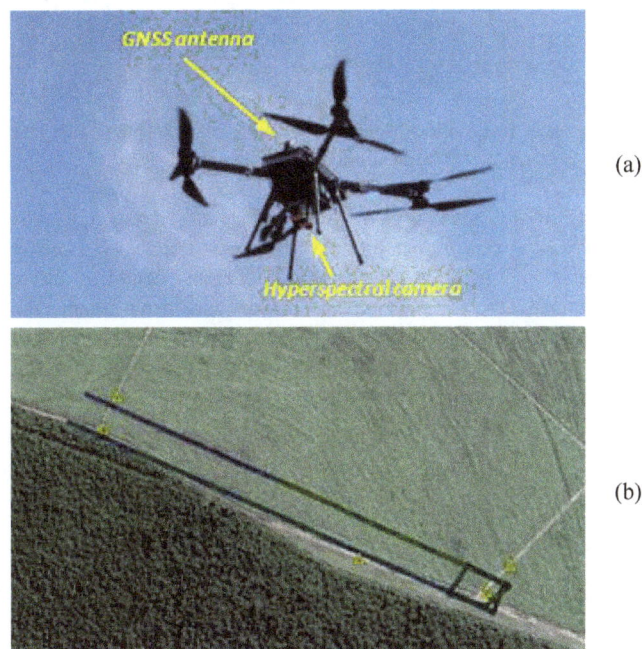

(a)

(b)

Figure 4. (a) UAV equipped with a hyperspectral camera and GNSS receiver. (b) Distribution of five GCPs (triangles) and trajectory followed by the UAV to acquire hyperspectral images.

4.3 Bundle adjustment

Considering the same two spectral bands selected for the camera calibration, two blocks were formed with the corresponding images and BBA was performed for each block using ERDAS Imagine – LPS software (v. 2015) with the following configuration:

- Initial positions of the camera PC were based on the data collected with GNSS receiver and processed by differential positioning technique with the base station on site (weighted constraints varying from 10 cm to 50 cm were used) and attitude angles were considered as unknowns;
- Calibrated IOPs were considered as absolute constraints;
- Image coordinates (automatically extracted) with standard deviation of 1/2 pixels;
- GCPs were surveyed in the midpoint between the two circular target centroids, and used with a standard deviation of 5 cm, based on the GNSS positioning accuracy.

Each GCP was located in one image and then transferred to homologue positions in the adjacent images using least squares matching via LPS software. Tie points were automatically generated using 25 points per model (default distribution).

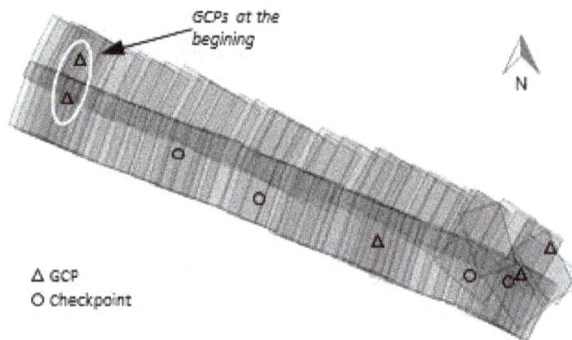

Figure 5. Image block geometry with 93 images, five GCPs, and four altimetry checkpoints.

One of the advantages in using presignalised targets is the automatic centre location that enables to achieve a better precision. In this project, the targets centroids were automatically located in the images and, from them, the midpoint corresponding to the GCP. This strategy was used to minimize measurement errors caused by image blurring and saturation.

It is important to note that all spectral bands are not simultaneously recorded at the same exposure time. The time interval from band to band is approximately 22 ms and image bands are grabbed sequentially with the platform displacement. Thus, the camera PC positions have to be interpolated for the selected spectral bands under study using the GPS time of a first acquired default band. The time interval for each band is determined by a sequence delay calculator provided by Rikola Company. The GPS time of the default band is grabbed by a navigation single frequency GPS integrated to the Rikola camera.

4.4 Results

Considering the band selected in the sensor 1 (689.56 nm), Figure 6 presents a graph generated with the root mean square error (RMSE) of GCPs resulting from the BBA for different weighted constraints in the camera PC position. As shown in Figure 5, five GCPs (in the corners) were used in the experiment to be compared when using two GCPs only at the beginning of the flying strips (see Figure 5).

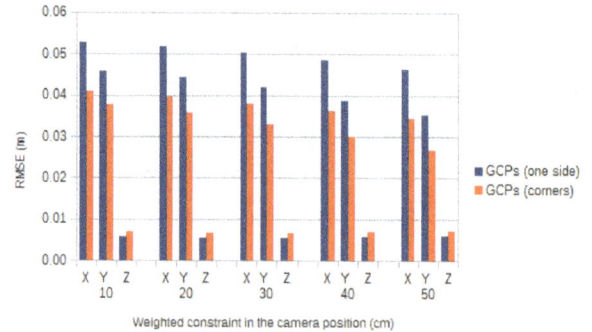

Figure 6. Sensor 1 – RMSE for GCPs resulting from the BBA considering GCPs in the corners and only on one side of the image block.

The results indicated larger RMSEs when GCPs were only used in one side of the image block, presenting values of 0.460-0.053 m in X and 0.035-0.046 m in Y. For the case with GCPs in the corners, the RMSEs were smaller varying 0.034-0.041 m in X e 0.027-0.038 m in Y. The Z coordinate presented approximate RMSEs < 0.01 m, which is better than XY and can be explained by the blurring affecting the measurements in the X direction. Another problem not yet assessed is the event logging error.

Figure 7 shows the RMSEs at GCPs with respect to the BBA for the sensor 2 (spectral band of 605.64 nm). When GCPs were only used on one side, the RMSEs were larger, 0.045-0.049 m in X and 0.069-0.076 m in Y. The larger errors presented in Y can be an effect caused by image blurring, which is more likely to occur in the sensor 2 (visible spectrum). With GPCs in the corners, the RMSEs were 0.031-0.38 m in X and 0.041-0.045 m in Y. In relation to Z coordinate, the errors were quite similar, being below 0.01 m.

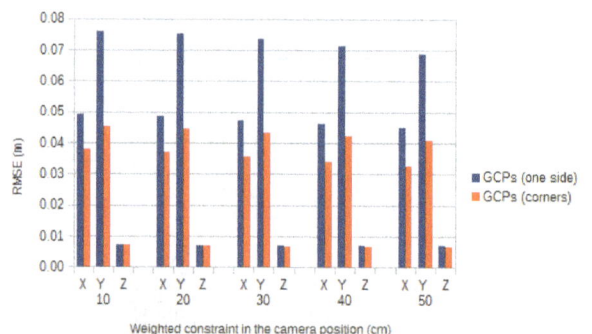

Figure 7. Sensor 2 – RMSE of GCPs resulting from the BBA considering GCPs in the corners and only on one side of the image block.

Comparing the results with GCPs in the corners and GCPs only on one side, the largest differences (in both graphs) were verified in the planimetry and the better results were obtained with GCPs in the corners, as expected. However, even using GCPs in one side, the RMSEs were smaller than 8 cm (< 1 GSD = 11 cm). Comparing the RMSEs in the five ranges of weighted constraints for the camera position, the RMSEs presented a slight decreasing value in planimetry following the ranges. This can be explained by errors in XY components of the PC coordinates which were probably caused by the time stamp provided by Rikola, blurring or by changes in the IOPs.

In the image space, the BBA produced sub-pixel residuals ranging between 0.25 and 0.40 pixels in all cases. In relation to the *a posteriori sigma*, Table 3 presents the sigma values for the several weighted constraints used in the camera position in both configurations of GCPs. Although the sigma values with GCP on one side are smaller in comparison with GCP in the corner, the differences are negligible.

Weighted constraint in the camera PC position		*A posteriori sigma*				
		10 cm	20 cm	30 cm	40 cm	50 cm
Sensor 1	GCP in the corner	0.57	0.57	0.55	0.54	0.53
	GCP in one side	0.56	0.55	0.54	0.53	0.52
Sensor 2	GCP in the corner	0.54	0.53	0.53	0.52	0.51
	GCP in one side	0.53	0.52	0.52	0.51	0.50

Table 3. *A posteriori sigma* in the BBA (*a priori sigma* = 0.50).

Another analysis can be made to assess the accuracy in altimetry. Four independent altimetry checkpoints were used to calculate the RMSE in Z coordinate. The planimetry errors were not assessed because these checkpoints were reused from a previous field survey and were not signalized for this hyperspectral flight. However, for forest application, the most critical errors are in heights, which can affect the modelling of forest canopy structure.

The graph in Figure 8 shows the RMSEs produced by the BBA using five weighted constraints in the camera position of the sensor 1. Three GCP configurations were used: without GCPs (only camera positions from GPS), integrated image orientation with GCPs in the corners, and integrated image orientation with GCPs at the beginning of the two flying strips.

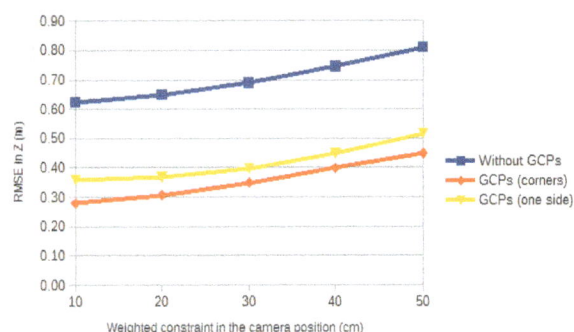

Figure 8. Sensor 1 – RMSE in Z coordinate calculated with four altimetry checkpoints in the BBA.

The RMSEs obtained without GCPs only using the camera PC positions generated altimetry discrepancies from 0.64 m up to 0.85 m depending on the weighted constraint. The use of GCPs showed to be important to improve the accuracy in Z. Using GCPs only on side yielded RMSEs from 0.34 m up to 0.69 m, while GCPs in the corner resulted in RMSEs from 0.29 m up to 0.43 m. The two curves (red and yellow) indicated smaller differences for the weighted restrictions below 30 cm, equivalent to differences around 1 GSD. Above the constraint of 30 cm, the RMSEs were larger and showed an increasing tendency.

Figure 9 presents the RMSEs at altimetry checkpoints for the sensor 2 which is similar to sensor 1. The BBA without GCPs generated RMSEs from 0.63 m to 0.81 m. Such RMSEs were improved using the GCPs in the corners, which presented altimetry discrepancies of 0.28-0.45 m. The GCPs only on one side resulted in RMSEs from 0.36 m to 0.52 m. In the comparison between the yellow and red curves, the differences were smaller than 1 GSD, showing less sharp curves for weighted constraints below 30 cm. Above 30 cm, a sharp increasing tendency of the RMSEs was observed.

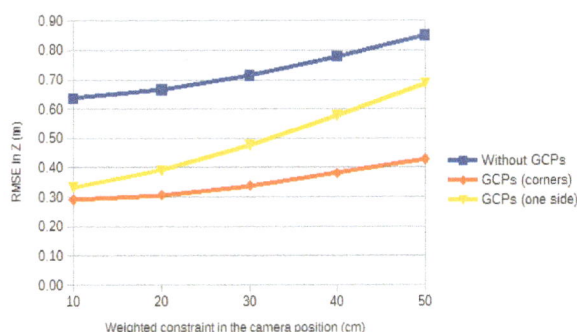

Figure 9. Sensor 2 – RMSE in Z coordinate calculated with four altimetry checkpoints in the BBA.

Based on the graphical analysis from Figures 8 and 9, the weighted restrictions below 30 cm presented the most suitable value for the configuration of the image block processing with sparse control, which resulted in altimetry discrepancies smaller than 40 cm.

The assessment based on RMSEs show that weighted constraints as it was expected are mandatory in the proposed technique and can produce acceptable results with few points in one strip side. Thus, a sparse ground control for forest applications can be used, when control targets can only be positioned outside the mapping area.

Figure 10 displays a mosaic generated with the images of a spectral band under study (605.64 nm). From the image orientation of this band, other spectral bands from the same sensor can be co-registered with relation to this one. Some radiometric effects can be observed in the mosaic, which was caused by illumination variations (cloud shadows) during the image acquisition.

Figure 10. Mosaic generated with the spectral band (605.64 nm, sensor 2) in study. Some radiometric effects can be realized due to the illuminarion variation during the flying survey.

5. CONCLUSIONS

This study presented a preliminary experiment on the hyperspectral frame image orientation for forest applications using UAV. The main objective was to assess the altimetry accuracy provided by using only sparse control outside the forest.

The experiments were based on the BBA of two hyperspectral bands (two sensors) using two flying strips collected with UAV which is the expected configuration for the future flights. Control targets were installed in a vegetation area to assess the feasibility of using only ground controls at the beginning of flying strips, simulating them outside the forest. The trials considered GCPs positioned in the image block corners and GCPs only on one border.

The analysis of results was based on altimetry discrepancies obtained from four altimetry checkpoints and discrepancies in the GCPs. The RMSEs showed that ground control is needed to improve accuracy for the case studied, which was verified in the comparison of RMSEs without using GCPs and RMSEs with GCPs in the BBA. When the BBA using GCPs in the corners was compared with GCPs in one border, the RMSE differences (Figures 8 and 9) presented approximately 1 GSD between the curves (yellow and red) for weighted constraints below 30 cm in the camera position.

In general, the preliminary results of both spectral bands presented a discrepancy of approximately 40 cm in Z coordinate, which is sufficient for photogrammetric applications in forests, such as DSM generation and vegetation mapping. For future studies, the planimetry accuracy in checkpoints should be assessed, as well as the optimum number of GCPs at the beginning of flying strips and other image block arrangements. The stability of the IOPs in the image orientation should also be investigated as well as suitable configurations of flight strips and GCP to enable on-the-job calibration.

ACKNOWLEDGEMENTS

The authors would like to thank the São Paulo Research Foundation (FAPESP) – grants 2014/05533-7 and 2013/50426-4 and Sensormap Company for providing aerial images.

REFERENCES

Aasen, H., Burkart, A., Bolten, A. and Bareth, G., 2015. Generating 3D hyperspectral information with lightweight UAV snapshot cameras for vegetation monitoring: from camera calibration to quality assurance. *ISPRS Journal of Photogrammetry and. Remote Sensing.* 108, pp. 245 – 259.

Colomina, I. and Molina, P., 2014. Unmanned aerial systems for photogrammetry and remote sensing: a review. *ISPRS Journal of Photogrammetry and. Remote Sensing.* 92(6), pp. 79 – 97.

Eisenbeiss, H., 2011. The potential of unmanned aerial vehicles for mapping. In: *Photogrammetrische Woche 2011.* Fritsch, D., pp. 135–145.

Fryer, J.G. and Brown, D.C., 1986. Lens distortion for close-range photogrammetry. *Photogrammetric Engineering & Remote Sensing.* 52(1), pp. 51–58.

Garrido-Jurado, S., Muñoz-Salinas, R., Madrid-Cuevas, F.J. and Marín-Jiménez, M.J., 2014. Automatic generation and detection of highly reliable fiducial markers under occlusion. *Pattern Recognition.* 47(6), pp. 2280 – 2292.

Honkavaara, E., Saari, H., Kaivosoja, J., Pölönen, I., Hakala, T., Litkey, P., Mäkynen, J. and Pesonen, L., 2013. Processing and assessment of spectrometric, stereoscopic imagery collected using a lightweight UAV spectral camera for precision agriculture. *Remote Sensing.* 5(10), pp. 5006–5039.

Kenefick, J. F., Gyer, M. S., Harp, B. F., 1972. Analytical self-calibration. *Photogrammetric Engineering*, 38, 1117–1126.

Kraus, K., 2007. *Photogrammetry: geometry from images and laser scans*, 2nd ed. de Gruyter, Berlin. 459 p.

Mäkeläinen, A., Saari, H., Hippi, I., Sarkeala, J. and Soukkamäki, J., 2013. 2D hyperspectral frame imager camera data in photogrammetric mosaicking. In: *The International Archives of the Photogrammetry, Remote Sensing and Spatial Information Sciences.* Rostoch, Germany. . XL-1/W2, pp. 263–267.

Remondino, F., Barazzetti, L., Nex, F., Scaioni, M. and Sarazzi, D., 2011. UAV photogrammetry for mapping and 3D modeling - current status and future perspectives, In: *The International Archives of the Photogrammetry, Remote Sensing and Spatial Information Sciences.* Zurich, Switzerland. Vol. XXXVIII-1/C22, 7 pages.

Ruy, R., Tommaselli, A.M.G., Galo, Maurício, M., Hasegawa, J.K. and Reis, T.T., 2009. Evaluation of bundle block adjustment with additional parameters using images acquired by SAAPI system In: *Proceedings of 6th International Symposium on Mobile Mapping Technology.* Presidente Prudente, Brazil.

Tommaselli, A. M. G., Berveglieri, A., Oliveira, R. A., Nagai, L. Y., Honkavaara, E., 2015. Orientation and calibration requirements for hyperpectral imaging using UAVs: a case study. In: *Proceeding of the Calibration and Orientation Workshop – EuroCOW 2015.* Lausanne, Switzerland, February 10-12, 8 pages.

Silva, S.L.A., Tommaselli, A.M.G. and Artero, A.O., 2014. Utilização de alvos codificados na automação do processo de calibração de câmaras. *Boletim de Ciências Geodésicas.* 20(3), pp. 636–656.

4

PRECISION ANALYSIS OF POINT-AND-SCALE PHOTOGRAMMETRIC MEASUREMENTS FOR CORRIDOR MAPPING: PRELIMINARY RESULTS

P. Molina[a,*], M. Blázquez, [a], J. Sastre[a], I. Colomina[a]

[a] GeoNumerics S.L., Parc Mediterrani de la Tecnologia, 08860 Castelldefels (Spain)
(pere.molina, marta.blazquez, jaume.sastre, ismael.colomina)@geonumerics.com

ICWG III/I

KEY WORDS: mapKITE, Unmanned Aerial Systems (UAS), Terrestrial Mobile Mapping Systems (TMMS), integrated sensor orientation (ISO), corridor mapping

ABSTRACT:

This paper addresses the key aspects of the sensor orientation and calibration approach within the mapKITE concept for corridor mapping, focusing on the contribution analysis of point-and-scale measurements of kinematic ground control points. MapKITE is a new mobile, simultaneous terrestrial and aerial, geodata acquisition and post-processing method. On one hand, the acquisition system is a tandem composed of a terrestrial mobile mapping system and an unmanned aerial system, the latter equipped with a remote sensing payload, and linked through a 'virtual tether', that is, a real-time waypoint supply from the terrestrial vehicle to the unmanned aircraft. On the other hand, mapKITE entails a method for geodata post-processing (specifically, sensor orientation and calibration) based on the described acquisition paradigm, focusing on few key aspects: the particular geometric relationship of a mapKITE network —the aerial vehicle always observes the terrestrial one as they both move—, precise air and ground trajectory determination —the terrestrial vehicle is regarded as a kinematic ground control point— and new photogrammetric measurements —pointing on and measuring the scale of an optical target on the roof of the terrestrial vehicle— are exploited.

In this paper, we analyze the performance of aerial image orientation and calibration in mapKITE for corridor mapping, which is the natural application niche of mapKITE, based on the principles and procedures of integrated sensor orientation with the addition of point-and-scale photogrammetric measurements of the kinematic ground control points. To do so, traditional (static ground control points, photogrammetric tie points, aerial control) and new (pointing-and-scaling of kinematic ground control points) measurements have been simulated for mapKITE corridor mapping missions, consisting on takeoff and calibration pattern, single-pass corridor operation potentially performing calibration patterns, and landing and calibration pattern. Our preliminary results show that the exterior orientation, interior orientation and tie points precision estimates are better when using kinematic control with few static ground control, and even with excluding the latter. We conclude then that mapKITE can be a breakthrough on the UAS-based corridor mapping field, as precision requirements can be achieved for single-pass operation with no need for traditional static ground control points.

1. INTRODUCTION

Unmanned Aerial Systems (UAS) have come to stay among the mapping community (Colomina and Molina, 2014). Small, low-cost and easy-to-operate aerial robots have blossomed since few years ago, and they are now a consolidated technology in our modern society at various levels (from professional to mass-market niches), just leaving room for questions about its future growth quotes and evolution potential. In the geoinformation sector, several UAS have established themselves as *must-have* surveyor tools (Sensefly's Swinglet or eBee, Trimble's UX-5 or Ascending Technologies' Falcon-8) and even some consumer-grade platforms are being integrated into the mapping field; e.g., Phantom DJI's via Pix4D tools. At the end, technical barriers have been broadly overcome as opposite to regulatory ones —the latter have leveraged the use of UAS in mining and archaeological site survey as well as agriculture, but delayed operations in infrastructure inspection or urban mapping.

Regarding urban and inter-urban scenarios, high-resolution mapping has been recently performed with Terrestrial Mobile Mapping Systems (TMMS) equipped with cameras and/or LiDAR sensors, able to produce 3D geoinformation with high resolution and accuracy relying on direct sensor orientation by means of navigation-grade inertial and GNSS systems. While high-resolu-

tion close-range applications e.g. building facades, have been quite resolved by TMMS (despite the challenge of navigation in harsh urban scenarios), the lack of a "total point of view," that is, aerial and ground view of the scene, might be a handicap.

To this respect, efforts in fusing Unmanned Aircraft (UA) images with terrestrial images (Püschel et al., n.d.), (Mayer, 2015), or combining 3D models based on UAS images with terrestrial LiDAR ones (Optech, 2015), indicate the feasibility of such an approach to achieve the total point of view from a mapping perspective. Yet, to our knowledge, there is no approach performing such thing in a simultaneous geodata acquisition mode. Moreover, with the current UAS architectures with static ground control station, limitations for the UA-to-station distance (usually, below visual line-of-sight) jeopardizes the mission productivity, specially in the context of long corridors.

In the context of corridor mapping, sensor orientation is a challenge itself. The weak geometry of single-strip blocks is usually compensated by a dense coverage of ground control points (GCPs) or by INS/GNSS aerial control. While current work by mapping companies (SenseFly, 2015), (Delair Tech, 2014) are based on the former, recent research explores the use of precise position and attitude aerial control to perform direct sensor orientation in corridor blocks (Rehak and Skaloud, 2015). The Fast AT concept was explored for corridor mapping in (Blázquez and Colomina, 2012), based again on precise position and attitude

---

*Corresponding author

aerial control and few image measurements of GCPs, and compared to traditional ISO and to direct sensor orientation (DiSO). The method does not rely on tie points between images, which avoids image measurements on ill-textured environments e.g. rivers, sand, trees, etc.

As previously highlighted, our work is devoted to the mapKITE paradigm for sensor orientation in corridors, which follows the trend of precise aerial control but introduces a new type of ground control, and aims at assessing the potential of the concept and identify its benefits and limitations. To do so, section 2 will provide the basics about the mapKITE concept; section 3 presents the modelling aspects that define the mapKITE sensor orientation approach and section 4 exposes the framework for the achieved results.

2. THE MAPKITE ESSENTIALS

The system. A mapKITE system consists of a TMMS and UAS tandem, equipped with remote sensing payloads, and linked by a "virtual tether," that is, a real-time supply of waypoints derived from the navigation solution of the TMMS and sent to the UAS ground control station to be executed by the UA. Basically, the effect is that of a UA following a terrestrial vehicle. (In other contexts, this technique is known as the "follow me" UA flying mode.) These waypoints are conveniently set to comply with the mapping mission requirements (image footprint size and overlap, etc.) and potentially perform INS/GNSS and/or camera calibration maneuvers. In mapKITE, the UAS ground control station is mobile, carried by the TMMS vehicle thus optimizing the line-of-sight conditions, necessary for the communication link and for safety aspects. In particular, this configuration enables UAS operation far beyond the starting point, in contrast to the usual [static] UAS configurations. The result is a "mapping kite" with a total point of view, from ground to air, targeting at corridor mapping applications such as mapping of roadways, railways, waterways and other linear structures. In these cases, the continuous-line-of-sight nature of the system is a productivity booster and a concept differentiator. More details about the concept and its current implementation can be found in (Molina et al., 2015).

New measurements within a new paradigm. One key element of a mapKITE system is an optical target placed on the roof of the terrestrial vehicle. Based on the concept of coded targets featuring fast and robust identification in images for target tracking (Cucci et al., 2015), the target enables image measurement in two different ways, *pointing* —classical photogrammetric measurement of the target— and *scaling* —metric measurement of the image projection of the target in relation to the original target or, put simply, measurement of the scale of a point in an image. Additionally, these image measurements can be associated to precise global coordinates by means of the post-processed navigation solution of the TMMS. Consequently, the target shall be interpreted as a Kinematic Ground Control Point (KGCP) that extends the classic concept of Static Ground Control Point (SGCP) by physically moving in solidarity with the terrestrial vehicle and, thus, being potentially measured in every image. Note also that, in these conditions, obtaining motionless KGCPs (an analogy to traditional GCP) is as simple as stopping the terrestrial vehicle yet much cheaper than surveying techniques. Precisely, the potential of this new type of ground control, the KGCP, through its point-and-scale image measurement is the object of study of the present work.

Stop-and-go calibration in missions. When considering multicopter platforms for mapKITE, one may think of a typical mission starting with an ascent to a nominal operational height and

end with a descent to ground. During these two phases, calibration maneuvers, e.g. bearing turns and short translations, can be done while the terrestrial vehicle in motionless. Additionally, a mapKITE operator may even stop the terrestrial vehicle (and consequently, the UA) during the mission and command such calibration maneuvers aiming at strengthening the measurement network on those points and help with the process of self-calibration.

Navigation approach. Precision, accuracy and reliability of the mapKITE aerial and ground trajectories is instrumental in many aspects, covering real-time to post-mission. To this respect, the use of modern and modernized GNSS systems is addressed by combining reliable GPS signals with the accuracy and multipath-resistant properties of the new GNSS like Galileo or Beidou. Regarding navigation of the UA, a hybrid INS/GNSS concept is envisioned, enhanced by using the European Geostationary Navigation Overlay Service (EGNOS) that provides real-time GPS differential corrections to improve accuracy and ensure integrity of the GNSS signals. In relation to the generation of the UA waypoints by the terrestrial vehicle, the use of the Galileo E5 AltBOC (15,10) signal and modulation is of particular interest because of its robustness against multipath errors. Additionally, its 2 cm noise pseudorange shall contribute to the quality of the post-mission ground trajectory determination, which is the source of the KGCP measurements.

3. SENSOR ORIENTATION AND CALIBRATION APPROACH

As already discussed, the mapKITE control configuration concept includes the usual control components of integrated sensor orientation —aerial control with a constant shift GNSS error model and ground control (the SGCPs)— plus the KGCPs. Moreover, since the KGCPs are materialized with targets of known size and shape, point-and-scale photogrammetric measurements can be made for the KGCPs. In short, a mapKITE ISO extends the usual ISO of aerial blocks by using KGCPs and their point-and-scale photogrammetric measurements. All this, together with calibration image sequences whose images are taken at different altitudes during landing and takeoff maneuvers, opens the door to the self-calibration of the interior orientation (IO) parameters as the geometry of the resulting network is rather strong.

In this article we report on preliminary research results rather than pursuing a comprehensive analysis on precision, accuracy —i.e., feasibility of calibration of systematic errors— and reliability [of the new type of measurements]. We have therefore made a number of assumptions on the precision of the input measurements and on the trajectory shape of the terrestrial-aerial vehicle tandem (confer section 4). Similarly, we have made assumptions on the stability properties of the aircraft camera(s) which, obviously, are camera dependent and that, in some cases, may totally deviate from our assumptions. More specifically, we have assumed that IO parameters are unstable from flight to flight but stable within a flight; i.e., a block-to-block variant and in-block invariant IO hypothesis. We have also assumed that radial and tangential distortions can be calibrated off-line, with lab- or field-calibration strategies, in a way that the parameters of their modeling functions —e.g., the k_1, k_2, k_3, p_1 and p_2 parameters of the Conrady-Brown model— are sufficiently decorrelated from the IO parameters. This can be achieved with an appropriate acquisition geometry.

With the above assumptions we try to emulate the situation of high-end, off-the-shelve cameras and high-end, off-the-shelve lenses of fixed focal length. This is a reasonable setup in the context

Coordinate Reference Frames

l	Cartesian local geodetic frame (Easting-Northing-height)
c	Cartesian camera instrumental frame (forward-left-up)

Parameters

$P^l = (E, N, h)^l$	ground point in l-frame
$P_0^l = (E_0, N_0, h_0)^l$	camera projection centre (l-frame)
$R_c^l(\omega, \phi, \kappa)$	camera attitude matrix (parameterized by the Euler angles $(\omega, \phi, \kappa)_c^l$
$(\Delta x_0, \Delta y_0, \Delta c)^c$	interior orientation biases (c-frame)

Observations

$\ell_x, \ell_y, \ell_\lambda$	image coordinates and scale observations of P^l
$\nu_x, \nu_y, \nu_\lambda$	image coordinates and scale residuals of P^l

Instrument constants

x_0, y_0, c	principal point coordinates and camera constant

Table 1: Coordinate reference frames, parameters, observations and instrument constants.

of the mapKITE missions and the targeted corridor mapping applications.

Last but not least in our orientation and calibration approach, is the mathematical model for point-and-scale photogrammetric measurements $\ell_x, \ell_y, \ell_\lambda$ that we will use for the KGCPs and that can be written in the obvious form

$$P^l = P_0^l + (\ell_\lambda + \nu_\lambda) \cdot R_c^l \cdot \begin{bmatrix} \ell_x + \nu_x - (x_0 + \Delta x_0) \\ \ell_y + \nu_y - (y_0 + \Delta y_0) \\ -(c + \Delta c) \end{bmatrix}^c \quad (1)$$

where ℓ_λ is the measured image scale factor at point P^l. The meaning of the rest of variables and reference frame symbols in the previous equation are described in Table 1.

4. SIMULATED CORRIDORS AND RESULTS

In order to understand the general properties and overall contribution of the new measurement types introduced for mapKITE, KGCPs and point-and-scale photogrammetric measurements, we have simulated single-strip corridor blocks by generating measurements for all required observables as detailed in table 2.

Seasoned researchers and practitioners know what simulations can tell and what they can be made to tell. Keeping this in mind and with the goal of developing a realistic —non-optimistic or, at least, not too optimistic— insight and intuition on the behavior of mapKITE blocks, we have "designed" a block which can be materialized from an operational standpoint and we have assigned measurement precisions which are attainable, rather pessimistic than optimistic, under operational conditions.

Table 2 summarizes the mission design parameters and the precision of the simulated measurements, and table 3 summarizes the results of the simulations for each simulated corridor. We discuss them in the next sections.

4.1 Block design

As described in table 2, we have simulated a 5 km, single-strip, rectilinear, 80% forward overlap block, acquired from an altitude of 100 m over flat terrain with camera parameters close to the popular Sony NEX-7 camera and a 20 mm camera constant lens resulting in a 2 cm GSD.

Observables	Precision	Units
Static GCP (SGCP)		
- $\sigma_{E,N}$	3	cm
- σ_h	5	cm
Kinematic GCP (KGCP)		
- $\sigma_{E,N}$	6	cm
- σ_h	10	cm
Tie Point Image Coordinates		
- $\sigma_{x,y}$	3	μm
	0.77	px
Point-and-scale coordinates of KGCP		
- $\sigma_{x,y}$	3	μm
	0.77	px
- σ_λ	100	ppm
GNSS aerial control		
- $\sigma_{E,N}$	3	cm
- σ_h	5	cm

Mission design parameters	Name/Value	Units
Equipment (UA camera)	Sony NEX-7	-
Equipment (TV navigation)	GNSS receiver geodetic grade	-
Equipment (UA navigation)	GNSS receiver geodetic grade	-
Sensor size	23.5 x 15.6	mm
	6000 x 4000	px
Pixel size	3.9	μm
Camera constant	20	mm
Corridor length	5	km
Corridor flying height	100	m
Base-to-height ratio	0.15	-
No. of bases	328	-
Calibration maneuvers	start & end	-
Ground Sampling Distance (GSD)	2	cm
Forward Overlap	80	%

Table 2: Precision of observables and mission design parameters.

Image tie point measurements are uniformly distributed along the image, and KGCPs correspond to the trajectory of a terrestrial vehicle "paired" to the UA trajectory, as in a mapKITE mission. SGCPs have been generated as pairs of points, aligned with the across-track direction, and distributed along the corridor with a separation of 900 metres. Since, in mapKITE, the aircraft is a multicopter, we have simulated maneuvers for image calibration; that is, sets of images taken during short ascent from or descent to 50 m, combined with horizontal translations and heading turns. These maneuvers are always feasible at the start and end of a mapKITE mission coinciding with the takeoff and landing phases. In fact, as already mentioned, it is always possible —and potentially beneficial— to stop the mission for a short period of time, in which the aircraft repeats the calibration maneuver while the terrestrial vehicle remains static. In this case, a KGCP becomes a SGCP.

4.2 Precision of measurements

We have given SGCPs the precision resulting from standard surveying best practices with geodetic-grade GNNS receivers ($\sigma_{E,N}$

$= 3$ cm, $\sigma_h = 5$ cm). We have assumed that KGCPs are measured with equivalent surveying equipment as the SGCPs. However, even with identical equipment, kinematic positioning is less precise than its static counterpart and, in a corridor mapping mission, GNSS signal fading —if not signal occlusions— may occur further resulting in less precise measurements because of a lower signal-to-noise ratio. Therefore, we have given KGCPs a larger random error ($\sigma_{E,N} = 6$ cm, $\sigma_h = 10$ cm). In both cases, SGCPs and KGCPs, we have assumed that they are accurate since short periods of static measurements for GNSS integer ambiguity resolution can always be implemented.

We have restricted aerial control to position control since, at this point in time of mapKITE research, we did not want to add an equipment requirement (IMU) to the unmanned aircraft payload that may prove to be not strictly necessary. (This is arguable. However, in any case, inertial measurements would just bring improvement upon the reported results.) Following the usual approach of airborne photogrammetry we have assumed that the aircraft is equipped with a geodetic grade GNSS receiver, which for state-of-the-art receivers, translates into a multi-constellation and multi-frequency receiver. GNSS measurements are assumed to be processed under the Precise Point Positioning (PPP) mode which is more precise but less accurate than differential carrier phase methods. Eventual systematic errors of PPP are modeled with the usual shift parameters. According to these we have assigned $\sigma_{E,N} = 3$ cm and $\sigma_h = 5$ cm to aerial control. (This is also arguable since differential processing might be also performed for short periods when the aircraft is on ground, to avoid large inaccuracies. Again, such an approach would just improve upon the reported results.)

Image coordinate measurements for natural and targeted tie points have been given an approximate precision of $\sigma_{x,y} \approx 1$ px, clearly on the conservative side. Last, we have set a precision of $\sigma_\lambda \approx 100$ ppm for the scale component of the point-and-scale measurements. The value is rather speculative, based on the expected precision of matching contours to extracted target edges. However, the simulations conducted so far, indicate that the parameter precision is rather insensitive to σ_λ. For instance, setting $\sigma_\lambda \approx 1000$ ppm deteriorates slightly ground point vertical accuracy (17%) —although augments its redundancy number, from 4% to 54%.

4.3 Software

Observation equation 1 has been implemented in the airVISION model toolbox of GeoNumerics' generic network adjustment platform GENA (Colomina et al., 2012). The rest of the mathematical models used in the simulations were already available in airVISION. The simulations themselves have been performed with GENA and airVISION.

4.4 Simulated configurations

In order to facilitate the understanding of the essential properties of the mapKITE orientation and calibration method, we have selected six plus one representative configurations of mapKITE missions. The selection aims also at the comparison between the traditional way of designing a corridor mapping mission (row 1), corridors featuring, long to too long GCP bridging (rows 2 and 3) and some candidate mapKITE configurations with KGCPs (rows 4, 5 and 6). The configurations correspond to the six rows of Table 3 and the single row of Table 4 respectively. All configurations include, at least, two camera calibration maneuvers, one for each end of the strip; one of them (Table 3, row 6) includes three additional ones.

The first configuration (Table 3, row 1) corresponds to a mapKITE flight whose orientation and calibration have been obtained in the traditional way, i.e., by using GNSS aerial control and SGCPs, at the ends of and within the strip. The bridging distance between pairs of GCPs is about 13 bases or 195 m. This configuration aims at being representative of non-mapKITE, traditional, as said, methods.

The second configuration (Table 3, row 2) is configuration 1 with less GCPs, with bridging distance between pairs of GCPs of about 60 bases or 900 m. It is an in-between configuration among the first and third configurations with the only purpose of illustrating the trade-offs between GCP density and performance.

The third configuration (Table 3, row 3) results from removing all GCPs but those at the strip ends. Bridging distance between pairs of GCPs is, therefore, 320 bases or the length of the strip. This configuration serves the purpose of illustrating the role of GCPs and the improvement brought by KGCPs. It is not representative of real-life situations and will not be used to draw conclusions.

The fourth configuration (Table 3, row 4) is our flagship. (It will result in being our recommended configuration.) This configuration includes just two pairs of GCPs at the strip ends and one KGCP for each image with its associated point-and-scale photogrammetric measure.

The fifth configuration (Table 3, row 5) is the fourth one without GCPs.

The sixth configuration (Table 3, row 6) adds three aircraft maneuvers with static terrestrial vehicle to the previous configuration, and in practice, also to the fourth one. Its goal is to quantify the improvement brought into the determinability of the calibration parameters, their redundancy numbers of the point-and-scale measurements and any possible significant improvement in the parameters of interest, ground points and exterior orientation parameters.

The last configuration (Table 4) is almost equivalent to configuration 4, the only difference being the precision of the scale measurement which, in this case is $\sigma_\lambda = 1000$ ppm instead of $\sigma_\lambda = 100$ ppm.

4.5 Simulation results

We provide Table 3 to present, on one side, a schematic view of the mission plan phases of the simulated blocks and, on the other hand, the estimated precision results for the relevant parameters of interest —Exterior Orientation (EO), Ground Points (GP) and Interior Orientation (IO)— together with the Internal Reliability (IR) for every measurement listed in 2. Yet, we provide additional explanations hereafter:

- The initial cell (first row, first column) depicts the symbols for SGCPs, KGCPs, calibration maneuvers and motionless KGCP symbols. Recall that aerial control and tie points are present in all corridors (thus not made explicit),

- The rest of the cells in the first row list the used estimators for precision (mean of the standard deviations) and internal reliability (mean of redundancy numbers) for each measurement,

- The rest of the rows describe the actual results following the model in the first row for the simulated blocks.

The main result is that with the mapKITE orientation and calibration method (the mapKITE ISO), we obtain the same results, if not slightly better, as with the traditional ISO method with dense GCPs. Moreover the redundancy numbers of the measurements in the mapKITE ISO are high with the exception of point-and-scale measurements that range between 4% and 54%.

In table 3, first, we observe the expected results for traditional ISO in the described corridor conditions: the precision of the EO position wanders around the precision of the aerial and ground control, a strong heading angle determination and a lower precision of the across-track component of the ground points. Measurement redundancy numbers are considered within expectations. Secondly, we observe also the effect of increasing the ground control bridging in rows 2 and 3, which is a degradation of the across-track ground point coordinates and of the EO attitude ω angle, while the EO position is still dominated by the aerial control. We observe a decrease of the SGCP redundancy, as we as we decrease the number of GCPs.

When KGCPs are introduced and measured, that is, performing mapKITE ISO, we immediately observe a precision improvement of all the estimated parameters as compared to traditional ISO, even with short bridging distances (row 1). In particular, a remarkable effect on ground point precision is observed. Additionally, we observe that the redundancy numbers of SGCP measurements are higher in presence of KGCPs. As a matter of fact, we observe by the results in row 5 that even without SGCP, mapKITE ISO leads to the same results as traditional ISO with high SGCP density. We do not recommend, though, this configuration and tend to favor configuration 4 as static GCPs have proven to be an excellent tool against GNSS receiver malfunctions, weak geometries, frequent datum mistakes, etc. Moreover, it is feasible from an operational standpoint as the corridor ends usually correspond to system set-up periods where surveying measurements can be taken confortably.

Regarding the calibration of the IO parameters, we observe some benefit with the addition of calibration maneuvers in the middle of the strip, as in row 6. Whether this improvement, in practice, is significant or not we shall see in the future. In any case, camera calibration maneuvers with a multicopter is a low-cost operation and supposes a low impact in aerial corridor mapping missions.

We also provide measurement redundancy numbers to analyze the contribution of these new measurements as compared to the traditional ones (table 3, right columns). First of all, we observe a high redundancy for KGCPs. This indicates a good condition for error identification through regression diagnostics mechanisms, which is convenient for mapKITE as we expect the terrestrial vehicle trajectory to be a major source of errors. Secondly, we observe (table 3) a low redundancy of the point-and-scale measurement, which indicates a strong influence of such measurements with respect to the rest. However, in table 4, for $\sigma_\lambda = 1000$ ppm, the scale measurement mean redundancy number goes up to 54% due to its lower weight at the expected expense of a less precise vertical component determination. Admittedly, at the current stage of this research, we have not rigorously characterized the precision of the scale measurement. In any case, double KGCP targeting results in significant higher redundancy numbers.

5. SUMMARY AND PRELIMINARY CONCLUSIONS

We have described the mapKITE orientation and calibration concept that we also call the "mapKITE ISO" and have performed simulations with rather conservative assumptions on the quality of the input measurements to understand its potential benefits.

The simulations indicate, that the mapKITE ISO can replace traditional ISO with great savings in time and money for corridor mapping projects.

The research leading to these results has been funded by the European Community Horizon 2020 Programme under grant agreement no. 641518 (project mapKITE, www.mapkite.com) managed by the European GNSS Agency (GSA).

MapKITE has been patented by GeoNumerics (Spanish patent 201231200) and further PCT applications have been filed to cover Europe, United States and Brazil.

REFERENCES

Blázquez, M. and Colomina, I., 2012. Performance analysis of Fast AT for corridor aerial mapping. ISPRS - International Archives of the Photogrammetry, Remote Sensing and Spatial Information Sciences XXXIX-B1, pp. 97–102.

Colomina, I. and Molina, P., 2014. Unmanned aerial systems for photogrammetry and remote sensing: A review. {ISPRS} Journal of Photogrammetry and Remote Sensing 92, pp. 79 – 97.

Colomina, I., Blázquez, M., Navarro, J. and Sastre, J., 2012. The need and keys for a new generation network adjustment software. ISPRS - International Archives of the Photogrammetry, Remote Sensing and Spatial Information Sciences XXXIX-B1, pp. 303–308.

Cucci, D., Constantin, D. and Rehak, M., 2015. Smile targets in aerial photogrammetry. In: Proceedings of the International Micro Air Vehicle Conference and Competition, Vol. 54, Aachen, Germany.

Delair Tech, 2014. Power lines inspection using mini-UAV. Online at http://goo.gl/NK5zZq; accessed 25-January-2015.

Mayer, H., 2015. From Orientation to Functional Modeling for Terrestrial and UAV Images. In: 55th Photogrammetric Week, Institut für Photogrammetrie, Universität Stuttgart, pp. 165–174.

Molina, P., Blázquez, M., Sastre, J. and Colomina, I., 2015. A method for simultaneous aerial and terrestrial geodata acquisition for corridor mapping. ISPRS - International Archives of the Photogrammetry, Remote Sensing and Spatial Information Sciences XL-1/W4, pp. 227–232.

Optech, 2015. ILRIS Terrestrial Laser Scanner Integrating UAV Camera Imagery. Online at http://goo.gl/ERIsA9; accessed 25-January-2015.

Püschel, H., Sauerbier, M. and Eisenbeiss, H., n.d. A 3D model of Castle Landenberg (CH) from combined photogrammetric processing of terrestrial and UAV-based images. Vol. XXXVII-B1, Beijing, China, pp. 963–970.

Rehak, M. and Skaloud, J., 2015. Fixed-wing micro aerial vehicle for accurate corridor mapping. ISPRS - International Archives of the Photogrammetry, Remote Sensing and Spatial Information Sciences.

SenseFly, 2015. Drones vs traditional instruments: corridor mapping in Turkey. Online at https://goo.gl/4rLmtZ; accessed 25-January-2015.

Legend: ● Static GCP   × Kinematic GCP   ⋈ Motionless KGCP   ↻ Calibration Manoeuver

Schematic mission plan	EO pos (mm) σ_E σ_N σ_h	EO att (mdeg) σ_ω σ_ϕ σ_κ	GP (mm) σ_{P_E} σ_{P_N} σ_{P_h}	IO (px) $\sigma_{\Delta c}$ $\sigma_{\Delta x_0}$ $\sigma_{\Delta y_0}$	IR SGCP (%) $\bar r_E$ $\bar r_N$ $\bar r_h$	IR KGCP (%) $\bar r_E$ $\bar r_N$ $\bar r_h$	IR ic (%) $\bar r_x$ $\bar r_y$ –	IR P&S (%) $\bar r_x$ $\bar r_y$ $\bar r_\lambda$	IR AC (%) $\bar r_E$ $\bar r_N$ $\bar r_h$
(mission plan diagram)	22 / 30 / 35	19 / 11 / 6	12 / 16 / 25	1.7 / 0.6 / 0.7	84 / 75 / 79	– / – / –	71 / 84 / –	– / – / –	62 / 50 / 94
(mission plan diagram)	23 / 32 / 37	25 / 11 / 7	16 / 30 / 31	1.8 / 0.6 / 0.7	77 / 56 / 75	– / – / –	71 / 83 / –	– / – / –	62 / 50 / 94
(mission plan diagram)	27 / 33 / 39	45 / 11 / 8	24 / 75 / 43	1.9 / 0.6 / 0.7	63 / 57 / 65	– / – / –	71 / 83 / –	– / – / –	62 / 50 / 93
(mission plan diagram)	21 / 29 / 31	19 / 11 / 6	12 / 14 / 24	1.5 / 0.6 / 0.7	89 / 86 / 87	92 / 91 / 98	71 / 84 / –	12 / 12 / 4	63 / 51 / 95
(mission plan diagram)	21 / 30 / 32	20 / 11 / 6	12 / 14 / 24	1.6 / 0.6 / 0.7	– / – / –	92 / 91 / 98	71 / 84 / –	12 / 12 / 4	63 / 51 / 95
(mission plan diagram, 3×)	19 / 24 / 21	17 / 11 / 6	12 / 14 / 23	0.9 / 0.4 / 0.5	– / – / –	92 / 91 / 98	72 / 84 / –	20 / 20 / 8	65 / 53 / 95

EO: Exterior Orientation; GP: Ground Point; IO: Interior Orientation; IR: Internal reliability; SGCP: static ground control point; KGCP: kinematic ground control point; ic: image coordinates; P&S: point-and-scale; AC: aerial control

Table 3: Precision and internal reliability results for the mapKITE corridors, varying SGCPs, KGCPs and calibration manoeuvers.

Schematic mission plan	EO pos (mm)	EO att (mdeg)	GP (mm)	IO (px)	IR SGCP (%)	IR KGCP (%)	IR ic (%)	IR P&S (%)	IR AC (%)
(mission plan diagram)	21 / 30 / 32	20 / 11 / 6	12 / 14 / 29	1.6 / 0.6 / 0.7	88 / 86 / 86	92 / 91 / 98	71 / 84 / –	12 / 12 / 54	63 / 50 / 94

Table 4: Precision and internal reliability results for the mapKITE corridor using $\sigma_\lambda = 1000$ ppm.

5

NEED FOR RELIABLE SENSOR CALIBRATION FROM THE PERSPECTIVE OF A NATIONAL MAPPING AGENCY

S. Baltrusch [a]

[a] LAiV M-V, NMCA M-V, Department for Photogrammetry, 19059 Schwerin, Germany – sven.baltrusch@laiv-mv.de

EuroCOW 2016

KEY WORDS: Calibration, Validation, Geometry, Radiometry, standardisation, Basic Geodata, NDVI

ABSTRACT:

The sensor calibration is one of the basic elements for getting effective and efficient production workflows out of airborne photogrammetry. Digital images and their orientations (interior and exterior) are the key to get the resulting products into the workflow of a national mapping agency (NMCA). Not only the geometric calibration is required meanwhile the radiometric calibration as well is used, for example in raster-based classification processes. In the paper the requirements are shown and examples are presented. At least open aspects and the outstanding debts are given.

1. INTRODUCTION

1.1 Current position

The NMCAs of the Federal Republic of Germany contract out aerial surveys since many years. In the past the most and nearly only aim was to update the information in the landscape model (Basis-DLM) as well as in the series of topographic maps.

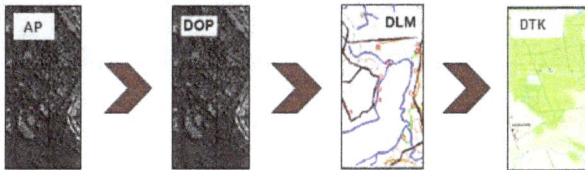

Figure 1. Traditional workflow for aerial survey results in NMCAs

The aerial photos still deliver fundamental information for the Authorative Topographic-Cartographic Information System ATKIS (Figure 2).

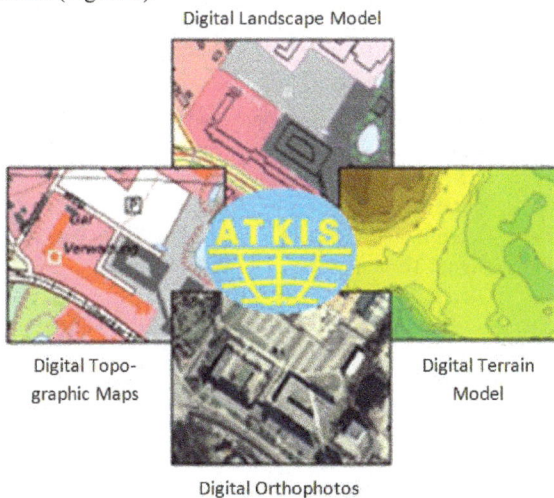

Figure 2. Authorative Topographic-Cartographic Information System ATKIS

With the change into digital photogrammetry workflows and the new basic approaches, deduced from the computer vision technologies, the aerial photos gained from aerial surveys have an increasing importance for a variety of applications. The quality demands have increased and the currency cycles were shortened parallel. These new use cases strengthen the need for reliable sensor calibration. Not only the geometric calibration is required meanwhile the radiometric calibration as well is used, for example in raster-based classification processes.

The NMCAs have to reach these requirements with respect to permanently shrinking number of staff, which requires automatic workflows, control steps and monitoring strategies. Only samples are possible. Sustained quality controls are inefficient and are extremely time-consuming. Figure 3 shows the controlpoint distribution for NMCA aerial survey projects (JRC, 2008).

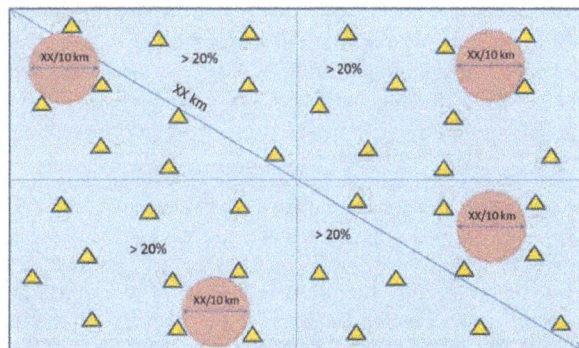

Figure 3: Controlpoint distribution for NMCA aerial survey projects

Other official agencies depend on a continuous delivery of qualified and contemporary datasets for their own purpose, commonly indentured by European or national laws.

1.2 Aerial Survey parameters of a NMCA

The German NMCAs have to tender their aerial survey projects with respect to the national and the European public procurement law. The Working Committee of the Surveying Authorities of the States of the Federal Republic Germany (AdV) published a guideline for tendering a digital aerial survey as a basis for the production of ATKIS-DOP and stereoscopic analysis (AdV, 2014).

The requirements reflected to the camera-system are described with the following facts:

- Large format digital camera systems
- Simultaneously recording of the panchromatic, the red, green and the blue channel as well as the near infrared with separated optics
- PAN-sharpening-ratio max. 1:4
- Illumination with a leaf shutter system
- Motion blurring has to be minimized with FMC/TDI and gyro techniques
- Orientation accuracy of the direct georeferencing shall be better than 0,5 m spatial variance

The following technical parameters complete the definition of an aerial project:

Parameter	Value
Geometric Resolution	GSD 10 – 20 cm
Spectral Resolution	PAN + RGBI
Radiometric Resolution	8 / 16 bit
Forward Overlap	70 – 80%
Side Overlap	30 – 50%
Flight cycle	2 – 3 years
Assignment period	Spring / Summer

Table 1. Aerial survey parameters of a NMCA

1.3 Calibration requirements

The camerasystem has to be calibrated geometrically and radiometrically. The requirements are based on the remarks given in the DIN 18740-4. The calibration has to be proven by a certificate of the constructor. The validity of the geometric calibration for the time of an aerial survey has to be attested by a validation flight (not older than one year) or a renewed complete calibration (not older than two years).

1.4 Orientation requirements

For the image orientation the direct georeferencing (GNSS, INS) has to be used. The correction parameters of the German SA*POS*-Service (DGNSS-permanent-services of the NMCAs) have to be considered. The orientation accuracy resulting from the direct georeferencing has to be better than 0,5 m spatial variance. The evidence has to be provided by the contractor at a spatial distributed selection of ground control points appropriated by the NMCAs. If the required accuracy cannot be provided the contractor has the duty to perform an aerotriangulation.

2. AUTHORITIVE PRODUCTS

2.1 The relevance of aerial photos

The basic product in the photogrammetric workflow in German NMCAs is the dataset of aerial photos. These photos have a fundamental relevance, because of

- no loss of information,
- no loss of projection,
- no loss of interpretation,
- no loss of dimensionality.

Thanks to the increasing accuracy of the direct georeferencing and the developments on the hardware consumer market as well as to the sinking prices for stereo visualisation techniques the aerial photos celebrated a renaissance.

2.2 Orientated aerial photos

Aerial photos with all necessary parameters for a stereoscopic evaluation (especially interior and exterior orientation) are called orientated aerial photos. The definitions, quality requirements and metadata definitions are based on the AdV-Standard for digital aerial photos (AdV, 2015).
The orientated aerial photos are divided in classes as a function of their orientation accuracy (1σ):

LB1:   image center (X, Y) approximated
LB2:   3-fold GSD[1] (resulting from direct georeferencing)
LB3:   2-fold GSD (resulting from aerotriangulation)
LB4:   1-fold GSD (resulting from aerotriangulation)

Accuracy-classes	GSD10	GSD20
LB1	1,00 m	1,00 m
LB2	0,30 m	0,60 m
LB3	0,20 m	0,40 m
LB4	0,10 m	0,20 m

Figure 4: Productclassification, defined by orientation accuracy

2.3 Production chain

The information of orientated aerial photos lead into a production chain with a series of following products like (Figure 5):

- Preliminary Orthophotos without interactive correction steps in geometry and radiometry (DOP-V)
- Orthophotos with interactive corrections (ATKIS-DOP)
- Rasterbased classification (Imageb. Classification)
- Dense Image Matching – Surface models (BDOM)
- Digital Terrain Models (ATKIS-DGM)
- Digital Surface Models (AdV-DOM)
- 3D-Building-Models (AdV-3DGbm)
- TrueOrthophotos (True-DOP)
- Timedifferenced Surface Models (tDOM)

---

[1] GSD = Ground Sample Distance

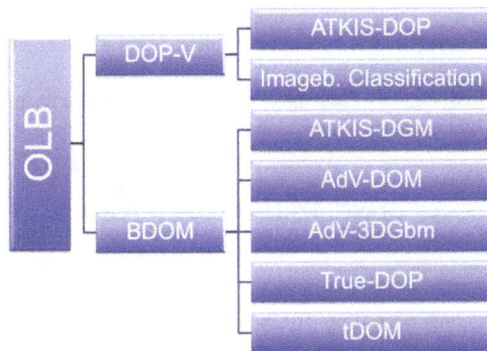

Figure 5: Product chain based on orientated aerial photos

These so called official basic geodatasets (Amtliche Geobasisdaten) are used in a lot of technical applications at other administrative bodies. Certified with the label "official product" they stand for neutrality on interests, reliability, unity in quantity and quality. These datasets have to be gapless and nationwide.

This induces the need for reliable sensor calibration from the perspective of a national mapping agency.

2.4 Additional requirements reasoned by European wide access to camera systems

The increasing accuracy requirements and the combined variety in the product chain lead to additional conditions of the camera system. The access and availability of the systems has to be made sure in the assignment periods.

Usually there are only a few days with weather conditions for aerial surveys so that the system has to guarantee stable calibration parameters for a complete assignment period. No additional camera calibrations are possible in "flight seasons".

The stability has to be made sure for a mission of several hours, only with stop-over for refueling.

The system parameters also have to be stable for varieties of weather conditions as well as in various flying heights caused by different target resolutions or restrictions of the aviation safety, for example at international airports. These parameters also depend on different conditions in different regions, for example depending on the changes of the relief.

3. EVERYDAY PROBLEMS

3.1 Default of standardised calibrations

In comparison to the analogue camera generation there is no standardisation of camera calibration anymore. The camera owner charges the camera constructor with the realisation of the camera calibration. Some owners do a calibration every two years, some owners perform a validation flight once a year.

The content of the calibration- or validation certificate varies between the camera constructors. Supplementary it has to be considered that the certificate is issued by the constructor and not an independent agency.

Especially the information resulting from a validation flight is commonly very sparse and not meaningful. The validation of the radiometric characteristics is completely missing. The procedure of the flight is not predefined. There are no specifications for standardised flying heights, scope and so on.

In the task force "ATKIS-DOP" of the AdV a requirement specification for the performance of a validation flight is

developed in association with scientists of TU Dresden (AdV, Prof. Maas, 2015).

- After every massive reconstruction at the camera system a new calibration has to be performed.
- The documentation of a validation check bases on the DIN 18740-4. The documentation has to give a replicable evidence, that the position- and height accuracy of the camera system worsens not more than 25% in comparison to the reference survey (burn-in flight).
- Precondition is a „burn-in flight" just after the last camera calibration. These results are used as a reference for the qualitive assessment of the validation flight. A comparison is factual, if both flights – burn-in flight and validation flight – are similar according to image scale, block configuration, use of control points and loading of all observation groups.
- Besides the full documentation of the bundel block adjustment, the graphical presentation of the flight structure, the control- and checkpointdistribution the following information have to be delivered:

  o   Flying height or GSD (max. deviation 20%)
  o   Forward overlap (max. deviation 10%)
  o   Side overlap (max. 5%)
  o   Number of photos (max. 50%)
  o   Number of flight strips (min. 3 strips)
  o   Number of cross strips (min. 1 strip)
  o   Number of control points (max. 8 points)
  o   Number of check points (min. 15 points)
  o   Identical configuration for GNSS/INS

  The deviation always refers to the comparison between the burn-in flight and the validation flight.
  The processes of the aerotriangulation including the strategies (loadings, camera calibration yes/no, boresight alignment yes/no) have to be similar.
  Finally the results from the statistics have to be compared. Both the internal values of the bundle block adjustment (for example: weight unit error) and the values of independent external reference measurements are expected to be similar with maximum deviations of 25%.

- For the assessment of the validation results and the compliance of the calibration values the documents belonging to the camera calibration as well as the results of the burn-in flight are needed.

These additional requirements are mandatory for the processing of validation flights for aerial surveys commissioned by a German NMCA.

3.2 Radiometric calibration

For the radiometric calibration a labour calibration is needed. In a camera validation these process is not included. Nevertheless the analysis and the use of the multispectral information growes constantly.

With respect to the multispectral information in a digital aerial survey with a large format camera system, the NMCAs try to adapt remote sensing algorithms used for satellite images on the analysis of aerial photos. Combined with other basic geodata information – mostly obtained from the same photo sources – it is possible to build up a multidimensional feature space for rasterbased classification applications.

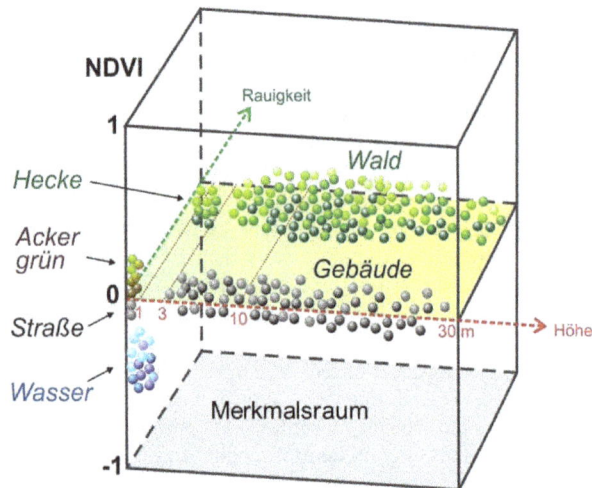

Figure 4. Multidimensional feature space (Rzeznik, 2015)

Current researches try to use these methods for updating the information in the landscape model, especially the updating of the effective landuse. This process demands a lot of man-power in interactive work.

For effective results the radiometric characteristic of the camera system has to be stable and reliable.

3.3 Requirements to post-processing-steps in camera software

Usually the task for the aerial survey contractors is to deliver aerial photos with a homogenous radiometric presentation for orthophoto productions. Therefore post-processing steps generate survey photographs from raw data. These production steps ignore the ratio between the red and the near infrared channel, which is a basic function in imagebased classification. This neglection leads to wrong assumptions based on the NDVI. In an aerial survey, flown at one day no constant NDVI values for the same objects are guaranteed.

3.4 Camera calibration in aerotriangulation steps

The NMCAs try to avoid camera calibration steps during an aerotriangulation. These calibration steps can be so powerful, that the final result, represented at a small number of control points, looks better than the projectwide quality really is.

Experiences of past projects show results, where the effects are not reasonable. A shift in the focal length or in the z-values lead to better results. But this is not justifiable.

The self-calibration is the last step to get better results. Normally geometric calibrated cameras are expected. If the camera self-calibration works effective enough there won´t be the demand for labour calibrations anymore.

3.5 No standardised exchange formats for camera parameters

Regardless on the discussions of the advantage and disadvantage of the common parametersets resulting from camera calibration in the aerotriangulation process there is no standardised exchange format for these parameters. This implicates that the accuracy improvements normally only can be used in the software solution similar to the aerotriangulation software. A change in the processing software with a reimport of the adjusted images is not solved at the moment.

But in the era of the renaissance of the orientated aerial photos non photogrammetric specialised software solutions want to use this information as well.

Additional analysis according to the stability of the radiometric sensor characteristic have to be implemented in the validation process.

4. CONCLUSIONS / OPEN TOPICS

The developments and requirements to digital photogrammetric products increased the last years very strong. This leads to high requirements according to the input data. The fundamental datasets are the orientated aerial photos with their geometric and radiometric information. The datasets have to be reliable calibrated so that the NMCAs can start their production workflows without attempts of camera improvements.

The NMCAs are looking forward to:

- Standardisation in camera calibration, preferably done by an independent agency,
- Implementing radiometric calibration into validation processes,
- Regard on the ratio (red / near infrared) in post-processing steps in camera software
- Standardised exchange-format for orientated aerial photos

REFERENCES

AdV, 2014. Guideline for tendering a digital aerial survey as a basis for the production of ATKIS-DOP and stereoscopic analysis (adv-online.de).

AdV, 2015. AdV-Standard for digital aerial photos (adv-online.de).

DIN18740-4:2007-09: Photogrammetric products — Part 4: Requirements for digital aerial cameras and digital aerial photographs.

JRC, 2008: Guidelines for Best Practise and Quality Checking of Ortho Imagery – Issue 3.0

Rzeznik, 2015: Change Detection from official remote sensing datasets (unpublished)

6

ODOMETRY AND LOW-COST SENSOR FUSION IN TMM DATASET

A. M. Manzino, C. Taglioretti

DIATI Department, Politecnico di Torino, Corso Duca degli Abruzzi 24, 10129 Torino, Italy –
(ambrogio.manzino, cinzia.taglioretti)@polito.it

KEY WORDS: Mobile Mapping, odometry, motion models, UKF, filtering techniques, sensor integration.

ABSTRACT:

The aim of this study is to identify the most powerful motion model and filtering technique to represent an urban terrestrial mobile mapping (TMM) survey and ultimately to obtain the best representation of the car trajectory. The authors want to test how far a motion model and a more or less refined filtering technique could bring benefits in the determination of the car trajectory.

To achieve the necessary data for the application of the motion models and the filtering techniques described in the article, the authors realized a TMM survey in the urban centre of Turin by equipping a vehicle with various instruments: a low-cost action-cam also able to record the GPS trace of the vehicle even in the presence of obstructions, an inertial measurement system and an odometer.

The results of analysis show in the article indicate that the Unscented Kalman Filter (UKF) technique provides good results in the determination of the vehicle trajectory, especially if the motion model considers more states (such as the positions, the tangential velocity, the angular velocity, the heading, the acceleration). The authors also compared the results obtained with a motion model characterized by four, five and six states.

A natural corollary to this work would be the introduction to the UKF of the photogrammetric information obtained by the same camera placed on board the vehicle. These data would permit to establish how photogrammetric measurements can improve the quality of TMM solutions, especially in the absence of GPS signals (like urban canyons).

1. INTRODUCTION

The majority of vehicles today have GPS instruments, digital maps or speedometers on board to measure their velocity. These instruments are characterized by low precision and are not able to integrate pieces of information from the different sources. The integration of the information obtained by these instruments with their accuracy and other visual data about the motion model could define a vehicle's future position more accurately (Lytrivis et al., 2010). The application fields of these studies are many: e.g. driver assistance systems, start-stop systems and adaptive cruise control (Schubert et al., 2008).

The Kalman filter, or its more modern derivatives, is the most commonly employed method to combine information about position, velocity, function of position and velocity and filtering techniques (Hartikainen and Särkkä, 2011).

The data filtering may take different form: if the stochastic model is represented by a Gaussian and the state equations between two successive epochs may be considered linear by leaving out the second order effects, a more suitable method is the Extended Kalman Filter (EKF). Alternatively, if these conditions are not verified, for example without the introduction of big approximations, it is possible to use other techniques such as the Unscented Kalman Filter (UKF).

However it is essential to adopt a motion model (Li and Jilkov, 2003) that can include in the state parameters and in the observation equations the measurements effectively realized during the survey.

2. CASE STUDY

In order to obtain an adequate dataset for the application of motion models and different filtering techniques, the authors realized a TMM survey in the urban centre of Turin. Therefore the path is characterized by rectilinear segments spaced out by

curves of up to 90 degrees, changes of velocity, some departures and stops and urban canyons, as shown in Figure 1.

Figure 1. In red: vehicle's trajectory, in green: portion of trajectory analyzed

With the aim of placing the instrumentation on the vehicle, the authors equipped its baggage rack with a structure reminiscent of a "cross". This is the name used (also in this article) to refer to this particular object.

The cross is able to host the whole instrumentation in known and unmovable positions.

In order to know the position of each instrument with a millimetric accuracy in respect of origin (O) and a direction (OA in Figure 2), the measurements were performed and adjusted depending on a tridimensional schema using a commercial software.

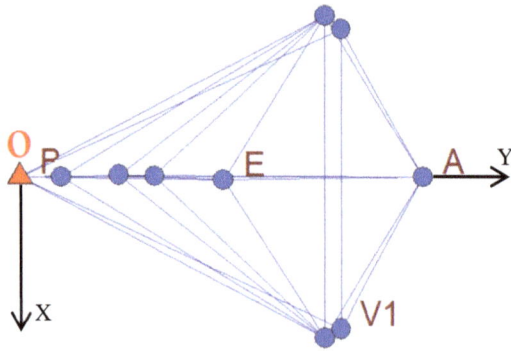

Figure 2. Schema of the positions of the various instruments on the cross

The authors equipped the vehicle with various instruments: a low-cost action-cam (position V1 in Figure 2, in orange in Figure 3) able to record the trace of the vehicle even in presence of obstructions, since it has an internal highly sensitive GPS; an inertial measurement system (INS, position E in Figure 2, in blue in Figure 3), which is useful for obtaining the reference trajectory especially in the absence of GPS signals; and a DMI (distance measurement indicator, located on the vehicle as shown in Figure 4).

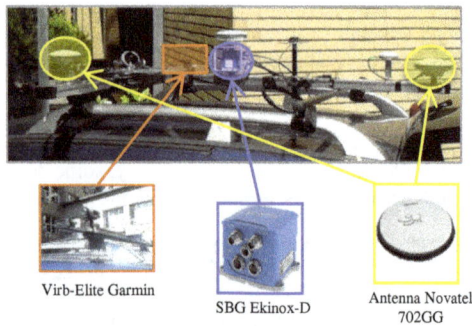

Figure 3. Instruments on board

Figure 4. The DMI in use

The Garmin Virb Elite Action-Cam 1080p HD used for the survey is able to acquire HD video (with 1920*1080 pixel resolution) and high velocity frames (25 fps), as well as to provide the GPS locations of the frames, while maintaining a low cost. In fact, even if the camera crosses into an obstructed area, its high sensibility GPS is able to determine its position: for this reason it is a very useful instrument, especially in urban canyons or near high buildings or trees. Moreover, the presence of the integrated GPS guarantees easy synchronization of the time scale of the camera with the UTC time.

Another instrument on board is the INS SBG Ekinox-D, a navigation grade sensor characterized by an internal GNSS receiver with two antennas (in yellow in Figure 3).
The use of this INS does not want to contradict the idea of analyzing an unfavourable scenario to understand how the use of different instruments can really improve the positioning solution, but it is necessary to define the "true trajectory" and it also constitutes one of the ways to interface with the DMI.

The authors decided to use the less precise GPS data (in fact are code points positions) acquired by the action-cam with the aim of testing how far a motion model and a more or less refined filtering technique could bring benefits in the determination of the vehicle trajectory and the position of the perspective centres of the camera.

Lastly, a DMI (Pegasem WSS) is also located on the vehicle, which permits to record vehicle speed information (and obviously the epochs of acquisition) and measurements of distance. The authors used these speeds information, and for this reason the instrument could be defined speedometer, but for brevity they use the term DMI.
The DMI permits the introduction of velocities into the analysis and also caters for the typical deficiencies of the motion models: e.g. failure to consider stops at traffic lights or zebra crossings, which are typically present in an urban road.
The DMI permits to "adjust" these models, realized for a continuous trend of the vehicle also to a discontinuous trend.

Obviously both camera and DMI were calibrated (Angelats and Colomina, 2014) before the survey, the first using the calibration tool of the commercial software Matlab® (Heikkilä and Silvén, 1997; Zhang, 1999), the second in the Laboratory of Topography of Politecnico di Torino, with the aim to determine the so-called "odometer gain" (in others words: the scale value of DMI). Figure 5 shows the instruments used for this latest calibration.

Figure 5. On the left: the DMI during calibration; on the right: the measure of the circumference of the wheel of the vehicle

The instruments used for the calibration of the DMI were a lathe, a speedometer and a chronometer, and obviously the value of measure of the circumference of the vehicle (shown in Figure 5). The instruments were used to define the correspondence between the mean value of velocity recorded by the DMI, when the odometer gain is in the amount of one pulse/m and the number of turns per second performed by the lathe.
The authors considered different measurements respectively corresponding more or less to three, five and seven turns per second of the lathe, and for each one they defined the mean value of velocity recorded by the DMI. This value was then divided by the specific number of turns per second corresponding to the specific measurement session. Finally, the results obtained by each measurement session were averaged (Table 1), for the purpose of defining the most reliable value useful for the determination of the scale of the DMI.

Measure session	Time of measure [min]	N° of turns	Turn/s	v_{MED} of DMI	v_{MED}/ (turn/s)
1	2	364	3.03	780	257.43
2	2	407	3.39	872	257.23
3	2	629	5.24	1343	256.30
4	2	802	6.68	1713	256.44

Table 1. Results of DMI calibration test

The final mean value of velocity related to turns per second, adopted for the determination of the scale value of DMI, was 256.85. This value, divided by the circumference of the wheel of the vehicle, permitted the determination of the scale value of the DMI (equation 1), used in the successive surveys:

$$Scale = \frac{256.85}{1.89} = 135.90 \tag{1}$$

The DMI data are recorded at a rate of about one second, but their acquisition times are not synchronized with the GPS camera positions: for this reason, resampling of the DMI data was necessary. The authors opted for a spline interpolation, finer than to the linear one. The residuals obtained by the linear interpolation with respect to the spline were at most 8 cm/s.

3. MOTION MODELS

It should be pointed out that each motion model is characterized by hypotheses and simplified assumptions, leaving out some errors, deliberately or otherwise. These un-modelled or causal effects relapse into the stochastic part of modern filters. We must remember that a simplified model, i.e. an "unpolished" model, even though it apparently works well, will be able to balance out only in a small part the simplified assumptions introduced in the motion errors.

Different motion models (Yuan et al., 2014) were studied for the analysis of the case study.
Linear models are the least complex and assume the vehicle velocity or its acceleration as constant: for this reason they are respectively called Constant Velocity or Constant Acceleration models (Schubert et al., 2008). Obviously their major advantage is the linearity of the state transition equation, but these models suppose a "straight" motion without changes of direction or curves, which are always present in an urban path.

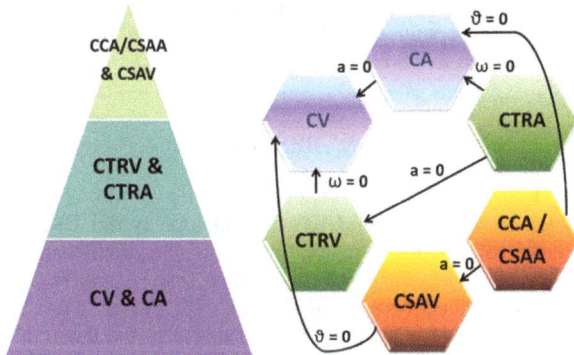

Figure 6. Scale of complexity of the motion models

Therefore, more complex curvilinear models were adopted to describe the case study in a realistic way.
These latest models could be classified using their unknown or constant parameters: the simpler is the so-called Constant Turn Rate and Velocity Model (CTRV), very useful in the description of the trajectories of aeroplanes; the second is the Constant Turn Rate and Acceleration (CTRA) (Altendorfer, 2009).
Besides yielding the best results in the definition of the vehicle trajectory and a realistic description of the motion, the CTRA method is also the only curvilinear method able to consider change of direction, the presence of rectilinear and curvilinear paths, and therefore passages in clothoids.

Figure 7 shows the variables used in the different motion models which are described in the next paragraphs.

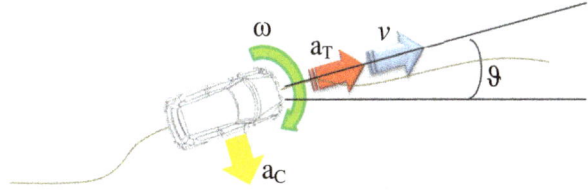

Figure 7. Variables in use in the motion models

3.1 Motion model with 4 states

This model considers the planimetric positions of the vehicle and its velocities in these directions.

$$\vec{x}(t) = (x, v_x, y, v_y) \tag{2}$$

where x, y = positions of the vehicle
v_x, v_y = constant velocities in x and y directions

This is a linear model and the state transition equation is shown in equation 3:

$$\vec{x}(t+T) = \begin{pmatrix} x(t) + T \cdot v_x \\ v_x \\ y(t) + T \cdot v_y \\ v_y \end{pmatrix} \tag{3}$$

where T = time interval between two successive epochs
x(t), y(t) = positions in the t-time
$\vec{x}(t+T)$ = states at i-time (t + T).

3.2 Motion model with 5 states: CTRV

In this model the states are five: two positions, the tangential velocity, the direction angle θ and the rotation angle ω.

$$\vec{x}(t) = (x, y, \theta, v, \omega) \tag{4}$$

where x, y = positions of the vehicle
v = tangential velocity
θ = direction angle
ω = rotation angle

The equation 5 shows the state transition equation:

$$\vec{x}(t+T) = \begin{pmatrix} \frac{v}{\omega} \cdot sin(\omega \cdot T + \theta) - \frac{v}{\omega} \cdot sin(\theta) + x(t) \\ -\frac{v}{\omega} \cdot cos(\omega \cdot T + \theta) + \frac{v}{\omega} \cdot cos(\theta) + y(t) \\ v \\ \omega \cdot T + \theta \\ \omega \end{pmatrix} \tag{5}$$

where T = time interval between successive epochs.

Here both the tangential velocity and the rotation angle are constants, in fact we have:

$$v(t+T) = v(t) \tag{6}$$

$$\omega(t+T) = \omega(t) \tag{7}$$

3.3 Motion model with 6 states: CTRA

This model "come from" the CTRV and includes also the acceleration of the vehicle.

$$\vec{x}(t) = (x, y, \theta, v, a, \omega) \tag{8}$$

The state transition equation is (Lategahn et al., 2012):

$$\vec{x}(t+T) =$$

$$(\ x(t) + \frac{1}{\omega^2}[(v(t)\omega + a\omega T)\cdot sin(\omega T + \theta(t)) + ...$$

$$+a\,cos(\omega T + \theta(t)) - v(t)\omega\,sin(\theta(t)) - a\,cos(\theta(t))]$$

$$y(t) + \frac{1}{\omega^2}[(-v(t)\omega - a\omega T)\cdot cos(\omega T + \theta(t)) + \cdots$$

$$... +a\,sin(\omega T + \theta(t)) + v(t)\omega\,cos(\theta(t)) - a\,sin(\theta(t))]$$

$$\omega\cdot T + \theta$$

$$+a\cdot T$$

$$0$$

$$0\ ) \qquad\qquad (9)$$

where   a = constant vehicle acceleration ($a(t + T) = a(t)$).

As the acceleration, also the rotation angle is constant: $\omega(t + T) = \omega(t)$.

4. DATA FILTERING

4.1 KF and EKF

The hypotheses of the Kalman filter (hereafter called "KF" for brevity; Kalman, 1960; Einicke, 2012) are:
- ✓ the system evolves linearly in the time;
- ✓ the measure equations are linear;
- ✓ both the state equations than the measure equations are normally distributed and uncorrelated.

In a lot of situations it is possible to linearize the measure equations and, if these are sufficiently precise, it is possible to consider these as normally distributed without falling into error and obtaining imprecise, less reasonable or divergent results.
Instead, less frequently the state equations are linear and indeed usually are differential equations.
However, if the time intervals between the measures are evaluated as "little" (a more detailed explication would request more space), it is usually possible to write non-linear state equations and then linearize these equations using a matrix of partial derivatives.

The initial state parameters, being non-linear equations - generally positions, velocities, accelerations, angular velocities, etc.-are obtained by the motion and measure equations of the previous epoch.
In this situation the filtering technique isn't known as Kalman Filter (KF), but is called Extended Kalman Filter (EKF).
The clearest difference is represented by the computation of the update of the state parameters, not by the classical computation of the parameters.
After the linearization of the state equations, the structure of the EKF follows the typical rules of the more traditional KF.
The KF is a recursive filter characterized by two steps: the filtering and the smoothing. The first step is divided into two parts: a time-update of the state equations (for this reason this passage is known as predictive estimation, or "prediction") and a following time-update of the measure equations (a process known as "correction").
At the successive epoch, in the EKF, the estimation just filtered is updated and then becomes the new starting point for the prediction of the successive instant.
In conclusion, the EKF permits the update of both the solution of the problem and the variance matrix.

4.2 UKF

The principal limitations that characterize both the KF and the EKF are the necessary linearization of the non-linear equations (Wan and Van der Merwe, 2000; Van der Merwe and Wan, 2001) and the hypothesis of the normality of the measures.
Thanks to the use of the Unscented Kalman Filter, it is possible to remove the first hypothesis and maintain the second, though with minor consequences (Tsogas et al., 2011).
The data filtering related to dynamic non-linear systems has in the recent years become a fascinating study area, considering that a lot of approaches are developed with the aim of solving the problem.
The Unscented Kalman Filter is based on unscented transformation, that is a mechanism for the propagation of the mean and the covariance, using non-linear transformations (Xiong et al., 2006; Terejanu et al., 2007).
The state vector is not unique, but instead constitutes a reduced number of state vectors (points in a space of 2L+1 dimensions, where L is the dimension of the variance-covariance matrix of the states). These state vectors (called sigma points; Van der Merwe, 2004; Zoeter et al., 2004) are accurately chosen to approximate both the 2L+1 dimensional mean, and the variance-covariance matrix of the variable computed a posteriori, which is still Gaussian, with a second order accuracy (the mean and the variance).
The improvement in respect of the EKF is clear if we consider that EKF permits only the obtainment of the first order of accuracy (we only know the mean value: the variance-covariance matrix is only obtained for a more or less linear rule and it refers to the only average value). Furthermore with the UKF it is not necessary to build the Jacobean matrix, and so in this respect the computational commitment is reduced.
The application fields for this method are various, for example, the fusion of the output data of different types of sensor; the determination of the position; or the training of a neural network (Haykin, 2001).
Therefore the literature suggests that this method is able to obtain better results than the EKF (Gustafsson and Hendeby, 2012), so the authors decided to apply the UKF to their case study.

5. RESULTS

The following tables and figures show the results obtainable using the two types of UKF method: one is the UKF not-augmented, described in the previous paragraphs, and the second method is the UKF augmented, that differs from the previous method because of its use of a greater number of states. This greater number of states is due to the fact that the state and measure equations are considered as non-linear with respect to the noise, which generally describes a real situation (Rutten, 2013). The number of states is not casual, it is justified by the introduction of the analysis of the noise that troubles the states (Guzzi, 2012).
The biggest advantage of the augmented method is the possibility of considering the influence of the noise of the measures in the computation, but its biggest disadvantage is represented by a computational commitment and by the possibility of not getting a positive weight matrix. The latter situation causes an inapplicability of the Cholesky factorization (a fundamental step in the UKF computational process) and the consequent end of the computation. In order to solve these problems, some authors proposed different alternative solutions to the factorization (Rutten, 2013).

In relation to both the procedures for implementation of the UKF method, whatever the number of states under examination, it is important define two parameters: the coefficient "α" (equation 10), which takes into account the spread of the sigma points (Turner and Rasmussen, 2010) and is generally inversely proportional to the considered number of states, and "β", which considers the a priori knowledge of the sigma points distribution (for Gaussian distribution generally is assumed equal to two).

$$\alpha\epsilon\ (0,1]\qquad(10)$$

In fact, for the case study, it was possible to observe that the assumption of values of α about its lower limit usually determined the impossibility of applying Cholesky factorization, and so the crash of the method.

Another element to take into account as input is the variance-covariance matrix of the process: the adjustment of this matrix depends on the case study, in particular on the number of states used. As an example the equation 11 and equation 12, show the variance-covariance matrix used for the motion model characterized by four states:

$$Q = \begin{bmatrix} [Q_1] & 0 & 0 \\ 0 & 0 & [Q_1] \end{bmatrix}\qquad(11)$$

where

$$Q_1 = q^2 \cdot \begin{bmatrix} \frac{1}{4}T^4 & \frac{1}{2}T^3 \\ \frac{1}{2}T^3 & T^2 \end{bmatrix}\qquad(12)$$

where
    q = standard deviation of the process
    T = time interval between successive epochs.

In the next tables, the authors show the means and the respective standard deviations (SQM) obtained by comparing the 2D positions of the reference INS (in black in the next figures) with the planimetric trajectories achieved by the UKF methods (augmented and not-augmented) applied to the GPS+DMI data. In these tables are shown the results obtained by the comparison between the reference and the smoothed values.
Obviously to compare these data, the authors applied the specific level-arm between the GPS+DMI data and the reference trajectory, using the heading extracted from the velocity vector.
In the related graphs, it is also possible to observe the trajectories: the reference in black; the trajectories obtained by the UKF filtering and smoothing applied to the GPS and DMI data in blue and green respectively, and the GPS data of the camera in red.
All the results shown in the next tables and graphs were obtained thanks to a script realized in Matlab® by the authors.

In Table 2 are shown the results related to the case study with 4 states (x, v_x, y, v_y).

N° states	UKF Augmented		UKF not-augmented	
	$\Delta 2D_{MEAN}$ INS-CASE [m]	$\Delta 2D_{SQM}$ INS-CASE [m]	$\Delta 2D_{MEAN}$ INS-CASE [m]	$\Delta 2D_{SQM}$ INS-CASE [m]
4	4.47	1.80	7.99	4.16

Table 2. Delta mean and delta standard deviation between INS and UKF augmented or not-augmented applied to the motion model with 4 states

The Figure 8 show the trajectory obtainable by the UKF, using four states:

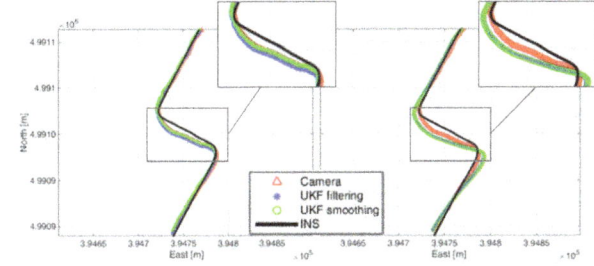

Figure 8. On the left: UKF augmented trajectory with 4 states; on the right: UKF not-augmented trajectories with 4 states

In Table 3 are shown the same results of Table 1, but related to the motion model with 5 states (x, y, θ, v, ω).

N° states	UKF Augmented		UKF not-augmented	
	$\Delta 2D_{MEAN}$ INS-CASE [m]	$\Delta 2D_{SQM}$ INS-CASE [m]	$\Delta 2D_{MEAN}$ INS-CASE [m]	$\Delta 2D_{SQM}$ INS-CASE [m]
5	4.56	2.16	5.29	1.63

Table 3. Delta mean and delta standard deviation between INS and UKF augmented or not-augmented applied to the motion model with 5 states

Figure 9 show the trajectory obtainable by the UKF methods considering the motion model characterized by 5 states:

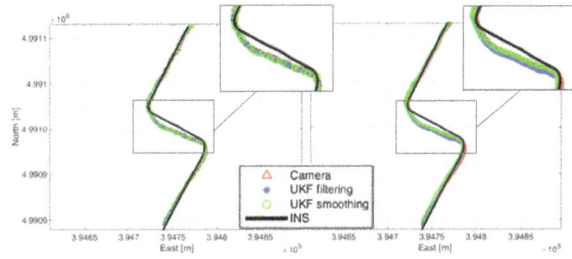

Figure 9. On the left: UKF augmented trajectory with 5 states; on the right: UKF not-augmented trajectories with 5 states

The trajectory shown in Figure 9 are quite the same.

Figure 10 shows also the comparison between the tangential velocity of DMI (in red), and the values obtained with UKF filtering (blue) or smoothing (green) augmented or not-augmented.

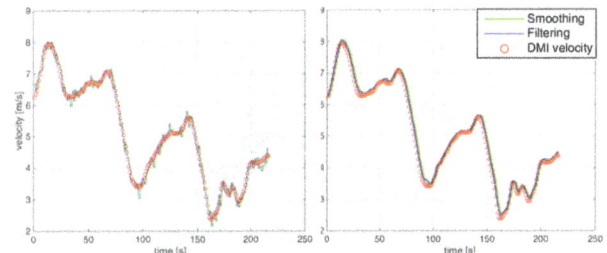

Figure 10. On the left the UKF augmented and on the right the not-augmented tangential velocities with 5 states

In Table 4 are shown the same results of the previous tables, but related to the motion model with 6 states ($x, y, \theta, v, a, \omega$).

N° states	UKF Augmented		UKF not-augmented	
	$\Delta 2D_{MEAN}$ INS-CASE [m]	$\Delta 2D_{SQM}$ INS-CASE [m]	$\Delta 2D_{MEAN}$ INS-CASE [m]	$\Delta 2D_{SQM}$ INS-CASE [m]
6	4.06	1.44	4.74	1.73

Table 4. Delta mean and delta standard deviation between INS and UKF augmented or not-augmented applied to the motion model with 6 states

Figure 11 shows the trajectory obtainable by the UKF methods:

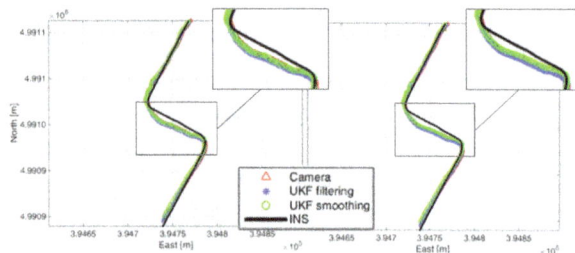

Figure 11. On the left: UKF augmented trajectory with 6 states; on the right: UKF not-augmented trajectories with 6 states

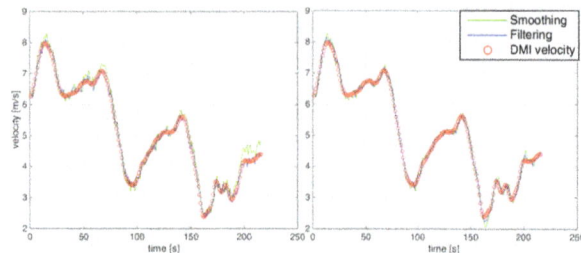

Figure 12. On the left the UKF augmented and on the right the not-augmented tangential velocities with 6 states

Figure 12 and the previous Figure 10 show results of tangential velocity a little bit more noised if it is considered the UKF augmented rather than the not-augmented.

With the aim of showing the results in a clearer scenario and to underline that a bigger number of states guarantees the best results in the definition of the trajectory, the authors show in Figure 13 the trajectories resulting from the UKF not-augmented method (with four, five and six states) in the urban canyon of Turin. They prefer to report the not-augmented results in a satellite view, because the differences in the various motion models in the definition of the final smoothed trajectory are clearer in this type of analysis than in the UKF augmented.

Figure 13. UKF not-augmented smoothed trajectories: in blue the INS, in red 4 states, in orange 5 states, in green 6 states

Figure 13 represents the results in the acquisition context, enabling us to see clearly that a greater number of states determines a more realistic definition of the trajectory (in green).

In fact, the "worst" trajectory corresponds to the least number of states: it is the motion model with four states represented in red. Here we can see that the trajectory is sinuous (that may be because of good UKF smoothing), but this trajectory is not always correctly located on the road: after the curves especially the positions are placed on trees or on the buildings.

An improvement is determined by the introduction of another state: the trajectory in orange corresponds to the motion model with five states, which is a little bit more realistic than the previous because it also follows the road after a curve.

However, it is the last analysis with the motion model based on six states that gives the best results: even if the final trajectory seems to be quite similar to the previous case, a visual analysis revels that after the curve this one is located in a more correct position than the other.

6. CONCLUSION

The results shown in the previous tables suggest that the UKF methods (especially the not-augmented), associated with a reduced number of states, are not able to consider all of the physics entities that have a role in the specific case study, and determines "dirty" results in the delta planimetric mean. In contrast, if the number of states considered is bigger, we can see an improvement (that is, a reduction) of the same value of the delta planimetric mean.

At the same time, the values of the delta planimetric SQM are almost unvaried, even if the number of states is augmented.

Moving from the most restricted number of states towards the biggest, we can see an improvement in the delta planimetric mean value for the UKF augmented, but the SQM does not change substantially.

Observing in a more detailed manner the case study characterized by five states, it possible to see that the augmented method apparently determines more noisy results, since it is characterized by a SQM 50 cm bigger than the equivalent not-augmented method. However it is important also to consider the delta planimetric mean, which is equal to 4.56 m for the augmented method and 5.29 m for the other: there is a worsening of about 70 cm between the two UKF filtering techniques.

With respect to the results shown in the previous paragraph, it is possible to conclude that if we consider a reduced number of states, the not considered entities in the analysis try to influence the results negatively, in particular "dirtying" the mean value; otherwise, if the number of states augments, it is possible to see a general improvement (a reduction) of the value of the planimetric mean.

It can therefore be concluded that the UKF augmented achieves better results than the not-augmented.

Following the integration of the GPS and DMI data, the available dataset permits another type of integration: GPS data with photogrammetric information (De Agostino et al., 2011).

Both the GPS data and the frames are obtained from the low-cost camera used for the survey.

The aim of this different type of integration is to see if the photogrammetric information is useful to improve the solution of positioning when there are not other technologies available.

Obviously it was possible using a software able to simulate the situation represented by the case study (Taglioretti and

Manzino, 2014), but in order to achieve this goal, the authors initially use a commercial software (frequently used in the field of photogrammetric research), able automatically to discover a lot of tie points (TP) between the frames and to compute the orientation between successive frames. The latest goal is the introduction of attitude parameters between the state parameters for the estimation of the camera coordinates with the respective errors of measure.

The authors decided to illustrate the results of a commercial software to permit an initial comparison with the values obtained using the UKF methods. This is because the authors are currently in a phase of implementation of a proprietary software: this software will be able to integrate GPS data with photogrammetric information, and will be the subject of future work.

It is important to underline that these results are obtained using a sort of "loosely coupled" integration.

The authors introduced in the commercial software the real values of position and accuracy of the camera obtained by the UKF methods, but they not identified any significant changes in respect of the introduction of the accuracy of the camera position. For this reason they believe more interesting understand how the errors change in respect of the reference trajectory modifying this parameter. They chose a value of camera accuracy equal to a low 1 meter, up to 3 meters and they decided to study the situation characterized by five states, because it represents a sort of "compromise" in the number of state at disposal.

The aim was to understand how photogrammetric measurements can improve positioning, in particular depending on the specific quality of the data available (and consequently depending on the camera quality) (Taglioretti et al., 2015).

The results obtained with the commercial software and compared with INS trajectory, are shown in the next Table 5.

Camera accuracy [m]	$\Delta 2D_{MEAN}$ INS-CASE [m]	$\Delta 2D_{SQM}$ INS-CASE [m]
1	2.44	1.42
2	2.57	2.11
3	5.63	3.20

Table 5. The results obtained by the commercial software

These values suggest that a loosely coupled integration permits an improvement of the mean value, but it is not sufficient to guarantee a real improvement in the positioning solution in respect of results obtained by the UKF. In fact the results shown in Table 5 obtained with the camera accuracy equal to two, show that there is an improvement in the mean value, but the SQM is quite similar to the previous case shown in Table 3.

For this reason, a more complete tightly-coupled integration is necessary (Cazzaniga et al., 2007) that permits the obtainment of better results, and which the authors will follow when developing their software.

ACKNOWLEDGEMENTS

The Authors would like to thank Eng. Horea I. Bendea for the support in the surveys and in the calibration tests.

REFERENCES

Altendorfer R., 2009. Observable dynamics and coordinate systems for automotive target tracking. Proceedings of the IEEE Intelligent Vehicles Symposium 2009, Xi'an, 3-5 June 2009, pp. 741-746. DOI: 10.1109/IVS.2009.5164369. ISSN :1931-0587.

Angelats E., Colomina I., 2014. One step mobile mapping laser and camera data orientation and calibration. *ISPRS - International Archives of the Photogrammetry, Remote Sensing and Spatial Information Sciences*, Vol. XL-3/W1.

Cazzaniga N. E., Forlani G., Roncella R., 2007. Improving the reliability of a GPS/INS navigation solution for MM vehicles by photogrammetry, Proceedings of the 5th International Symposium on Mobile Mapping Technology, Padua.

De Agostino M., Lingua A., Marenchino D., Nex F., Piras M., 2011. GIMPHI: a new integration approach for early impact assessment. Applied Geomatics, Vol. 3(4), pp. 241-249.

Einicke G. A., 2012. Smoothing, Filtering and Prediction - Estimating The Past, Present and Future. InTech, DOI: 10.5772/2706. ISBN: 978-953-307-752-9. Pages 286.

Gustafsson F., Hendeby G., 2012. Some relations between extended and unscented Kalman filters. IEEE Transactions on Signal Processing, (60), 2, pp. 545-555.

Guzzi R., 2012. Introduzione ai meodi inversi. Con applicazioni alla geofisica e al telerilevamento. Springer-Verlag Italia, Milano. ISSN: 2038-5730.

Hartikainen J., Solin A., Särkkä S., 2011. Optimal Filtering with Kalman Filters and Smoothers - a Manual for the Matlab toolbox EKF/UKF Version 1.3. Aalto University.

Haykin S., 2001. Kalman filtering and Neural Networks. John Wiley & Sons, Inc. ISBNs: 0-471-36998-5 (Hardback); 0-471-22154-6 (Electronic), 304 pages.

Heikkilä J. and Silven O., 1997. A Four-step Camera Calibration Procedure with Implicit Image Correction. Proceedings of the 1997 Conference on Computer Vision and Pattern Recognition (CVPR '97), pp. 1106-1113, June 17-19 1997. ISBN:0-8186-7822-4.

Kalman R. E., 1960. A new approach to linear filtering and prediction problems. Transactions of the ASME, Journal of Basic Engineering, vol. 82, pp. 35–45.

Lategahn H., Geiger A., Kitt B., Stiller C., 2012. Motion-without-structure: real time multipose optimization for accurate visual odometry. IEEE Intelligent Vehicles Symposium (IV).

Li X. R. and Jilkov V. P., 2003. Survey of Maneuvering Target Tracking. Part I: Dynamic Models. IEEE Transactions on Aerospace and Electronic Systems , Oct. 2003. vol.39 n°4, pp.1333-1364. doi: 10.1109/TAES.2003.1261132.

Lytrivis P.,Tsogas M., Thomaidis G., Karaseintanidis G. e Amditis A., 2010. IEEE Intelligent Vehicles Symposium University of California, San Diego, CA, USA, 21-24 June 2010.

Rutten M. G., 2013. Square-Root Unscented Filtering and Smoothing. IEEE Eighth International Conference on Intelligent Sensors, Sensor Networks and Information Processing, Melbourne, VIC, 2-5 April 2013, pp. 294-299. Doi: 10.1109/ISSNIP.2013.6529805. ISBN: 978-1-4673-5499-8.

Schubert R., Richter E., Wanielik G., 2008. Comparison and Evaluation of Advanced Motion Models for Vehicle Tracking. Proceedings of the 11th International Conference on Information Fusion , pp. 730-735.

Taglioretti C., Manzino, A. M., 2014. Terrestrial Mobile Mapping: photogrammetric simulator. *ISPRS - International Archives of the Photogrammetry, Remote Sensing and Spatial Information Sciences*, XL-3, 333-339, doi:10.5194/isprsarchives-XL-3-333-2014.

Taglioretti C., Manzino A. M., Bellone T., Colomina I., 2015. On outlier detection in a photogrammetric mobile mapping dataset. *ISPRS-International Archives of the Photogrammetry, Remote Sensing and Spatial Information Sciences*, 40 (3W2), pp. 227-233.

Terejanu G., Singh T., Scott P. D., 2007. Unscented Kalman Filter/Smoother for a CBRN Puff-Based Dispersion Model. 10th International Conference on Information Fusion, 9-12 July 2007, pp.1-8. DOI: 10.1109/ICIF.2007.4408076.

Turner R., Rasmussen C. E., 2010. Model based learning of sigma points in unscented Kalman filtering. In Samuel Kaski, David J. Miller, Erkki Oja, and Antti Honkela, editors, *Machine Learning for Signal Processing (MLSP 2010)*, pp. 178-183, Kittilä, Finland, August 2010.

Van der Merwe R., Wan E. A., 2001. The square-root Unscented Kalman Filter for state and parameter-estimation. Proceeding of the IEEE International conference on Acoustics, Speech, and Signal Processing, Salt Lake City, UT, 07 May 2001-11 May 2001, Volume:6, pp. 3461 – 3464. DOI:10.1109/ICASSP.2001.940586. ISSN : 1520-6149.

Van der Merwe R., 2004. Sigma-Point Kalman Filters for Probabilistic Inference in Dynamic State-Space Models. A dissertation submitted to the faculty of the OGI School of Science & Engineering at Oregon Health & Science University in partial fulfillment of the requirements for the degree Doctor of Philosophy in Electrical and Computer Engineering.

Wan E.A., Van der Merwe R., 2000. The unscented Kalman filter for nonlinear estimation. Adaptive Systems for Signal Processing, Communications, and Control Symposium 2000, Lake Louise, Alta, 01-04 October 2000, pp.153-158, DOI: 10.1109/ASSPCC.2000.882463, ISBN: 0-7803-5800-7.

Xiong K., Zhang H.Y., Chan C. W., 2006. Performance evaluation of UKF-based nonlinear filtering. Automatica Elsevier 42(2), 261-270.

Yuan X., Lian F., Han C., 2014. Models and Algorithms for Tracking Target with Coordinated Turn Motion. Mathematical Problems in Engineering, vol. 2014, Article ID 649276, 10 pages, 2014. doi:10.1155/2014/649276.

Zhang Z., 1999. Flexible Camera Calibration By Viewing a Plane From Unknown Orientations. International Conference on Computer Vision (ICCV'99), Corfu, Greece, September 1999, pp. 666-673.

Zoeter O., Ypma A. and Heskes T., 2004. Improved unscented Kalman smoothing for stock volatility Estimation. Proceeding of the 2004 14th IEEE Workshop on Machine Learning for Signal Processing Society Workshop, Sao Luis, Sept. 29 2004-Oct. 1, pp. 143 – 152, DOI: 10.1109/MLSP.2004.1422968, ISSN: 1551-2541.

7

SPATIAL DATA QUALITY AND A WORKFLOW TOOL

M.Meijer [a*] L.A.E. Vullings [a], J.D. Bulens [a], F.I. Rip [a], M. Boss [a], G. Hazeu [a], M.Storm [a,]

[a] Alterra, Wageningen University and Research Centre, Wageningen, The Netherlands

Commission II, WG II/4

KEY WORDS: Spatial Data Quality, Communication, Producer, Consumer, Fitness for Use, Open Data, LPIS, ETS

ABSTRACT:

Although by many perceived as important, spatial data quality has hardly ever been taken centre stage unless something went wrong due to bad quality. However, we think this is going to change soon. We are more and more relying on data driven processes and due to the increased availability of data, there is a choice in what data to use. How to make that choice? We think spatial data quality has potential as a selection criterion.

In this paper we focus on how a workflow tool can help the consumer as well as the producer to get a better understanding about which product characteristics are important. For this purpose, we have developed a framework in which we define different roles (consumer, producer and intermediary) and differentiate between product specifications and quality specifications. A number of requirements is stated that can be translated into quality elements. We used case studies to validate our framework. This framework is designed following the fitness for use principle. Also part of this framework is software that in some cases can help ascertain the quality of datasets.

1. INTRODUCTION

1.1 General Instructions

Although spatial data quality for many years has been considered important, it has not always been a priority. However, times are changing. In our day to day lives we are more and more depending on applications and services in which data plays a crucial role. And as more data becomes available we also have different datasets to choose from. But how do we make that choice? How do we make sure that we end up using the best possible data set for our application?

It is our view that spatial data quality has the potential to help users choose the correct dataset. Specifically the fitness for use approach (Devillers et al., 2007) can contribute in facilitating the choice in what data to use for a specific type of application. Fitness for use of course isn't a new concept within the scope of spatial data quality. There is already a great deal of research related to fitness for use and spatial data (e.g. Vasseur et al., 2003, Frank et al. 2004 and Devillers et al., 2007).

In June 2014 we organized a symposium titled 'Why Spatial data quality?' More than eighty Dutch scientist and policymakers shared their thoughts on this subject. It was concluded that spatial data quality has indeed the potential to become a selection criterion and that fitness for use should be the guiding principle, but in order to reach its full potential more attention is needed to subjects such as the definition of spatial data quality, validation, communication, business case development, and means of determining spatial data quality.

Based on the outcome of the before mentioned symposium and our broad experiences dealing with spatial data in a number of cases (Meijer & Vullings, 2012; Storm et al., 2012a; Storm et al., 2012b; Hazeu et al., 2014; Meijer et al., in press) we have defined a framework for spatial data quality. This framework is validated by case studies from a consumer as well as a producer perspective.

1.2 Background

There are many ways to describe and categorise the quality of spatial data. According to ISO 8402 (1994) quality is defined as "the totality of characteristics of an entity that bear upon its ability to satisfy stated and implied needs." And "The purpose of describing the quality of geographic data is to facilitate the comparison and selection of the dataset best suited to application needs or requirements' (ISO 19157:2013).

Not only do we have many ways of describing quality from different perspectives, we also have numerous standards describing quality. However since we are converging towards the ISO standards for spatial data we focus on the ISO 19157 (2013) standard describing spatial data quality by the following six groups of elements: Completeness, Logical consistency, Positional accuracy, Thematic accuracy, Temporal accuracy and Usability element (ISO 19157:2013). According to Devillers et al. (2006) there's a general consensus about these criteria.

All the elements before mentioned refer to the internal quality of a dataset. With internal quality we refer to *the level of similarity between the data produced and the 'perfect' data that should have been produced* (Devillers et al., 2006). However quality of spatial data in isn't absolute. It differs in accordance with the intended use and the ability to satisfy stated and implied needs (ISO, 1994). This type of quality is perhaps better known as external quality or fitness for use.

---

[*] Corresponding author

2. FRAMEWORK

Based on our own experience with spatial quality projects (specifying criteria and auditing datasets (Meijer & Vullings, 2012; Storm et al., 2012a; Storm et al., 2012b; Hazeu et al., 2014; Rip & Bulens, 2013; Meijer et al., in press)) and existing literature related to spatial data quality (Devillers et al., 2006), we defined a framework for assessing spatial data quality. In order to verify and specify the framework further we used case studies.

For the basic setup of the framework we take the user as a starting point. By describing the use case we identify the relevant context to be the universe of discourse. Based on this information we define the product that is wanted by the user. This can vary from plain data provisioning to automated procedures like an App to providing human services. In this stage we will limit the functionality of the framework to data and the requirements of processes. For now we leave the quality aspect of the processes themselves as well as the institutional aspect out of our scope. They are to be included in a next stage.

In the framework the user as a consumer plays a central role, since the consumer and the context of the actual use determines the necessary quality (fitness for use). In order to get a good understanding of what the user actually needs it's important to unravel the information question into criteria with the help of spatial data quality expertise. Depending on the case this can be done by a structured questionnaire or, if more complex, with the help of mediators and using interview techniques. The mediators need to be good communicators with a good understanding of spatial data quality. In figure 2 we present a workflow that will help guide us to get a better understanding of the quality needed.

The first phase is focussed on revealing all relevant product characteristics and, where applicable, relating them to quality requirements using the standardised quality elements. So, we distinguish between product characteristics and quality characteristics. For instance if the consumer is looking for data of trees in his municipality and he needs to know what type of tree is located where, a product characteristic is that the data set needs to include data on the type of tree. A quality characteristic can be that 95% of the records in the dataset provide species information. At the end of this process the user/consumer has a complete list of appropriate product characteristics and quality characteristics.

Next to the consumer there is a producer and an intermediary or broker role defined. The broker is defined as a service provider of spatial data between the consumer and producer and as such can add value to the data of the producer or supply services that provide the data as a product. Both the broker and producer should specify the product characteristics and the value of the quality characteristics of their dataset(s) and in case of data set transformations of all the 'in between' products as well. This information has to be comprehensible and easily accessible for the consumer. Only then a consumer can judge whether a dataset is 'good enough' to fit the intended use. The framework is visualised in figure 1.

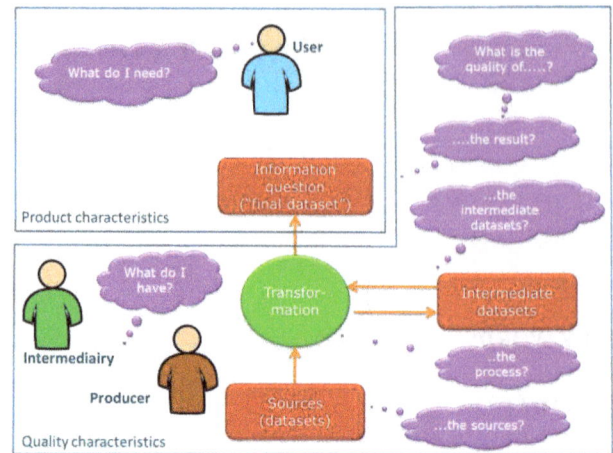

Figure 1: framework spatial data quality

In the past producers were the ones creating datasets, usually initiated by a specific need for that data. In time more use cases can evolve that have a need for the same type of data. Common practice was to use that data or when necessary transform the data to be useful for the case. It is conceivable that a producer from a business point of view will market his data. In that case the data should be multipurpose, fitting more than one need. Instead of just one use case, the producer should make an inventory of possible use cases for which the data could be needed and for each use case the product and quality characteristics should be defined. This process is visualised in figure 1.

Software like for example webbased workflow tool can facilitate processes to ascertain the quality of datasets for producers. Firstly it will help them to establish that the quality of the dataset is in accordance with the set of standards for consumers. Secondly it will help the producer to find out whether the quality of the dataset is conform the criteria defined by the information question. In the next chapter the use of a workflow tool is illustrated by an example about the land parcel information system in the Netherlands.

3. USE CASE: LAND PARCEL INFORMATION SYSTEM

In Europe farmers receive around €50 BLN of subsidies a year. Most of this money is reserved for so-called area based subsidies. Meaning for example that for every hectare of arable land or square meter of hedgerows farmers receive a certain amount. These subsidies are part of the Common Agricultural Policy (CAP).

To make sure that subsidies are paid for the correct area all member states have setup a Land Parcel Identification System (LPIS). One of the most important datasets of the LPIS is the reference layer. The reference layer is used as a control instrument to check area applications made by farmers.

In the Netherlands this reference layer was based on a product called TOP10NL. This is a digital topographical dataset produced by the Dutch Cadastre. It is the most detailed product within the national topographical base registration. It is generated from aerial photo interpretation, combined with information from cyclorama photographs. Cyclorama's are high-quality 360° panoramic images with high accuracy. They provide current and clear views of street level environments readily from the web. Cycloramas have a number of unique features, including metric accuracy and geo-referencing.

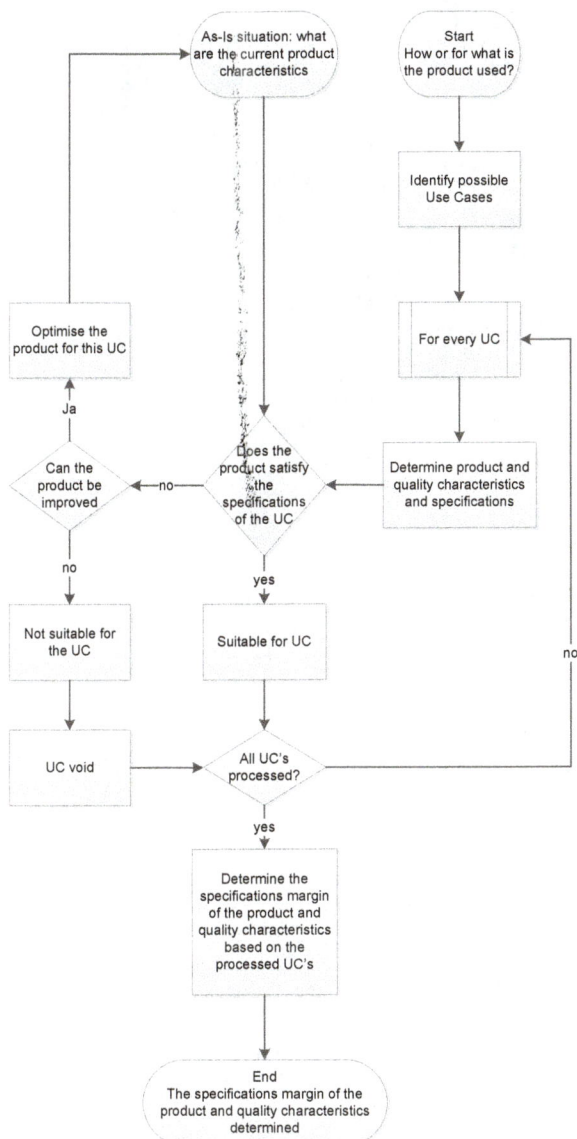

Figure 2: process flow of identifying customer product and quality requirements for producer

For many years the Dutch paying agency responsible for controlling and paying subsidies to farmers considered the TOP10NL fit for use to control area based payments. But a couple of years ago the Dutch government was sanctioned by the European Commission because the Dutch LPIS was unable to perform two explicit functions;

- unambiguous localisation of all declared agricultural parcels by farmer and inspectors,
- the quantification of all eligible area for crosschecks during the administrative controls by the paying agency.

In figure 3 and 4 two examples are given illustrating the problems related to the Dutch LPIS.

Figure 3: In this example the land use is unclear and not all parcel boundaries correspond to features in the field.

Figure 4: This example show an incorrect interpretation of the operator. Non agricultural area (a roadside verge) was included in the reference parcel.

Failure of an LPIS in the unambiguous localisation induces risks for double declaration of land and for ineffective inspections; inadequate quantification of eligible area renders the crosschecks ineffective for preventing and identifying over-declarations by farmers. Both failures involve financial risks for the EU Funds.

Based on decision process similar to the flowchart presented in figure 1 the Dutch paying agency looked for an alternative. Eventually it was decided to create a new reference layer from scratch. But how can you be sure that this new reference layer is able to perform the two functions previously mentioned? And not only now but also in the future?

The Dutch landscape is dynamic in nature. Land changes for example from agricultural to residential or grassland changes into arable land. If these changes are not recorded properly and/or quick enough this will have an impact on the quality of a spatial dataset. But how to decide if the quality is below a certain threshold? When is the quality so bad that the LPIS can't perform its two explicit functions? The first step is to relate the two explicit functions to quality criteria and find the best fitness for use. In paragraph 1.2 we already presented six groups of elements we believe are important. We can also use these elements in the case of the LPIS. For example thematic accuracy can be linked to the eligibility of a reference parcel or the correct land cover classification. Completeness can be linked to the coverage of all agricultural land in a specific region or country. Temporal quality can be linked to changes of the land and the processing of these changes in the reference layer. In the Netherlands this has been used to setup a quality framework for the LPIS.

For each of the different criteria the European Commission has indicated which quality should be. To give an example. One of the quality is related to the correct quantification of the area eligible for subsidies. To check this an operator has to digitize the same area covered by an existing reference parcel and compare both areas. If the area difference is less than 3% the existing reference parcel is considered to be conform.

4. WORKFLOW TOOL

In order to facilitate the user but also the producer in assessing the quality of the reference layer we built a web-based workflow tool that guides for example the quality control expert through a number of steps which will eventually lead to an overview of the quality of the product.

The first phase consists of taking a representative sample of the reference layer and collecting recent orthoimagery. The orthoimagery can be a recent aerial image or high resolution satellite imagery. After sampling the reference parcels the selected parcels are inspected. The first step in inspecting the reference parcels consists of ascertaining if it's feasible to inspect the selected reference parcels. In some cases it's impossible to check a particular parcel because of issues with the imagery (see figure 5). If a parcel can't be inspected the parcel is skipped and a new parcel is added to the sample.

Figure 5. The selected reference parcel can't be inspected because of clouds.

If the reference parcel can technically be checked the operator has to digitise the parcel again and label the parcel in accordance with the land cover visible on the image.

After the entire sample is digitised again and all errors and deviations are labelled the results need to analysed and reported to the European Commission (see also figure 7).

Figure 6: In this image in red the original boundaries of the reference parcel are shown. In yellow the boundaries of the parcel digitised by the operator.

The workflow tool that is used to guide the operator is called the ETS Manager. ETS refers to Executable Test Suite which is often used in conformance testing (Sagris et al., 2013).

Currently this workflow tool is mainly used for checking the quality of the LPIS. The main advantage of the ETS Manager is that it follows the workflow of the LPIS Quality Assurance Framework (Sagris et al., 2013). All operators follow the same set of rules which makes standardisation and reporting the results of the LPIS QAF to the European Commission a lot easier. The tool has led to a reduction of operator time but, also increased the quality of the actual reporting, and made it easier to share information about the results between different departments.

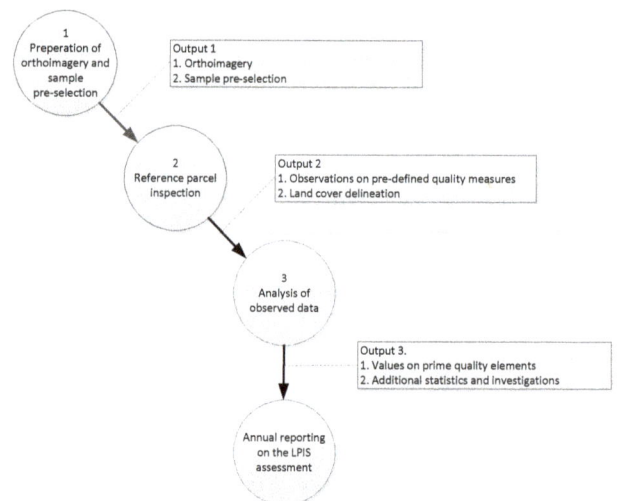

Figure 7: an overview of the different phases of the inspection of the quality of the reference layer (source: https://marswiki.jrc.ec.europa.eu/wikicap/index.php/Main_Page)

5. DISCUSSION AND FUTURE RESEARCH/ RECOMMENDATIONS

In order to use spatial data quality as a selection criterion when choosing a data set for usage in an application, the determination and communication of spatial data quality between consumers, brokers and producers needs to improve. A workflow tool as the one presented in the article can play a role in this process.

More specifically the use of a workflow tool can lead to a harmonisation of the way quality of a dataset is assessed. It also makes reporting more easy. Additionally the use of a web-based workflow tool makes it possible to visually and remotely discuss specific issues between for example the consumer and producer.

Although the ETS tool is already used by a number of different consumers and producers we want to develop the workflow tool further. To get a better understanding of how your specific dataset is performing it will be interesting to compare the results with the quality inspection results of a similar dataset. So one of the features we want to add is benchmarking.

In the framework we focussed firstly on extracting and defining characteristics based on its intended use. Future work will be to extent the framework to other relevant data quality properties like for example temporal aspects of spatial data. One extension will be to assess the information published besides the data itself. One can think of the availability of feature catalogues containing commonly, standardized and excepted definitions of spatial features and their attributes. The level of compliance with existing standards (INSPIRE). Proper documentation and metadata using standards, availability of managed code lists accessible through registries based on described standardized hierarchies as for example Simple Knowledge Organization System (SKOS). Furthermore we like to continue focussing on the communication aspects, so all parties involved can find and know what is meant by quality information.

It all matters when one has the luxury to choose what data to use; it will improve use of spatial data and avoid capital mistakes.

6. REFERENCES

Devillers, R. and Jeansoulin, R. (eds), 2006. Fundamentals of spatial data quality. IST.

Devillers, R., Bédard, Y., Jeansoulin, R. and Moulin, B., 2007. Towards spatial data quality information analysis tools for experts assessing the fitness for use of spatial data. In: *International Journal of Geographical Information Science*, 21:3, 261-282.

Frank, A.U., Grum, E. and Vasseur, B., 2004, Procedure to select the best dataset for a task. In: *Proceedings of the Third International Conference on Geographic Information Science* (GIScience 2004), Adelphi, USA, pp. 81–93.

Hazeu, G.W., Schuiling, C., Dorland, G.J., Roerink, G.J., Naeff, H.S.D., and Smidt, R.A., 2014. Landelijk Grondgebruiksbestand Nederland versie 7 (LGN7); Vervaardiging, nauwkeurigheid en gebruik. Wageningen,

Alterra Wageningen UR (University & Research Centre), Alterra-rapport 2548. 86 blz.; 16 fig.; 12 tab.; 15 ref.

ISO 19157:2013: Geographic information - Data quality.

ISO 8402:1994: Quality management and quality assurance – Vocabulary.

Justice, C.O., Belward, A., Morisette, J., Lewis, P., Privette, J. & Baret, F., 2000. Developments in the 'validation' of satellite sensor products for the study of land surface. In: *International Journal of Remote Sensing*, 21, 3383-3390.

Meijer, M., Rip, F.I., Van Benthem, R., Clement, J. and Van der Sande, C, 2015. Boomkronen afleiden uit het Actueel Hoogtebestand Nederland. Alterra rapport (in prep.). Wageningen University and Research Centre.

Meijer, M. and Vullings, L.A.E., 2012. Kwaliteit van ruimtelijke data in relatie tot het LPIS; kwaliteitsaspecten rondom het beheer van ruimtelijke data. Wageningen, Alterra, Alterra-rapport 2285. 124 blz; 51 fig.; 4 tab.; 40 ref.

Sagris, V., Wojda, P., Milenov, P., and Devos, W., 2013. The harmonised data model for assessing Land Parcel Identification Systems compliance with requirements of direct aid and agri-environmental schemes of the CAP. In: *Journal of environmental management*, 118, 40-48.

Storm, M.H., Knotters, M. and Brus, D.J., 2012a. Controlemethodiek Basisregistratie Topografie. Wageningen, Alterra.

Storm, M.H., Knotters, M. and Brus, D.J., 2012b. Audit Basisregistratie Topografie, resultaten van een eerste wettelijk vereiste externe controle op de kwaliteit van de BRT.

Vasseur, B., Devillers, R. and Jeansoulin, R., 2003, Ontological approach of the fitness of geospatial datasets. In: *Proceedings of 6th Agile Conference on Geographic Information Science*, Lyon, France, pp. 497–504.

8

TOWARDS EFFICIENCY OF OBLIQUE IMAGES ORIENTATION

W. Ostrowski [a]*, K. Bakuła [a]

[a] Department of Photogrammetry, Remote Sensing and Spatial Information Systems, Faculty of Geodesy and Cartography, Warsaw University of Technology, Poland - (w.ostrowski, k.bakula)@gik.pw.edu.pl

KEY WORDS: oblique, oblique images, orientation, adjustment, bundle, block

ABSTRACT:

Many papers on both theoretical aspects of bundle adjustment of oblique images and new operators for detecting tie points on oblique images have been written. However, only a few achievements presented in the literature were practically implemented in commercial software. In consequence often aerial triangulation is performed either for nadir images obtained simultaneously with oblique photos or bundle adjustment for separate images captured in different directions. The aim of this study was to investigate how the orientation of oblique images can be carried out effectively in commercial software based on the structure from motion technology. The main objective of the research was to evaluate the impact of the orientation strategy on both duration of the process and accuracy of photogrammetric 3D products. Two, very popular software: Pix4D and Agisoft Photoscan were tested and two approaches for image blocks were considered. The first approach based only on oblique images collected in four directions and the second approach included nadir images. In this study, blocks for three test areas were analysed. Oblique images were collected with medium-format cameras in maltan cross configuration with registration of GNSS and INS data. As a reference both check points and digital surface models from airborne laser scanning were used.

1. INTRODUCTION

Oblique airborne imagery become more and more popular photogrammetric datasets which is obviously related with rising market of their applications. The major advantage of oblique images from the consumer point of view is their 'natural' view which is much easier to interpret for non-expert users (Remondino and Gerke, 2015). These type of images have been successfully used for years in applications which do not demand high accuracy (Höhle, 2008). Because of that for many years enough accurate way of determining elements of exterior orientation parameters for oblique images was direct referencing. However, such a solution is insufficient for some applications (Grenzdorffer et at. 2008, Rupnik et al. 2015), and recently there has been a growing number of applications which need higher accuracies of measurement, like 3D City Modelling (Haala et al. 2015) and another verification or extraction information about urban environment (Rau et al. 2015, Nex et al. 2013, Nyaruhuma, et al. 2012a, Nyaruhuma, et al. 2012b).

The most popular way of reliably determining orientation of images is aerial triangulation by bundle block adjustment, unfortunately classical photogrammetric workflow is not suitable for oblique images. Experiments with orientation of this type of data with digital photogrammetric workstation would not provide fully satisfactory results (Jacobsen, 2008; Gerke and Nyaruhuma, 2009). As a consequence the multi-step orientation method was developed and used, these methods assuming separation images into sub-block by looking direction and orientation oblique images directly to adjusted nadir images (Wiedemann and More, 2012) or separate and independent adjustment of each view direction (Hu, et al. 2015).

One of the major reasons why the classical photogrammetric software is not suitable for orientation of oblique images are different orientation angels. Photogrammetric workflow for years was optimized to work with nadir images with similar scale.

Oblique images have very different view angle, tilting of optical axis means that in different parts of image there is a different scale. Traditional algorithms used in photogrammetry in order to extract tie points are not able to deal with this type of image distortion. Furthermore, most of this algorisms are not invariant to affine distortion whose character have differences of view between nadir and oblique images (Xiao et al. 2013).

In recent years interesting studies have been published on the proposed methods for oblique image orientation (Rupnik et al. 2013 Rupnik et al. 2015), using existing algorithms such as SIFT, or proposed new algorithms dedicated to matching oblique images. Searching for ways to improve automatic aerial triangulation for oblique images Yang et al. (2012) proposed a multi-stage algorithm based on SIFT matching. Another solution proposed Xiao et al. (2013) is the using approximated exterior orientation elements in NAIF algorithm (Nicer Affine Invariant Feature),. Another method is proposed by Hu et al. (2015).

Unfortunately, the availability of tools described in the above-mentioned studies is still limited and therefore the oblique image orientation still causes trouble in everyday applications. However, in recent years a whole variety of programs successfully used in other segments of market have arisen. They have been developed to some extent regardless of the methods used in large-format (classical) photogrammetry. This software is related to orientation of images from Unmanned Aerial Vehicles (UAV) and Close-Range Photogrammetry. This software based usually on the structure from motion algorithms. Imaging captured in those branches of photogrammetry has often a very diverse geometry - similar to the oblique images. A major problem when using these types of methods to the orientation of oblique aerial images can be computation performance (Karel and Pfeifer, 2015).

The aim of this study was to verify whether it is possible to effectively orientation oblique images in the software used for the

---

* Corresponding author

orientation of images from UAVs and Close-Range Photogrammetry. For this purpose, two quite popular software AgiSoft PhotoScan and Pix4D Mapper Pro were tested in presented research

2. TEST AREAS

During experiments three test areas was used, all data was acquired by Polish Photogrammetric Company – MGGP Aero. Cameras which were used are combination of five cameras IGI DigiCam in maltan cross, with tilt angle of 45 degrees. Three of them were 39 Mpx (nadir one has 50 mm focal length, forward and backward looking cameras have 100 mm focal length), another two were 50 Mpx (left and right looking also with 100 mm focal length). All cameras were calibrated and calibration parameters were used as precalibrated values during self-calibration process. The platform on which cameras are placed is equipped with a GNSS/INS system. The accuracy of the f initial value of the EO by these systems has been defined a posteriori of 10 cm for translation and 0.5 degrees for the rotation.

Two of the datasets used in the research were blocks for urban areas. They covered the centers of the two Polish cities of Katowice and Wrocław. Data for Katowice was acquired in May 2014, and included the 1070 block shots (854 oblique and 216 nadir) with an average GSD of 9 cm and overlap of 60% in line and 30% between the strips. For the test area 56 ground control points were measured in photogrammetric intersection (Fig. 5). The measurement was performed on large-format Images collected with DMCII 230 camera with GSD of 10 cm. As control points selected identifiable manhole covers or rarely road markings were selected.

The flying mission for second test area - Wrocław was carried out in May 2015. 3430 oblique from four cameras were acquired with overlap of 60% in line and between strips. Avarage GSD was 7 cm and for the block 12 control points was signalled (painted on the road) and measured using GNSS.

The third of the used blocks included the rural areas, in the northern part of Poland . It consisted of 1108 oblique images on the GSD 10 cm and characterized by a much smaller overlaps of approx. 30% both in lines and between strips.

3. EXPERIMENTS AND ANALYSIS

Experiments were divided into two groups. Firstly, settings and specific options for individual software are tested in order to determine their influence on orientation process. Next, all other factors which are common for any aerial triangulation process

(tie points number per image, GCP distribution, usage of nadir images, overlap between images and number of images in block) were examined.

3.1 Settings exclusive for one of used software

The first group of experiments was related to configuration parameters unique to each test programs. Agisoft PhotoScan offers two options of preselection pairs during matching. The first of them (Reference) is based on the approximate EO and the average height of the terrain to find common overlaps of images. The second option (Generic) finds the corresponding images using preliminary matching and pyramid images. Tests have been conducted on the block images from Katowice. Time and effectiveness of these adjustments are presented in Table 1. It can be noticed that the use of options Reference accelerates the process of orientation. Despite a good approximation of exterior orientation that use them in tested software decreases the effectiveness of orientation. Parts of the image blocks remain non-oriented. Therefore after further experiments it was decided not to use the approximate angular exterior orientation of both programs.

No.	Key points	Pair Preselection	Time*	Oriented images
1	90 000	Generic	245 min	1065/1070
2	90 000	Reference	132 min	1046/1070
3	60 000	Generic	257 min	1070/1070
4	60 000	Reference	190 min	1018/1070

Table 1. Time and effectiveness of orientation in Agisoft PhotoScan regarding to Pair Preselection Method. *For 1, 2 and 3, 4 experiments different workstations were used so results in time cannot be directly compared between these pairs.

Pix4D has two configuration options which, without modifying the strategy of matching images may have an impact on the orientation results. The first of them is the use of additional geometry verification of matching and the second is subsequent re-matching after the first calculation of the image orientation.

To verify the effect of these two options on the orientation, a series of experiments with a reduced number of key points (to 10 000) was carried out. Firstly, data from Katowice (Tab. 2) where oriented. The block consisted of both oblique and nadir images. Dense and numerous control points allow to 47 check-points and 9 GCP used in orientation

No.	Dataset	Keypoints	Matches (median)	Geometry Verified Matching	Rematch	Time [min]	Oriented images	Check Point RMS X/Y/Z [m]
1	Katowice	10 000	2318	Yes	Yes	109	1068/1070	0.05/0.08/0.09
2	Katowice	10 000	2043	Yes	No	86	1068/1070	0.05/0.08/0.09
3	Katowice	10 000	1955	No	Yes	446	1026/1070	0.31/0.12/1.25
4	Katowice	10 000	1078	No	No	333	1023/1070	0.05/0.09/0.10

Table 2. Time and effectiveness of orientation in Pix4D regarding to usage of Geometry Verified Matching and Rematch settings.

No.	Dataset	Keypoints	Matches (median)	GVM	ReM	Time [min]	Oriented images	GCP RMS X/Y/Z [m]	Check Point RMS X/Y/Z [m]
1	Wrocław	10 000	1947	Yes	No	409	3427/3430	0.04/0.03/0.03	0.02/0.04/0.02
2	Wrocław	10 000	2323	Yes	Yes	872	3427/3430	0.04/0.04/0.04	0.02/0.02/0.01

Table 3. Time and effectiveness of orientation in Pix4D regarding to usage of Rematch option.

The results obtained for the block from Katowice, clearly show that the use of Geometry Verified Matching improves the results which is particularly noticeable in the approach 3 and 4, where due to the reduction of the number of tie points Pix4D had trouble orienting all images (more than 40 remained not-oriented). This also resulted in the extension of computation time. Differences in effectiveness or accuracy between approach 1 and 2 are not noticed.

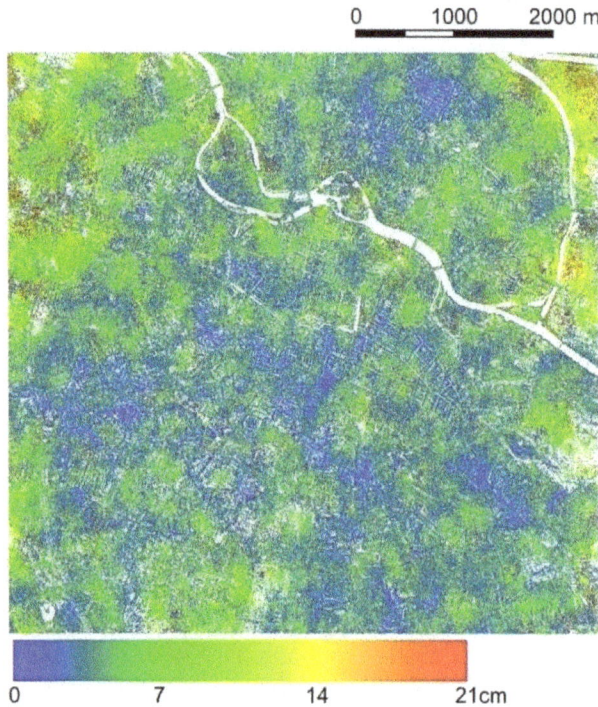

Figure 1. Map of spatial distribution of deviation in tie points height computed with and without Geometry Verified Matching in Pix4D (Wrocław dataset).

Additional experiments were performed for block from Wrocław, which consisted of only the oblique images and had more accurate controls. It could be more likely that the impact of a rematch option would be better seen here. Control was much rarer - during orientation 9 GCPs 4 check points were used. Due to the small number of controls it was decided to further compare the height difference between the tie points extracted in both orientations (Fig. 1). The average height difference was 0.05 m with a standard deviation (STD) of 0.05 m. In this figure on the distribution minimal systematic differences on the edge of the block can be noticed.

3.2 Number of tie points

One of the crucial factors which should have the direct impact on both the time of processing and results of aerial triangulation is number of tie poinst. Decreasing number of tie points should reduce the time of points extraction as well as matching them. However,a reduction of tie point might lead to poor connection between images in block which can have an influence on achieved accuracies. In case of Pix4D the user cannot define desired number of tie points but there is the possibility to define number of key points per image which will be used during matching. Agisoft Photoscan gives both these possibilities.

To evaluate the influence of the reduction of key point numbers a series of aerial triangulation process was carried out. Each

process were carriedout fully independently and consists of all steps of processing from key point extraction by image matching to the final adjustment of the block with self-calibration. Dataset from Katowice consist of 1070 images (both oblique and nadir) with 9 GCP and 47 Check Points was used.

Results from Pix4D software clearly show that there is a high potential of increased efficiency in reduction of key point number. Especially because the automatic number of them, which are default settings, is really high (median 86 000 per image). The number of key point have almost linear influence on time processing (Fig. 2) and do not influence the achieved accuracy (Fig. 3), It is also visible that the decreasing number of key points below 20 000 might influence the effectiveness of orientation (Fig. 4).

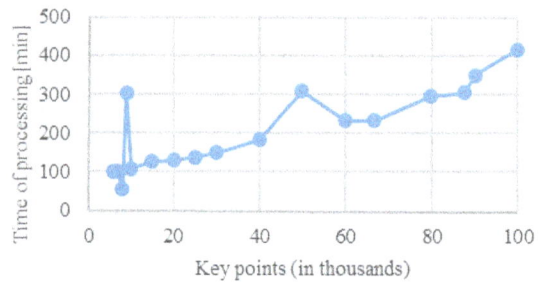

Figure 2. Time of processing (in minutes) in Pix4D of Katowice dataset with different number of key points

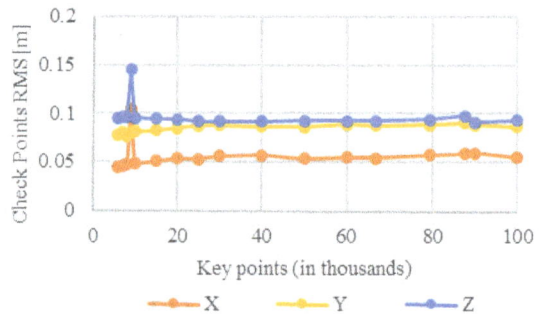

Figure 3. RMS of errors (in meters) achieved on Check Points in Pix4D (Katowice dataset) with different number of key points

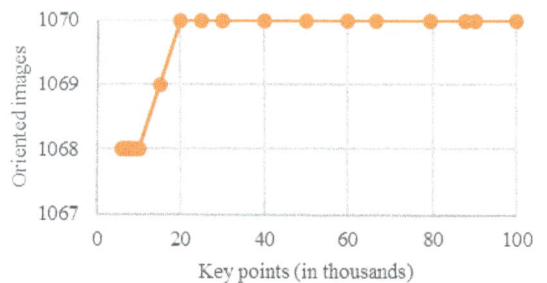

Figure 4. Effectiveness of orientation (Katowice dataset) in Pix4D with different number of key points

Experiments with reduction of the tie point number in Agisoft Photoscan (Tab. 4) shows that any reduction of tie point number below the default (40 000) has direct impact into orientation process stability and effectiveness. There is no visible change in the duration of processing.The only changes involve the number of successfully oriented images.

No.	Key points	Time [min]	Oriented images
1	10 000	260	872/1070
2	15 000	322	865/1070
3	20 000	260	959/1070
4	30 000	278	967/1070
5	40 000	273	1047/1070

Table 4. Time and effectiveness of orientation in AgiSoft PhotoScan regarding to number of key point per image.

3.3 GCP distribution

Another factor that significantly affects the results of aerial triangulation is the number and distribution of ground control points. Due to the large number of tie points and block geometry, it is expected that for a block of oblique images the number of control points can be significantly limited. Reducing the number of GCP is so important that their acquisition is time-consuming, and the number of them can be increased with the division into sub-blocks.

The block from Katowice, due to smaller overlap between the strips can be more sensitive to the reduction of the GCP number Experiments were performed in Pix4D in two variants. The first one used 9 GCP and the second only 4 (Fig. 5). Both variants made two alignments, first using 25 000 key points per image and the second of 10 000 key points per image. RMS value analysis on Check Points (Tab. 5) showed no significant differences in alignment between using 4 or 9 GCPs. To further verify the impact of ground control points distribution the distance between tie points for both examined variants was created. Such comparison is presented in Fig.5 and this analysis was performed with 10 000 key points per image, assuming that the less tie poits you have the impact of analysis should be more visible. The average difference in distance was 4 cm (with STD of 4 cm) and it can be noted that the spatial distribution of the distance between the position of tie points from both alignment does not show any significant systematic component.

3.4 Adjustment together with nadir images

The process of relative orientation of images is strongly dependent on their configuration, especially since the occurrence of nadir images is considered. Using them allows it to be much easier to match corresponding points on oblique images with different view direction. However, not for all objectives have oblique images synchronously captured nadir images. In such cases, resignation of nadir images could also improve efficiency.

0   1000   2000 m

0   4   8   12 cm

Figure 5. Map of GCP distribution in Katowice dataset. Markers: Yellow – Check points, Green – GCP in both variants, Pink – GCP only in variant with 9 GCPs. Spatial distribution of distances between tie points computed in variant with four and nine GCPs.

To test the impact of nadir images in oblique images orientation two alignments of blocks in Katowice were carried out. In the first all images (oblique and nadir) were used in adjustment and in the second only oblique images were a subject of aerial triangulation. Adjustment was carried out using 25 000 key points per image and 4 GCP.

No.	Dataset	Keypoints	Matches (median)	Oriented images	GCP	Check points	RMS GCP X/Y/Z [m]	Check Point RMS X/Y/Z [m]
1	Katowice	10 000	2318	1068/1070	9	47	0.06/0.06/0.05	0.05/0.09/0.09
2	Katowice	25 000	5844	1070/1070	9	47	0.06/0.07/0.05	0.05/0.09/0.09
3	Katowice	10 000	2309	1068/1070	4	47	0.05/0.04/0.05	0.05/0.09/0.10
4	Katowice	25 000	5843	1070/1070	4	47	0.04/0.05/0.06	0.06/0.09/0.09

Table 5. The results for GCP number and distribution analysis in Pix4D

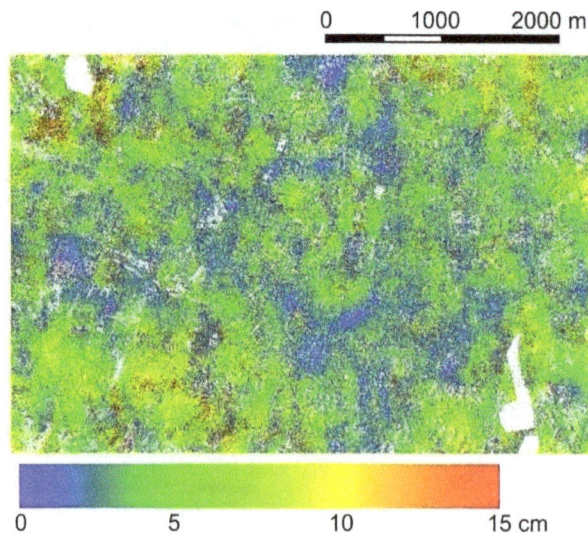

Figure 6. Map of spatial distribution of deviation in tie points height computed with and without nadir images (Katowice dataset).

Fig. 6 presents height differences for tie points extracted in adjustment for blocks where and images were and were not included. As it can be seen using nadir images it is not necessary for an effective orientation, and it does not substantially affect the precision and effectiveness of the orientation.

3.5 Overlap between images

A factor that may have a significant impact on the time of matching images is overlaps in a block. More overlaps means that more imagery is considered in adjustment. It should also be noted that tested programs were designed with the aim of processing of images with high overlaps.

A Comparison that was carried out between the time for orientation of the block in Katowice (60/30 overlap) and Wroclaw (60/60) showed no significant increase in computation time omitting an increasing number of images. There are also no differences in the effectiveness of orientation - comparable errors.

Experiments were also carried out on the block with much less overlap of approx. 30/30%. Attempts at orientation were successful both in Pix4D nor in Agisoft. Major part of block has remained non0oriented or EO parameters were wrong. However, due to other characters of the area, which was a rural area, it is difficult to clearly determine whether the direct cause of such low efficiency of orientation was caused by the lack of full stereoscopic overlap (which is definitely not an advantage) orwhether it was mostly caused by the type of terrain.

3.6 Number of images – division into sub-blocks

The last of the examined parameters, which can be crucial for the orientation process, is the number of images used in one alignment. To determine how the increase of the number influences the time of processing in orientation of blocks two datasets (Katowice and Wroclaw) were divided into sub blocks.

For a block from Katowice linear increase was observed for series of adjustments with different number of photos taken to orientation. This caused by small number of in whole block in

contrary to a block from Wroclaw where a significant increase in duration time it was observed when large amount of images is considered - 3000 images (Fig. 7) in a single alignment.

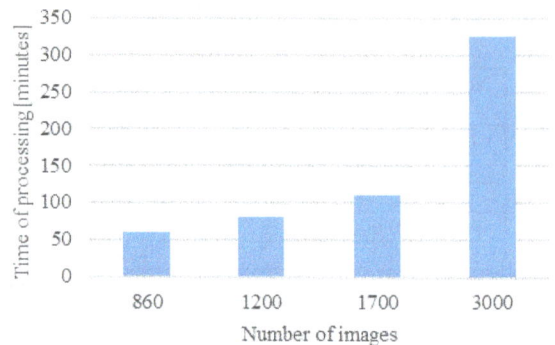

Figure 7. Time of processing (in minutes) in Pix4D depend of number of images (Wroclaw dataset).

It can be expected that the decline in performance orientation depends on how many photos and will also largely depend on the workstation used

4. SUMMARY AND CONCLUSIONS

The presented research proves that it is possible to process bundle adjustment of oblique imagery in commercial software. Many variants of adjustment lead to conclusions as follows:

(1) The performed experiments performed using Pix4D and Agisoft have shown that it is possible to use commercial software for effective orientation of the oblique images. However, considering two test programs, Pix4D has a much greater stability. Moreover, in this software orientation of oblique image block can be performed with nadir images simultaneously.

(2) Due to the fact that these programs today cannot effectively use the angular exterior orientation of images, registration with the use of expensive IMU system seems to be pointless. This situation could be changed with the implementation of other algorithms of matching suggested ,for example, by Xiao et al. (2013).

(3) A significant influence on the duration of the aerial triangulation is the number key points which may be significantly reduced in Pix4D with respect to the default settings without having a negative impact on the results.

(4) An important element in the efficient development of large blocks is their division into sub-blocks which improves the productivity of software. In researchan increase of duration was observed when a large number of images was used in aerial triangulation. This can avoid a situation in which it effective alignment will be possible (Karel and Pfeifer, 2015). This solution is much better than with conventional photogrammetry. A block of oblique images is characterized by a more resistant geometry (more tie points on many images, higher overlaps), and can afford a significant reduction in the number GCP without affecting the orientation results.

ACKNOWLEDGEMENTS

The authors wish to thank MGGP Aero for their friendly attitude and for the data from this company's own projects

REFERENCES

Gerke, M. & Nyaruhuma, P., 2009. Incorporating scene constraints into the triangulation of airborne oblique images. *ISPRS - International Archives of the Photogrammetry, Remote Sensing and Spatial Information Sciences*, 38(part 1-4-7/W5),

Grenzdorffer, G.J., Guretzki, M. & Friedlander, I., 2008. Photogrammetric Image Acquisition and Image. *The Photogrammetric Record*, 23(December), pp.372–386.

Haala, N., Rothermel, M. & Cavegn, S., 2015. Extracting 3D urban models from oblique aerial images. *In 2015 Joint Urban Remote Sensing Event (JURSE). IEEE*, pp. 1–4.

Höhle, J., 2008. Photogrammetric measurements in oblique aerial images. *Photogrammetrie, Fernerkundung, Geoinformation, 1*, pp.7–14.

Hu, H. et al., 2015. Reliable Spatial Relationship Constrained Feature Point Matching of Oblique Aerial Images. *Photogrammetric Engineering & Remote Sensing*, 81(1), pp.49–58.

Jacobsen, K., 2008. Geometry of vertical and oblique image combinations. In *Remote Sensing for a Changing Europe: Proceedings of the 28th Symposium of the European Association of Remote Sensing Laboratories*, Istanbul, Turkey.

Karel W., Pfeifer N., 2015. Analysis of oblique image datasets with OrientAL. In *EUROSDR/ISPRS workshop on 'oblique cameras and dense image matching'* 19 - 20 October Southampton, UK.

Nex, F., Rupnik, E. & Remondino, F., 2013. Building footprints extraction from oblique imagery. *ISPRS Annals of the Photogrammetry, Remote Sensing and Spatial Information Sciences*, II(November), pp.61–66.

Nyaruhuma, P., Gerke, M. & Vosselman, G., 2012. Verification of 3D Building Models Using Mutual Information in Airborne Oblique Images. *ISPRS Annals of the Photogrammetry, Remote Sensing and Spatial Information Sciences*, I(September), pp.275–280.

Nyaruhuma, A.P. et al., 2012. Verification of 2D building outlines using oblique airborne images. ISPRS *Journal of Photogrammetry and Remote Sensing*, 71, pp.62–75.

Rau, J., Jhan, J. & Hsu, Y., 2015. Analysis of Oblique Aerial Images for Land Cover and Point Cloud Classification in an Urban Environment. *IEEE Transactions on geoscience and remote sensing*, 53(3), pp.1304–1319.

Remondino, F. & Gerke, M., 2015. Oblique Aerial Imagery – A Review. *Photogrammetric Week 2015*, pp.75–83.

Rupnik, E., Nex, F. & Remondino, F., 2013. Automatic orientation of large blocks of oblique images. *ISPRS - International Archives of the Photogrammetry, Remote Sensing and Spatial Information Sciences*, XL(May), pp.21–24.

Rupnik, E. et al., 2015. Aerial multi-camera systems: Accuracy and block triangulation issues. *ISPRS Journal of Photogrammetry and Remote Sensing*, 101(60), pp.233–246.

Wiedemann, A. & Moré, J., 2012. Orientation Strategies for Aerial Oblique Images. *ISPRS - International Archives of the Photogrammetry, Remote Sensing and Spatial Information Sciences*, XXXIX-B1(September), pp.185–189.

Xiao, X. et al., 2013. Robust and rapid matching of oblique UAV images of urban area. In *Proc. SPIE 8919, MIPPR 2013: Pattern Recognition and Computer Vision*,. p. 89190Y..

Yang, H., Zhang, S. & Wang, Y., 2012. Robust and precise registration of oblique images based on scale-invariant feature transformation algorithm. *IEEE Geoscience and Remote Sensing Letters*, 9(4), pp.783–787.

9

CHANGE DETECTION AND LAND USE / LAND COVER DATABASE UPDATING USING IMAGE SEGMENTATION, GIS ANALYSIS AND VISUAL INTERPRETATION

Jean-François Mas and Rafael González

Centro de Investigaciones en Geografia Ambiental (CIGA)
Universidad Nacional Autónoma de México (UNAM)
Antigua Carretera a Patzcuaro No. 8701
Col. Ex-Hacienda de San José de La Huerta
C.P. 58190 Morelia Michoacan MEXICO
jfmas@ciga.unam.mx; jrgonzalez@pmip.unam.mx

KEY WORDS: Land cover database, Updating, Uncertainty, Image segmentation, Visual interpretation

ABSTRACT:

This article presents a hybrid method that combines image segmentation, GIS analysis, and visual interpretation in order to detect discrepancies between an existing land use/cover map and satellite images, and assess land use/cover changes. It was applied to the elaboration of a multidate land use/cover database of the State of Michoacán, Mexico using SPOT and Landsat imagery. The method was first applied to improve the resolution of an existing 1:250,000 land use/cover map produced through the visual interpretation of 2007 SPOT images. A segmentation of the 2007 SPOT images was carried out to create spectrally homogeneous objects with a minimum area of two hectares. Through an overlay operation with the outdated map, each segment receives the "majority" category from the map. Furthermore, spectral indices of the SPOT image were calculated for each band and each segment; therefore, each segment was characterized from the images (spectral indices) and the map (class label). In order to detect uncertain areas which present discrepancy between spectral response and class label, a multivariate trimming, which consists in truncating a distribution from its least likely values, was applied. The segments that behave like outliers were detected and labeled as "uncertain" and a probable alternative category was determined by means of a digital classification using a decision tree classification algorithm. Then, the segments were visually inspected in the SPOT image and high resolution imagery to assign a final category. The same procedure was applied to update the map to 2014 using Landsat imagery. As a final step, an accuracy assessment was carried out using verification sites selected from a stratified random sampling and visually interpreted using high resolution imagery and ground truth.

1. INTRODUCTION

Due to its latitudinal position, its topography, climate and geology diversity, Mexico presents a high biodiversity. In particular, there are many types of vegetation, including temperate and tropical forests (Toledo, 1994). Besides, Mexico presents high rates of land use/cover change (LUCC) including important processes of deforestation and forest degradation (Mas et al., 2004, Velázquez et al., 2010). LUCC results in complex mosaics of land use and forest patches. For these reasons, mapping land use/cover is not an easy task.

As a result of this large diversity, land use/cover (LUC) map's classification schemes are complex. For instance, LUC maps from the National Institute of Statistics and Geography (INEGI), the Mexican mapping agency, have 57 different types of vegetation and 20 types of land use (INEGI, 2011). Different types of covers present similar spectral response and the same type of cover can present different spectral responses depending on the phenology, the conservation state and the density of vegetation. Due to LUCC, frequent updating and the elaboration of multidate cartographic databases are required to assess change. Different approaches can be used to elaborate and update existing LUC maps. On the one hand, visual interpretation, often computer-aided, has been widely used to elaborate LUC cartography including cartography over large areas such as Europe (Feranec et al., 2007), Africa (Disperati and Virdis, 2015) and, China (Zhang et al., 2014). It enables map producers to include many classification criteria such as texture, shape, pattern, size of object and proximity between object, interpreter's knowledge, etc. and has been shown to achieve more accurate results than spectral-based digital approaches (Sader et al., 1990, Mas and Ramirez, 1996,

Palacio Prieto and González, 1994, van den Broek et al., 2004). When visual interpretation is used to update existing cartography, LUCC are extracted accurately (Zhang et al., 2014, Disperati and Virdis, 2015). On the other hand fully automatic processing approaches based on digital classification of spectral data eventually combined with ancillary information (Gebhardt et al., 2014) permit a faster analysis but they present often more classification errors than visual interpretation.

This study aims at developing a hybrid method (semi-automatic processing and computer-aided visual interpretation) which combines image segmentation, GIS analysis, and visual interpretation to elaborate a multidate cartographic database from remote sensing data, to produce an updated cartography and assess LUCC (deforestation).

2. STUDY AREA AND MATERIALS

The State of Michoacán (Fig.1), which encompasses about 60,000 square kilometers, is one of the most diverse State of Mexico with different types of tropical and temperate forests. It also presents important processes of land use/cover change (Bocco et al., 2001).

We used 32 SPOT 5 images of 2007 along with an outdated LUC cartography at scale 1:100,000 obtained by visual interpretation of a Landsat image dated 2003 (Figure 3). Image processing was carried out using the spatial modeling platform DINAMICA EGO (Soares-Filho et al., 2002), BIS Cloud (Berkeley Image Segmentation on the cloud) (Berkeley Image Segmentation, 2015) and R (R Core Team, 2013).

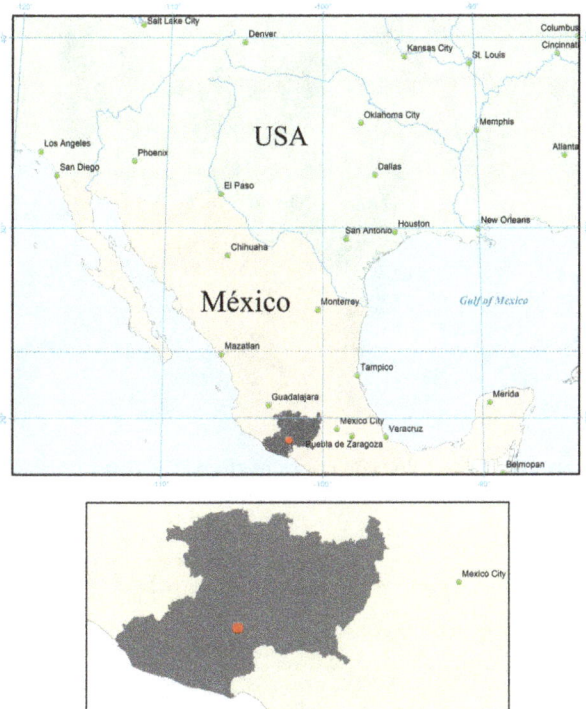

Figure 1: State of Michoacán, Mexico. The red square is the area represented in the subsequent figures

imagery and ground truth.

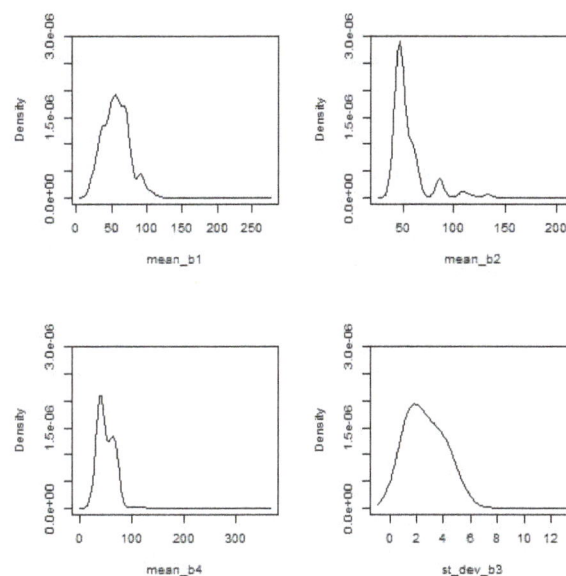

Figure 2: Density function per analysed band.

3. METHODS

Method was modified from the approach proposed by Radoux & Defourny (2010) and Radoux et al., (2014) to update cartography and/or detect changes. This approach consists of using an existing but outdated LUC map to classify a recent image. In this case, a segmentation of the 2007 SPOT (Système Probatoire d'Observation de la Terre) images was carried out using a region growing algorithm, creating spectrally homogeneous objects with a minimum area of two hectares. Then the segmented image and the outdated map were overlaid in order to assign a label (category) at each segment: Each segment was labeled based on the class from visual interpretation-based map covering the largest proportion of its area. The spectral response of each segmented was also computed. Therefore each segment was associated to a category (from the map) and spectral indices (mean and standard deviation of digital number values for each spectral band). Each category was then defined by a density function (figure 2), which describes the relative likelihood for a pixel with a given spectral value to belong to a particular LUC category. Segments whose label from the map did not match the "typical" spectral response of their category were identified as uncertain segments (outliers) by trimming (removal of extreme "outliers" values). These segments were classified using spectral response by means of the tree C5 classifier as a help for the visual interpreter. However, the final category was assigned by visual interpretation of a false color composite of the SPOT images. Finally, an accuracy assessment was carried out using a stratified random sampling and panchromatic fused SPOT images following the method proposed by Card (1982).

The same procedure was applied to update the map to 2014 using Landsat imagery. In the update of 2007 map to 2014, it can be reasonably assumed that many discrepancies were effective land use/cover change. As a final step, an accuracy assessment was carried out using verification sites selected from a stratified random sampling and visually interpreted using high resolution

4. RESULTS

A classification scheme of 25 LUC categories including 15 forest categories was defined (Table 1) In this 1:100,000 2003 LUC map, objects representation depends on cartographic rules as minimum cartographic area and polygon generalization (Figure 3). The segmentation of the 2007 SPOT images created spectrally homogeneous objects with a minimum area of two hectares (Figure 4). The segmented image was therefore spatially more detailed than the 2003 map obtained through visual interpretation (Figure 5).

The accuracy assessment from the 2007 updated segmentation indicated an overall accuracy of 79.5 % ± 3.3 and user and producer accuracies between 50 and 100% (Table 2 in Appendix).

The 2007 map was used to update the information using 2014 Landsat imagery (Figures 6 and 7). In summary, this method permits to update maps from different imagery inputs.

5. DISCUSSION AND CONCLUSION

Visual interpretation is widely used to update existing cartography. According to Zhang et al. (2014) processes of LUCC are extracted more accurately by visual interpretation than by digital classification. However, in case of the elaboration of new cartography (without existing previous map) or when the outdated existing map is represented at a coarser scale, visual delimitation of polygons, a time consuming task, has to be carried out. In these cases, digital image segmentation is a way to capture the limits between the different LUC types. In addition, hybrid approach allowed optimizing the work of the interpreter by identifying areas with likely errors or changes. Spectral classification is used but the "last word" is given to the visual interpretation because, due to spectral confusion, we considered expert opinion more accurate.

The obtained accuracy is higher than accuracy commonly obtained by digital approaches in Mexico. This method will be used

LUC category	code
Cropland (Irrigated)	1
Cropland (rainfed)	2
Cropland (perennial)	3
Human settlements	4
Grassland	5
Oak forest (primary)	6
Oak forest (secondary)	7
Oyamel forest (primary)	8
Oyamel forest (secondary)	9
Pine forest (primary)	10
Pine forest (secondary)	11
Mountain cloud forest	12
Mountain cloud forest (secondary)	13
Pine-oak forest (primary)	14
Pine-oak forest (secondary)	15
Scrubland	16
Scrubland (secondary)	17
Tropical deciduous forest	18
Tropical deciduous forest (secondary)	19
Medium deciduous forest	20
Medium deciduous forest (secondary)	21
Water body	22
Mangrove	23
Popal-tular (Wetland vegetation)	24
Areas without apparent vegetation	25

Table 1: Classification scheme.

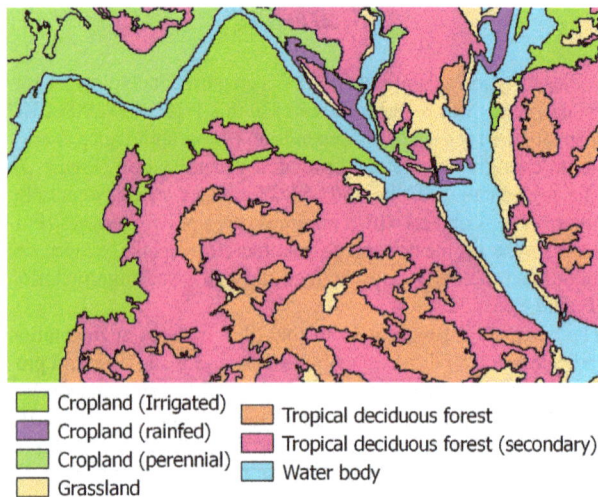

Figure 4: False color composite of the SPOT 5 Image, 2007. © CNES.

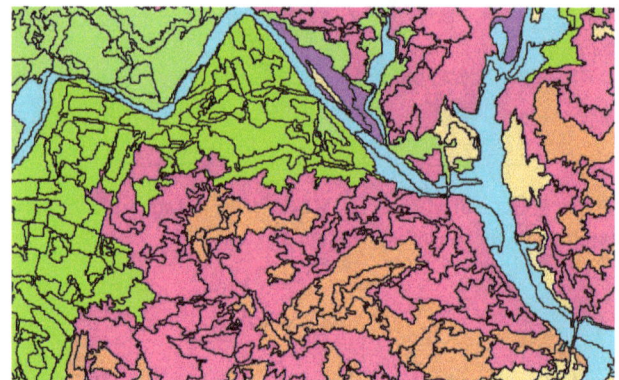

Cropland (Irrigated)
Cropland (rainfed)
Cropland (perennial)
Grassland
Tropical deciduous forest
Tropical deciduous forest (secondary)
Water body

Figure 5: SPOT 5 based updated segmentation (2007).

Cropland (Irrigated)
Cropland (rainfed)
Cropland (perennial)
Grassland
Tropical deciduous forest
Tropical deciduous forest (secondary)
Water body

Figure 3: 2003 LUC map.

for the elaboration of a whole time series to monitor LUCC in Michoacan. Further work is necessary to deeply evaluate classification errors. the LUC map will be evaluated by interpreters from INEGI and verification sites used in the first accuracy assessment will be interpreted again by a second interpreter and, eventually verified in field in case of incongruence between the first and second interpretations. Finally, information from field plots from the National Forest Inventory will be integrated in the database.

ACKNOWLEDGEMENTS

This study was supported by the project *"Monitoreo de la cubierta del suelo y la deforestación en el Estado de Michoacán: un análisis de cambios mediante sensores remotos a escala regional"* (Fondo Mixto Conacyt - Gobierno del Estado de Michoacán, grant number 192429). We would like to thank Luis

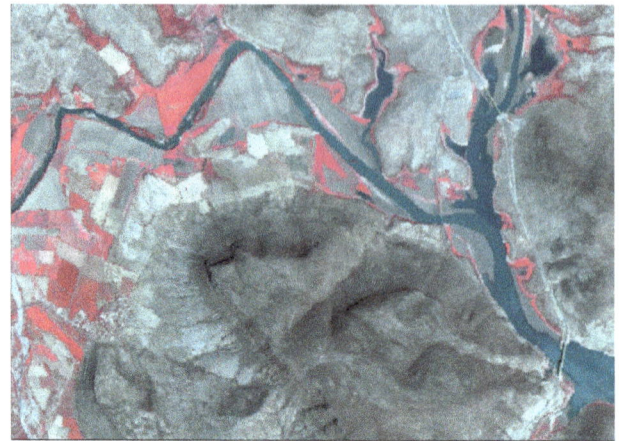

Figure 6: False color composite of the Landsat 8 Image, 2014. © USGS/NASA Landsat. Red objects were the "uncertain" segments.

Giovanni Ramírez Sánchez, González Rodríguez Marisol, Evelyn Herrera-Flores, Jairo López-Sánchez, Andrés Piña Garduño and Richard Lemoine Rodríguez for their contributions. SPOT images were obtained through the ERMEXS-UNAM agreement.

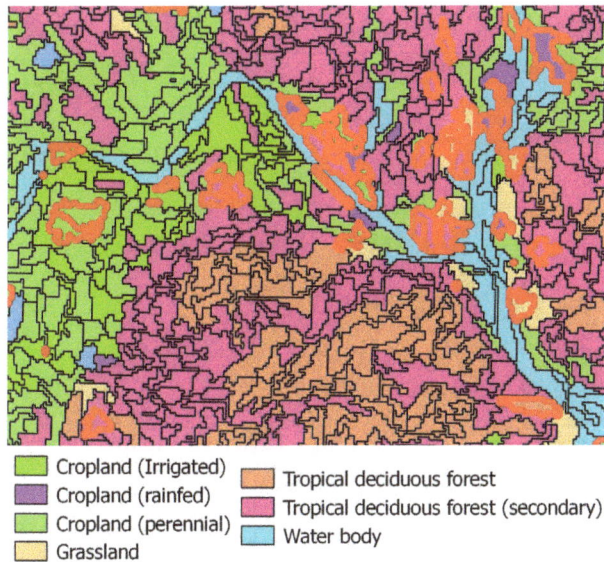

Cropland (Irrigated)
Cropland (rainfed)
Cropland (perennial)
Grassland
Tropical deciduous forest
Tropical deciduous forest (secondary)
Water body

Figure 7: Landsat 8 based updated segmentation (2014) Probable classification by C5 algorithm. Red objects were the "uncertain" segments before visual interpretation.

REFERENCES

Berkeley Image Segmentation, 2015.

Bocco, G., Méndoza, M. and Masera, O. R., 2001. La dinámica del cambio del uso del suelo en Michoacán. Una propuesta metodológica para el estudio de los procesos de deforestación. Investigaciones Geográficas 44, pp. 18–38.

Card, D. H., 1982. Using known map category marginal frequencies to improve estimates of thematic map accuracy.

Disperati, L. and Virdis, S. G. P., 2015. Assessment of land-use and land-cover changes from 1965 to 2014 in tam giang-cau hai lagoon, central vietnam. Applied Geography 58(0), pp. 48 – 64.

Feranec, J., Hazeu, G., Christensen, S. and Jaffrain, G., 2007. Corine land cover change detection in europe (case studies of the netherlands and slovakia). Land Use Policy 24(1), pp. 234 – 247.

Gebhardt, S., Wehrmann, T., Ruiz, M. A. M., Maeda, P., Bishop, J., Schramm, M., Kopeinig, R., Cartus, O., Kellndorfer, J., Ressl, R. et al., 2014. Mad-mex: Automatic wall-to-wall land cover monitoring for the mexican redd-mrv program using all landsat data. Remote Sensing 6(5), pp. 3923–3943.

INEGI, 2011. Metodología para la Generación y Actualización de la Información de Uso de Suelo y Vegetación, escala 1:250,000, Serie IV. Technical report.

Mas, J. and Ramirez, I., 1996. Comparison of land use classifications obtained by visual interpretation and digital processing. ITC Journal (3), pp. 278–283.

Mas, J.-F., Velázquez, A., Díaz-Gallegos, J., Mayorga-Saucedo, A., Alcántara, C., Bocco, G., Castro, R., Fernández, T. and Pérez-Vega, A., 2004. Assessing land use/cover changes: A nationwide multidate spatial database for Mexico. International Journal of Applied Earth Observation and Geoinformation 5(4), pp. 249–261.

Palacio Prieto, J. and González, L. L., 1994. Clasificación espectral automática vs. clasificación visual: un ejemplo al sur de la ciudad de méxico. Investigaciones Geográficas Boletín 29, pp. 25–40.

R Core Team, 2013. R: A Language and Environment for Statistical Computing. R Foundation for Statistical Computing, Vienna, Austria.

Radoux, J. and Defourny, P., 2010. Automated image-to-map discrepancy detection using iterative trimming. Photogrammetric Engineering & Remote Sensing 76(2), pp. 173–181.

Radoux, J., Lamarche, C., Van Bogaert, E., Bontemps, S., Brockmann, C. and Defourny, P., 2014. Automated training sample extraction for global land cover mapping. Remote Sensing 6(5), pp. 3965–3987.

Sader, S. A., Stone, T. A. and Joyce, A. T., 1990. Remote sensing of tropical forests: an overview of research and applications using non-photographic sensors. Photogrammetric Engineering and Remote Sensing 56(10), pp. 1343–1351.

Soares-Filho, B. S., Cerqueira, G. C. and Pennachin, C. L., 2002. Dinamica—a stochastic cellular automata model designed to simulate the landscape dynamics in an amazonian colonization frontier. Ecological modelling 154(3), pp. 217–235.

Toledo, V. M., 1994. La biodiversidad de México. Ciencias 34, pp. 43–57.

van den Broek, A. C., Smith, A. J. E. and Toet, A., 2004. Land use classification of polarimetric sar data by visual interpretation and comparison with an automatic procedure. International Journal of Remote Sensing 25(18), pp. 3573–3591.

Velázquez, A., Mas, J.-F., Bocco, G. and Palacio-Prieto, J. L., 2010. Mapping land cover changes in Mexico, 1976-2000 and applications for guiding environmental management policy. Singapore Journal of Tropical Geography 31(2), pp. 152–162.

Zhang, Z., Wang, X., Zhao, X., Liu, B., Yi, L., Zuo, L., Wen, Q., Liu, F., Xu, J. and Hu, S., 2014. A 2010 update of national land use/cover database of china at 1:100000 scale using medium spatial resolution satellite images. Remote Sensing of Environment 149(0), pp. 142 – 154.

APPENDIX

LUC class	code	User's accuracy %	Confidence interval %	Product's accuracy %	Confidence interval%
Cropland (Irrigated)	1	47.83	20.42	70.3	17.39
Cropland (rainfed)	2	86.84	10.75	62.7	8.19
Cropland (perennial)	3	84.00	10.16	93.7	6.08
Human settlements	4	95.83	5.65	99.6	0.72
Grassland	5	85.42	9.98	74.0	8.02
Oak forest (primary)	6	55.10	13.93	86.9	16.47
Oak forest (secondary)	7	58.00	13.68	75.4	13.24
Oyamel forest (primary)	8	94.59	7.29	100.0	0.00
Oyamel forest (secondary)	9	90.00	18.59	72.0	39.21
Pine forest (primary)	10	86.27	9.44	88.3	17.02
Pine forest (secondary)	11	81.25	11.04	77.3	12.62
Mountain cloud forest	12	50.00	17.89	100.0	0.00
Mountain cloud forest (secondary)	13	40.00	42.94	0.9	1.14
Pine-oak forest (primary)	14	85.71	9.80	88.0	7.98
Pine-oak forest (secondary)	15	77.08	11.89	78.2	10.46
Scrubland	16	93.88	6.71	92.3	9.98
Scrubland (secondary)	17	83.33	13.34	66.9	19.88
Tropical deciduous forest	18	80.00	11.09	86.8	9.23
Tropical deciduous forest (secondary)	19	89.80	8.48	81.2	7.33
Medium deciduous forest	20	70.00	14.20	96.5	6.56
Medium deciduous forest (secondary)	21	70.73	13.93	85.5	10.47
Water body	22	100.00	0.00	99.6	0.70
Mangrove	23	65.00	20.90	100.0	0.00
Popal-tular (Wetland vegetation)	24	93.10	9.22	100.0	0.00
Areas without apparent vegetation	25	93.33	8.93	100.0	0.00

Table 2: User's and Product's accuracies along with their confidence interval

10

TRAJECTORY ADJUSTMENT OF MOBILE LASER SCAN DATA IN GPS DENIED ENVIRONMENTS

P. Schaer [a, *], J. Vallet [b]

[a] GEOSAT SA, Route du Manège 59b, 1950 Sion, Switzerland – schaer@geosat.ch
[b] HELIMAP SYSTEM SA, Le Grand-Chemin 73, 1066 Epalinges, Switzerland –julien.vallet@helimap.ch

KEY WORDS: MLS, mobile mapping, LiDAR, long tunnel survey, GPS denied environment

ABSTRACT:

Mobile mapping often occurs in environments with poor or no GNSS signal reception (tunnels, urban canons, canopy …). Most mobile scanning systems are equipped with tactical grade inertial systems, GNSS outages longer than 30-60 seconds may lead to a rapid decrease in absolute positioning accuracy that might be below the client's expectations. In order to guarantee sufficient positioning accuracy even in such environments the data must be corrected. This can either happen by readjusting the final point cloud using control points (3D translations) or by directly correcting the trajectory by adding external position updates. The approach presented in this paper consists in detecting targets (either existing or placed before measurement) that can easily be identified in the point cloud and that have been measured independently. By identifying the 3D point closest to the target's center, computing the coordinate difference to the corresponding GCP and retrieving its GNSS-timestamp and the internal scanner coordinates, an external position update at a given time can be computed. This procedure has proven its efficiency in several projects including scanning tunnels up to 5km length, where the positional error (in 3D) of the resulting point cloud could be reduced from 5m (in the middle of the tunnel) down to better than 5cm using GCPS only every 400m.

1. INTRODUCTION

The acquisition of laser point clouds for road environments by mobile mapping has become very popular, due to the fact that compared to terrestrial static laser scanning (TLS), kinematic acquisition of 3D data has substantial benefits in terms of cost efficiency (less time spent measuring, no closure of road necessary) and security (no workers exposed to traffic directly). Mobile mapping often occurs in environments where GNSS signals are partly (urban canons, canopy) or completely (tunnels, galleries) masked. Even though most mobile scanning systems are equipped with tactical or navigation grade IMUs, GNSS outages longer than 30-60 seconds may lead to a rapid decrease in absolute positioning accuracy that might be below the client's expectations (see table 1). In order to guarantee sufficient positioning accuracy even in such environments the data must be corrected.

Often these corrections are carried out by measuring control points that are used to create a local 3D deformation model which is applied to readjust directly the final point cloud generated by direct georeferencing (a posteriori adjustment). However this technique requires the presence of many control points (at least every 50meters), a task that may lead to high costs due to road closure and difficulty in access. As no correction on the trajectory data itself is applied, such procedure also neglects the non-linear behavior of position errors in trajectories derived from inertial data. It is therefore preferable to perform the corrections directly to the trajectory by adding external position updates (a priori adjustment).

In this paper we discuss the methodology for generating such external position updates directly form signalized targets in the point cloud. We also present results from a test in a tunnel of 5km length that validate our approach.

2. PROPOSED APPROACH

2.1 Mathematical model

IMU Grade	MEMS low cost	MEMS tactical	FOG tactical	FOG navigation
Angular drift	5-10°/h	1-3°/h	0.1-1°/h	<0.01°/h
Position drift after 60sec	1-10m	20-80cm	15-30cm	10-15cm
Position accuracy GPS/INS	10 cm	5-10cm	5cm	5cm
Angular accuracy GPS/INS	0.1-0.2°	0.05°	<0.02°	<0.005°
Typical usage	Leisure	Short range ALS	ALS, MLS	ALS, MLS
Price	50-1'000 €	5-10'000 €	30'000 €	100'000 €

Table 1: Different types of IMU, their accuracies and usage

Compared to airborne laser scanning, where typical ranges reach from 200m up to 3500m, the measured laser ranges in mobile mapping systems (denoted MLS hereafter) are much shorter (typically 2-50m). In confined environments, such as tunnels, the maximal ranges often do not even exceed 5m. If the MLS is equipped with an IMU with an angular drift of less than 0.1°/h, the point cloud error due to the angular drift after one hour of dead reckoning does not exceed 1cm and can therefore be neglected for the readjustment of the data. The positional errors however increase much faster and may reach easily decimeter level after only 1 minute of dead reckoning even for navigation grade IMUs (see table 1). Therefore adding external position updates into the Kalman Filter (KF) at regular intervals

---

* Corresponding author

is of uppermost importance in order to guarantee a sub-decimeter absolute position accuracy of the computed point cloud.

Now we can consider the direct georeferencing formula (Baltsavias, 1999) that expresses the computation of a point cloud coordinate **p** in a mapping frame m in function of scanner measurements, the GPS/INS measurements and system calibration parameters:

$$\mathbf{p}(t)^m = \mathbf{s}(t)^m + \mathbf{R}_b^m(t)\left[\mathbf{R}_s^b \rho(t)\begin{pmatrix} \sin\theta(t) \\ 0 \\ \cos\theta(t) \end{pmatrix} + \mathbf{a}^b\right] \quad (1)$$

Where

$\mathbf{p}(t)^m$ is a point in the mapping frame computed at time t

$\rho(t), \theta(t)$ are the scanner range and the encoder angle measured by the scanner at time t

$\mathbf{s}(t)^m$ is the position of the scanner center in the arbitrary mapping frame m at time t

$\mathbf{R}_b^m(r(t),p(t),y(t))$ is the attitude matrix from the IMU body frame to the mapping frame at time t

$\mathbf{R}_s^b(e_x,e_y,e_z)$ is the boresight matrix describing the angular offsets between the body frame and the scanner frame

$\mathbf{a}^b$ is the lever-arm offset between the IMU and scanner frame origins expressed in the body frame

In the case a point was measured independently in 3D (denominated **gcp**^{m}) that materializes a clearly identifiable target, we can find a point $\mathbf{p}(t_i)^m$ in the point cloud generated by direct georeferencing that is closest in 3D to this point

$$\left\| \mathbf{gcp}^m - \mathbf{p}(t_i)^m \right\| = min \quad (2)$$

Assuming that the boresight $\mathbf{R}_s^b$ and leverarm $\mathbf{a}^b$ are perfectly calibrated, that the attitude angles (rpy) at time t_i have been estimated by GPS/INS integration with an accuracy better than 0.1°, the only unknown in equation 1 is the position of the scanner $\mathbf{s}(t_i)^m$ at a given time t_i. Accordingly equation 1 can be reformulated in order to compute scanner positions at a time t_i replacing the 3D coordinates of $\mathbf{p}(t_i)^m$ with the ones observed by independent control measures **gcp**^{m}.

2.2 Procedure

Based on the mathematical model explained before the following calibration procedure has been developed (figure 1):

- The trajectory is first computed by GPS/INS integration without any external position updates
- A point cloud (containing intensity value and the precise timestamp for every record) is generated based on this initial trajectory
- Within the initial point cloud the target center coordinates corresponding to a given GCP are extracted
- The timestamp t_i to the point closest to the target center is extracted and the 3D difference between the target center in the point cloud and the GCP coordinates is computed in order to compute position updates at the time t_i

- An adjusted trajectory is recomputed including the position updates at the times $t_1 \dots t_n$
- Finally the adjusted trajectory is used to compute the final point cloud where the targets centers within the point cloud match the GCP coordinates.

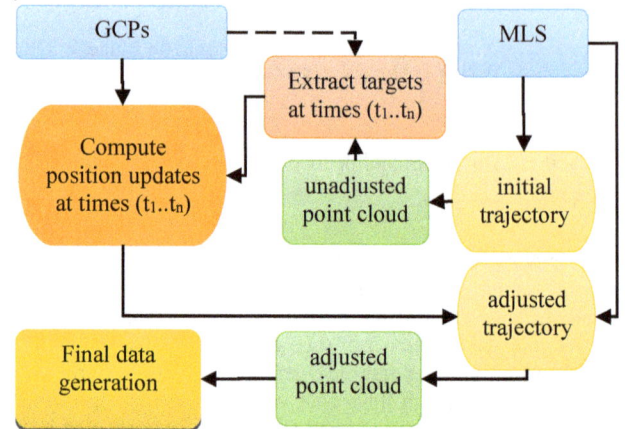

Figure 1: Schematic representation of trajectory adjustment procedure

3. VALIDATION OF CONCEPT

3.1 Test setup

In the frame of an airborne and mobile data acquisition campaign near Martigny (Switzerland) a more than 5km long section with tunnels and galleries (from Bourg St. Pierre to the north entry of the Great St. Bernhard tunnel) was acquired by mobile mapping in September 2013. The section was acquired in in both directions driving at 40-50km/h, thus recovering a GNSS signal on each side after approximately 6 minutes of driving.

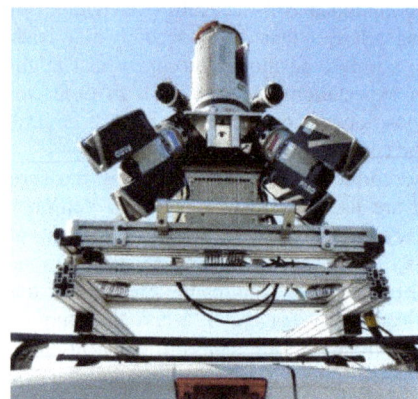

Figure 1: Mobile mapping system IGI-SAM used for the test in the tunnel section

The measures were performed using the IGI-SAM system that is composed of the following elements (see figure 2):

- Two scanners Faro Focus 3D (scanning rate up to 250'000pt/sec) tilted by 45° both in roll and pitch in order to measure a full 360° profile of any road section and also any perpendicular faces of road infrastructure
- One scanner Riegl VZ400 (scanning rate up to 150'000pt/sec) to scan the soft shoulder of the road

- Two cameras 8Mpix for documentation purposes
- L1/L2 GNSS receiver and a navigation grade IMU (IGI IMU-IIe: FOG Bias 0.03°/h) yielding an integrated position accuracy of 0.05m, roll /pitch accuracy of 0.003° and a heading accuracy of 0.007° (Minten, 2009).

Within the tunnel section GCPs were measured by a line traverse approximately every 200m with an accuracy of 1cm in 3D. The points were materialized as circular white targets with a diameter of 30cm (see figure 3a). Following the procedure described in paragraph 2.2, first the unadjusted point cloud was computed (figure 3b), then the circular targets were detected using the intensity values and the 3D displacement vectors to the GCP coordinates computed (figure 3c). Finally the timestamp of the laser point closest to the target center was extracted. These values were used to generate external observations to be inserted into to the Kalman Filter. The latter described steps were performed semi-automatically by a tool specially developed for this purpose by the contractors (partly shown in figure 3c).

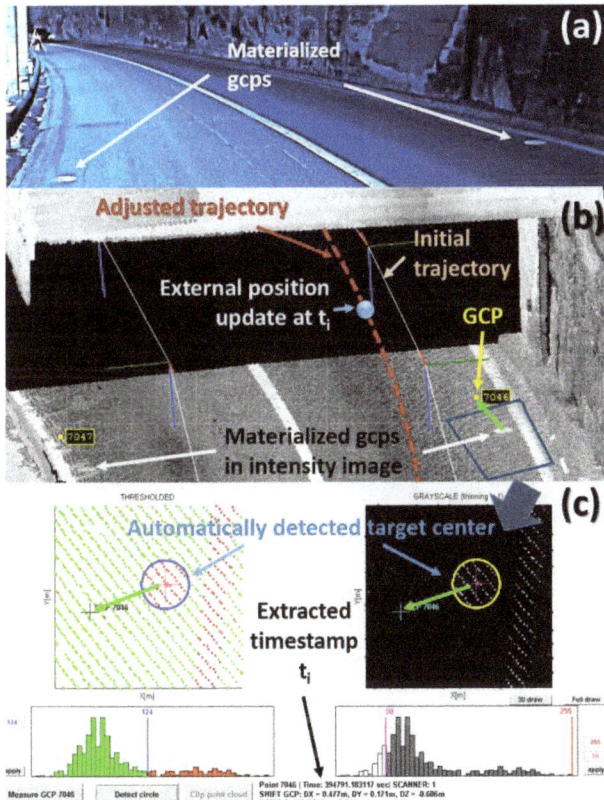

Figure 3: (a) Measure and materialization of targets with circular spots, (b) Initial point cloud color-coded by intensity, (c) Tool for detection of target centers, displacement vectors and timestamp extraction

3.2 Results

To evaluate what external update rate is required in order to reach a desired accuracy (5cm in this case), the computation of the trajectory and point cloud was performed using no GCPS (figure 4), GCPS every 1000m (figure 5) and GCPS every 400m (figure 6). For all scenarios, the target coordinate differences (in XY and Z) were computed for all GCPs available.

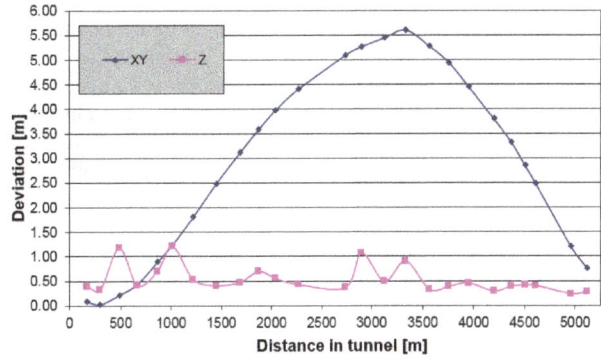

Figure 4: Deviation on GCPs without any adjustment

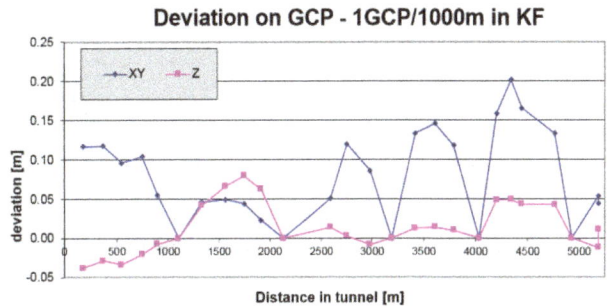

Figure 5: Deviation on GCPs using GCPS every 1000m for KF update

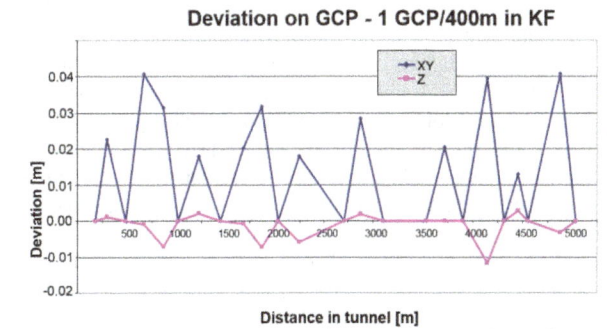

Figure 6: Deviation on GCPs using GCPS every 400m for KF update

The plot in figure 4 (pure inertial navigation for 6mins) shows that the position drift reaches its maximum (5m in XY and 1.2m in Z) approximately in the middle of the section. If the GNSS outage time is reduced by a factor 5 (update every 70sec), the maximum observed position drift does not exceed 20cm in XY and 7cm in Z (figure 5) thus improving the position accuracy by a factor of 25. If the GNSS outage time is reduced even further (update every 400m/35sec), the position accuracy can improved by a factor of 100 (max 5cm in XY and 1cm in Z), approaching the integrated position accuracy without GNSS outages (figure 6). The measures show that the absolute position accuracy degrades at the square-root of the GNSS outage time. This reflects the fact that in inertial navigation position estimates are computed from accelerometer observations and that errors which arise in the accelerometers propagate through the double integration (Woodman, 2007).

3.3 Interpretation of results

Figure 7 shows the trend curve for the position drift in function of the GNSS outage for the XY and Z component. It can be stated that for the category of IMU used (navigation grade

FOG) an external update every ~30 seconds (or every 400m at 40-50km/h) is enough to guarantee an absolute positioning better than 5cm in XY and 3cm in Z. This allows to reduce drastically the number of target points that have to be measured in comparison to the method where the point cloud is adjusted a posteriori (no correction of the trajectory). However, if the MLS is equipped with a less performant IMU (tactical grade IMU), the update rates must probably be increased.

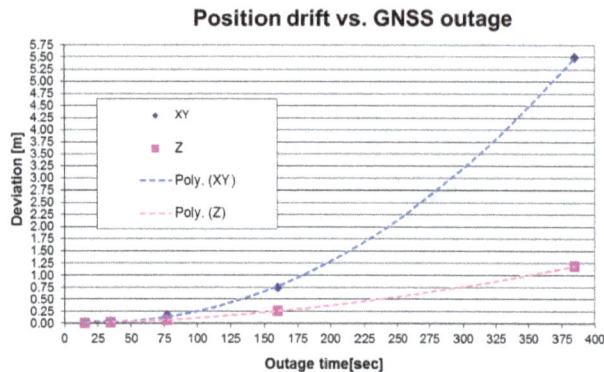

Figure 7:   Computed trend curve for position drift in function of GNSS outage time for a navigation grade IMU

4. CONCLUSION

In this paper we have presented a methodology capable of delivering MLS data with absolute positioning better than 5cm in XY and 3cm even in long tunnels. The methodology consists in measuring targets as ground control points (GCPS), identifying the 3D point closest to the target's center and computing the coordinate difference to the corresponding GCP. By retrieving the timestamp of the closest laser measurement, an external position update at a given time can be computed. In comparison to an a posteriori point cloud adjustment, the usage of GCPS to generate external time-tagged observations for the KF allows to reduce the number of GCPs by a factor of 5 to 10. The proposed methodology requires that the trajectory and the point cloud have to be computed twice, but the gain in accuracy and the economy in time on the field (less GCPs needed) clearly counterbalances this inconvenient. Additionally, if the GCPs are materialized by clearly identifiable patterns in the intensity color-coded point cloud (e.g. circular patterns), the target detection and external update computation can be highly automated.

REFERENCES

Baltsavias, E.P., 1999. Airborne laser scanning: basic relations and formulas. ISPRS Journal of Photogrammetry and Remote Sensing, 54(2-3): 199-214.

Minten, H. 2009. The Modular System Concept of IGI. In: FRITSCH, D. (Ed.): Photogrammetric Week '09, Wichmann, pp. 41-47.

Woodman J, 2007. An Introduction to Inertial Navigation ,Technical Report Number 696, Computer Laboratory UCAM-CL-TR-696 ISSN 1476-2986

11

LOW-LEVEL TIE FEATURE EXTRACTION OF MOBILE MAPPING DATA (MLS/IMAGES) AND AERIAL IMAGERY

P. Jende, Z. Hussnain, M. Peter, S. Oude Elberink, M. Gerke, G. Vosselman

University of Twente, Faculty of Geo-Information Science and Earth Observation (ITC),
Department of Earth Observation Science, The Netherlands
{p.l.h.jende,s.z.hussnain,m.s.peter,s.j.oudeelberink,m.gerke,george.vosselman}@utwente.nl

KEY WORDS: Mobile Mapping, Feature Extraction, Feature Matching, Image Orientation

ABSTRACT:

Mobile Mapping (MM) is a technique to obtain geo-information using sensors mounted on a mobile platform or vehicle. The mobile platform's position is provided by the integration of Global Navigation Satellite Systems (GNSS) and Inertial Navigation Systems (INS). However, especially in urban areas, building structures can obstruct a direct line-of-sight between the GNSS receiver and navigation satellites resulting in an erroneous position estimation. Therefore, derived MM data products, such as laser point clouds or images, lack the expected positioning reliability and accuracy. This issue has been addressed by many researchers, whose aim to mitigate these effects mainly concentrates on utilising tertiary reference data. However, current approaches do not consider errors in height, cannot achieve sub-decimetre accuracy and are often not designed to work in a fully automatic fashion. We propose an automatic pipeline to rectify MM data products by employing high resolution aerial nadir and oblique imagery as horizontal and vertical reference, respectively. By exploiting the MM platform's defective, and therefore imprecise but approximate orientation parameters, accurate feature matching techniques can be realised as a pre-processing step to minimise the MM platform's three-dimensional positioning error. Subsequently, identified correspondences serve as constraints for an orientation update, which is conducted by an estimation or adjustment technique. Since not all MM systems employ laser scanners and imaging sensors simultaneously, and each system and data demands different approaches, two independent workflows are developed in parallel.
Still under development, both workflows will be presented and preliminary results will be shown. The workflows comprise of three steps; feature extraction, feature matching and the orientation update. In this paper, initial results of low-level image and point cloud feature extraction methods will be discussed as well as an outline of the project and its framework will be given.

1. INTRODUCTION

Mobile Mapping is on the verge of becoming a substantial addition to the family of geo-data acquisition techniques. Airborne or satellite data cover large areas, but have limited capabilities when it comes to the density of data postings and high accuracy, whereas classical terrestrial techniques are expensive and often impractical. Particularly in urban areas, MM shapes up to be an extraordinarily useful technique not just to complement airborne or satellite coverage, but to enable a completely new array of possibilities. MM imaging systems and laser scanners collect high-resolution data, but have to rely on external georeferencing by GNSS. As GNSS being intermittently available, INS provides relative measures between position fixes and compensates for measurement noise and errors. Although GNSS carrier-phase measurements allow highly accurate positioning, urban areas remain problematic regarding the measurement reliability due to multipath effects and occlusions. When these phenomena persist over longer periods, accurate positioning cannot be maintained, and consequently data accuracy will be diminished (Godha, Petovello et al. 2005). This paper presents a method to detect and extract low-level image and point cloud features as a prerequisite for the rectification of MM data using aerial imagery. First, a brief outline of the project will be given. In section 2, a literature overview on similar work will be presented, and applied feature detection and extraction methods will be shortly introduced, followed by section 4 addressing low-level feature extraction for images as well as for point clouds. Section 5 discusses initial results of low-level feature extraction methods of both aerial and MM images as well as point cloud data. Lastly, section 6 concludes the work presented

in this paper as well as gives an outlook on future developments.

2. PROJECT OVERVIEW

The aim of our research project is to enable a reliable localisation pipeline for MM data obtained in urban areas, and to verify existing data sets according to their localisation accuracy in order to economise the acquisition of ground control. Due to apparent differences in the sensor setup and data, two workflows for Mobile Laser Scanning (MLS) and Mobile Mapping Imaging (MMI) are being developed. The common basis is the utilisation of high-resolution aerial nadir and oblique imagery as an external reference to compensate for vertical as well as for horizontal errors. In a first stage, common features between the ground data and aerial nadir imagery are sought. Based on the imprecise, but approximate exterior orientation of the MM data, more reliable and efficient matching techniques can be employed. For instance, a confined search for correspondences and their verification in the other image can be inferred even from coarse orientation parameters. The next stage will be the integration of oblique images into the pipeline to yield common features on the vertical axis in order to better detect errors in height, and to increase the overall number of tie features considerably. Façades and other vertical objects, such as street lights and traffic signs, are potential objects which can be used for that purpose in the future. In a last step, this tie information allows for either a re-computation of the trajectory or, alternatively, an adjustment of the data as such.

3. RELATED WORK

3.1 Previous Approaches

Coping with poor localisation of mobile platforms in urban areas has been addressed by many authors. Mostly by employing tertiary data as an external reference, either the data itself (Tournaire, Soheilian et al. (2006); Jaud, Rouveure et al. (2013); Ji, Shi et al. (2015)) or the platform's trajectory (Kümmerle, Steder et al. (2011); Levinson and Thrun (2007); Leung, Clark et al. (2008)) has been corrected. Depending on the data input and type (e.g. aerial imagery, digital maps or ground control points), different registration methods were utilised to impose unaffected, reliable and precise orientation information from external data on MM data sets. Subsequently, yielded correspondences were used as a constraint within a filter or adjustment solution. Even though many authors achieved a successful localisation based on an external reference, errors in height were not corrected, and a consistent sub-decimetre accuracy could not be reached.

3.2 Low-Level Feature Extraction

Both, low- and high-level feature extraction methods, are relevant for this research project. Whereas low-level features allow a great flexibility towards the selection of suitable correspondences, the registration of data originating from different sensors (i.e. Mobile Laser Scanning and aerial imagery) may demand an extension of that concept. Although MLS intensity information enables the derivation of corner features, an abstract representation by identifying common objects in both data sets can facilitate determining thorough and reliable transformation parameters. Hence, high-level feature extraction methods will be highlighted in the future. In this paper, however, emphasis will be placed on low-level feature extraction which is still an active field of research as real-time applications have been gaining more attention in the last few years. Classic feature detection algorithms, such as the Förstner-Operator (Förstner and Gülch 1987) or the Harris Corner Detector (Harris and Stephens 1988) are accompanied by state-of-the-art approaches like AKAZE (Alcanterilla, Nuevo et al. 2013) or FAST (Rosten and Drummond 2006). Although many improvements have been made in this field, the most important property of a feature detector remains to identify the same keypoints over a set of images.

Once features have been detected in the image, they have to be described unambiguously to increase their distinctiveness among other features in order to match them correctly. Low-level feature description approaches can be divided into two categories – binary and float description. Whereas float descriptors, such as SIFT (Lowe 2004), are based on a Histogram of Oriented Gradients (HoG), binary descriptors (e.g. BRIEF (Calonder, Lepetit et al. 2010)) are analysing the neighbourhood of a feature keypoint with a binary comparison of intensities according to a specific sampling pattern. Float descriptors are typically more expensive to compute, and need more memory to store their output than binary descriptors. However, depending on the application, robustness of these two categories varies (Heinly, Dunn et al. (2012); Miksik and Mikolajczyk (2012)).

In this paper, different feature detection as well as float and binary description methods will be compared taking the example of aerial nadir, MM panoramic imagery and intensity images derived from MLS data. Feature keypoints across the data sets will be computed with SIFT (Lowe 2004), KAZE (Alcanterilla, Bartoli et al. 2012), AKAZE (Alcanterilla, Nuevo

et al. 2013) and the Förstner Operator (Förstner and Gülch 1987).

SIFT detects blobs with a Difference-of-Gaussian method at different scaled instances of the image. KAZE computes a non-linear scale space using an additive operator splitting technique, where keypoints are detected at locations with a maximum response of the determinant of the Hessian matrix. Similarly, AKAZE also relies on keypoint detection based on the Hessian matrix, but computes a non-linear scale space with fast explicit diffusion. Förstner detects corners based on the search for local minima of eigenvalues of a covariance matrix of image gradients. Except for Förstner, all aforementioned procedures allow for an additional feature description. SIFT utilises a HoG in a local neighbourhood to describe a keypoint. KAZE's keypoints are described with the SURF descriptor (Bay, Ess et al. 2008) modified to be compatible with the detector's non-linear scale space. AKAZE uses a binary description based on an adapted version of Local Difference Binary (Yang and Cheng 2012) where sample patches around the keypoint are averaged and then compared in a binary manner. For Förstner keypoints, LATCH (Levi and Hassner 2015) has been used for a binary feature description. LATCH compares sample-triplets around a keypoint, where the sampling arrangement is learnt. Respective results will be discussed in section 5.

4. LOW-LEVEL FEATURE EXTRACTION

4.1 MMI & Aerial Nadir Images

Aerial nadir ortho-images with a ground sampling distance of approximately 12 centimetres serve as the reference data set in this project. The MM images are 360*180 degrees panoramic images (Figure 1) acquired every 5 metres along the platform's trajectory. For more details and specifications, please see (Beers 2011).

Figure 1 Mobile mapping panoramic image in equirectangular projection

In order to successfully use the aerial images' exterior orientation for the rectification of MM data, respective tie information has to be reliable and accurate. Although ground and aerial nadir data have a different perspective on the scene, low-level feature correspondences can be identified in all data sets. For example, corners of road markings, centres of manholes and building corners resemble each other across all sensors.

4.1.1 Pre-processing

In order to simplify and optimise feature matching, the panoramic images are projected onto an artificial ground plane to increase the resemblance to the aerial images. The ground plane is computed based on the location of the MM imaging sensor and the fixed height of the sensor above ground. Especially in areas where the actual ground is not exactly flat, this approximation can lead to certain distortions (see Figure 2). In the future, the rather reliable relative orientation between two recording locations will be used to compute a more accurate plane. Since this paper focuses solely on feature detection and description, and the aerial images used are ortho-projected, this fact can be neglected for now.

MM panoramic images are stored in an equirectangular projection, encoding directly spherical coordinates for every image pixel. Therefore, no projection matrix or other intrinsic parameters are needed to reproject the panoramic image. The quadratic ground plane is centred at the dropped perpendicular foot of the respective recording location. Analogue to the aerial imagery's resolution of 12 centimetres, the ground plane is rasterised holding a world coordinate for every cell. Subsequently, each raster cell's coordinate is back-projected into the panoramic image in order to extract the respective RGB value, and transfer the information back onto the ground plane.

Since every back-projected ray will pierce the image plane of the panoramic image, and thus every raster cell will contain an RGB value, an interpolation of the resulting projected image seems dispensable. However, the geometric representation of the pixels of both grids varies, leading to multiple assignments of the same RGB value especially at the edge of the projected image appearing as blur. Hence, a bilinear interpolation of the extracted value according to the pixel neighbourhood of the panoramic image is conducted. Consequently, every pixel in the projected image is composed of an individual set of grey values.

Figure 2 Panoramic projected onto an artificial ground plane

4.1.2 Feature Extraction

The only overlapping area for feature detection induced by different original perspectives between aerial ortho-images and MM images is the road surface and its immediate vicinity. Therefore, road markings, such as zebra crossings or centre lines are being targeted on for feature detection. Resulting from atmospheric conditions and motion blur (esp. cameras without forward motion compensation), the image quality of the aerial photographs can be affected. To compensate for these effects, the projected panoramic images might need to be blurred even though sharing the same resolution with the aerial image. In the process of projecting the panoramic images onto the ground, not just the projection but also the approximate scale and rotation of the aerial image have been retrieved simultaneously. In particular, this circumstance simplifies the matching process considerably, but also renders to be useful for the step of feature description as less invariances and therefore fewer ambiguities have to be considered by the descriptor; i.e. the descriptor does not have to account for scale and rotational invariance since the panoramic image is north oriented and has got the same resolution.

On the other hand, the images have not been acquired at the same time and with different sensor systems. Consequently, this fact is resulting in another category of a description problem.

For instance, changes in illumination and contrast may affect the computation of the descriptor.

Moreover, repetitive patterns of road markings (e.g. zebra crossings) cannot be ignored as they may result in false feature matches. Either this issue has to be tackled on the descriptor level or during the matching stage. Introducing rules, such as ordering constraints (Egels and Kasser (2001), p. 198) or perceptual grouping (Lowe (1985), p. 4), to describe a chain or group of adjacent features may prevent misassignment. Additionally, approximate camera parameters can be exploited within the matching procedure. By back-projecting identified keypoints into the other image, a window can be defined to constrain the search for correspondences. These methods are currently under development or labelled future work. Aforementioned feature detection and description procedures will be applied to our data sets and results will be discussed in section 5.

4.2 Mobile Laser Scanning

The Mobile laser scanning point cloud (MLSPC) is acquired from one or more lidar sensors mounted on a moving car. The car's trajectory is estimated by GNSS and IMU, where a GNSS based position is retrieved after one second intervals. The IMU is used to interpolate all intermediate positions. A particular mobile mapping car moving at a speed of 36 km/h covers an area of 10m in 1 second. During this 1 second interval, the IMU provides relatively accurate positions which favours to crop MLCPC patch-wise, where the size of each patch is 10 by 10 m. State of the art laser scanning systems claim to achieve a relative accuracy of 10 mm, when a control point is provided within 100 m of scanning. Thus, even if the scanning is conducted at a slower acquisition speed, the 10 by 10 m patch would not be affected by (IMU-based) distortions to an extent that would hamper feature extraction. Moreover, the point cloud which has been used in this project, already has an absolute accuracy in sub-metre range for roughly 25 km of scanning, which means that the relative accuracy of the point cloud is still within a 10 mm range.

Thereafter, each cropped point cloud patch is converted to an ortho-image by assigning a barycentric interpolation of laser intensities to its corresponding image pixel. A particular point cloud patch and the generated ortho-image is shown in Figure 3. The proposed method detects low-level features from ortho-image gradients using SIFT, KAZE, AKAZE and Förstner feature detectors. The feature point description is obtained from SIFT, KAZE, AKAZE and LATCH feature descriptors.

Figure 3 Point cloud patch (left) to an orthoimage (right).

5. RESULTS

In this section, feature detection and description methods will be compared according to their potential for deriving significant tie features and correspondences between aerial nadir and mobile mapping panoramic images as well as between aerial nadir and MLS intensity images. First, a comparison between SIFT[1], KAZE, AKAZE and Förstner[2] on each of the three data sets will be conducted. Subsequently, acquired keypoints will be described with their corresponding method except for Förstner where a LATCH description will be used. Although still under development, feature matching will be utilised to compare the quality of each descriptor. To this end, simple descriptor matching to yield correspondences and a homography estimation to detect outliers will be used. As the focus of this project is on urban areas, four subsets with each 15 m side length of a typical road scene between two intersections have been selected for this experiment (Figure 4).

Figure 4 Four subsets of a typical urban scene (coloured tiles from scene 1 on the left to scene 4 on the right)

5.1   Feature detection

In urban areas, road markings and other prominent objects, such as kerbstones or manholes, identifiable among all data sets are favoured for feature detection. However, due to noise and different original perspectives, it is considered to be a challenging task for the step of feature detection to maintain a comparable detection rate over the entire data set.

Depending on the scene, this detection rate varies. The number of road markings and the detector itself, highly influence the results. For instance, due to its scale invariance SIFT detects keypoints on different blurred instances of the same image, and thus yields a lot more potential features than a corner detector, such as Förstner. As it will be shown in section 5.2, a potent feature detection alone is not sufficient for a successful registration.

	Aerial Image	Panoramic Image	MLS Intensity	Total
SIFT	234	379	810	1423
KAZE	119	304	458	881
AKAZE	29	68	175	272
Förstner	40	75	153	268

Table 1 Number of combined keypoints over all subsets per detection method

5.1.1   SIFT

SIFT yields more keypoints than any other method used in this paper (Table 1). It detects 60% keypoints more than KAZE and even 5 times more keypoints than AKAZE or Förstner. Being very sensitive to image noise and detecting keypoints on different image scales, the detected features are not always

useful. In particular, this comes into effect for both types of MM images as they have a higher original resolution and therefore a higher entropy (Figure 5).

Figure 5 SIFT keypoints detected in aerial image (left), panoramic image (centre) and MLS intensity image (right)

5.1.2   KAZE

KAZE detects fewer keypoints than SIFT, but still considerably more than AKAZE or Förstner. However, road markings are very well preserved, and especially their corner features, which are the most important image entity in our case, were mostly detected (Figure 6).

Figure 6 KAZE keypoints detected in aerial image (left), panoramic image (centre) and MLS intensity image (right)

5.1.3   AKAZE

Although, AKAZE and KAZE are quite similar in the way how a feature is detected (determinant of Hessian), their main difference lies in the computation of image pyramids to detect keypoints at different image scales. AKAZE detects fewer keypoints than KAZE, but these keypoints are most often important corners of road markings (Figure 7). Nonetheless, in two of four aerial images, AKAZE only detected one single keypoint which turns out to be too few for matching purposes.

Figure 7 AKAZE keypoints detected in aerial image (left), panoramic image (centre) and MLS intensity image (right)

5.1.4   Förstner Operator

The Förstner Operator is the only feature detector without the consideration of scale. This, and the fact that Förstner detects features solely at corners and centres of small image objects, leads to a very deliberate keypoint detection. However, almost every detected feature can be regarded as significant for the registration process. Due to its capability of sub-pixel localisation of keypoints, the same object point can be represented slightly shifted among different image sources which shapes up as a challenge for feature description.

---

[1] For SIFT, KAZE, AKAZE and LATCH, their respective OpenCV implementation has been used

[2] Implementation of the Förstner-Operator by Marc Luxen, University of Bonn

Figure 8 Förstner keypoints detected in aerial image (left), panoramic image (centre) and MLS intensity image (right)

5.2 Feature description

Identified keypoints need to be described unambiguously to enable feature matching between two images. In general, difficulties arise if there is a change in perspective, illumination, coverage, or scale between the images as well as ambiguities resulting from repetitive patterns. In order to obviate apparent difficulties for registering the images, the MM data has been projected onto the ground to increase the resemblance to aerial imagery. As a consequence, scale and perspective are more similar among the data sets, but differences in illumination and coverage cannot be mitigated easily. Thus, different description methods are evaluated with regard to their ability to cope not just with the aforementioned changes but also to their performance to bridge sensor-induced differences.

SIFT, KAZE, AKAZE and LATCH will be used for feature description. As mentioned earlier, LATCH will be used for keypoints detected with the Förstner-Operator. To measure the descriptor quality of each method, the images have to be matched. The number of matches, inliers classified by RANSAC as well as the actual number of correct correspondences will be compared among different descriptors. Two out of four scenes (scene 1 and scene 2) will be discussed in detail[3]. Moreover, for every test scene, four different settings have been tested. To this end, MM data has been blurred with a Gaussian filter to increase the resemblance to the aerial data set. Moreover, a resampling of all data sets has been conducted as it has been shown that increasing the sampling size can facilitate a feature's distinctiveness considerably (Köthe 2003).

1st run	No modification of source images
2nd run	Gaussian blurring of source images
3rd run	Resampling to 150% of original size
4th run	Blurring and subsequent resampling of source images

5.2.1 Aerial images and panoramic images

5.2.1.1 Scene 1

The first scene comprises of a zebra crossing and dotted road markings aggravating correct matching due to possible descriptor ambiguity. If enough correct correspondences are found, RANSAC converges to a correct solution. In the first run, however, none of the methods was able to achieve a good result (see e.g. Figure 9). The derived keypoint descriptors were not distinct enough to be matched accordingly. By blurring the images with a Gaussian low-pass filter in the second iteration, results slightly improved for KAZE (see Table 2). Yet, by resampling the source images to 150% of their original size, results got significantly better especially for KAZE, but also a

bit for SIFT and AKAZE (see Figure 10). In the fourth run, a Gaussian blurring followed by a resampling did not have an impact on the matching quality of this scene (see Table 3). Furthermore, LATCH yielded very poor results regardless of the iteration.

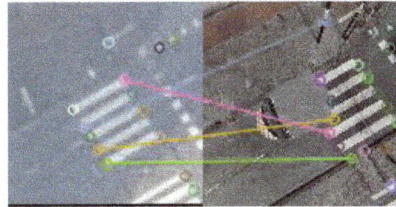

Figure 9 Matched LATCH keypoints in the first scene and first iteration

	1st run			2nd run		
	Matches	Inliers	Correct Matches	Matches	Inliers	Correct Matches
SIFT	61	15	1	61	13	0
KAZE	41	12	1	41	12	2
AKAZE	14	6	0	14	6	0
LATCH	9	4	0	9	0	0

Table 2 Matching results of scene 1 between aerial and panoramic image of the 1st and 2nd iteration

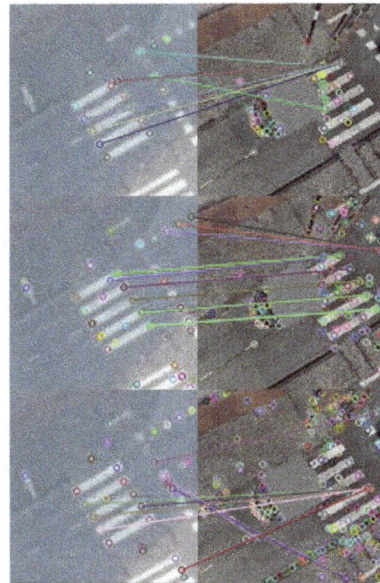

Figure 10 Comparison of matching results of AKAZE (top), KAZE (centre) and SIFT (bottom) in 3rd run of the 1st scene

	3rd run			4th run		
	Matches	Inliers	Correct Matches	Matches	Inliers	Correct Matches
SIFT	59	15	2	59	16	0
KAZE	78	17	10	78	16	10
AKAZE	24	9	2	24	9	0
LATCH	7	4	0	7	0	0

Table 3 Matching results of scene 1 between aerial and panoramic image of the 3rd and 4th iteration

---

[3] More results are provided on
https://www.researchgate.net/profile/Phillipp_Jende

5.2.1.2   Scene 2

The second scene shows linear road markings and parts of a zebra crossing. Whereas major parts of the zebra crossing and the dotted road markings were visible in the first scene, large parts of the road markings are covered by the mobile mapping car itself in the second scene which may impede the matching process. Similarly to the first scene, unmodified imagery was difficult to match and LATCH nor AKAZE found a single correspondence. With SIFT, however, a couple of keypoints could be matched, even though just one correct correspondence has been identified (see Figure 11).

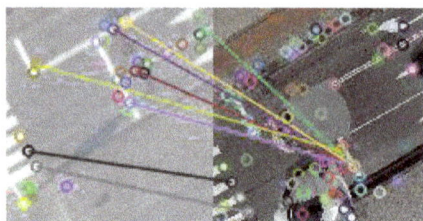

Figure 11 Matched SIFT keypoints in the second scene and first iteration (correct correspondence is light purple)

	1st run			2nd run		
	Matches	Inliers	Correct Matches	Matches	Inliers	Correct Matches
SIFT	54	12	1	54	11	2
KAZE	42	8	1	42	12	0
AKAZE	1	0	0	1	6	0
LATCH	1	0	0	1	0	0

Table 4 Matching results of scene 2 between aerial and panoramic image of the 1st and 2nd iteration

By blurring the images with a Gaussian filter, especially SIFT returns a better result. Albeit only two correct correspondences have been identified, results got considerably better (see Figure 12). Apparently, RANSAC removed a couple of outliers, and was able to stabilise the estimation of the homography. Without ground truth, the matched bars of the zebra crossing might appear as correct correspondences. These descriptor ambiguities have to be tackled on another processing level.

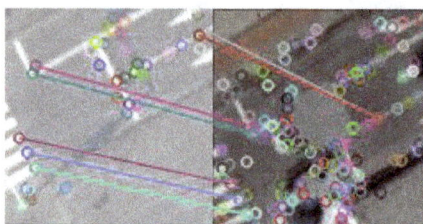

Figure 12 Matched SIFT keypoints in the second scene and second iteration

Now, by resampling the images to 150% of their original size in the 3rd iteration, KAZE benefitted the most, although yielding only 4 correct matches out of 14 matches classified as inliers (see Figure 13).

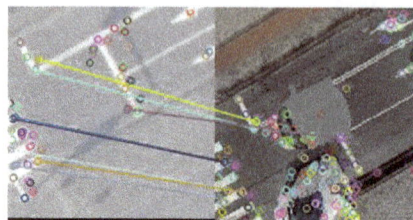

Figure 13 Matched KAZE keypoints in the second scene and third iteration

Blurring the images prior to resampling them further improves the results for KAZE and AKAZE while decreasing the quality of SIFT's output (see Figure 14). Again, LATCH did not show any improvement.

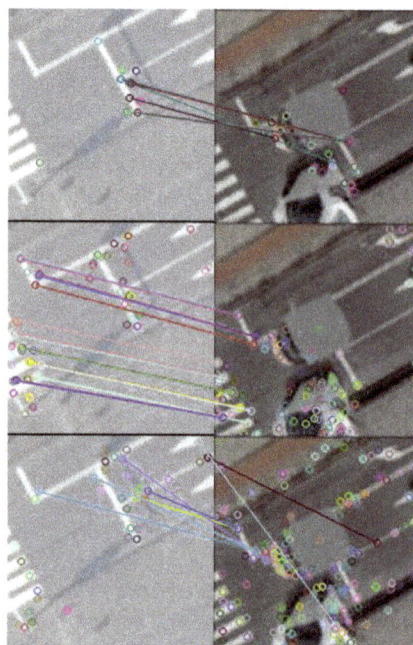

Figure 14 Comparison of matching results of AKAZE (top), KAZE (centre) and SIFT (bottom) in 4th run of the 2nd scene

	3rd run			4th run		
	Matches	Inliers	Correct Matches	Matches	Inliers	Correct Matches
SIFT	51	12	0	51	11	1
KAZE	79	14	4	79	20	6
AKAZE	15	6	0	15	8	2
LATCH	3	0	0	3	0	0

Table 5 Matching results of scene 2 between aerial and panoramic image of the 3rd and 4th iteration

5.2.2   Aerial images and MLS intensity images

5.2.2.1   Scene 1

In the 1st run (Table 6), KAZE yielded the best results with only few mismatches, and those occurred due to descriptor ambiguity. AKAZE has a competitive result, however obtained fewer matches than KAZE. SIFT and LATCH both equally failed to achieve a reliable number of matches. In the 2nd run, blurring the images, dramatically improved the number of correct matches from SIFT descriptor. Similarly, KAZE's result also improved moderately. AKAZE has performed consistently and results did not improve. The image blurring did not have an effect on the poor results of the LATCH descriptor. In the 3rd run (Table 7), resizing the images to 150%, results of SIFT and KAZE have improved. Interestingly, all calculated matches are correct and there is no mismatch. The results from AKAZE have improved slightly as well.

The 4th run, blurring and resizing the images, increased the number of inliers from KAZE, while there is no mismatch as shown in Figure 15. The total number of matches even decreased in case of AKAZE.

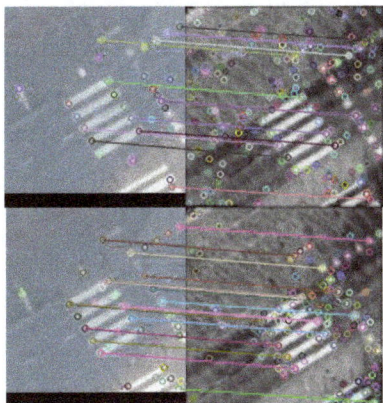

Figure 15 Comparison of SIFT (top) and KAZE (bottom) in 4th run on 1st scene.

	1st run			2nd run		
	Matches	Inliers	Correct Matches	Matches	Inliers	Correct Matches
SIFT	61	15	1	61	14	12
KAZE	41	12	8	41	13	12
AKAZE	14	7	5	14	7	4
LATCH	9	4	0	9	4	0

Table 6 Matching results of scene 1 between aerial and MLS ortho-image of the 1st and 2nd iteration

	3rd run			4th run		
	Matches	Inliers	Correct Matches	Matches	Inliers	Correct Matches
SIFT	59	14	14	59	14	14
KAZE	78	19	19	78	20	20
AKAZE	24	9	6	24	8	3
LATCH	7	4	0	7	0	0

Table 7 Matching results of scene 1 between aerial and MLS ortho-image of the 3rd and 4th iteration

5.2.2.2 Scene 2

In the 1st run (Table 8), on this difficult scene, all descriptors totally failed except KAZE, which also performed poorly due to descriptor ambiguity. SIFT also seemed to struggle with the descriptor ambiguity and therefore yielded no match. Blurring the images did not change anything, except that KAZE's result slightly improved. The 3rd run (Table 9) did not lead to any significant improvements. Contrarily to the 1st scene, the total number of matches from KAZE even decreased due to descriptor ambiguity. Although AKAZE was able to derive some matches, it cannot be considered as a significant improvement. SIFT and LATCH also failed to achieve a single

match in the 3rd run. In the 4th run, the number of matches from KAZE improved significantly. Interestingly, not a single mismatch could be identified. Results are shown in Figure 16. SIFT, however, could not improve and yielded only a single match.

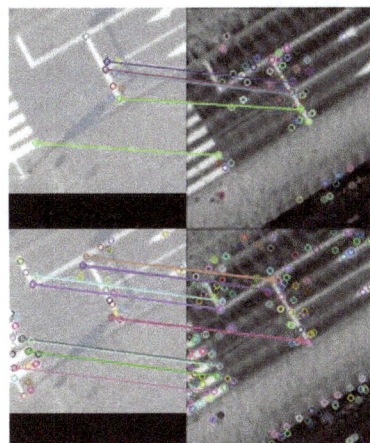

Figure 16 Matching results of AKAZE (top) and KAZE (bottom) in 4th run on scene 2.

	1st run			2nd run		
	Matches	Inliers	Correct Matches	Matches	Inliers	Correct Matches
SIFT	54	10	0	54	11	0
KAZE	42	10	4	42	11	7
AKAZE	1	0	0	1	0	0
LATCH	1	0	0	1	0	0

Table 8 Matching results of scene 2 between aerial and MLS ortho-image of the 1st and 2nd iteration

	3rd run			4th run		
	Matches	Inliers	Correct Matches	Matches	Inliers	Correct Matches
SIFT	51	9	0	51	9	1
KAZE	79	17	5	79	21	21
AKAZE	15	6	3	15	8	5
LATCH	3	0	0	3	0	0

Table 9 Matching results of scene 2 between aerial and MLS ortho-image of the 3rd and 4th iteration

6. DISCUSSION

6.1 Conclusion

This paper addressed the topic of tie feature extraction within the framework of the registration of aerial nadir images, mobile mapping panoramic images and MLS data. The aim of the overall project is to develop an automatic pipeline to correct the trajectory of mobile mapping platforms, especially in urban areas where reliable GNSS localisation is scarce. As a prerequisite for an orientation update of the platform's trajectory, precise tie information is needed. In this paper, feature-based extraction techniques have been evaluated. It could be shown that the outcome highly depends on the algorithm itself and data pre-processing. KAZE seems to be the most reliable feature extraction method in both cases – mobile laser scanning intensity and panoramic imagery. SIFT and AKAZE only yield mediocre results, and do not benefit from resampling and blurring the images to the same extent as KAZE does. Although the Förstner-Operator detects good and significant features, LATCH failed to describe them accordingly to allow for a successful matching. In our scenario, binary descriptors are not as powerful coping with changes in illumination and contrast as float descriptors, and cannot manage to handle features originating from different sensors that well. However, due to this specific setup, and a very generic feature matching, further tests have to be conducted to draw a thorough conclusion on binary descriptors' performance. Interestingly, the discussed techniques show a better performance in conjunction with MLS and aerial data than with MMI data.

6.2 Outlook

Evidently, repetitive patterns of road markings are the biggest obstacle for a successful registration of the data sets. To efficiently tackle this issue, spatial information has to be introduced. Although the positioning accuracy of mobile platforms may be diminished in urban areas, their exterior orientation could support feature matching. In other words, they can be utilised to introduce search constraints as they allow for the localisation of individual keypoints. Besides that, using contextual information and shape knowledge can augment feature description to prevent mismatches. Additionally, nonessential descriptor invariances or capabilities could be removed from the respective original implementations (e.g. rotational & scale invariance, sub-pixel localisation etc.).

As far as MLS data is concerned, utilising high-level feature extraction methods could further facilitate the registration process also in areas with a lower point density and therefore fewer distinct keypoints. For this reason, entities, such as kerbstones or entire zebra crossings, can be utilised to accomplish this task.

ACKNOWLEDGEMENTS

This project is funded and supported by the Dutch Technology Foundation STW, which is part of Netherlands Organisation for Scientific Research (NWO), and which is partly funded by the Ministry of Economic Affairs.

REFERENCES

Alcantarilla, P. F., A. Bartoli and A. J. Davison (2012). "KAZE Features."

Alcanterilla, P. F., J. Nuevo and A. Bartoli (2013). "Fast Explicit Diffusion for Accelerated Features in Nonlinear Scale Spaces."

Bay, H., A. Ess, T. Tuytelaars and L. Van Gool (2008). "Speeded-up robust features (SURF)." Computer vision and image understanding 110(3): 346-359.

Beers, B. (2011). "Collection and Application of 2D and 3D Panoramic Imagery."

Calonder, M., V. Lepetit, C. Strecha and P. Fua (2010). "Brief: Binary robust independent elementary features." Computer Vision–ECCV 2010: 778-792.

Egels, Y. and M. Kasser (2001). Digital Photogrammetry, Taylor & Francis, Inc.

Förstner, W. and E. Gülch (1987). "A fast operator for detection and precise location of distinct points, corners and circular features."

Godha, S., M. G. Petovello and G. Lachapelle (2005). "Performance Analysis of MEMS IMU/HSGPS/Magnetic Sensor Integrated System in Urban Canyons." ION GNSS 2005.

Harris, C. and M. Stephens (1988). "A Combined Corner and Edge Detector." Proceedings of Fourth Alvey Vision Conference: 147-151.

Heinly, J., E. Dunn and J.-M. Frahm (2012). Comparative evaluation of binary features. Computer Vision–ECCV 2012, Springer: 759-773.

Jaud, M., R. Rouveure, P. Faure and M.-O. Monod (2013). "Methods for FMCW radar map georeferencing." ISPRS Journal of Photogrammetry and Remote Sensing 84(0): 33-42.

Ji, S., Y. Shi, J. Shan, X. Shao, Z. Shi, X. Yuan, P. Yang, W. Wu, H. Tang and R. Shibasaki (2015). "Particle filtering methods for georeferencing panoramic image sequence in complex urban scenes." ISPRS Journal of Photogrammetry and Remote Sensing 105(0): 1-12.

Köthe, U. (2003). "Edge and junction detection with an improved structure tensor." Pattern Recognition, Proceedings 2781: 25-32.

Köthe, U. (2003). Gradient-Based Segmentation Requires Doubling of the Sampling Rate.

Kümmerle, R., B. Steder, C. Dornhege, A. Kleiner, G. Grisetti and W. Burgard (2011). "Large scale graph-based SLAM using aerial images as prior information." Auton. Robots 30(1): 25-39.

Leung, K. Y. K., C. M. Clark and J. P. Huissoon (2008). Localization in urban environments by matching ground level video images with an aerial image. Robotics and Automation, 2008. ICRA 2008.

Levi, G. and T. Hassner (2015). "LATCH: Learned Arrangments of Three Patch Codes."

Levinson, J. and S. Thrun (2007). "Map-Based Precision Vehicle Localization in Urban Environments." Robotics: Science and Systems.

Lowe, D. G. (1985). Perceptual Organization and Visual Recognition, Kluwer Academic Publishers.

Lowe, D. G. (2004). "Distinctive image features from scale-invariant keypoints." International Journal of Computer Vision 60(2): 91-110.

Miksik, O. and K. Mikolajczyk (2012). "Evaluation of Local Detectors and Descriptors for Fast Feature Matching." 2012 21st International Conference on Pattern Recognition (Icpr 2012): 2681-2684.

Rosten, E. and T. Drummond (2006). Machine learning for high-speed corner detection. Computer Vision–ECCV 2006, Springer: 430-443.

Tournaire, O., B. Soheilian and N. Paparoditis (2006). "Towards a Sub-Decimetric Georeferencing of Ground-Based Mobile Mapping Systems in Urban Areas: Matching Ground-Based and Aerial-based Imagery Using Roadmarks." ISPRS Commission I Symposium "From Sensors to Imagery".

Yang, X. and K.-T. Cheng (2012). LDB: An ultra-fast feature for scalable augmented reality on mobile devices. Mixed and Augmented Reality (ISMAR), 2012 IEEE International Symposium on, IEEE.

12

CALIBRATING CELLULAR AUTOMATA OF LAND USE/COVER CHANGE MODELS USING A GENETIC ALGORITHM

J.F. Mas[a]*, B. Soares-Filho[b], H. Rodrigues[b]

[a] Centro de Investigaciones en Geografía Ambiental, Universidad Nacional Autónoma de México,
58190 Morelia, Mexico - jfmas@ciga.unam.mx
[b] Centro de Sensoriamento Remoto, Universidade Federal de Minas Gerais, Belo Horizonte
31270-900, MG, Brasil - britaldo@csr.ufmg.br

Commission II, WG II/4

KEY WORDS: Stochastic spatial simulation, Genetic algorithm, Amazon deforestation, landscape pattern, fragmentation, connectivity

ABSTRACT:

Spatially explicit land use / land cover (LUCC) models aim at simulating the patterns of change on the landscape. In order to simulate landscape structure, the simulation procedures of most computational LUCC models use a cellular automata to replicate the land use / cover patches. Generally, model evaluation is based on assessing the location of the simulated changes in comparison to the true locations but landscapes metrics can also be used to assess landscape structure. As model complexity increases, the need to improve calibration and assessment techniques also increases. In this study, we applied a genetic algorithm tool to optimize cellular automata's parameters to simulate deforestation in a region of the Brazilian Amazon. We found that the genetic algorithm was able to calibrate the model to simulate more realistic landscape in term of connectivity. Results show also that more realistic simulated landscapes are often obtained at the expense of the location coincidence. However, when considering processes such as the fragmentation impacts on biodiversity, the simulation of more realistic landscape structure should be preferred to spatial coincidence performance.

1. INTRODUCTION

Spatially explicit land use / land cover change (LUCC) models aim at simulating the patterns of change on the landscape (Paegelow et al., 2013). Many of the models are based on a inductive pattern-based approach: In this approach, LUCC is modelled empirically using past LUCC spatial distribution and rate to develop a mathematical model that estimates the change potential as a function of a set of explanatory spatial variables and the expected amount of change (Paegelow and Olmedo, 2005; Mas et al., 2014). In prospective modelling, allocation procedures are used to simulate the projected amount of change in the most likely locations. In order to simulate landscape structure and fragmentation pattern, the simulation procedures of most computational LUCC models use a cellular automata (CA) that intends to replicate the land use/cover (LUC) patches.

Generally, the assessment of model performance is based on the spatial coincidence between a simulated map and an observed LUC map for the same date and does not evaluate the model ability to simulate the landscape pattern as, for instance, the size, the shape and the distribution of patches. Landscapes metrics can be used to assess simulated landscape structure (Mas et al., 2012). As model complexity increases, the need to improve calibration techniques also increases. This study aims at applying a genetic algorithm to optimize cellular automata's parameters to simulate deforestation in a region of the Brazilian Amazon.

2. MATERIALS

Dinamica EGO freeware (hereafter DINAMICA) is an environmental modeling platform for the design of space-time models. It has been applied to a variety of studies, such as modeling tropical deforestation (Soares-Filho et al., 2001, 2002, 2006; Cuevas and Mas, 2008), urban growth (Almeida et al., 2005), fire regimes (Silvestrini et al., 2011) and, landscape patterns (Pe'er et al., 2013; Soares Filho et al., 2003), among others. We chose it due to its flexibility and computing eficiency (Mas et al., 2014).

We used a portion of one of the 12 case-study areas from Soares-Filho et al. (2013), comprising a TM-Landsat scene map from the PRODES project (INPE, 2011). The study area is located in the State of Pará along the road between Santarém and Cuiabá. Deforestation maps encompassing the years 1997 and 2001 were rasterized into a 250-m raster. As spatial drivers of deforestation, we selected only three variables from the dataset: distance to previously deforested lands, proximity to roads and, elevation in order to have a simple model easier to interpret.

3. METHODS

DINAMICA uses transition probability maps that are based on a Bayesian method of conditional probability known the weight of evidence method. These maps of probability are used to simulate landscape dynamics using both Markov matrices to project the quantity of change and a cellular automata (CA) approach to reproduce spatial patterns (Soares-Filho et al., 2002, 2010). Two complementary CA are available: the Expander, that simulates the expansion of previously formed patches and, the Patcher that generates new patches through a seeding mechanism. The behaviour of the CA is controlled by four main parameters: the mean patch size, the patch size variance, the isometry and the prune factor. Increasing patch size value leads to simulated maps with a less fragmented landscape; increasing the patch size variance leads to a more diverse landscape in term of size of the patches. Setting the isometry value greater than one leads to create more isometric patches. Increasing the prune factor allows simulated changes to occur in less likely areas. With a prune factor of one, the model becomes almost deterministic, that is changes are restricted to the areas with higher change probability (Soares-Filho et al., 2002; Mas et al., 2014).

DINAMICA has also a genetic algorithm tool which has been

---

used to calibrate the weights of evidence (Soares-Filho et al., 2013). Genetic algorithms are based on the Darwinian mechanisms of evolution and attempt to mimic the natural evolution of a population by selection, combination, and mutations (random changes in genes). Each chromosome is composed of genes that define its characteristics. In computational models, a chromosome is a string of numbers encoding the parameters that the genetic algorithm attempts to optimize. First, a population of chromosomes is created randomly. Subsequently, the genetic algorithm generates new individuals from the existing ones by means of processes of selection, crossover and mutation. The parents are selected from the best chromosomes, according to a fitness criterion. To increase genetic heterogeneity, mutation and crossover operators randomly changes or interchange some genes on a chromosome sequence. A new generation is created by copying the most successful individuals, by crossing-over them and by mutating some chromosome sequences. These processes generate a new population of chromosomes with a greater chance of including a near-optimum solution for the problem. This evolution process iterates until the fitness-stopping criterion is satisfied. Genetic algorithms have been shown to be able to optimize multi-parameter function Wang (1997). In the present study, the genetic algorithm optimizes the four main parameters of the Parcher CA: the mean patch size, the patch size variance, the isometry and, the prune factor.

Forest Maps of 1997 and 2001 were overlaid in order to map deforested and conserved forest area. This deforestation map allowed to compute a matrix of Markov, the weights of evidence and a map of probability using the three explanatory variables (see appendix). Then deforestation was simulated from 1997 to 2001 using the 1997 as initial LUC map, the matrix of Markov to calculate expected annual deforested area and the set of weights of evidence to compute the probability of change. The variable distance from previously deforested area is a dynamic variable, that is computed at each annual iteration of the simulation. It is worth noting that, in the present case, the training (or calibration) period is the same than the simulation period because the objective of the study is fitting the model and not testing its prospective ability.

The CA calibration aimed at fitting the parameters in order to simulate a landscape similar to true landscape with regards to the size of the patches of deforestation and its general spatial distribution (e.g. avoiding that simulated changes concentrate in most likely areas, near previous deforested area when in true landscape there are also a little quantity of changes in remote areas). We calculated three indices which depict the landscape characteristics: The mean area of the deforestation patches (DPMA), the standard deviation of the deforestation patches area (DPASD) and the mean distance to deforested patches in remote areas ($MDFP_r$) which is the mean distance of forest cells to deforested area taking into account only cells located at a larger distance than the average distance. Fitness was assessed through the difference of the indices for the true change and the simulated change maps. The fitness criterion was computed by the weighted sum of three fitness components (equation 1).

$$F = \frac{w_1}{\delta DPMA} + \frac{w_2}{\delta DPSD} + \frac{w_3}{\delta MDFP_r} \qquad (1)$$

where w_1, w_2, w_3 = pondering weights,
$\delta DPMA$ = Absolute value of the difference of index DPMA between simulated and true map,
$\delta DPSD$ = Absolute value of the difference of index DPSD between simulated and true map,
$\delta MDFP_r$ = Absolute value of the difference of index $MDFP_r$ between simulated and true map.

Final simulated map was assessed by visual inspection and through the computing of the spatial coincidence between true change and simulated changes during 1997-2001.

4. RESULTS

Figures 1, 2 and 3 present the forest in 1997, 2001 and the deforested areas during the period 1997-2001 respectively. Figure 4 shows the map of probability of deforestation according to the weights of evidence based on the three explanatory variables (maps of explanatory variables are in appendix). It can be observed that change presents a clear spatial pattern (patches) and occurred mainly on high probability areas but also in less likely areas (e.g. remote areas).

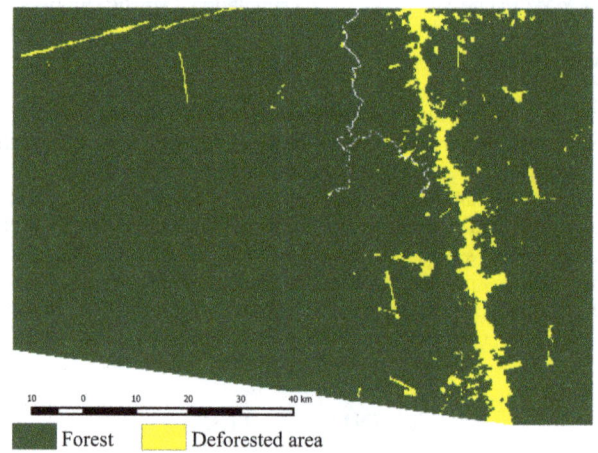

Figure 1: Land use / cover map (1997)

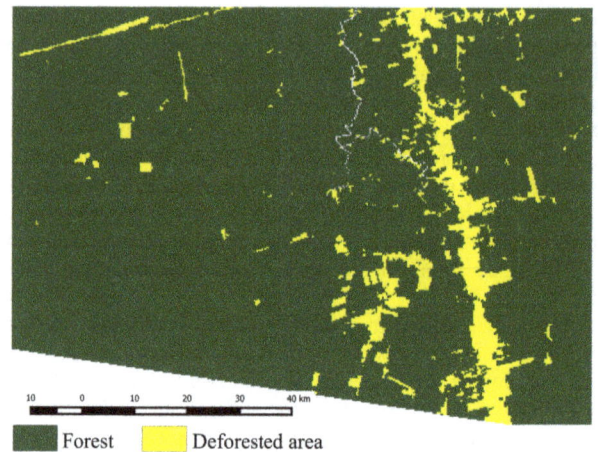

Figure 2: Land use / cover map (2001)

Figures 5 represents the simulated LUCC map from the modelled which CA was calibrated by the genetic algorithm. As comparison, figure 6 shows the simulated LUCC map thresholding the cells with the highest probability to simulate the change. It can be observed that the map obtained using CA is more realistic in term of landscape structure because it presents patches of deforestation of broadly the same size and distribution than the true map of change (Figure 3). However, the unrealistic map obtained without CA has a higher coincidence with true map (29%) than the more realistic map obtained with the CA calibrated by the genetic algorithm (22%).

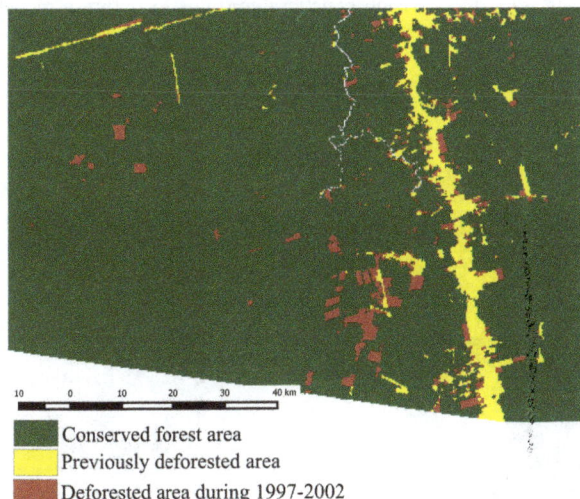

Figure 3: LUCC map (1997-2002). Previously deforested area refers to forest area cleared before 1997. Some deforestation patches are located in remote areas

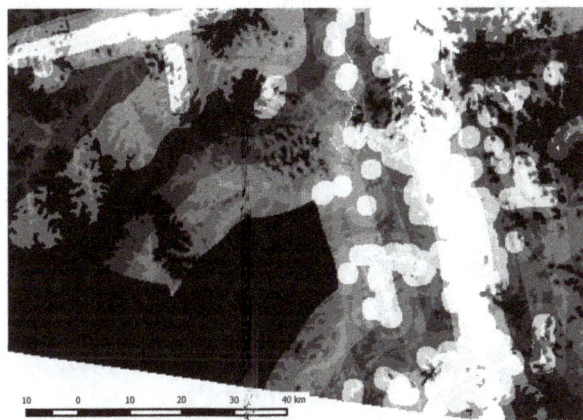

Figure 4: Map of probability of deforestation (1997)

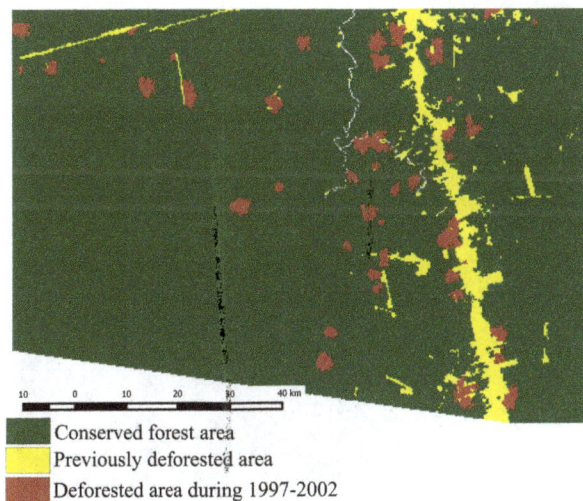

Figure 5: Simulated LUCC map (1997-2002) using the CA calibrated by the genetic algorithm. Previously deforested area refers to forest area cleared before 1997

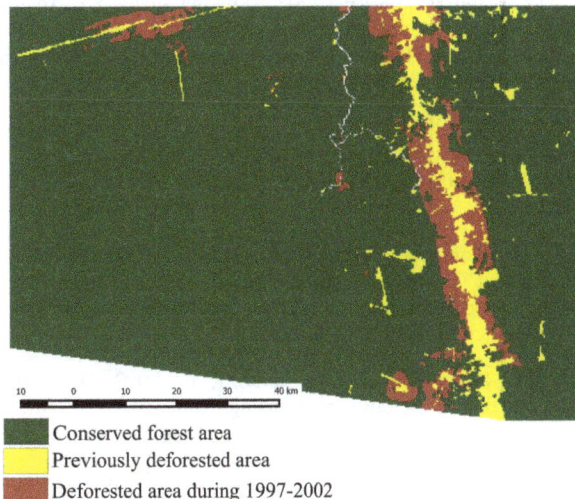

Figure 6: Simulated LUCC map (1997-2002) applying a threshold to the probabilities. Previously deforested area refers to forest area cleared before 1997

5. CONCLUSION

We found that the genetic algorithm was able to calibrate the model's CA to simulate more realistic landscape in term of connectivity. Different spatial patterns can be observed depending on human activities contributing to deforestation as cattle ranching, shifting cultivation, commercial agriculture or logging (Anwar and Stein, 2015; Lorena and Lambin, 2007). This approach can be used to calibrate LUCC models and other types of models aiming at simulating landscape patterns. Results show also that more realistic simulated landscapes are often obtained at the expense of the location coincidence. However, when LUCC modelling is used to assess processes such as the fragmentation impacts on biodiversity, the simulation of more realistic landscape structure and change dynamics should be preferred to spatial coincidence performance (Malanson et al., 2007; Mas et al., 2012).

ACKNOWLEDGEMENTS

This study was supported by the *Consejo Nacional de Ciencia y Tecnología* (CONACYT) and the *Secretaría de Educación Pública* through the project *¿Puede la modelacin espacial ayudarnos a entender los procesos de cambio de cobertura/uso del suelo y de degradacin ambiental?*-Fondos SEP-CONACyT 178816.

References

Almeida, C. M. D., Monteiro, A. M. V., Câmara, G., Soares-Filho, B. S., Cerqueira, G. C., Pennachin, C. L. and Batty, M., 2005. GIS and remote sensing as tools for the simulation of urban landuse change. International Journal of Remote Sensing 26(4), pp. 759–774.

Anwar, S. and Stein, A., 2015. Spatial pattern development of selective logging over several years. Spatial Statistics 13(0), pp. 90–105.

Cuevas, G. and Mas, J., 2008. Land use scenarios : a communication tool. In: M. Paegelow and M. T. Camacho-Olmedo (eds), Modelling Environmental Dynamics, Springer Berlin Heidelberg, Berlin, chapter Land use s, pp. 223–246.

INPE, 2011. PRODES: Assessment of Deforestation in Brazilian Amazonia.

Lorena, R. and Lambin, E., 2007. Linking spatial patterns of deforestation to land use using satellite and field data. In: Geoscience and Remote Sensing Symposium, 2007. IGARSS 2007. IEEE International, pp. 3357–3361.

Malanson, G. P., Wang, Q. and Kupfer, J. A., 2007. Ecological processes and spatial patterns before, during and after simulated deforestation. Ecological Modelling 202(34), pp. 397–409.

Mas, J.-F., Kolb, M., Paegelow, M., Camacho Olmedo, M. T. and Houet, T., 2014. Inductive pattern-based land use/cover change models: A comparison of four software packages. Environmental Modelling & Software 51, pp. 94–111.

Mas, J.-F., Pérez-Vega, A. and Clarke, K. C., 2012. Assessing simulated land use/cover maps using similarity and fragmentation indices. Ecological Complexity 11, pp. 38–45.

Paegelow, M. and Olmedo, M. T. C., 2005. Possibilities and limits of prospective GIS land cover modellinga compared case study: Garrotxes (France) and Alta Alpujarra Granadina (Spain). International Journal of Geographical Information Science 19(6), pp. 697–722.

Paegelow, M., Camacho Olmedo, M. T., Mas, J.-F., Houet, T. and Pontius Jr., R. G., 2013. Land change modelling: moving beyond projections. International Journal of Geographical Information Science 27(October), pp. 1691–1695.

Pe'er, G., Zurita, G. A., Schober, L., Bellocq, M. I., Strer, M., Müller, M. and Pütz, S., 2013. Simple Process-Based Simulators for Generating Spatial Patterns of Habitat Loss and Fragmentation: A Review and Introduction to the G-RaFFe Model. PLoS ONE 8(5), pp. e64968.

Silvestrini, R. A., Soares-Filho, B. S., Nepstad, D., Coe, M., Rodrigues, H. and Assunção, R., 2011. Simulating fire regimes in the Amazon in response to climate change and deforestation. Ecological Applications 21(5), pp. 1573–1590.

Soares-Filho, B., Moutinho, P., Nepstad, D., Anderson, A., Rodrigues, H., Garcia, R., Dietzsch, L., Merry, F., Bowman, M., Hissa, L., Silvestrini, R. and Maretti, C., 2010. Role of Brazilian Amazon protected areas in climate change mitigation. Proceedings of the National Academy of Sciences of the United States of America 107, pp. 10821–10826.

Soares-Filho, B., Rodrigues, H. and Follador, M., 2013. A hybrid analytical-heuristic method for calibrating land-use change models. Environmental Modelling and Software 43, pp. 80–87.

Soares-Filho, B. S., Assunção, R. M. and Pantuzzo, A. E., 2001. Modeling the Spatial Transition Probabilities of Landscape Dynamics in an Amazonian Colonization Frontier. BioScience 51(12), pp. 1059.

Soares Filho, B. S., Corradi Filho, L., Coutinho Cerqueira, G. and Leite Araújo, W., 2003. Simulating the spatial patterns of change through the use of the DINAMICA model. In: INPE (ed.), Anais do XI Simpósio Brasileiro de Sensoriamento Remoto - SBSR, Belo Horizonte, MG, Brasil, pp. 721–728.

Soares-Filho, B. S., Coutinho Cerqueira, G. and Lopes Pennachin, C., 2002. DINAMICA - A stochastic cellular automata model designed to simulate the landscape dynamics in an Amazonian colonization frontier. Ecological Modelling 154, pp. 217–235.

Soares-Filho, B. S., Nepstad, D. C., Curran, L. M., Cerqueira, G. C., Garcia, R. A., Ramos, C. A., Voll, E., McDonald, A., Lefebvre, P. and Schlesinger, P., 2006. Modelling conservation in the Amazon basin. Nature 440, pp. 520–523.

Wang, Q., 1997. Using genetic algorithms to optimise model parameters. Environmental Modelling & Software 12, pp. 27–24.

APPENDIX

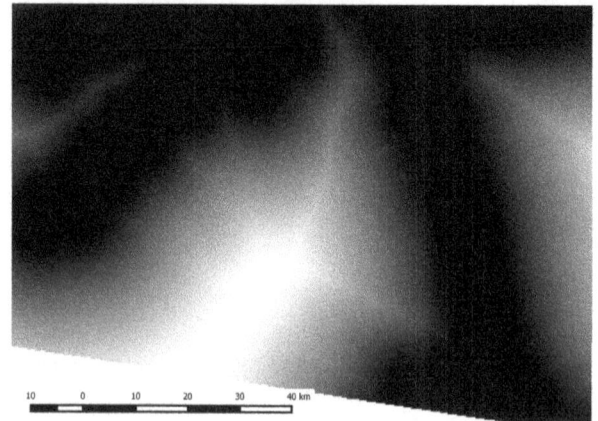
Figure 7: Map of distance from roads used as explanatory variable

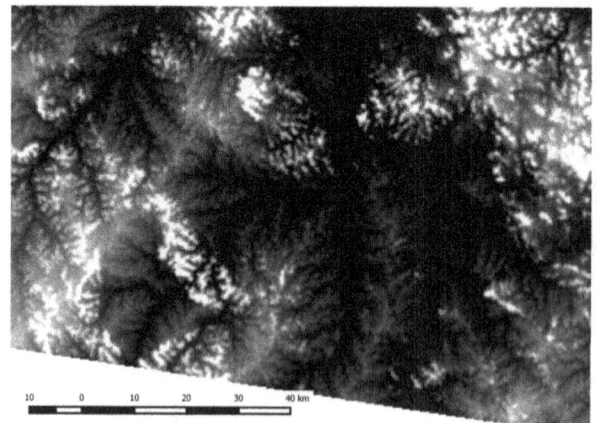
Figure 8: Digital model elevation used as explanatory variable

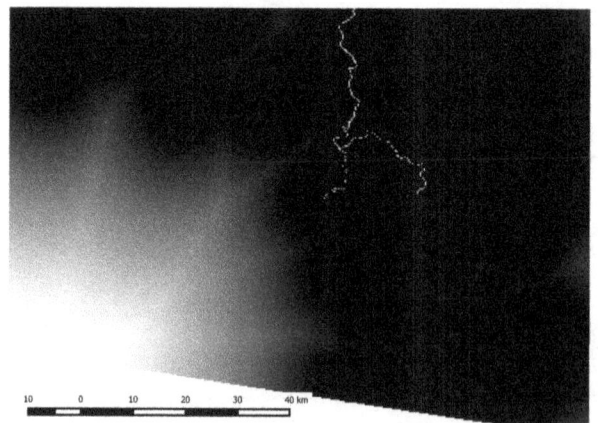
Figure 9: Map of distance from previously deforested area (1997). This map is updated at each modeling step

13

STUDY OF LEVER-ARM EFFECT USING EMBEDDED PHOTOGRAMMETRY AND ON-BOARD GPS RECEIVER ON UAV FOR METROLOGICAL MAPPING PURPOSE AND PROPOSAL OF A FREE GROUND MEASUREMENTS CALIBRATION PROCEDURE

M. Daakir[a,d], M. Pierrot-Deseilligny[b,d], P. Bosser[c], F. Pichard[a], C. Thom[d], Y. Rabot[a]

[a]Vinci-Construction-Terrassement, 1, Rue du docteur Charcot, 91421 Morangis, France - (mehdi.daakir, francis.pichard, yohann.rabot)@vinci-construction.com
[b]Université Paris-Est, IGN, ENSG, LOEMI , 6-8 Avenue Blaise Pascal, 77455 Champs-sur-Marne, France - marc.pierrot-deseilligny@ensg.eu
[c]ENSTA Bretagne-OSM Team, 2 rue Francois Verny, 29806 Brest, France - pierre.bosser@ensta-bretagne.fr
[d]Université Paris-Est, IGN, SRIG, LOEMI, 73 avenue de Paris, 94160 Saint-Mande, France - christian.thom@ign.fr

KEY WORDS: UAV, photogrammetry, GPS, lever-arm, calibration, direct-georeferencing, synchronisation, MicMac

ABSTRACT:

Nowadays, Unmanned Aerial Vehicle (UAV) on-board photogrammetry knows a significant growth due to the democratization of using drones in the civilian sector. Also, due to changes in regulations laws governing the rules of inclusion of a UAV in the airspace which become suitable for the development of professional activities. Fields of application of photogrammetry are diverse, for instance: architecture, geology, archaeology, mapping, industrial metrology, etc. Our research concerns the latter area. *Vinci-Construction-Terrassement* is a private company specialized in public earthworks that uses UAVs for metrology applications. This article deals with maximum accuracy one can achieve with a coupled camera and GPS receiver system for direct-georeferencing of Digital Surface Models (DSMs) without relying on Ground Control Points (GCPs) measurements. This article focuses specially on the lever-arm calibration part. This proposed calibration method is based on two steps: a first step involves the proper calibration for each sensor, i.e. to determine the position of the optical center of the camera and the GPS antenna phase center in a local coordinate system relative to the sensor. A second step concerns a $3d$ modeling of the UAV with embedded sensors through a photogrammetric acquisition. Processing this acquisition allows to determine the value of the lever-arm offset without using GCPs.

1. INTRODUCTION

In the context of direct-georeferencing of UAV aerial images using an on-board GPS receiver, lever-arm calibration is an important step in the data processing pipeline to ensure maximum accuracy when generating a DSM. The lever-arm vector is the offset between the GPS receiver antenna phase center and the optical center of the camera. In our configuration, we rely on processing carrier-phase raw embedded GPS data to get the camera position while the photogrammetric image processing based on tie points extraction (Lowe, 2004) provides the camera orientation. The conventional method to calibrate this vector is generally based on external measurements as well as the mechanical stability of the mounting of the sensors on the UAV is required. Indeed, if one of the sensors, the camera or the GPS receiver, exchange position/orientation, calibration results are obviously no longer valid. We propose in the following a study of the impact of lever-arm determination on direct-georeferencing accuracy and a different calibration method requiring only photogrammetric measurements without any special mechanical sensors mounting configuration constraint.

2. HARDWARE

As part of our research we developed a UAV prototype which can perform, by embedding a camera and a GPS receiver, direct-georeferencing of camera centers in an absolute coordinate system. This UAV uses home made IGN^1 $LOEMI^2$ developed sensors. The choice of the UAV model is intended to be low-cost as for the embedded sensors. The total mass should not exceed $2\ Kg$. Figure 1 shows the UAV used for experiments as well as on-board sensors:

Figure 1: DJI-F550 UAV with IGN GeoCube (left) and on-board light camera (right)

The UAV is the F-550 hexacopter model developed by DJI[3]. It allows to take-off within $0.8\ Kg$ payload but its autonomy still relatively low, about 10 min. The GPS receiver is the *GeoCube* devel-

---

[1]French Mapping Agency
[2]Opto-Electronics, Instrumentation and Metrology Laboratory
[3]www.dji.com/product/flame-wheel-arf/

oped at *IGN*. The *GeoCube* (Benoit et al., 2015) is a multi-sensor instrument that embeds a LEA-6T-001 u-blox (u-blox, 2014) GPS module chip. It records carrier-phase raw measurements on L1[4] band of GPS constellation. The light camera used (Martin et al., 2014) is also developed at *IGN* and is specially designed for aerial photogrammetry applications using UAVs. This camera has a full frame sensor with a fixed focal length. Its mass does not exceed $300\ g$.

3.   LEVER-ARM CALIBRATION

3.1   Auto-calibration method

A conventional method for calibrating the lever-arm offset is to acquire GPS data coupled with images over a scene containing GCPs. These GCPs are used to estimate the unknown parameters of the $3d$ similarity to convert camera poses from bundle block adjustment relative frame to an absolute one. Figure 2 shows an image acquired by a UAV of a scene provided with GCPs which are used for in-flight classical lever-arm calibration.

Figure 2: Scene containing GCPs used for lever-arm calibration

These GCPs allow simultaneously to constrain the estimation of the lever-arm offset. In theory one image is needed to estimate the value of this vector. However, in order to qualify the accuracy of this estimation, and supposing that time synchronisation is perfect between both sensors, a least squares adjustment is performed with several dozen images by minimizing the following system:

$$\begin{cases} \cdots\cdots \\ \mathcal{S}(\vec{\mathcal{C}}_k|_r) = \vec{\mathcal{G}}_k|_t + \mathcal{R}_k \cdot \vec{\mathcal{O}} \\ \cdots\cdots \end{cases} \qquad (1)$$

where: $\mathcal{S}$ = 7 parameters $3d$ similarity $(\vec{\mathcal{T}},\mu,\mathcal{R})$
              $\mathcal{R}_k$ = orientation of the image k
              $\vec{\mathcal{O}}$ = lever-arm vector
              $\vec{\mathcal{G}}_k|_t$ = GPS position of the image k

For aerial acquisitions using UAVs, vertical component of global translation $\vec{T}|_z$ and vertical component of lever-arm vector $\vec{O}|_z$ are correlated. This correlation is introduced due to the fact that the camera maintains the same orientation during a UAV flight.

---

[4] 1575.42 MHz

We need at least one GCP measurement in order to apply a constraint during the adjustment. This constraint is applied to the estimation of $3d$ similarity parameters as following:

$$\begin{cases} \cdots\cdots \\ \mathcal{S}(\vec{\mathcal{C}}_k|_r) = \vec{\mathcal{G}}_k|_t + \mathcal{R}_k \cdot \vec{\mathcal{O}} \\ \cdots\cdots \\ \mathcal{S}(\vec{\mathcal{P}}_i|_r) = \vec{\mathcal{P}}_i|_t \\ \cdots\cdots \end{cases} \qquad (2)$$

3.2   Pseudo-materialization method

Phase and optical centers can not be directly measured. This calibration method is to make a pseudo-materialization of these. The idea is to determine their relative position with respect to a network of known points on the GPS receiver and the camera in a local frame. Figure 3 shows the targets on sensors.

Figure 3: Targets fixed on the sensors

The acquisition of the calibration data protocol, which is carried out only once, is based on the following photogrammetric measurements:

1. Camera optical center calibration:

   - a photogrammetric acquisition is performed using the camera to calibrate
   - the camera to calibrate is positioned on a stable surface
   - the camera to calibrate takes a single image when positioned on the stable surface
   - a second camera performs a photogrammetric acquisition around the first one

   Figure 4 shows a photogrammetric acquisition around the camera to be calibrated while positioned on a stable surface.

Figure 4: Photogrammetric acquisition around the UAV on-board camera

2. GPS antenna phase center calibration:

- the GPS receiver is positioned on a scene that contains already known targets in an absolute reference system

- for each GPS observations session, a photogrammetric acquisition is performed

- the GPS receiver performs a $\sim 90°$ rotation and a new session is started coupled to a new photogrammetric acquisition

Figure 5 shows a photogrammetric acquisition of the GPS receiver for a given position.

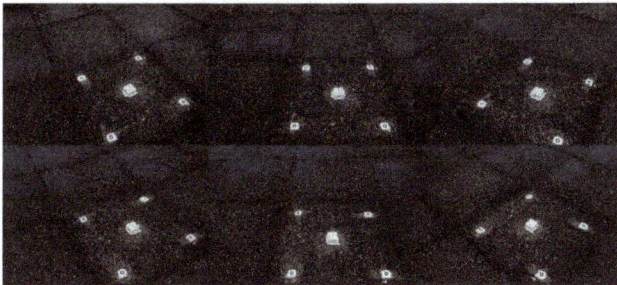

Figure 5: Photogrammetric acquisition of the UAV on-board GPS receiver

3.2.1 Camera optical center calibration

As a result of data acquisition step, we have taken two converging photogrammetric acquisitions of the same scene. The first performed with the camera to be calibrated and the second one performed with another camera around the first. A ruler is disposed on the scene because the results of cameras poses estimation must be expressed in a scaled frame. Bundle block adjustment of all acquired images is achieved using Apero (Pierrot-Deseilligny and Clery, 2011) module of the IGN MicMac photogrammetric suite. A convergent acquisition, in the form of a circle, is the most suitable to ensure high overlap rates. Figure 6 shows images acquisition geometry.

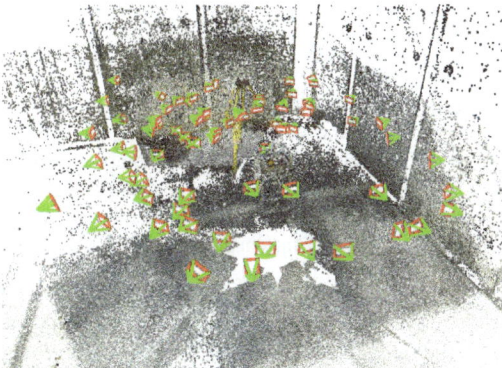

Figure 6: The circular geometry of the acquisition

After performing the bundle block adjustment, the targets $\vec{p}_i|_k$ located on the camera 3 to be calibrated are measured in the images k and their $3d$ coordinates $\vec{P}_i$ estimated by pseudo-intersections. This estimate is carried out by a least squares adjustment of the following system of equations:

$$\begin{cases} \cdots\cdots \\ \vec{p}_i|_k - \zeta(\pi(\mathcal{R}_k(\vec{P}_i - \vec{\mathcal{C}}_k))) = \vec{0} \\ \cdots\cdots \end{cases} \quad (3)$$

where:

- $\vec{p}_i|_k$ vector of $2d$ image coordinates of target i on image k

- ζ is a $\mathbb{R}^2 \to \mathbb{R}^2$ application describing internal parameters of the camera

- π is a $\mathbb{R}^3 \to \mathbb{R}^2$ projective application

- $(\vec{\mathcal{C}}_k, \mathcal{R}_k)$ position and orientation of image k

- $\vec{P}_i$ target $3d$ position parameter to be estimated

Using the external parameters of the single image s acquired by the camera to be calibrated while placed on the stable surface, the $3d$ targets positions $\vec{P}_i|_r$ are expressed in the camera frame $\vec{P}_i|_{cam}$ using:

$$\vec{P}_i|_{cam} = \mathcal{R}_s^T \cdot (\vec{P}_i|_r - \vec{\mathcal{C}}_s|_r) \quad (4)$$

where:

- $\vec{P}_i|_{cam}$ position of target i expressed in the camera frame

- $\vec{\mathcal{C}}_s|_r, \mathcal{R}_s$ position and orientation of image s expressed in the relative frame

- $\vec{P}_i|_r$ previous estimated $3d$ position of target i expressed in the relative frame

3.2.2 GPS antenna phase center calibration

The calibration of the GPS antenna phase center is based on the same principle. Here, we need to express the targets in an absolute reference system[5]. Photogrammetric acquisitions are performed on a scene where are installed known targets in the same system determined with topometric measurements.

We perform 4 rotations of $\sim 90°$ to highlight the relative displacement of the antenna phase center position expressed in the local frame of the receiver. For each rotation $3\ h$ GPS raw observations sampled at $1\ Hz$ are recorded. RTKlib (Takasu, 2011) open-source software is used to perform differential static postprocessing of L1 carrier-phase measurements. Each session gives an estimate of the GPS antenna phase center position. For each rotation, a photogrammetric acquisition is performed. Figure 7 shows images acquisition geometry.

Figure 7: Images acquired around the GPS receiver

The estimated camera poses from bundle block adjustment are expressed in the GPS absolute reference system using 4 GCPs. The targets $\vec{g}_i|_k$ located on the GPS receiver 3 are measured in

---

[5]GPS World Geodetic System 1984

GIS APPROACHES FOR REMOTE SENSING AND PHOTOGRAMMETRY

the images k and their $3d$ coordinates $\vec{G}_i$ estimated using system of equations 3 . Targets positions and the position of the antenna phase center after the first session of measurements are arbitrarily considered as a reference state. Successive positions are expressed in the frame of the first session by estimating the 6 parameters $(R_i, \vec{T}_i)$ of the rigid transformation $\mathcal{D}_i$ such that:

$$\mathcal{D}_i(\vec{P}_i) = \vec{P}_0 \qquad (5)$$

where:

- $\vec{P}_i$ position of targets and antenna phase center after session of observations i

- $\vec{P}_0$ position of targets and antenna phase center after the first session of observations

Figure 8 shows the planimetric displacement of the antenna phase center expressed in the frame of the first session of observations.

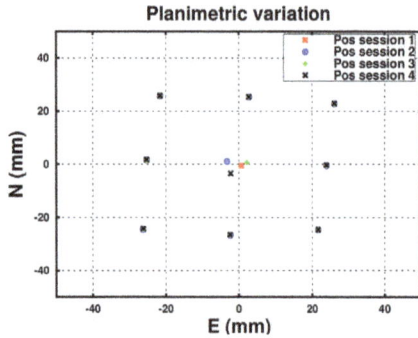

Figure 8: Planimetric variation of the antenna phase center

Symbols at the center of Figure 8 represent the predictions of the antenna phase center for each session. The mean deviation between all sessions is $0.7\ cm$ and the maximum deviation is $1.1\ cm$. Final values for the calibration of targets and antenna phase center positions are the average of estimated positions including all observations sessions.

3.2.3 Computing lever-arm offset value To get the value of the lever-arm vector after the calibration of optical and phase centers performed, it is necessary to conduct a photogrammetric acquisition of the UAV, with the sensors mounted, before take-off. Figure 9 shows images and geometry acquisition of the UAV on the ground.

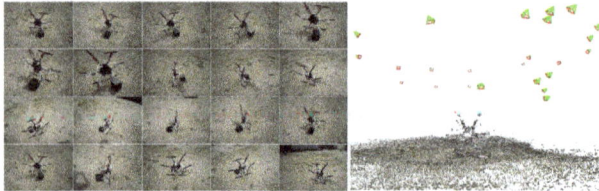

Figure 9: Photogrammetric acquisition around UAV before take-off (left) and images geometry acquisition (right)

To calculate the lever-arm offset value only image measurements are needed. Let N be the number of targets measured on the camera and N' the number of targets measured on the GPS receiver. Let:

- $\vec{v}|_{R_r} = [\vec{P}_1|_r, \ldots, \vec{P}_N|_r]$ camera targets estimated positions in the bundle block adjustment relative frame

- $\vec{v}|_{R_{cam}} = [\vec{P}_1|_{cam}, \ldots, \vec{P}_N|_{cam}]$ camera targets estimated positions from calibration expressed in the camera frame

- $\vec{v'}|_{R_r} = [\vec{G}_1|_r, \ldots, \vec{G}_{N'}|_r]$ GPS targets estimated positions in the bundle block adjustment relative frame

- $\vec{v'}|_{R_l} = [\vec{G}_1|_l, \ldots, \vec{G}_{N'}|_l]$ GPS targets estimated positions from calibration expressed in the GPS receiver local frame

- $\vec{v'}|_{R_c} = [\vec{G}_1|_c, \ldots, \vec{G}_{N'}|_c]$ GPS targets estimated positions expressed in the camera frame

- $\vec{C}_{gps}|_{R_l}$ antenna phase center from calibration expressed in the GPS receiver local frame

- $\vec{O}$ lever-arm vector by definition expressed in the camera frame

To estimate the value of the lever-arm offset $\vec{O}$, one must realize:

1. estimate the 7 parameters $(\lambda, R_1, \vec{T}_1)$ of the $3d$ similarity $\mathcal{S}$ such that camera poses are expressed in the calibrated camera frame:

$$\mathcal{S}(\vec{v}|_{R_r}) = \lambda \cdot R_1 \cdot \vec{v}|_{R_{cam}} + \vec{T}_1 \qquad (6)$$

2. express GPS targets in the camera frame:

$$\vec{v'}|_{R_c} = \mathcal{S}(\vec{v'}|_{R_r}) \qquad (7)$$

3. estimate the 6 parameters $(R_2, \vec{T}_2)$ of the rigid transformation $\mathcal{D}_i$ such that:

$$\mathcal{D}(\vec{v'}|_{R_l}) = R_2 \cdot \vec{v'}|_{R_c} + \vec{T}_2 \qquad (8)$$

4. express the lever-arm offset value as:

$$\vec{O} = R_2 \cdot \vec{C}_{gps}|_{R_l} + \vec{T}_2 \qquad (9)$$

4. RESULTS

A data acquisition with the system as presented in 1 has been made. Two photogrammetric acquisitions are performed as shown in Figure 9 . The first one before UAV take-off and the second one after landing. Each acquisition provides a lever-arm value with respect to the second calibration methodology 3.2 and allows us at least to control the mechanical stability between the start of the acquisition and its end.

Before comparing the values of the lever-arm vector provided by this method and the one obtained by classical in-flight self-calibration method based on GCPs 3.1, we can compare the calculated values before the UAV take-off and after the UAV landing. Table 1 gives lever-arm values for both measurements from two photogrammetric acquisitions of the UAV on the ground and the in-flight estimated value using all available GCPs (17) during this experiment.

Method	X_c (mm)	Y_c (mm)	Z_c (mm)
Before UAV take-off	86	-52	-82
After UAV landing	85	-51	-81
In-flight calibration	98	-70	-91

Table 1: Different estimates of values of the lever-arm offset

Both photogrammetric acquisitions of the UAV give two lever-arm values within $2\ mm$ difference. It proves firstly that the mechanical assembly is stable and the sensors keep a fixed relative positions while the difference between the two calibration techniques is 26 mm and is more significant.

As stated above, it is possible to estimate the lever-arm offset without adding a constraint equation using one or several GCPs. Figure 10 shows the variation of the lever-arm vector estimation value based on the number of used GCPs during a least squares adjustment for the in-flight self-calibration method.

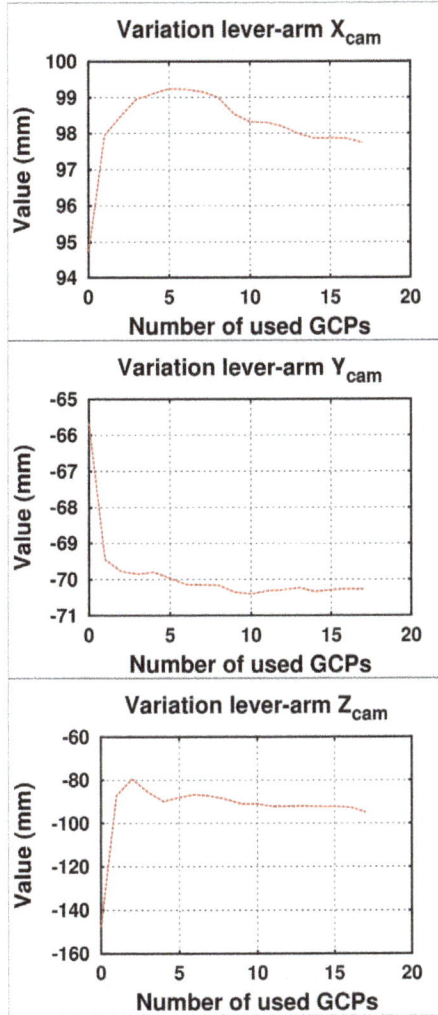

Figure 10: Variation of the lever-arm estimation value using 0 to 17 GCPs

The curves of the axial components show a convergence of the parameter estimation with a stability which increases with the number of GCPs used. However, the difference between the two lever-arm calibration methods seems high (26 mm).

To study the impact of this difference a second flight is made to have independent data sets. The first estimated lever-arm values given in Table 1 are used to correct GPS post-processed trajectory to perform a direct-georeferencing of camera centers. The residual on check points is calculated using camera poses based on GPS data. Figure 11 gives a comparison on residuals computed on check points for each lever-arm calibration method.

Figure 11 shows that residuals are most important for the calibration method based on a photogrammetric acquisition of the UAV on the ground. Residuals for this method give an accuracy of 2 cm ± 1.4 cm while the in-flight calibration method gives an

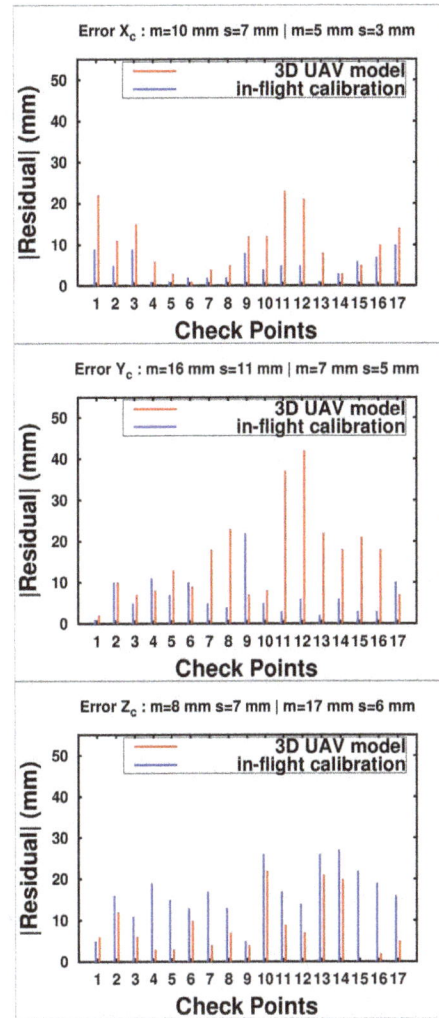

Figure 11: Comparison of absolute residuals using two different lever-arm calibration techniques

accuracy of 1.8 cm ± 0.8 cm. However, the second method requires GCPs and a stable mechanical mounting of sensors on the UAV to be reused while the first one needs only image measurements and requires no special mechanical configuration.

The difference in accuracy between the two methods is due to the fact that, for the photogrammetric acquisition of the UAV method, the calibration of the antenna phase center is not entirely correct. Indeed, this calibration, which requires making static observations and photogrammetric acquisitions of the GPS receiver, is performed on the ground. However, the antenna phase center position is correlated with the antenna environment through the multipath effect. The position of the antenna phase center in the local frame of the GPS receiver finds itself biased because of the multipath effect.

Given the fact that the multipath effect is negligible for an on-board antenna on a UAV during a flight, this bias introduced during phase antenna center calibration can be estimated by computing the difference between the values of the lever-arm, the one estimated in-flight and the one estimating with a photogrammetric acquisition of the UAV on the ground.

5. FURTHER WORK

In order to validate the lever-arm proposed calibration method, it is necessary to conduct a second experiment where two independent datasets will be acquired. The difference here is that the position and the orientation of the camera must be different. The first flight will be used to correct the value of the measured lever-arm on the ground from the multipath effect and the second flight will determine whether this correction improves the accuracy on check points in comparison to the results provided above.

6. CONCLUSIONS

We propose in this article a study about a method of estimating the lever-arm offset in the context of direct-georeferencing of a close-range photogrammetric acquisition performed with a UAV. The proposed method is flexible because requires no mechanical constraint for the mounting of the sensors on the UAV on each flight. Once the optical and antenna phase centers calibrated, the estimation of the lever-arm value is based solely on photogrammetric measurements across a photogrammetric images acquisition of the UAV on the ground. Also, this method allows to control the stability of the sensors mounting between different flights. The accuracy achieved by this method is lower than the conventional in-flight estimation method. During our experiments we achieved an absolute accuracy of $2\ cm \pm 1.4\ cm$ on check points while the classical method achieved an absolute accuracy of $1.8\ cm \pm 0.8\ cm$. However, this second method requires having external measurements (GCPs) and that the mechanical configuration remains the same from one flight to another that it may be still valid.

REFERENCES

Benoit, L., 2014. Positionnement GPS précis et en temps-réel dans le contexte de réseaux de capteurs sans fil type Geocube : application à des objets géophysiques de taille kilométrique. PhD thesis, École Normale Supérieure.

Benoit, L., Briole, P., Martin, O., Thom, C., Malet, J.-P. and Ulrich, P., 2015. Monitoring landslide displacements with the geocube wireless network of low-cost gps. Engineering Geology 195, pp. 111–121.

Bláha, M., Eisenbeiss, H., Grimm, D. and Limpach, P., 2011. Direct georeferencing of uavs. ISPRS - International Archives of the Photogrammetry, Remote Sensing and Spatial Information Sciences XXXVIII-1/C22, pp. 131–136.

Chiang, K.-W., Tsai, M.-L. and Chu, C.-H., 2012. The development of an uav borne direct georeferenced photogrammetric platform for ground control point free applications. Sensors 12(7), pp. 9161–9180.

Cramer, M., Bovet, S., Gültlinger, M., Honkavaara, E., McGill, A., Rijsdijk, M., Tabor, M. and Tournadre, V., 2013. On the use of rpas in national mapping ; the eurosdr point of view. ISPRS - International Archives of the Photogrammetry, Remote Sensing and Spatial Information Sciences XL-1/W2, pp. 93–99.

Daakir, M., Pierrot-Deseilligny, M., Bosser, P., Pichard, F. and Thom, C., 2015. Uav onboard photogrammetry and gps positionning for earthworks. ISPRS - International Archives of the Photogrammetry, Remote Sensing and Spatial Information Sciences XL-3/W3, pp. 293–298.

DJI, 2015. F550 user manual.

Fraser, C. S., 1997. Digital camera self-calibration. ISPRS Journal of Photogrammetry and Remote sensing 52(4), pp. 149–159.

Fraser, C. S., 2013. Automatic camera calibration in close range photogrammetry. Photogrammetric Engineering & Remote Sensing 79(4), pp. 381–388.

Gruen, A., 2012. Development and status of image matching in photogrammetry. The Photogrammetric Record 27(137), pp. 36–57.

Klobuchar, J. et al., 1987. Ionospheric time-delay algorithm for single-frequency gps users. Aerospace and Electronic Systems, IEEE Transactions on (3), pp. 325–331.

Lowe, D. G., 2004. Distinctive image features from scale-invariant keypoints. International journal of computer vision 60(2), pp. 91–110.

Martin, O., Meynard, C., Pierrot-Deseilligny, M., Souchon, J. and Thom, C., 2014. Réalisation d'une caméra photogrammétrique ultralégère et de haute résolution. http://drone.teledetection.fr/.

Pfeifer, N., Glira, P. and Briese, C., 2012. Direct georeferencing with on board navigation components of light weight uav platforms. ISPRS - International Archives of the Photogrammetry, Remote Sensing and Spatial Information Sciences XXXIX-B7, pp. 487–492.

Pierrot-Deseilligny, M., 2015. MicMac, Apero, Pastis and Other Beverages in a Nutshell!

Pierrot-Deseilligny, M. and Clery, I., 2011. Apero, an open source bundle adjustment software for automatic calibration and orientation of set of images. ISPRS - International Archives of the Photogrammetry, Remote Sensing and Spatial Information Sciences XXXVIII-5/W16, pp. 269–276.

Pierrot-deseilligny, M. and Paparoditis, N., 2006. A multiresolution and optimization-based image matching approach: An application to surface reconstruction from spot5-hrs stereo imagery. In: In: Proc. of the ISPRS Conference Topographic Mapping From Space (With Special Emphasis on Small Satellites), ISPRS.

Rehak, M., Mabillard, R. and Skaloud, J., 2013. A micro-uav with the capability of direct georeferencing. ISPRS - International Archives of the Photogrammetry, Remote Sensing and Spatial Information Sciences XL-1/W2, pp. 317–323.

Rieke, M., Foerster, T., Geipel, J. and Prinz, T., 2011. High-precision positioning and real-time data processing of uav-systems. International Archives of Photogrammetry, Remote Sensing and Spatial Information Sciences 38, pp. 1–C22.

Saastamoinen, J., 1972. Atmospheric correction for the troposphere and stratosphere in radio ranging satellites. The use of artificial satellites for geodesy pp. 247–251.

Stempfhuber, W. and Buchholz, M., 2011. A precise, low-cost rtk gnss system for uav applications. ISPRS - International Archives of the Photogrammetry, Remote Sensing and Spatial Information Sciences XXXVIII-1/C22, pp. 289–293.

Takasu, T., 2011. Rtklib: An open source program package for gnss positioning.

Tournadre, V., Pierrot-Deseilligny, M. and Faure, P. H., 2014. Uav photogrammetry to monitor dykes - calibration and comparison to terrestrial lidar. ISPRS - International Archives of the Photogrammetry, Remote Sensing and Spatial Information Sciences XL-3/W1, pp. 143–148.

Tournadre, V., Pierrot-Deseilligny, M. and Faure, P. H., 2015. Uav linear photogrammetry. ISPRS - International Archives of the Photogrammetry, Remote Sensing and Spatial Information Sciences XL-3/W3, pp. 327–333.

Tsai, M., Chiang, K., Huang, Y., Lin, Y., Tsai, J., Lo, C., Lin, Y. and Wu, C., 2010. The development of a direct georeferencing ready uav based photogrammetry platform. In: Proceedings of the 2010 Canadian Geomatics Conference and Symposium of Commission I.

Turner, D., Lucieer, A. and Wallace, L., 2014. Direct georeferencing of ultrahigh-resolution uav imagery. Geoscience and Remote Sensing, IEEE Transactions on 52(5), pp. 2738–2745.

u-blox, 2009. GPS Antennas RF Design Considerations for u-blox GPS Receivers. u-blox AG.

u-blox, 2013. u-blox 6 Receiver Description Including Protocol Specification. u-blox AG.

u-blox, 2014. LEA-6 u-blox 6 GPS Modules Data Sheet Objective Specification. u-blox AG.

14

SUB-CAMERA CALIBRATION OF A PENTA-CAMERA

K. Jacobsen [a], M. Gerke[b]

[a]Leibniz University Hannover, Institute of Photogrammetry and Geoinformation, Germany;
- jacobsen@ipi.uni-hannover.de
[b]Faculty of Geo-Information Science and Earth Observation of the University of Twente, Netherlands
- m.gerke@utwente.nl

KEY WORDS: self calibration, penta camera, bundle block adjustment

ABSTRACT:

Penta cameras consisting of a nadir and four inclined cameras are becoming more and more popular, having the advantage of imaging also facades in built up areas from four directions. Such system cameras require a boresight calibration of the geometric relation of the cameras to each other, but also a calibration of the sub-cameras.

Based on data sets of the ISPRS/EuroSDR benchmark for multi platform photogrammetry the inner orientation of the used IGI Penta DigiCAM has been analyzed. The required image coordinates of the blocks Dortmund and Zeche Zollern have been determined by Pix4Dmapper and have been independently adjusted and analyzed by program system BLUH. With 4.1 million image points in 314 images respectively 3.9 million image points in 248 images a dense matching was provided by Pix4Dmapper. With up to 19 respectively 29 images per object point the images are well connected, nevertheless the high number of images per object point are concentrated to the block centres while the inclined images outside the block centre are satisfying but not very strongly connected. This leads to very high values for the Student test (T-test) of the finally used additional parameters or in other words, additional parameters are highly significant.

The estimated radial symmetric distortion of the nadir sub-camera corresponds to the laboratory calibration of IGI, but there are still radial symmetric distortions also for the inclined cameras with a size exceeding 5μm even if mentioned as negligible based on the laboratory calibration. Radial and tangential effects of the image corners are limited but still available. Remarkable angular affine systematic image errors can be seen especially in the block Zeche Zollern. Such deformations are unusual for digital matrix cameras, but it can be caused by the correlation between inner and exterior orientation if only parallel flight lines are used. With exception of the angular affinity the systematic image errors for corresponding cameras of both blocks have the same trend, but as usual for block adjustments with self calibration, they still show significant differences.

Based on the very high number of image points the remaining image residuals can be safely determined by overlaying and averaging the image residuals corresponding to their image coordinates. The size of the systematic image errors, not covered by the used additional parameters, is in the range of a square mean of 0.1 pixels corresponding to 0.6μm. They are not the same for both blocks, but show some similarities for corresponding cameras.

In general the bundle block adjustment with a satisfying set of additional parameters, checked by remaining systematic errors, is required for use of the whole geometric potential of the penta camera. Especially for object points on facades, often only in two images and taken with a limited base length, the correct handling of systematic image errors is important. At least in the analyzed data sets the self calibration of sub-cameras by bundle block adjustment suffers from the correlation of the inner to the exterior calibration due to missing crossing flight directions. As usual, the systematic image errors differ from block to block even without the influence of the correlation to the exterior orientation.

1. INTRODUCTION

Multiple lens and multiple camera arrangements for aerial purposes are in use more than 100 years (Manual of Photogrammetry 1952, Jacobsen 2008). They became a revival with digital cameras supported by direct sensor orientation (Remondino and Gerke 2015). Especially for urban mapping penta cameras are in use since years as e.g. Pictometry camera, Track'Air MIDAS, UltraCam Osprey, Leica RCD30 oblique, IGI penta camera and several individual mid-format and small camera combinations. Due to the difficult and time consuming orientation of multi head cameras in most cases a direct sensor orientation is preferred, using pre-calibration of the sub-cameras together with a calibration of the camera system and a boresight calibration in relation to the combination of GNSS and giros.

The pre-calibration of such multi-head cameras is required for operational use of these systems due to too time consuming inner and relative self calibration. In most cases the accuracy requirement is limited to presentation scale of the generated product. Several solutions have been published as e.g. Jacobsen 2008, Madani 2012, Rupnik et al. 2014. With progress in Unmanned Aerial Systems (UAS) several self arranged multi-head systems came in use requiring a system calibration e.g. Niemeyer et al. 2013, Li et al. 2013, Detchev et al. 2014.
For this paper the geometry of the sub-cameras of the IGI penta camera with one nadir and four inclined cameras with 45° nadir viewing direction (figure 1) and the relation of the sub-cameras to each other has been investigated. Especially the exact camera geometry and geometric stability has been analyzed.

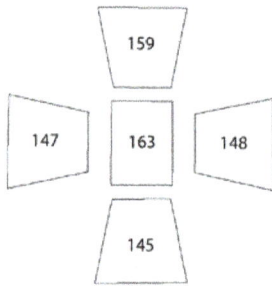

Figure 1: Sub-camera footprints of used penta-camera
Focal length for sub-camera
163: 50mm
145 – 159: 82mm
Image size: 49.056 x 36.792mm
Pixel size: 6μm
View directions for 145 – 159: 45° from nadir

The precise image geometry was determined by bundle block adjustment with self calibration by program system BLUH (Jacobsen 2007, Jacobsen et al. 2010). By self calibration the details of inner orientation can be determined, while the boresight misalignment can be extracted from exterior orientation. As input for the bundle block adjustment image coordinates taken by Pix4Dmapper have been used. Pix4Dmapper is operational software for the determination of tie points also for penta cameras with 45° nadir angle for the view direction of the side looking cameras. The bundle block adjustment with BLUH has not been handled with the same projection centre for all sub-cameras from one imaging instant. This is weakening the block adjustment, but it allows a boresight calibration without any pre-condition as for example caused by not exactly simultaneous imaging.

2. DATA SET

An image flight with the penta camera was made over the ISPRS benchmark test fields Dortmund and Zeche Zollern (http://www2.isprs.org/commissions/comm1/icwg15b/benchmark_main.html, Nex et al. 2015). For the nadir camera block Dortmund has 60% endlap and sidelap, while this is 80% for Zeche Zollern (figure 2). In Zeche Zollern the flight lines have been flown twice in opposite direction.

The image flight Dortmund was made 905m above ground, while it was 860m for Zeche Zollern. Corresponding to this the ground sampling distance (GSD) for the nadir camera is 10.4cm respectively 9.9cm and for the inclined cameras in block Dortmund 7.5cm x 8.5cm up to 14cm x 29cm for horizontal objects, for facades, perpendicular to view direction, in the far range 14cm x 16cm. For Zeche Zollern the GSD is 5% smaller.

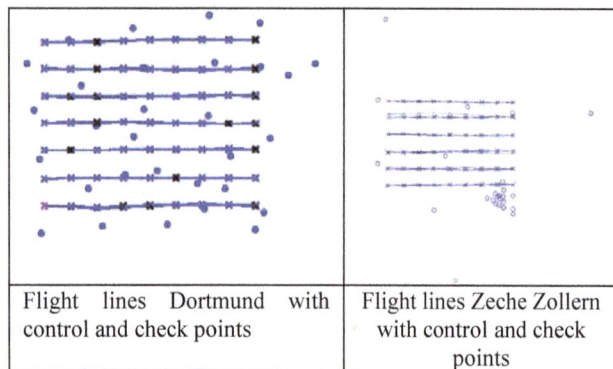

Flight lines Dortmund with control and check points	Flight lines Zeche Zollern with control and check points

Figure 2. Flight lines, projection centres and ground control and check points

The strong overlay of all images is shown by the foot print plots, colour coded for the used sub-cameras (figures 3 and 4). It also shows that only in the block centre the images are strongly overlapped. Crossing flight lines, having some advantages for self calibration, are not available. The possible image connections are shown in figures 6 and 7 colour coded

corresponding to the number of overlapping images. By theory it would be possible to have the same image point in 25 respectively 41 images, but in reality it is only 19 respectively 29 and this only for one object point in both blocks.

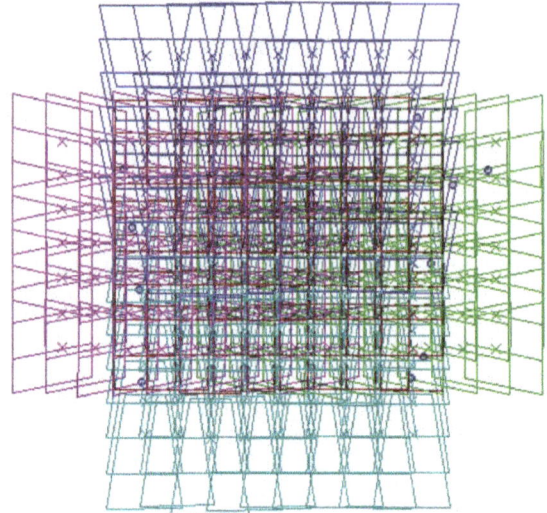

Figure 3: foot prints of all images Dortmund

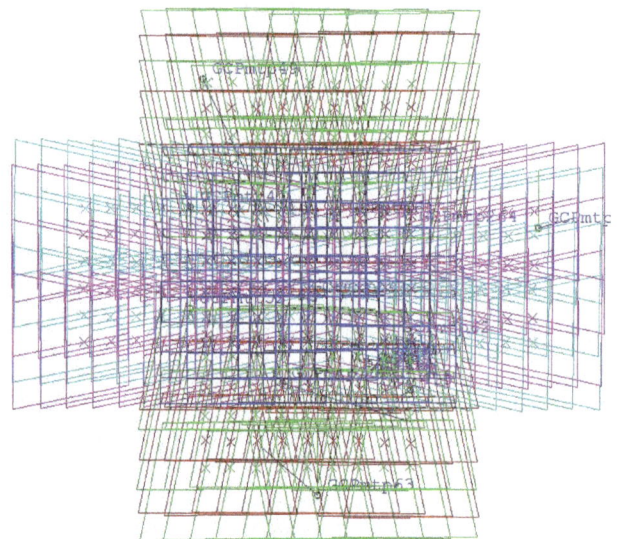

Figure 4: Foot prints of all images Zeche Zollern

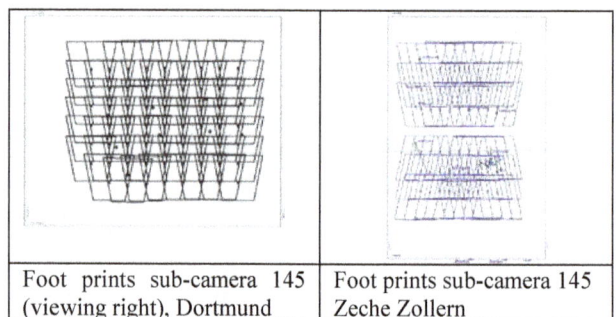

Foot prints sub-camera 145 (viewing right), Dortmund	Foot prints sub-camera 145 Zeche Zollern

Figure 5: Foot prints for inclined sub-camera 145

The frequency distribution of number of images per object point is shown in figures 8 and 9. Even if object points just measured in one image are excluded from the adjustment and the influence of object points, located just in two images, to the block tie is limited, the image tie leads to a satisfying image connection, only in the periphery the blocks are a little week. The distribution of the real image tie can be seen in figure 10.

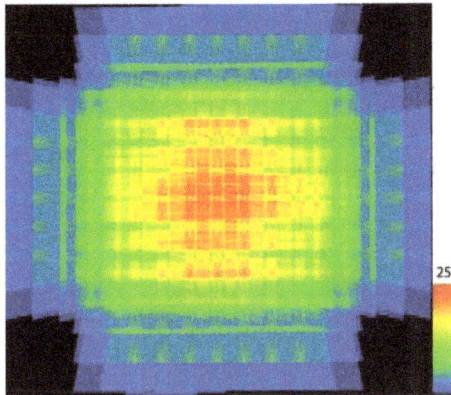

Figure 6: overlap of images, Dortmund

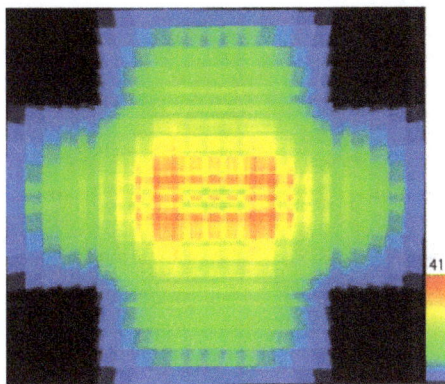

Figure 7: overlap of images, Zeche Zollern

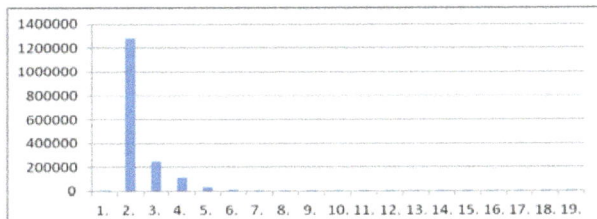

Figure 8. Number of images per object point, block Dortmund

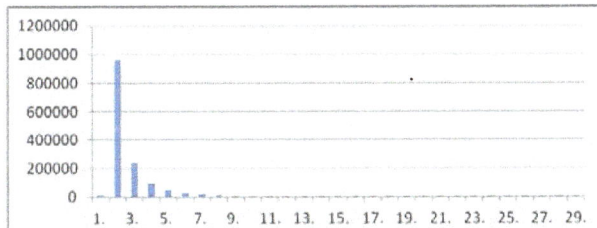

Figure 9. Number of images per object point, block Zeche Z.

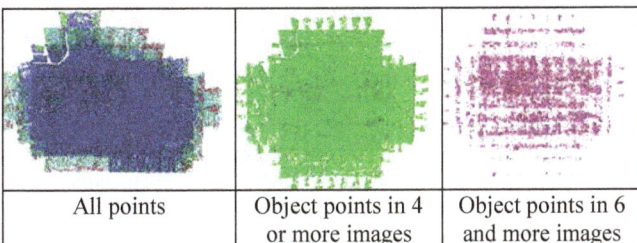

| All points | Object points in 4 or more images | Object points in 6 and more images |

Figure 10. Object points measured at least in the named number of images, Dortmund (blue=2 images, green 3 and 4 images, red >5 images)

3. BUNDLE BLOCK ADJUSTMENT

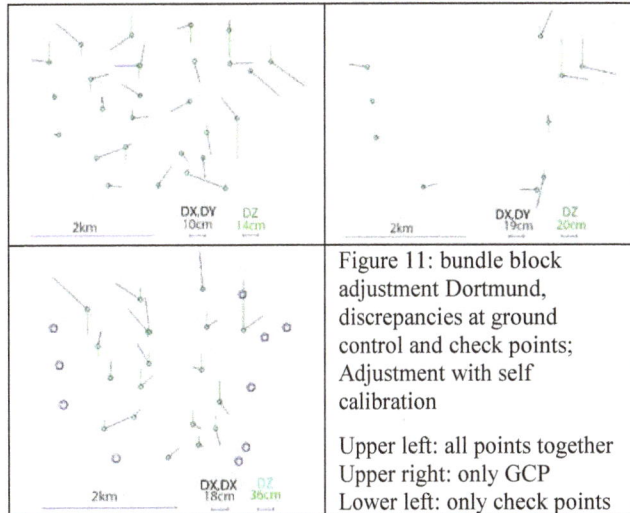

Figure 11: bundle block adjustment Dortmund, discrepancies at ground control and check points; Adjustment with self calibration

Upper left: all points together
Upper right: only GCP
Lower left: only check points

Based on 314 images of block Dortmund and 268 images of the block Zeche Zollern the results shown in tables 1 and 2 have been reached by bundle block adjustment.

Program BLUH is using a standard set of 12 additional parameters for standard self calibration (Jacobsen 2007). In addition digital cameras often have systematic effects at image corners, mainly due to not satisfying flatness of the CCD-matrix. Depending on the fixing of the CCD-arrays in the cameras, the flatness may depend on the temperature. This effect can be determined with the special parameters 81 up to 88 (Jacobsen et al. 2010). Also the principal point (x and y) as well as the focal length can be included as unknown. The use of the focal length as unknown was not successful; also the principal point did not lead to realistic results. This is not surprising in an object area with only limited undulation in height and same view direction for all images of a sub-camera. In general the additional parameters are checked for justification (Student test, correlation, total correlation), not justified parameters are automatically eliminated by the program. So the number of originally chosen additional parameters is reduced to the required set. Approximately 50% up to 70% of the additional parameters have been significant.

	$\sigma 0$	SX	SY	SZ	N
At ground control points					
Dortmund	9.5μm	0.20m	0.49m	0.60m	10
Zeche Zollern	8.6μm	0.12m	0.34m	0.43m	9
At check points					
Dortmund	-	0.26m	0.35m	0.51m	19
Zeche Zollern	-	0.07m	0.09m	0.82m	15

Table 1. Standard deviation of block adjustment without self calibration

	$\sigma 0$	SX	SY	SZ	N
At ground control points					
Dortmund	8.7μm	0.29m	0.16m	0.24m	10
Zeche Zollern	7.2μm	0.04m	0.14m	0.34m	9
At check points					
Dortmund	-	0.16m	0.19m	0.36m	19
Zeche Zollern	-	0.06m	0.06m	0.81m	15
All GCP and check points used as control points					
Dortmund	8.7μm	0.12m	0.08m	0.28m	29
Zeche Zollern	7.2μm	0.10m	0.08m	0.42m	23

Table 2. Standard deviation of block adjustment with self calibration by additional parameters

The improvement of the block adjustment by self calibration is obvious. As usual, especially the height accuracy depends upon the self calibration which can reduce block deformation.

The accuracy of the bundle block adjustments is typical for such type of images and the used tie point generation. The sigma0 exceeds the pixel size of 6µm, this is not unusual for such camera systems with matching of images taken with quite different view direction, being often limited to the pixel address, not reaching sub-pixel accuracy. The standard deviation of the horizontal coordinates X and Y based on all ground control points (GCP) is in the range of the GSD. For the height the accuracy is not as good due to not always good intersections. The ground control points are measured in 2 up to 11 images in case of block Dortmund, for Zeche Zollern up to 22 images per control point are available. With only 9 not optimal distributed GCP (figure 2) the block Zeche Zollern is not very stable in the height.

4. IMAGE GEOMETRY

The image geometry is determined by self calibration with additional parameters. By analysis of image coordinate residuals also remaining systematic image errors can be identified. For this all image residuals for one camera are overlaid corresponding to the image positions. By averaging all overlaid residuals in a chosen image matrix good information about remaining systematic image errors is achieved. Only residuals of object points measured determined at least in 4 images have been used. An averaging in 25 x 25 sub-areas was selected. Due to the high number of respected image points in average any vector represents the average of 275 residuals for block Dortmund and for Zeche Zollern even more (figure 14).

The determined systematic image errors are not the same for both blocks, flown with the same camera. Due to the parallel flight lines the systematic image errors are correlated to discrepancies at GCP, so only the same trend of systematic image errors can be expected (figure 12). In addition aerial cameras are changing the geometry from image flight to image flight, so finally only by self calibration optimal results can be achieved. In most cases the radial symmetric distortion dominates the systematic image errors. Also for the investigated data set the radial symmetric parameters are very important even if the used image coordinates are pre-corrected by information included in the calibration certificate. Again, the radial symmetric distortion has the same tendency for the sub-cameras of both project areas, but it is not the same (figure 13). The remaining systematic image errors (figure 14) are even more different in both blocks.

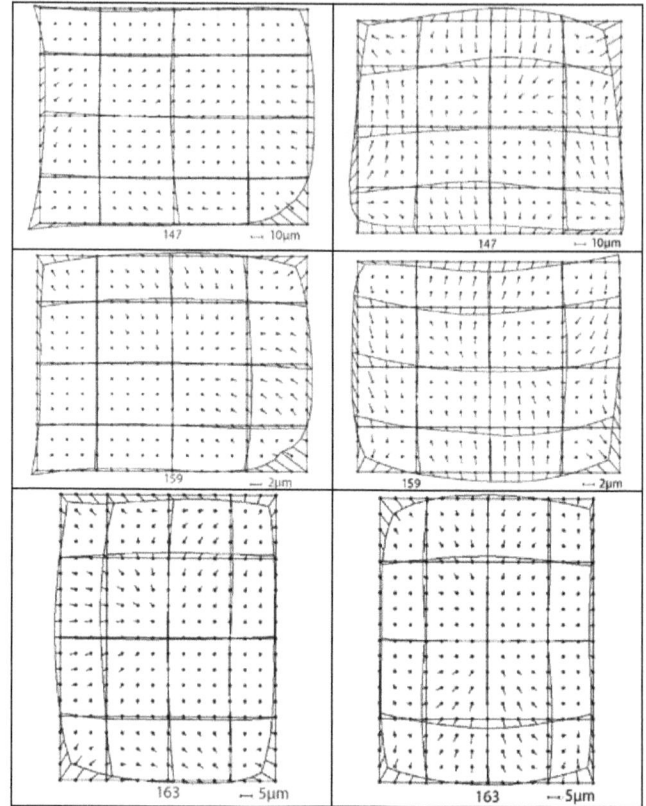

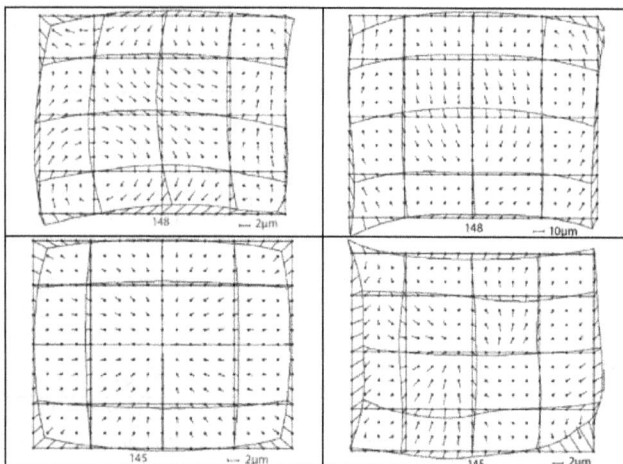

Figure 12: Systematic image errors for the 5 sub-cameras, left: Dortmund, right: Zeche Zollern (! different vector scale !)

Sub-camera 147 (backward looking camera) shows some problems in both data sets – the systematic image errors are quite large (note different vector scale in figure 12) and also the remaining systematic image errors are large – the vector scale for 147 in figure 14 is 10µm instead of 3 µm for the other sub-cameras.

	σo	SX	SY	SZ	N
Block adjust.	8.2µm	0.10m	0.19m	0.47m	9
At check points	-	0.05m	0.06m	0.78m	15

Table 3: Block adjustment with systematic image errors from block Dortmund as pre-correction for block Zeche Zollern (no self calibration)

Nevertheless a pre-correction of the image coordinates by systematic image errors determined in block Dortmund leads to a small improvement of the block adjustment Zeche Zollern (table 3), vice versa this is not the case, which may be explained by instability of block Zeche Zollern. Table 3 has to be compared with the results for Zeche Zollern in table 1.

With exception of sub-camera 147 the systematic image errors are not so large and may only lead to small improvement by pre- calibration of other data sets taken by this penta-camera, so the sigma0 of the block adjustment is reduced from 9.5µm by self calibration to 8.7µm for block Dortmund and from 8.6µm to 7.2µm for block Zeche Zollern. Nevertheless even such small influence to sigma0 may lead to block deformation if the block is not supported by direct sensor orientation and the number of GCP is limited, but the handled blocks with 60 respectively 54 nadir images are not very large. In addition especially object points determined just in two or few images with not optimal intersections may be strongly influenced also by such limited change of the image coordinates.

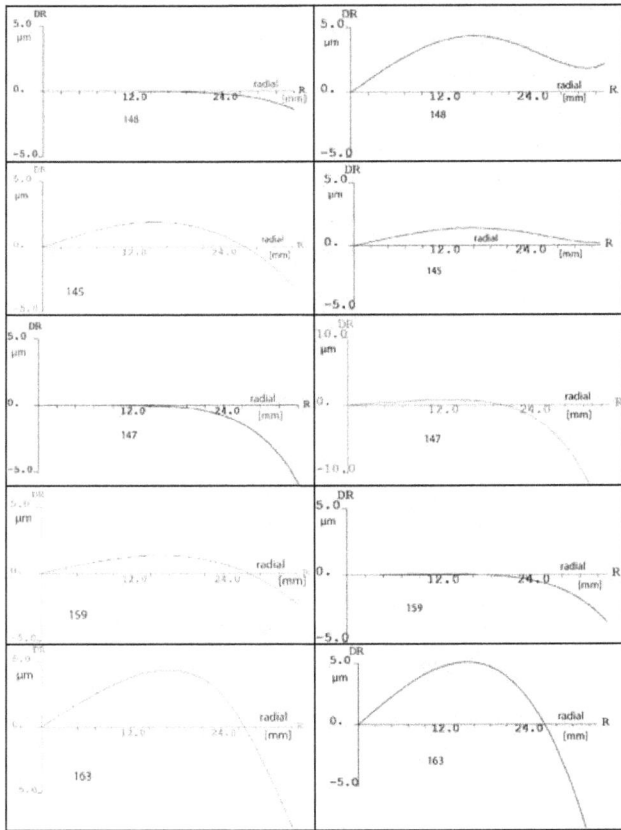

Figure 13: Radial symmetric distortion, left: Dortmund, right Zeche Zollern

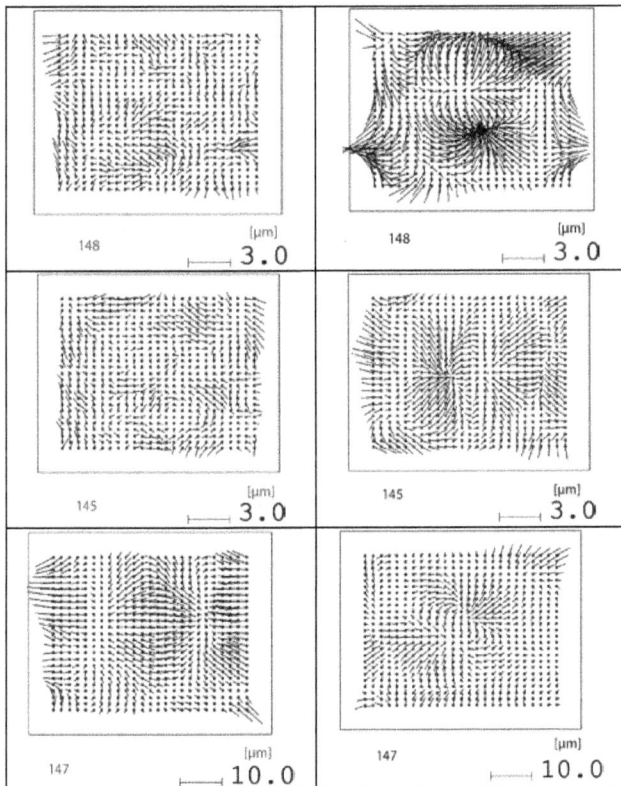

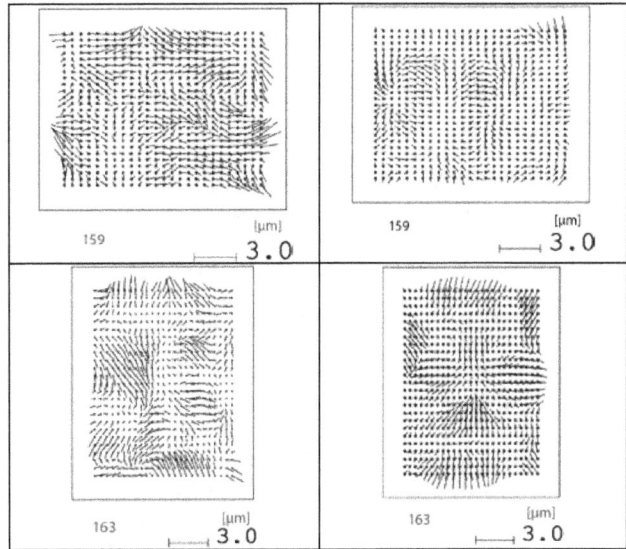

Figure 14: Remaining systematic image errors, left: Dortmund, right: Zeche Zollern

5. BORESIGHT MISALIGNMENT

As mentioned, the sub-cameras have not been fixed to the same projection centres and also no direct sensor information was used to allow an independent calibration.

camera	Dlong	Dlat	DZo	Dlong	Dlat	DZo
	Without self calibration			With self calibration		
148	36cm	-31cm	82cm	-343cm	20cm	134cm
145	66cm	1cm	24cm	-376cm	11cm	54cm
147	42cm	20cm	115cm	-179cm	200cm	299cm
159	39cm	-77cm	56cm	-422cm	-100cm	161cm

Table 4. Discrepancies of averaged projection centres – related to nadir camera, block Dortmund – in flight direction (Dlong), across flight direction (Dlat) and Height (DZo)

camera	Dlong	Dlat	DZo	Dlong	Dlat	DZo
	Without self calibration			With self calibration		
148	-9cm	-17cm	100cm	-413cm	-110cm	39cm
145	33cm	-115cm	31cm	-168cm	-113cm	47cm
147	-77cm	16cm	91cm	-403cm	71cm	139cm
159	44cm	-110cm	90cm	-190cm	-202cm	157cm

Table 5. Discrepancies of averaged projection centres – related to nadir camera, block Zeche Zollern

	Slong	Slat	SZo	Slong	Slat	SZo
	Without self calibration			With self calibration		
DO	57cm	66cm	56cm	24cm	49cm	34cm

Table 6. Root mean square variation of projection centre position in relation to nadir camera, block Dortmund

The discrepancies of the projection centre positions against the nadir camera (tables 4 and 5) are larger as possible for boresight. In flight direction (Dlong) differences may be explained by discrepancies in synchronization of the sub-cameras, but this is not the case across flight direction and in height. In addition there is no correlation between results of block Dortmund to block Zeche Zollern. The projection centre offsets are clearly larger for the results based on adjustment with self calibration, demonstrating the influence of the self calibration to the exterior orientation. Nevertheless the variations of the projection centres for the individual camera poses are clearly smaller for the block adjustment with self calibration (table 6). This is not shown for Zeche Zollern because of limited number of projection centres where all sub-images have been used. The listed results cannot be used for a boresight determination.

Also the attitudes have been compared with the calibration certificate from IGI. The discrepancies especially in roll and yaw are not realistic; they are also correlated with the offsets in table 4 and 5. Here joint adjustments with the same projection centres for all sub-cameras of the camera system are required.

6. CONCLUSION AND FUTURE STEPS

The details of image deformations, computed by block adjustment with self calibration by additional parameters could be determined. The self calibration clearly improves the results of the block adjustment even if the systematic image errors are limited in size. Nevertheless for both blocks they only show the same tendency. The systematic image errors of block Dortmund used as pre-correction for the images from block Zeche Zollern are improving the block adjustment without self calibration, but not vice versa. The high number of in the average 13200 respectively 14600 points per image reduce the standard deviation of the single additional parameters, but in relation to larger blocks of usual nadir cameras, handled by the author, they are not extremely high. It has to be respected that 31% respectively 25% of the object points are just available in two images - they have only a very limited effect to the self calibration. Of course large Student test values only can be reached if a corresponding image deformation exist and they are also depending upon the sigma0 of the block adjustment which is high in relation to standard blocks of nadir cameras.

The exterior orientations, based on independent adjustment of the sub-cameras, are not useful for a boresight determination.

In future additional analysis shall be made by joining the sub-camera images together with a single projection centre. For this however, the exact camera synchronisation remains a challenge. In addition the correlation of the additional parameters with exterior orientation will be improved by changing the formulas of the additional parameters without changing the influence in image space. This will improve the determination of the angular relation of the sub-cameras to each other.

REFERENCES

Detchev, I., Mazaheri, M., Rondeel, S., Habib, A., 2014: Calibration of Multi-Camera Photogrammetric Systems, IAPRS Volume XL-1, 2014

Jacobsen, K., 2007: Geometric Handling of Large Size Digital Airborne Frame Camera Images, Optical 3D Measurement Techniques VIII, Zürich 2007, pp 164 – 171

Jacobsen, K., 2008: Calibration of Camera Systems, Annual ASPRS Annual Conference Portland, 2008

Jacobsen, K., Cramer, M., Ladstätter, R., Ressl, C., Spreckels, V., 2010: DGPF Project: Evaluation of Digital Photogrammetric Camera Systems – Geometric Performance, PFG 2010, 2, pp 83-97

Li, B., Heng, L., Koeser, K., Pollefeys, M., 2013: A Multiple-Camera System Calibration Toolbox Using A Feature Descriptor-Based Calibration Pattern, IROS 2013.

Madani, M., 2012: Accuracy Potential and Applications of MIDAS Aerial Oblique Camera System, IAPRS Volume XXXIX-B1, 2012

Manual of Photogrammetry, 2nd edition, chapter II, part1, Aerial cameras and accessories, ASPRS, 1952

Nex, F., Gerke, M., Remondino, F., Przybilla, H.-J., Bäumker, M., Zurhorst, A., 2015: ISPRS benchmark for multi-platform photogrammetry. ISPRS Annals of the Photogrammetry, Remote Sensing and Spatial Information Sciences, Vol. II-3/W4, pp. 135-142. PIA15+HRIGI15 – Joint ISPRS conference, 25–27 March 2015, Munich, Germany

Niemeyer, F., Schima, R., Grenzdörffer, G., 2013: Relative and absolute Calibration of a multihead Camera System with oblique and nadir looking Cameras for a UAS, IAPRS Volume XL-1/W2, 2013

Remondino, F., Gerke, M., 2015: Oblique Aerial Imagery – a Review, Photogrammetric Week 2015, Wichmann/VDE Verlag, Belin & Offenbach, 2015

Rupnik, E., Nex, F., Remondino, F., 2014: Oblique multi-Camera Systems – Orientation and Dense Matching, EuroCOW 2014, IAPRS Volume XL-3/W1

15

AUTONOMOUS NAVIGATION OF SMALL UAVS BASED ON VEHICLE DYNAMIC MODEL

M. Khaghani[a]*, J. Skaloud[a]

[a]EPFL, Geodetic Engineering Laboratory TOPO, Route Cantonale,1015 Lausanne, Switzerland - (mehran.khaghani, jan.skaloud)@epfl.ch

KEY WORDS: Autonomous Navigation, UAV, Vehicle Dynamic Model, Integration, GNSS outage

ABSTRACT:

This paper presents a novel approach to autonomous navigation for small UAVs, in which the vehicle dynamic model (VDM) serves as the main process model within the navigation filter. The proposed method significantly increases the accuracy and reliability of autonomous navigation, especially for small UAVs with low-cost IMUs on-board. This is achieved with no extra sensor added to the conventional INS/GNSS setup. This improvement is of special interest in case of GNSS outages, where inertial coasting drifts very quickly. In the proposed architecture, the solution to VDM equations provides the estimate of position, velocity, and attitude, which is updated within the navigation filter based on available observations, such as IMU data or GNSS measurements. The VDM is also fed with the control input to the UAV, which is available within the control/autopilot system. The filter is capable of estimating wind velocity and dynamic model parameters, in addition to navigation states and IMU sensor errors. Monte Carlo simulations reveal major improvements in navigation accuracy compared to conventional INS/GNSS navigation system during the autonomous phase, when satellite signals are not available due to physical obstruction or electromagnetic interference for example. In case of GNSS outages of a few minutes, position and attitude accuracy experiences improvements of orders of magnitude compared to inertial coasting. It means that during such scenario, the position-velocity-attitude (PVA) determination is sufficiently accurate to navigate the UAV to a home position without any signal that depends on vehicle environment.

1. INTRODUCTION

This paper is a shortened version of (Khaghani and Skaloud, 2016). For more details and complementary results and discussions, readers are encouraged to refer to (Khaghani and Skaloud, 2016) here and on several occasions throughout the text.

1.1 Motivation

The dominant navigation system for small UAVs today is based on INS/GNSS integration (Bryson and Sukkarieh, 2015). GNSS provides absolute time-position-velocity TPV data for the platform, at relatively low frequency of only a few Hz, while INS provides relative position and attitude data at much higher frequencies than GNSS, typically at few hundreds of Hz. The integration of these data types can provide solutions with enough sort-term and long-term accuracy. However, the problem arises when GNSS outage happens (Lau et al., 2013), which is not a rare situation and can happen due to intentional corruption of GNSS signals (jamming and spoofing), or loss of line of sight to the satellites or interference in satellite signal reception (Groves, 2008). In such cases, the navigation solution is based on standalone INS with possible aiding from navigation aids such as barometric altimeters. The accuracy of the data provided by INS is directly determined by the quality of the IMU that is used in the system. The long-term accuracy of 3D inertial coasting based on small and low-cost IMUs available for small UAVs is so low that after only a minute of GNSS outage, the position uncertainty is too far from being of practical use. In other words, unless this drift is controlled by other means, the UAV is completely lost in space or may even become dynamically unstable. This may cause serious problems especially in non-line-of-sight flights and can lead to loss of the UAV with possible damages to objects or people on ground.

1.2 Available solutions

Many researches have been conducted to address the problem of rapid drift of navigation solution during GNSS outage conditions of minutes. Some have tried to improve INS error modeling using advanced techniques (Noureldin et al., 2009), and some have chosen to employ additional sensors to aid the system (Yun et al., 2013). The first approach does not still provide sufficiently good improvements for aerial vehicles. The solutions related to the second approach add cost and complexity to the system, and more importantly, their performance depends on environmental conditions that are not met all the times, which challenge the autonomy of the system. A widely used (yet partial) solution of the second category is employing vision based methods that provide relative or absolute measurements to inertial navigation (Wang et al., 2008) (Angelino et al., 2012). Apart from adding extra weight and hardware and software complications, their correct function requires some prerequisites on light, visibility, and terrain texture. For example, they might not work at night or in foggy conditions or over ground with uniform texture (vegetation, water, snow, etc.).

Therefore, it is a challenge to find a solution that mitigates the quick drift of low-cost inertial navigation during GNSS outage in airborne environment while preserving the autonomy of the system, and avoiding extra cost and additional sensors. Finding a suitable solution can be extremely beneficial for increasing the reliability of autonomous navigation of small UAVs.

There have been some previous research activities related to VDM integration to improve the navigation accuracy, especially in GNSS outage conditions. However, most of the proposed solutions employ the kinematic modeling (INS) as the main process model within the navigation filter (Bryson and Sukkarieh, 2004) (Vasconcelos et al., 2010) (Dadkhah et al., 2008), while using the VDM output either in the prediction phase or in the update phase

---
*Corresponding author

within this filter. Such approach relies totally on IMU and therefore is not robust if IMU failure occurs. Some authors have considered both INS and VDM at the same level within the filter (Koifman and Bar-Itzhack, 1999), but still the navigation solution at the end is based on filtered INS output. Therefore, IMU failure disables navigation in this case, as well. In many cases the presence of wind is not considered (Bryson and Sukkarieh, 2004) (Vasconcelos et al., 2010) (Dadkhah et al., 2008) (Crocoll et al., 2014) (Sendobry, 2014), or the capability of correcting the parametric errors in dynamic model on-flight is not provided (Bryson and Sukkarieh, 2004) (Vasconcelos et al., 2010) (Dadkhah et al., 2008), or the VDM is included only partially within the filter (Crocoll et al., 2014) (Crocoll and Trommer, 2014) (Müller et al., 2015). Some researchers also consider IMUs of higher accuracy (Koifman and Bar-Itzhack, 1999), which is impractical for small UAVs in terms of size/weight and cost.

1.3 Proposed approach

In this research, a solution is proposed that integrates VDM with inertial navigation for its autonomous part, and GNSS or other aiding when available. It improves the accuracy of navigation and significantly mitigates the drift of navigation uncertainty during GNSS signal reception absence. The main idea of this concept is to benefit from the available information of vehicle dynamic modeling and control input within the navigation system that implicitly rejects the physically impossible movements suggested by the IMU. Significant improvements in navigation solution accuracy in case of GNSS outages are reported via simulation studies. The proposed solution requires no additional sensors, so it preserves the autonomy of the navigation system when GNSS outages happen. Adding no extra sensors also means no additional cost and weight on the platform, which is an important aspect in small UAVs.

A key feature in the proposed solution is VDM acting as the main process model within the navigation filter, where its output is updated with raw IMU observations and if available, GNSS measurements. Such architecture avoids the complications of multi-process model filters (Bryson and Sukkarieh, 2004) (Koifman and Bar-Itzhack, 1999) and thus leads to simpler filter implementation, smaller state vector, and less computation cost. It is also preferred over the architectures in which INS is the main process model that gets updated by VDM, due to the following reasons. In case of IMU failure, the proposed architecture can simply stop using IMU observations, while the architecture with INS as the main process model will fail. On the other hand, the high frequency measurement noise in IMU data causes divergence in navigation solution when integrated within the navigation filter, as analytically shown in (Schwarz and Wei, 1994). The mechanical vibrations on the platform also affect the IMU measurements, but not the VDM output. Therefore, treating the IMU data as observations and avoiding integration of them as a process model is expected to improve the error growth.

The VDM needs to be fed with the control input to the UAV. This information is already available in the control/autopilot system, but it needs to be put in correct time relations with IMU and other measurements. The other input to VDM is the wind velocity. The proposed solution makes it possible to estimate the wind velocity within the navigation filter itself, even in absence of air pressure sensors. This adds certain redundancy to the system in case of air pressure sensor misbehavior when employed.

The VDM requires a proper structure based on the host platform type (fixed wing, copter, etc.) and its control actuators, which is well described in the literature (Cook, 2013) (Ducard, 2009).

Some parameters in the model depend on the specific platform. They can be either identified and pre-calibrated, or estimated in-flight. This capability for online parameter estimation (dynamic model identification) that does not require pre-calibration, minimizes the effort required in design and operation.

Proof of the proposed concept is performed via Monte Carlo simulation study in several different situations. To make the simulations realistic, errors are introduced to all the a-priori information available to the navigation system, such as initial values of states, dynamic model parameters, and error statistics of IMU and GNSS measurements. Also, real 3D wind velocity data (KNMI and Alterra, 2012) is used in simulations. Results of Monte Carlo simulations on a sample trajectory are presented and analyzed in the paper. More detailed results, along with observability and accuracy prediction discussions can be found in (Khaghani and Skaloud, 2016).

2. DEFINITIONS

Three coordinate frames are considered in this research, "navigation", "body", and "wind" frames. The navigation frame is a local frame oriented in north, east, and down directions denoted by (x_n, y_n, z_n) or (x_N, x_E, x_D), and considered to be inertial. Definitions of body frame and wind frame (x_b, y_b, z_b) can be perceived from Figure 1. The rotation matrix to transform vectors from body frame to navigation frame is defined as

$$R_n^b = R_1(\phi)\, R_2(\theta)\, R_3(\psi)$$
$$= \begin{bmatrix} 1 & 0 & 0 \\ 0 & \cos\phi & \sin\phi \\ 0 & -\sin\phi & \cos\phi \end{bmatrix} \begin{bmatrix} \cos\theta & 0 & -\sin\theta \\ 0 & 1 & 0 \\ \sin\theta & 0 & \cos\theta \end{bmatrix} \begin{bmatrix} \cos\psi & \sin\psi & 0 \\ -\sin\psi & \cos\psi & 0 \\ 0 & 0 & 1 \end{bmatrix}.$$
$$(1)$$

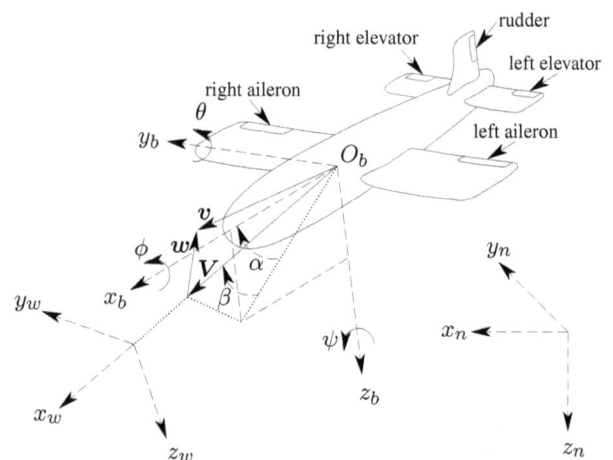

Figure 1: Navigation, body, and wind frames with airspeed (V), wind velocity(w), and UAV velocity (v), as weel as roll (ϕ), pitch (θ), and yaw (ψ)

The wind frame has its first axis in direction of airspeed V, and is defined by two angles with respect to body frame, angle of attack α and sideslip angle β. Velocity of airflow that is due to UAV's inertial velocity v and wind velocity w is denoted by airspeed vector V as

$$v = V + w. \qquad (2)$$

The rotation matrix from body frame to wind frame is defined as a function of Euler angles.

$$R_b^w = R_3(\beta) \, R_2^T(\alpha)$$

$$= \begin{bmatrix} \cos\beta & \sin\beta & 0 \\ -\sin\beta & \cos\beta & 0 \\ 0 & 0 & 1 \end{bmatrix} \begin{bmatrix} \cos\alpha & 0 & \sin\alpha \\ 0 & 1 & 0 \\ -\sin\alpha & 0 & \cos\alpha \end{bmatrix} \quad (3)$$

The density of the air is calculated based on the International Standard Atmosphere model as a function of local pressure and temperature, which can be expressed as functions of the altitude as detailed in (Khaghani and Skaloud, 2016).

3. VEHICLE DYNAMIC MODEL

The VDM employed in this research is based on rigid body dynamics for a fixed wing UAV and follows from (Ducard, 2009). It considers polynomial models for aerodynamic forces and moments. A brief description of the model is presented here, along with the key definitions and equations. More details can be found in (Ducard, 2009) and (Khaghani and Skaloud, 2016). The states vector $\boldsymbol{X}_n = \begin{bmatrix} x_N, x_E, x_D, v_x^b, v_y^b, v_z^b, \phi, \theta, \psi, \omega_x, \omega_y, \omega_z, n \end{bmatrix}^T$, control input vector $\boldsymbol{U} = [n_c, \delta_a, \delta_e, \delta_r]^T$, and wind velocity vector $\boldsymbol{w} = [w_N, w_E, w_D]^T$ are related via the dynamic model of the form $\dot{\boldsymbol{X}}_n = f(\boldsymbol{X}_n, \boldsymbol{U}, \boldsymbol{w})$. Components of UAV velocity vector $\boldsymbol{v}$ are represented by v_x^b, v_y^b, and v_z^b, while ω_x^b, ω_y^b, and ω_z^b denote the rate of change for roll, pitch, and yaw, respectively. Deflections of aileron, elevator, and rudder are denoted by δ_a, δ_e, and δ_r, respectively. Propeller speed is denoted by n, where n_c shows the commanded value for that, and τ_n is the time constant for its dynamics. Kinematic equations, Newton's equations of motion, and a first order model for propeller dynamics form the vehicle dynamic model as follows.

$$\begin{bmatrix} \dot{x}_N \\ \dot{x}_E \\ \dot{x}_D \end{bmatrix} = R_b^n \begin{bmatrix} v_x^b \\ v_y^b \\ v_z^b \end{bmatrix} \quad (4)$$

$$\begin{bmatrix} \dot{v}_x^b \\ \dot{v}_y^b \\ \dot{v}_z^b \end{bmatrix} = \begin{bmatrix} -g\sin\theta \\ g\sin\phi\cos\theta \\ g\cos\phi\cos\theta \end{bmatrix} + \frac{1}{m}\left[\begin{pmatrix} F_T \\ 0 \\ 0 \end{pmatrix} + R_w^b \begin{pmatrix} F_x^w \\ F_y^w \\ F_z^w \end{pmatrix} \right]$$

$$- \begin{bmatrix} \omega_y v_z^b - \omega_z v_y^b \\ \omega_z v_x^b - \omega_x v_z^b \\ \omega_x v_y^b - \omega_y v_x^b \end{bmatrix} \quad (5)$$

$$\begin{bmatrix} \dot{\phi} \\ \dot{\theta} \\ \dot{\psi} \end{bmatrix} = R_\omega \begin{bmatrix} \omega_x \\ \omega_y \\ \omega_z \end{bmatrix}, \quad R_\omega = \begin{bmatrix} 1 & \tan\theta\sin\phi & \tan\theta\cos\phi \\ 0 & \cos\phi & -\sin\phi \\ 0 & \sin\phi/\cos\theta & \cos\phi/\cos\theta \end{bmatrix} \quad (6)$$

$$\begin{bmatrix} \dot{\omega}_x \\ \dot{\omega}_y \\ \dot{\omega}_z \end{bmatrix} = (I^b)^{-1} \left(\begin{bmatrix} M_x^b \\ M_y^b \\ M_z^b \end{bmatrix} - \begin{bmatrix} \omega_x \\ \omega_y \\ \omega_z \end{bmatrix} \times I^b \begin{bmatrix} \omega_x \\ \omega_y \\ \omega_z \end{bmatrix} \right) \quad (7)$$

$$\dot{n} = -\frac{1}{\tau_n}n + \frac{1}{\tau_n}n_c \quad (8)$$

The four aerodynamic forces and the three aerodynamic moments are expressed as polynomial functions of navigation states, control inputs, wind velocity, and physical properties of the UAV called dynamic model parameters hereafter. The aerodynamic forces include:

- "thrust force" as $F_T = f(\rho, V, D, n, C_{F_T}...)$
- "drag force" as $F_x^w = f(\rho, V, S, C_{F_x}..., \alpha, \beta)$
- "lateral force" as $F_y^w = f(\rho, V, S, C_{F_y}..., \beta)$
- "lift force" as $F_z^w = f(\rho, V, S, C_{F_z}..., \alpha)$

The aerodynamic moments include:

- "roll moment" as $M_x^b = f(\rho, V, S, b, C_{M_x}..., \delta_a, \beta, \omega_x, \omega_z)$
- "pitch moment" as $M_y^b = f(\rho, V, S, \bar{c}, C_{M_y}..., \delta_e, \alpha, \omega_y)$
- "yaw moment" as $M_z^b = f(\rho, V, S, b, C_{M_z}..., \delta_r, \omega_z, \beta)$

Propeller diameter is denoted by D, and S, b, and $\bar{c}$ represent wing surface, wing span, and mean aerodynamic chord, respectively. Density of air is shown by ρ, and $C_{...}$'s represent aerodynamic coefficients for associated force and moment components. The vehicle dynamic model parameters are collected in (9).

4. FILTERING METHODOLOGY

An extended Kalman filter (Gelb, 1974) is chosen to serve as the navigation filter in this research, which is detailed in this section.

4.1 Scheme

The proposed navigation system utilizes VDM as main process model within a differential navigation filter. As depicted in Figure 2, VDM provides the navigation solution, which is updated by the navigation filter based on available measurements. Hence, IMU output is treated as a measurement within the navigation filter, just the way GNSS observations are, whenever they are available. Any other available sensory data, such as altimeter or magnetometer output, can also be integrated within the navigation filter as additional observations. VDM is fed with the control input of the UAV, which is commanded by the autopilot and therefore available. Other needed input is the wind velocity as an input, which can be estimated either by the aid of airspeed sensors, or within the navigation system with no additional sensors needed. The latter approach is implemented here. Finally, the parameters of VDM are required within the navigation filter. Pre-calibration of these parameters as fixed values is an option. However, to increase the flexibility, as well as the accuracy of the proposed approach while minimizing the design effort, an online parameter estimation/refinement is implemented. Last but not least, IMU errors are also modeled and estimated within the navigation system as additional filter states.

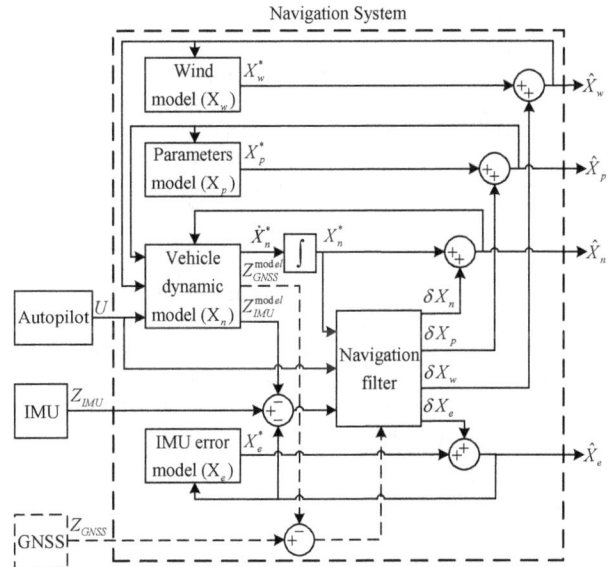

Figure 2: Navigation system architecture

4.2 State space augmentation

The augmented state vector includes the navigation states $\boldsymbol{X}_n$, the UAV dynamic model parameters $\boldsymbol{X}_p$ excluding mass (m) and moments of inertia (I_x, I_y, I_z, I_{xz}), the IMU error terms $\boldsymbol{X}_e$, and the wind velocity components $\boldsymbol{X}_w$.

The dynamic model parameters are included in a 26×1 state vector as in (9), and modeled as constant parameters with initial uncertainties. Description of these parameters is provided in the nomenclature, and the numerical values used in simulation can be found in (Ducard, 2009).

$$\boldsymbol{X}_p = \begin{bmatrix} S, & \bar{c}, & b, & D, & C_{F_T 1}, & \dots \\ \dots C_{F_T 2}, & C_{F_T 3}, & C_{F_z 1}, & C_{F_z \alpha}, & C_{F_x 1}, & \dots \\ \dots C_{F_x \alpha}, & C_{F_x \alpha 2}, & C_{F_x \beta 2}, & C_{F_y 1}, & C_{M_x a}, & \dots \\ \dots C_{M_x \beta}, & C_{M_x \tilde{\omega}_x}, & C_{M_x \tilde{\omega}_z}, & C_{M_y 1}, & C_{M_y e}, & \dots \\ \dots C_{M_y \alpha}, & C_{M_y \tilde{\omega}_y}, & C_{M_z \delta_r}, & C_{M_z \beta}, & C_{M_z \tilde{\omega}_z}, & \dots \\ \dots \tau_n \end{bmatrix}^T$$

(9)

Mass and moments of inertia are not included in this vector, since they appear as scaling factors in equations of motion and therefore they are completely correlated with the already included coefficients of aerodynamic forces and moments.

The error in each accelerometer and gyroscope inside the IMU is modeled as a random walk ($b_{rw}^{\dots}$) process. Therefore, the IMU error terms vector is defined as

$$\boldsymbol{X}_e = \begin{bmatrix} b_{rw}^{a1} & b_{rw}^{a2} & b_{rw}^{a3} & b_{rw}^{g1} & b_{rw}^{g2} & b_{rw}^{g3} \end{bmatrix}^T, \quad (10)$$

where ai and gi superscripts denote the i-th accelerometer and gyroscope, respectively. This model has been found sufficient for the low-cost IMU in consideration, but can be extended as needed.

The wind velocity is stated as a vector in local (navigation) frame consisting of the three components of wind velocity in north, east, and down directions.

$$\boldsymbol{X}_w = [w_N, w_E, w_D]^T \quad (11)$$

Wind velocity is also modeled as a random walk process.

4.3 Errors and uncertainties

For the purpose of simulation, a MEMS-grade IMU is considered. Random biases with standard deviations of 8 mg for accelerometers and 720 $^\circ/hr$ for gyroscopes are considered, accompanied by white noise and first order Gauss-Markov processes. GNSS error is modelled as independent white noise signals for each channel (north, east, down), with standard deviations 1 m. The sampling frequency is 100 Hz for IMU and 1 Hz for GNSS measurements.

In terms of initialization errors, random errors are considered for different runs of the Monte-Carlo simulations with standard deviations of 1 m for each position component, 1 m/s for each velocity component, 3° for roll and pitch, 5° for yaw, $1^\circ/s$ for rotation rates, and 15 rad/s for propeller speed. The errors considered for the UAV dynamic model parameters ($\boldsymbol{X}_p$) are randomly distributed with a standard deviation of 10%.

More details on the process model noise, observation noise, and initial uncertainties can be found in (Khaghani and Skaloud, 2016)

5. MONTE CARLO SIMULATIONS

To evaluate the performance of the navigation system, a Monte Carlo simulation has been performed with 100 runs, using real 3D wind velocity data (KNMI and Alterra, 2012). While the trajectory and the wind has been kept the same in each realization, the error in observations, initialization, and VDM parameters has changed randomly for each individual run. Figure 3 depicts the reference trajectory, as well as the solution from a sample run.

For more comprehensible presentation, the results are presented in ENU frame, instead of navigation NED frame. The trajectory has a footprint bigger than $2\ km \times 2\ km$ on the ground and $1km$ change in altitude. The results of Monte Carlo simulations are presented in sections 5.1 and 5.2.

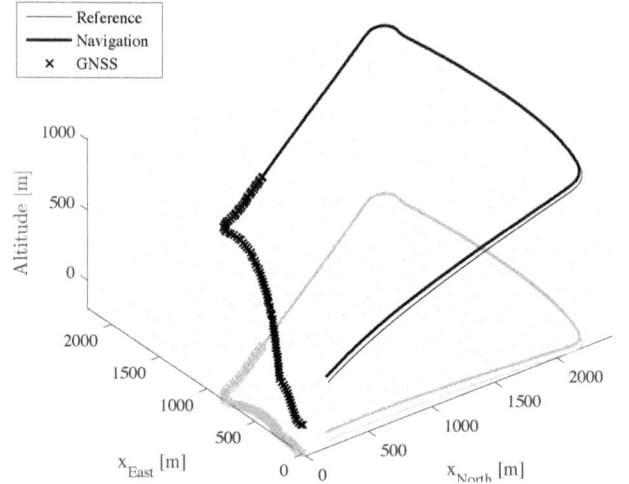

Figure 3: Reference trajectory and the solution from a sample rum with GNSS signals available during first $100s$ only

5.1 Navigation States

Figure 4 shows comparison of RMS of position and yaw errors for all the 100 runs between proposed VDM/INS/GNSS approach and classical INS/GNSS navigation algorithm over the whole interval. While the RMS of position error is $11.7km$ for classical INS coasting after 5 minutes of autonomous navigation during GNSS outage, this is reduced to less than $110m$ with the proposed VDM/INS/GNSS integration under exactly the same situations, which means an improvement of two orders of magnitude in position accuracy. The attitude determination also shows an improvement of 1 to 2 orders magnitude, which will be detailed shortly. It is worth mentioning here that the improved performance of the proposed filter is noticeable also during the availability of GNSS in estimating yaw.

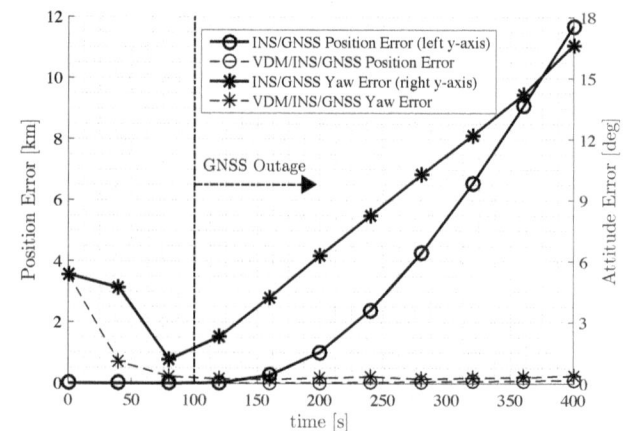

Figure 4: Comparison between INS/GNSS and VDM/INS/GNSS: RMS of position and yaw errors from 100 Monte Carlo runs

The position error for all the 100 Monte Carlo runs is presented in Figure 5. The graphs show how the error grows as time passes after GNSS outage starts, and how the overall behavior is similar for individual runs. An empirical RMS is calculated from these individual errors and plotted against the predicted confidence level (1σ). It is important to notice that how closely (and slightly conservatively) the error is predicted within the filter, which reveals the correctness of filter setup.

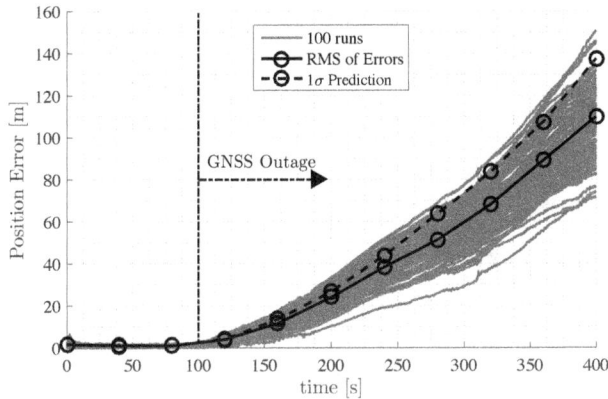

Figure 5: Position errors for all the 100 Monte Carlo runs

Figure 6 depicts the empirical RMS of attitude errors for all the 100 runs, with associated predicted values of confidence (1σ). The results are very satisfactory in terms of preserved navigation accuracy, with the RMS of error to be only $0.0072°$ for roll, $0.020°$ for pitch, and $0.38°$ for yaw after 5 minutes of autonomous navigation during GNSS outage. In comparison, the classical INS coasting would result in errors of $2.6°$ for roll, $1.5°$ for pitch, and $16.6°$ for yaw under exactly the same situations.

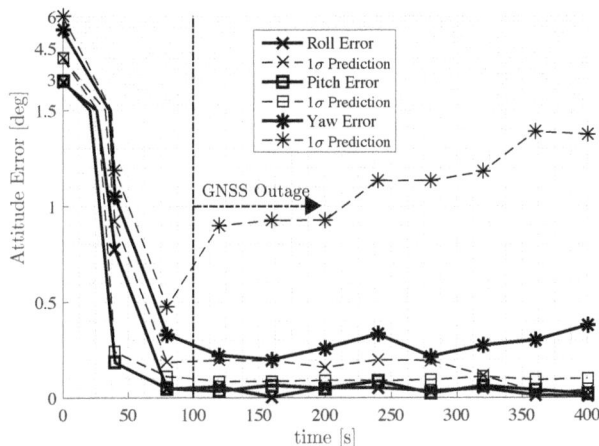

Figure 6: RMS of position and attitude errors from 100 Monte Carlo runs

5.2 Auxiliary States

Successful estimation of auxiliary states is a key enabler of navigation improvement within the filter. The results are briefly presented in this section, with all the reported values calculated as an RMS of the values for all the 100 Monte-Carlo runs. More detailed results on auxiliary states estimation and associated plots are presented in (Khaghani and Skaloud, 2016).

The time correlated part of the IMU error gets estimated quickly during the first tens of seconds of navigation and remains rather unchanged afterwards, even during the GNSS outage period. The estimation error has an average of 6.0% for the three accelerometers and an average of 14.8% for the three gyroscopes at the end of the whole navigation period.

The mean error in estimation of VDM parameters shows a sharp decrease from the initial value of 10% to 6% during the first 40

seconds with GNSS available, which is followed by a slowly decreasing trend until the end. The reason behind the second regime is the correlation between some parameters within the set. In such situations, the groups of parameters are estimated rather than individual parameters, and those individual errors contribute to increasing the mean error for the whole set.

Finally, the wind velocity is estimated very well during GNSS availability period, reaching an error of only 3.9% for wind speed after 100 seconds. The estimation error starts to grow when GNSS outage begins. However, the rate of this growth is well controlled, and the error is still below 9.6% after 5 minutes of GNSS outage.

6. CONCLUSION

In this work, a novel method was presented to perform autonomous navigation and sensor integration for unmanned aerial vehicles. The key concept of this method is employing vehicle dynamic model in navigation system. Unlike the traditional method of kinematic modeling in which IMU serves as navigation process model within navigation filter, here the navigation filter features dynamic model of UAV as navigation process model whose output gets updated using IMU measurements. GNSS measurements can also be used within the filter whenever they are available, as well as other measurements such as those from a barometric altimeter.

In addition to navigation states and error terms related to inertial sensors, UAV dynamic model parameters and wind velocity components can also be estimated within the filter. This is all possible with no extra sensors. The designed filter is thus polyvalent as it can accommodate changes in the platform and/or environmental conditions, such as the wind acting on the platform.

The scenario of GNSS signal reception disruption (e.g., due to electromagnetic interference) is a situation where this method can be most useful. In case of GNSS outage of 5 minutes, the presented Monte Carlo simulations show an improvement of more than 2 orders of magnitudes in position accuracy compared to inertial coasting, for random initial errors of 10% in UAV dynamic model parameters. Such gain of navigation autonomy is internal to UAV, therefore very suitable for limited or zero-visibility operations. Attitude estimation also shows a major improvement, reducing the error on yaw from $16.61°$ to only $0.38°$ for example. The parameters of the dynamic model are calibrated in flight and such calibration is possible even if GNSS signal reception is not available. The time-correlated errors in accelerometers and gyroscopes are also estimated within the filter as a random walk process, so are the wind velocity components.

Further development of current work will include studies on proposed navigation system in real scenarios. Technical and perhaps scientific challenges can be expected in real implementation. Proper time stamping of all sensor observations and scaling the control input signals are examples of technical challenges. On the scientific part, the main challenges will probably be related to unmodeled dynamics and disturbances, and the inclusion of additional effects, such as sensor misalignments, actuator dynamics, UAV body elasticity, and asymmetric mass distribution.

ACKNOWLEDGEMENTS

This paper is a shortened version of (Khaghani and Skaloud, 2016). The authors thank the Royal Netherlands Meteorological Institute (KNMI) and ALterra research institute for permission to use real wind data in this research, originally provided under the consortium agreement on Cabauw Experimental Site for Atmospheric Research (CESAR).

REFERENCES

Angelino, C. V., Baraniello, V. R. and Cicala, L., 2012. UAV position and attitude estimation using IMU, GNSS and camera. In: 15th International Conference on Information Fusion (FUSION), IEEE, pp. 735–742.

Bryson, M. and Sukkarieh, S., 2004. Vehicle model aided inertial navigation for a UAV using low-cost sensors. In: Proceedings of the Australasian Conference on Robotics and Automation, Citeseer.

Bryson, M. and Sukkarieh, S., 2015. UAV Localization Using Inertial Sensors and Satellite Positioning Systems. In: Handbook of Unmanned Aerial Vehicles, Springer Netherlands, pp. 433–460.

Cook, M. V., 2013. Flight dynamics principles a linear systems approach to aircraft stability and control. Butterworth-Heinemann, Waltham, MA.

Crocoll, P. and Trommer, G. F., 2014. Quadrotor Inertial Navigation Aided by a Vehicle Dynamics Model with In-Flight Parameter Estimation. In: Proceedings of the 27th International Technical Meeting of The Satellite Division of the Institute of Navigation (ION GNSS+ 2014), Tampa, Florida, pp. 1784–1795.

Crocoll, P., Seibold, J., Scholz, G. and Trommer, G. F., 2014. Model-Aided Navigation for a Quadrotor Helicopter: A Novel Navigation System and First Experimental Results. Navigation 61(4), pp. 253–271.

Dadkhah, N., Mettler, B. and Gebre-Egziabher, D., 2008. A model-aided ahrs for micro aerial vehicle application. In: Proceedings of the 21st the International Technical Meeting of the Satellite Division of The Institute of Navigation ION GNSS 2008, pp. 545–553.

Ducard, G. J., 2009. Fault-tolerant flight control and guidance systems: Practical methods for small unmanned aerial vehicles. Springer, London.

Gelb, A. (ed.), 1974. Applied optimal estimation. M.I.T. Press, Cambridge, Massachusetts, London.

Groves, P. D., 2008. Principles of GNSS, inertial, and multisensor integrated navigation systems. GNSS technology and applications series, Artech House, Boston.

Khaghani, M. and Skaloud, J., 2016. Autonomous Vehicle Dynamic Model Based Navigation for Small UAVs. NAVIGATION, Journal of the Institute of Navigation (In Print).

KNMI and Alterra, 2012. Data provided by KNMI and Alterra, under the Consortium agreement on Cabauw Experimental Site for Atmospheric Research (CESAR), www.cesar-database.nl.

Koifman, M. and Bar-Itzhack, I. Y., 1999. Inertial navigation system aided by aircraft dynamics. IEEE Transactions on Control Systems Technology 7(4), pp. 487–493.

Lau, T. K., Liu, Y.-H. and Lin, K. W., 2013. Inertial-Based Localization for Unmanned Helicopters Against GNSS Outage. IEEE Transactions on Aerospace and Electronic Systems 49(3), pp. 1932–1949.

Müller, K., Crocoll, P. and Trommer, G., 2015. Wind Estimation for a Quadrotor Helicopter in a Model-Aided Navigation System. In: 22nd Saint Petersburg International Conference on Integrated Navigation Systems.

Noureldin, A., Karamat, T., Eberts, M. and El-Shafie, A., 2009. Performance Enhancement of MEMS-Based INS/GPS Integration for Low-Cost Navigation Applications. IEEE Transactions on Vehicular Technology 58(3), pp. 1077–1096.

Schwarz, K. P. and Wei, M., 1994. Modeling INS/GPS for attitude and gravity applications. Proc. High Precision Navigation 95, pp. 200–18.

Sendobry, A., 2014. Control System Theoretic Approach to Model Based Navigation. Ingenieurwiss. Verlag, Bonn.

Vasconcelos, J., Silvestre, C., Oliveira, P. and Guerreiro, B., 2010. Embedded UAV model and LASER aiding techniques for inertial navigation systems. Control Engineering Practice 18(3), pp. 262–278.

Wang, J., Garratt, M., Lambert, A., Wang, J. J., Han, S. and Sinclair, D., 2008. Integration of GPS/INS/vision sensors to navigate unmanned aerial vehicles. The International Archives of the Photogrammetry, Remote Sensing and Spatial Information Sciences 37, pp. 963–970.

Yun, S., Lee, Y. J. and Sung, S., 2013. IMU/Vision/Lidar integrated navigation system in GNSS denied environments. In: Aerospace Conference, IEEE, pp. 1–10.

16

ROLE OF TIE-POINTS DISTRIBUTION IN AERIAL PHOTOGRAPHY

S. Kerner [a], I. Kaufman [a], Y. Raizman [a]

[a] VisionMap, 19D Habarzel, Tel Aviv, Israel - {slavak, itay, yuri}@visionmap.com

KEY WORDS: Photogrammetry, Image matching, Tie-Points, Feature-based matching, RANSAC, Von Gruber positions

ABSTRACT:

Automatic image matching algorithms, and especially feature-based methods, profoundly changed our understanding and requirements of tie points. The number of tie points has increased by orders of magnitude, yet the notions of accuracy and reliability of tie points remain equally important. The spatial distribution of tie points is less predictable, and is subject only to limited control. Feature-based methods also highlighted a conceptual division of the matching process into two separate stages – feature extraction and feature matching.

In this paper we discuss whether spatial distribution requirements, such as Von Gruber positions, are still relevant to modern matching methods. We argue that forcing such patterns might no longer be required in the feature extraction stage. However, we claim spatial distribution is important in the feature matching stage.

We will focus on terrains that are notorious for difficult matching, such as water bodies, with real data obtained by users of VisionMap's A3 Edge camera and LightSpeed photogrammetric suite.

1. INTRODUCTION

Image matching is a key stage in photogrammetric adjustment. It is a problem of great interest in computer vision. Specific matching algorithms were developed in the context of aerial photography. One of the methods discussed in this paper is specifying tie-points' spatial patterns. Von Gruber positions are the standard method used in photogrammetry.

We start by giving a brief historic overview of image matching, from the manual to the automated era, specifically in aerial photogrammetry.

We proceed to analyze different use-cases, emphasizing the most difficult scenarios for matching.

Finally we discuss tie-points' spatial distribution patterns in modern aerial triangulation, focusing on feature-based matching as the currently dominant method. We argue that forcing such patterns might be no longer required in feature extraction stage. However, we claim spatial distribution is important in the feature matching stage.

2. IMAGE MATCHING OVERVIEW

2.1 Manual Matching

In the manual matching era, identifying tie-points was the most time-consuming process in aerial triangulation. Consequently, minimizing the number of tie-points, while preserving photogrammetric accuracy and stability, was a major challenge. Much research was carried out, and several distribution patterns were advised, with Von Gruber positions being most popular. A typical pattern had several tie-points spread in a gridded manner across image. One of the underlying assumptions was an accurate flight with guaranteed high overlaps, both along and across-track. Another assumption was that most of the image was "matcheable", that is – one could reliably find tie-points throughout the whole image.

2.2 Automated Matching

Starting in the 1990s, the matching process became automated. Various approaches were employed: spatial vs. frequency domain, featured points vs. lines vs. contours, area correlation based vs. feature based, etc.

Extracting tie-points became "cheap", yet it was still desirable to limit the number of tie-points, as the run-time of matching, and subsequent photogrammetric adjustment, was of great importance.

2.2.1 Area-based matching

Probably the most common method during the 2000s was area-based: identifying highly featured points in a source image (e.g. (Shi, 1994)), extracting image patches around the points, and recognizing these patches in the target image, most often using NCC (e.g. (Lewis, 1995)). It was common to force source points to lie on a grid, or to be close to the grid vertices. An image pair typically yielded dozens of tie-points.

One of the limitations of this approach was performance: if relative transformation between the source and target image was not well estimated, one would have to choose between a long processing and pyramid approach, running the risk of mismatch. Additionally, while being stable over radiometric changes, this approach wasn't robust enough for scale and rotation.

2.2.2 Feature-based matching

Prior to the 2000s, aerial photography was one of the driving forces behind computer vision. During the 2000s, an abundance of digital cameras (and later cellular phones) and overall computerization led computer vision to focus on general image matching problems. Much progress was made over the last decade, particularly in feature-based matching, as this approach proved to be most robust and efficient for the generic setting. Several new matching methods were proposed (e.g. (Bay, 2006) and (Lowe, 2004))

Also during the 2000s, hardware improvements, the continued growth of clusters, and particularly the introduction of

(GP)GPU led to a dramatic runtime reduction. As runtime decreases rapidly, storage/network IO becomes the bottleneck. Consequently, it is highly desirable to read each image only once. Feature-matching proves suitable: at stage 1, each image is read, and its featured points are extracted alongside descriptors. These are stored in DB, typically requiring x10-100 less storage than the original imagery. At stage 2, featured points of each overlapping image pair are matched. A common practice is to fit a transformation (e.g. homography) based on the matches (typically using RANSAC (Fischler, 1981) or RANSAC-like majority voting scheme), and remove matches that don't agree with the transformation. This encouraged developing numerous variations and improvements on RANSAC (Torr, 2000), (Torr, 2002), (Chum, 2004), (Feng, 2003), (Choi, 2008).

Hundreds of tie-points become the norm. Tie-point abundance also leads to improved robustness and accuracy of following photogrammetric adjustment.

3. IMAGE MATCHING IN AERIAL PHOTOGRAPHY

3.1 Problem is simple

We argue that image matching in aerial photography is usually an easier problem than image matching in the general settings of computer vision. Some of the reasons:
- Images are captured in a controlled manner, and one knows which pairs of images actually overlap,
- Images are taken with significant overlaps,
- All objects are of similar depth,
- Images come from the same camera,
- Images are large, high-quality,
- Images are taken in good light conditions,
- Scene is large, diverse,
- Most of the scene is static.

On the other hand, some of image matching challenges in the context of aerial photography are well met by robust feature descriptors:
- Invariance to rotation,
- Invariance to scale,
- Invariance to lightning conditions, exposure time,
- Little or no prior knowledge of transformation between images (that's usually the case when IMU is not present).

As a result, feature-based image matching in aerial photography is now a very accurate and reliable procedure, most of the time (fig. 1).

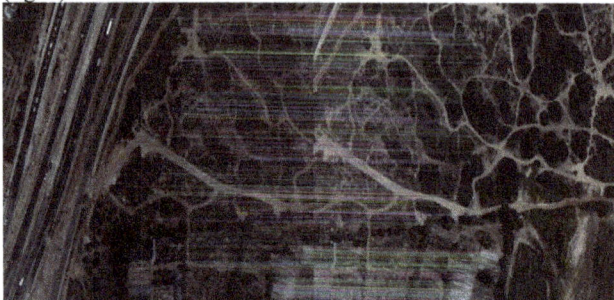

Figure 1: Typical problem; numerous correct tie-points

3.2 Problem is complicated

3.2.1 Clouds
While cloudy images have been somewhat difficult to match using area-based methods, this is less the case when feature-based matching is used. Assuming some parts of the image are not covered by clouds (otherwise the image is of no use anyway) - these are easily identified and matched by any of the common feature-based methods (fig. 2).

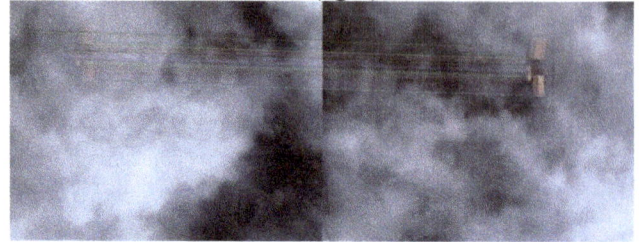

Figure 2: Matching clouded images

3.2.2 Repetitive areas
Areas of repetitive pattern are difficult to match, as the correct match area is almost identical to that of the outlier. That's sometimes the case with perfectly gridded agricultural areas, large solar fields etc.

Using a ratio test (e.g. (Lowe, 2004)) is critical; it allows ignoring featured points at high risk of being outliers, and choosing less-featured points that are unique. Any prior estimation of transformation between images is of great use, for example – based on GPS, gyro accelerometers or IMU.

Based on our experience, the absolute majority of repetitive images are reliably and accurately matched (fig. 3).

Figure 3: Matching images with repetitive areas

3.2.3 Desert
While desert was always notorious for difficult matching, that's not necessarily true with modern matching. Almost all desert images we encounter have grains of featureness scattered across the image (stones, dunes, plants etc.). Most importantly – these features are stable during a flight.

3.3 Problem is difficult

In our experience, water bodies with small objects like small islands or rocks, stones or natural or manmade objects surrounded by water are the biggest challenge in aerial photography image matching. It is necessary that every image containing any information (bridge, pierce, rocks, islet etc.) will be robustly matched (fig. 4A, 5A). A large river or strait often breaks a project into two disconnected areas, leading to inconsistent adjustments of both areas, or to the need for additional GCPs. It is therefore crucial to match all images that allow connecting both components. Also, often images with ground features or structures surrounded by water might be lost.

Unlike desert, images of water bodies contain numerous pixels of high featureness (such as reflections, waves, sea foam, ripple etc.), that move over time. These are bound to become outliers (fig. 4b, 5b). That risk is especially high between consecutive images: with a typical difference of 1 sec, a wave might move these points consistently, leading a RANSAC-like scheme to select these as compatible matches. In practice, this might lead to a mismatch of a few pixels.

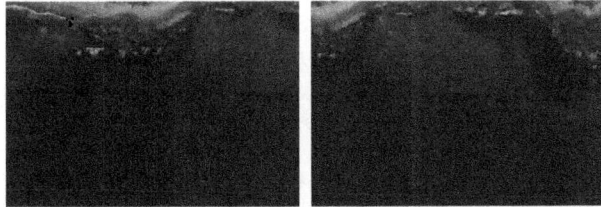

Figure 4a: Water with small objects - source (l.) and target (r.).

Figure 4c: Candidate matches

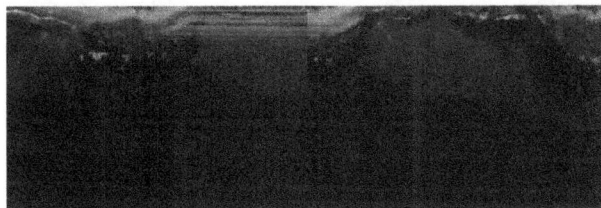

Figure 4c: Matches filtered by RANSAC-fitted homography

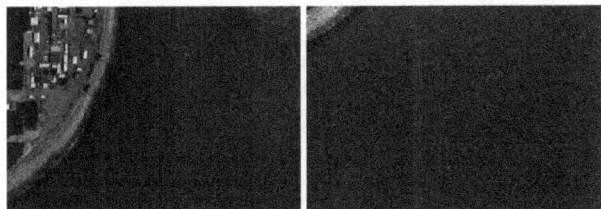

Figure 5a: Water with small objects: source (l.) and. target (r.).

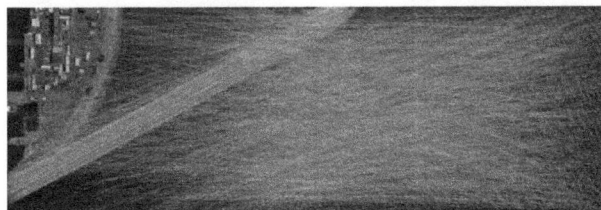

Figure 5b: Candidate matches

Figure 5c: Matches filtered by RANSAC-fitted homography

4. TIE-POINTS' SPATIAL DISTRIBUTION

4.1 Forcing spatial constraints in the feature extraction stage

As claimed above, the challenging scenarios occur when there are few small matchable objects in image. We argue that in these cases, forcing tie-points to behave according to some spatial distribution is detrimental.

A typical scheme to which most spatial constraints can be reduced is:
- Split the image into N regions,
- Select the best M features in each region.

In the case in question, it often occurs that all features fall within one region. Spatial constraints might produce two negative effects:
- Selecting only M points from the whole image – might result in too few points,
- Forcing or encouraging the extraction of featured points in regions where there are none – increases the risk of incorrect matches.

4.2 Forcing spatial constraints in the feature matching stage

We argue that forcing spatial constraints may be beneficial at the stage of fitting transformation (RANSAC-like step). Spatial distribution of tie-points becomes critical when most tie-points in the image are clustered in a spatially degenerate configuration: either in a small part of the image, or in a linear alignment. In that case, fitted transformation is unstable and unreliable. As a result, correct tie-points outside the cluster will be filtered out.

We suggest giving preference, at the RANSAC-like step, to transformation agreeing on spatially scattered points. There are two steps in RANSAC causing the above phenomenon:
- The fitting subset is small (often 4 candidates), and selected randomly. As a result, when one cluster contains the vast majority of candidate matches in an image, most fitting subsets will be based on that cluster only; therefore most transformations will be degenerate.
- The voting step gives identical weight to all candidates. One could say RANSAC prefers quantity over quality. RANSAC might choose transformation supported by all points in a cluster, rather than transformation supported by part of cluster, and other points outside the cluster.

We tried several modifications to the RANSAC step:
- Apply RANSAC iteratively, with converging tolerance threshold. That allows keeping, at least in the first iterations, tie-points outside the cluster, even when fitted transformation is degenerate.
- Encourage RANSAC to construct a fitting subset from spatially distributed points. For example, after three points are selected randomly, the fourth point is required to be distant.
- Recognize the degenerate cluster that RANSAC produced, and try to re-match points outside the cluster. The cluster is modelled as ellipse, and is

declared degenerate if its minor radius is below threshold.

- Run RANSAC on each cluster separately, and compare estimated transformations. The K-means method was used to separate clusters. Majority transformation was chosen.

- In the RANSAC voting step, give larger weight to points not coming from a dominant cluster (if such exists).

In all cases of clustered tie-points, where standard RANSAC filtered out tie-points outside the cluster – the spatial RANSAC variations we tested produced better spread of tie-points (as in fig. 6). However, each method had partial success; none was decisively best.

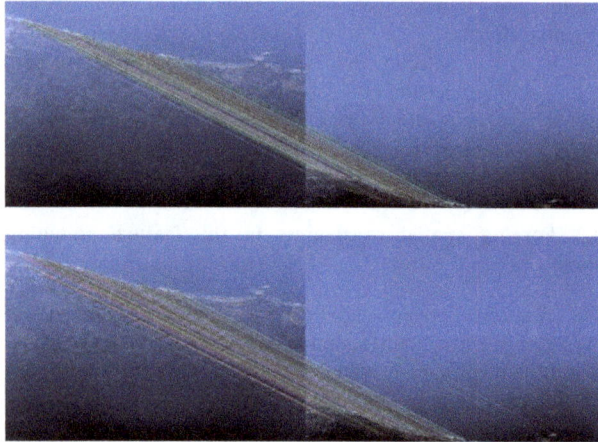

Figure 6: Matches using standard RANSAC (top) vs. matches using variation on RANSAC that prefers even spatial distribution (bottom). Note the linear cluster of tie-points in top, and additional tie-points outside the cluster bottom

5. CONCLUSIONS AND FUTURE WORK

Over the last 20 years, algorithmic and hardware advances made image matching in aerial photography easy, robust, reliable and fast. The biggest challenge is posed by images with few true features (often coming in clusters), and many outliers. A typical scenario is a water body with small objects.

Feature-based matching, currently the dominant method, naturally assumes no pattern for tie-points' spatial distribution. Furthermore, we believe forcing tie-points' spatial distribution patterns at the extraction stage is no longer required. However, we believe encouraging spatial distribution patterns at matching stage may be beneficial in the challenging scenarios, preventing degenerate configurations.

More work is needed to develop a robust algorithm, forcing stable spatial distribution in the matching stage. We believe the best direction is to develop a variation of RANSAC preferring a consensus set spread across the image.

REFERENCES

Bay, H., Tuytelaars, T., and Van Gool, L., 2006. Surf: Speeded up robust features. In European Conference on Computer Vision

Choi, S. and Kim, J.H. 2008. Robust regression to varying data distribution and its application to landmark-based localization. In Proceedings of the IEEE Conference on Systems, Man, and Cybernetics

Chum, O., Matas, J. and Obdrzalek, S. 2004. Enhancing RANSAC by generalized model optimization. In Proceedings of the Asian Conference on Computer Vision (ACCV)

Feng, C.L. and Hung, Y.S. 2003. A robust method for estimating the fundamental matrix. In Proceedings of the 7th Digital Image Computing: Techniques and Applications, number 633–642,

Fischler, M.A., and Bolles, R. C., 1981. Random Sample Consensus: A paradigm for model fitting with applications to image analysis and automated cartography. Communications of the ACM, 24(6):381–395

Lewis, J.P. 1995. "Fast normalized cross-correlation," Tech. Rep., Industrial Light & Magic

Lowe, D. G. 2004. Distinctive image features from scale-invariant keypoints. International Journal of Computer Vision, 60(2):91–110

Shi, J. and Tomasi, C., 1994. "Good Features to Track", Proc. IEEE Conf. on Computer Vision and Pattern Recognition,

Torr, P.H.S. and Zisserman, A. 2000. MLESAC: A new robust estimator with application to estimating image geometry. Computer Vision and Image Understanding, 78:138–156

Torr, P.H.S., 2002. Bayesian model estimation and selection for epipolar geometry and generic manifold fitting. International Journal of Computer Vision, 50(1):35–61

17

COMPARING NATIONAL DIFFERENCES IN WHAT PEOPLE PERCEIVE TO BE *THERE*: MAPPING VARIATIONS IN CROWD SOURCED LAND COVER

A. Comber [a], P. Mooney [b], R.S. Purves [c], D. Rocchini [d], A. Walz [e]

[a] School of Geography, University of Leeds, Leeds, LS2 9JT, UK, a.comber@leeds.ac.uk
[b] Department of Computer Science, National University of Ireland Maynooth, Ireland Peter.Mooney@nuim.ie
[c] Department of Geography, University of Zurich, 8057 Zurich, Switzerland ross.purves@geo.uzh.ch
[d] Fondazione Edmund Mach, 38010 S. Michele all'Adige, Italy duccio.rocchini@fmach.it
[e] Potsdam Institute for Climate Impact Research, 14412, Potsdam, Germany ariane.walz@pik-potsdam.de

Commission VI, WG VI/4

KEY WORDS: Semantics, Geo-Wiki, VGI, Land cover, Citizen Science

ABSTRACT:

This paper describes a simple comparison of the distributions of land cover features identified from volunteered data contributed by different social groups – in this case comparing two groups of Geo-Wiki campaigns. Understanding the impacts on analyses of citizen science data contributed by different groups is critical to ensure robust scientific outputs and to fully realise the potential benefits to formal scientific research. It is well known that different people, with different backgrounds and subject to different cultural factors, hold varying landscape conceptualisations. This paper analyses volunteered geographical information on land cover to generate land cover maps. It uses a geographically weighted approach to generate land cover mappings. The mappings generated by different groups (in this case a from a specific unnamed country) are compared and the results show how the predicted land cover distributions vary, with large differences in some classes (e.g. Barren land, Shrubland, Wetland) and little difference in others (e.g. Tree cover). This suggests that for some landscape features cultural and national differences matter when it comes to using crowdsourced data in formal scientific analyses and highlights the potential problems of *not* considering contributor backgrounds in citizen science. This is important because such data re now routinely being used to develop global land cover data, to generate uncertainty estimates of existing global land cover products and to generate global forest inventories. These in turn are being suggested as suitable inputs to such things as global climate models. A number of critical research directions arising from these findings are discussed.

1. INTRODUCTION

There is much interest in using crowd-sourced data, data generated through citizen science activities or what Goodchild (2007) referred to as *volunteered geographical information* to support formal scientific endeavours. As a result the scientific community has explored different opportunities arising from crowd-sourced data collection and analysis (Cohn, 2008; Coleman, 2010; Haklay et al, 2010; Hand, 2010) and there has been an explosion of applications underpinned by crowd-sourced data in nearly all areas of scientific investigation: from astronomy (Raddick et al, 2010) to zoology (Silvertown, 2009). In the domain of land cover / land use, the European Commission has funded a number of projects considering how such data may be used to help manage crises and emergencies[1], to develop Citizen Observatories for Land Cover and Land Use[2] and to monitor deforestation[3]. The reasons for these initiatives in the context of land cover are various but include uncertainties over future funding of remote sensing in Europe[4] and cost benefits (for example LUCAS sampling is expensive, costing €6.42m)[5]. As a result a number of crowd-sourced land cover data collection systems have been initiated and perhaps the best known is the Geo-Wiki system developed by Perger et al (2012) at IIASA, Austria although others exist (Pistorius & Poona, 2014; Vaz & Arsanjani 2015; Kinley, 2013). Geo-Wiki has been used for a number of campaigns and has seen some system development in order to increase data collection and contribution.

The rise of activities such as Geo-Wiki within mainstream scientific investigation provides a critical research context to the research presented in this paper. Geo-Wiki collects land cover data from volunteers and a number of applications have been developed to, for example, assess the quality of existing land cover products (Fritz et al, 2009), determine their uncertainties (Fritz et al., 2011) and generate hybrid global land cover maps (See et al., 2015). However, one of the critical issues associated with the use of citizen data relates to its quality (Foody et al, 2013; Comber et al, 2013a). One key problem is that different contributors or volunteers may have different underlying conceptualisations of the features that observed and thus recorded in crowd-sourced data. In the context of land cover, variation in concepts result in different interpretations of the boundaries between classes and so in the land cover data that recorded. This issue is illustrated by ethnophysiography (Derungs et al, 2013; Mark and Turk 2003) and by linguistic and cultural factors (Smith and Mark, 1998) and is well known in the context of *formal* land cover creation (for example, Comber et al, 2005; Comber et al, 2008). It is important to note that some differences may reflect real variation in the way landscapes are perceived, and there may be no "right" answer in terms of land cover or land use, without consideration of context. In formal land cover creation, divergent conceptualisations are mitigated by the inclusion of experimental designs: data collection protocols, training, sampling designs, QA procedures etc. These ensure that the inferences from any data analysis are statically robust.

This paper evaluates the potential impacts of using volunteered data by comparing the land cover data contributed by two groups, one formed of volunteers solely from one country,

---

1 http://projects.jrc.ec.europa.eu/jpb_public/act/publicexportworkprogramme.html?actId=453
2 https://ec.europa.eu/research/participants/portal/desktop/en/opportunities/h2020/topics/2196-sc5-17-2015.html
3 http://www.gmes-masters.com/sites/default/files/media/inline/gmes_results_booklet_2012.pdf
4 http://news.eoportal.org/policy/121001_pol1.html
5 http://epp.eurostat.ec.europa.eu/cache/ITY_SDDS/en/lan_esms.htm

named *Gondor* to avoid making inferences based on national stereotypes, the other containing all other nationalities.

2. DATA AND CASE STUDY

The research uses data collected through the Geo-Wiki project. Geo-Wiki is an open, web and app interface. As part of the registration process, volunteers are asked to describe their experience and where they are from. Once registered, volunteers contribute to different campaigns in which they describe the land cover at a series of randomly selected locations with Google Earth providing background imagery. In this research, data from two of these, capturing land cover using the same 10 classes were combined and the data for North and South America selected. The distributions of the data are summarised in Table 1.

Class	Group	
	Others	*Gondor*
Tree	3100	4144
Shrub	673	1860
Grass	1607	1554
Crop	1305	1176
Wetland	543	245
Urban	91	107
Snow	368	256
Barren	856	593
Water	555	364

Table 1. The land cover data collected by different volunteers

It is evident that despite a random sample of locations there are large differences between Gondor and Others especially in the number of images classified as *Shrub* and *Barren* land cover. The distribution of the data points is shown in Figure 1.

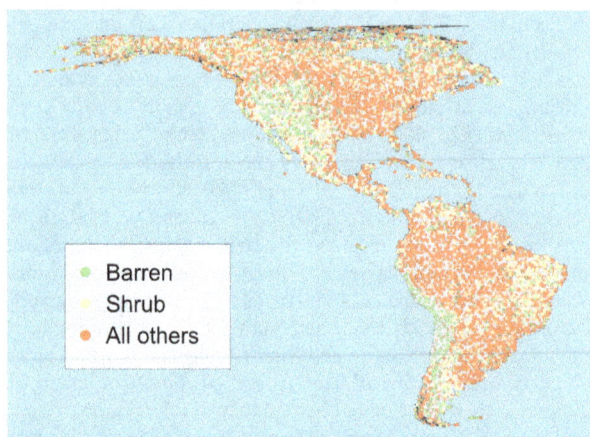

Figure 1. The distribution of the data

A Kernel Density Estimation (KDE) was used to generate surfaces for visualising the data trends and to provide a visual reference for later analyses. The KDEs in Figure 2 describe the probability of a class being present at each location.

In order to examine the potential impacts of using data operationally that was contributed by volunteers from different countries, with no consideration of the number of volunteers, their experience, background and training, the data were separated into 2 subsets each with groups. The first contained *Professional* and *Non-Professional*, the second contained contributed volunteers from *Gondor* and *Others* (i.e. data from volunteers from all other countries).

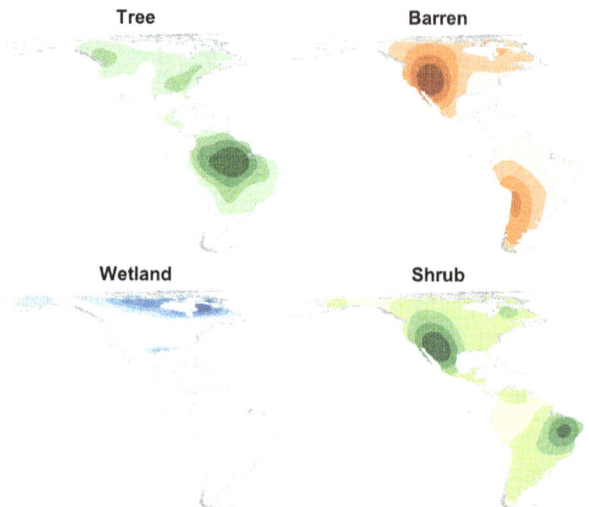

Figure 2. The KDE surfaces for 4 classes

To compare the impacts of different groups, geographically weighted averages were computed for each class at each point on 50km grid under a 50km kernel. This generated a value in the range [0, 1] at each of the 80,073 grid locations for each class, and the class with the greatest value was assigned. This approach is a smoothing approach similar to that used by Comber (2013a) to determine fuzzy memberships distributions.

3. RESULTS

Generating KDEs of the data contributed by different groups provides a convenient way to summarise the potential impacts. Figure 3 compares the surfaces generated for 2 land covers of data contributed volunteers from *Gondor* and data contributed by *Others*. Evidently different features are mapped in different locations (*Wetland* in Louisiana) and to different degrees (*Barren* land in Chile) by different groups.

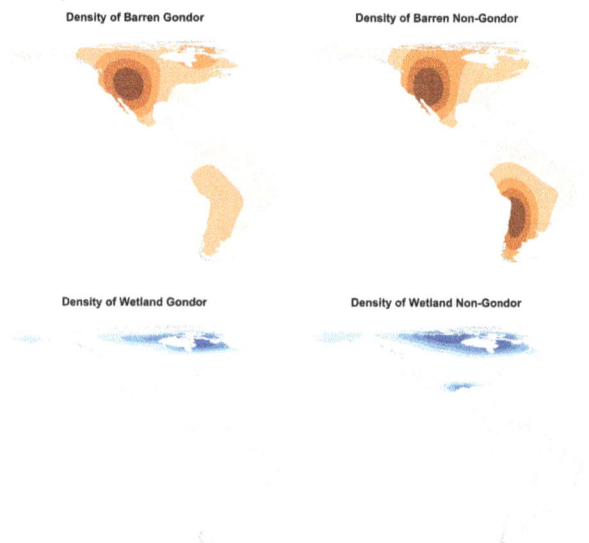

Figure 3. KDEs from data contributed by different groups

The Geo-Wiki data was analysed in the following way. At each location in a 50km grid and for each class, a local probability measure was generated, describing the probability of that class being present at that location. The local probability was calculated using a geographically weighted regression model that analysed the number of data points of that class under a 50km kernel, weighting each point's contribution to the model by its distance to the kernel centre. This generates probability surfaces for each class and at each location, the class with the greatest probability was assigned. Figure 4 shows the land cover map generated in this way using data from all contributors. This land cover mapping is generated in the same way the operational data described by See et al (2015) and Schepaschenko et al (2015). Figure 4 provides a baseline against which to compare the impacts of only using data from specific groups (Figure 5 and Figure 6).

All Data

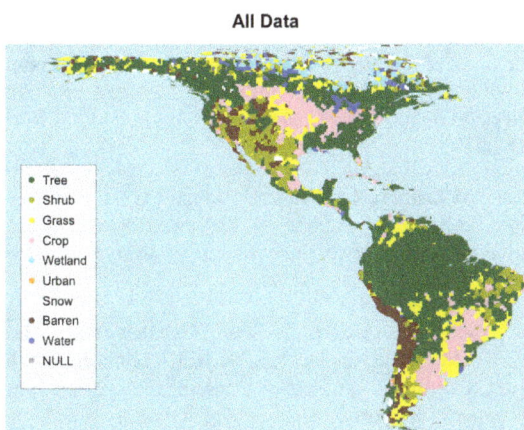

Figure 4. The land cover data generated by all contributors

Comparing Figures 4, 5, and 6 there are clear differences between the groups, although the maps generated from the data contributed by *Others* has much smaller differences to the map in Figure 4. Interesting and potentially significant differences are the subtle but important differences in the distributions of the *Wetland* class, *Shrub* and *Barren* and *Crop* and *Grass*.

Gondor

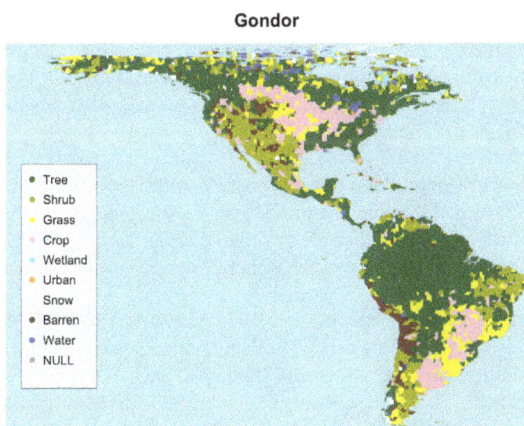

Figure 5. The land cover data generated by contributors from Gondor

Non-Gondor

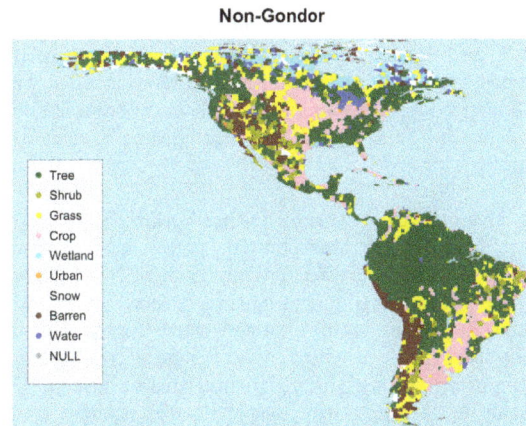

Figure 6. The land cover data generated by people from other countries

4. DISCUSSION

The analysis in this paper presents an initial analysis comparing differences between data contributed by 2 groups of volunteers, 21 from *Gondor* and 70 people of other nationalities or whose background was unknown. These numbers are very typical for a Geo-Wiki campaign. These preliminary results for a North and South American case study suggest that the well known differences in *what is perceived to be there* by different groups **matters**, even with a very simple 10 class nomenclature. This has profound implications for a number of on-going research activities that are using crowdsourced data to generate hybrid global land cover datasets from existing (but uncertain) global datasets and from crowdsourced data (See et al., 2015; Schepaschenko et al., 2015). These researches have not considered the impacts of contributor cultural or national background but their datasets they are creating are being suggested as suitable and improved inputs to global climate change models.

The methods used to develop the land cover maps apply a simplified approach. For example, a bandwidth of 50km was chosen rather than the optimal bandwidth being determined, as recommended in the GWR literature, and no comparisons with formal data were made. This objective of this work was to evaluate the potential impacts of contributors with different backgrounds. The number of contributors does allow for the possibility that the any observed differences (for example comparing Figure 5 and Figure 6) may be statistical noise / variation rather than representing any underlying Additionally, the land cover was allocated to the single class with the greatest probability at each location. In many locations many classes have similar levels of (high) probability suggesting the need for spatially distributed measures of uncertainty in the class allocation.

However, the geographically weighted approach to analysing Geo-Wiki data is at the core of many current activities which are using the data to construct land cover maps for operational usage (See et al., 2015; Schepaschenko et al., 2015). The research presented in this paper suggests that, as well thinking about developing measures of crowd-sourced data quality (Comber, 2013a; Foody et al 2014), there is a critical need to consider some of the 'COSIT' considerations related to spatial cognition[6] and who the volunteers are, where they come from,

---

[6] http://www.cosit.info

what their background (cultural and professional) is and so on. There is also a need to consider how volunteers are recruited and whether that can be done in a more representative way or even a targeted way, where for example, data contributed by an individual who fails to meet some criteria are excluded from analysis.

There are a number of areas for further work. These include the need to compare the interaction of professional experience, identified as an important factor (Comber et al, 2013b; Foody et al, 2014), with nationality and the degree to which differences between the groups are ameliorated when this is considered. Second, to explore whether the differences and similarities between groups persists for different classes in different areas: this study focused on a particular study area where one of the groups may have greater knowledge and experience. Third, to develop methods to consider how to integrate citizen science data with formal data and to develop quantitative measures related to citizen semantics. For example, there may be large difference in the way that different groups of citizens resolve any ambiguity they may have around the labelling of features – consider the example of Forest illustrated by Comber et al (2005). The Geo-Wiki volunteers attached measures of their confidence in the class labels that were attached to each data point. This may capture their uncertainty for example around their understanding of the affordances associated with the different land cover types. Fourth, recent activity in citizen science is using data from an increasingly wide range of data sources. Features may be labelled in different and novel ways depending on how the crowd-sourced data are captured or mined. Fifth, there is a need to consider the impact of digital and other divides on the nature of the information that is contributed and the potential for biases towards western, developed populations in recording and describing the world and the formalisation of that set of cultural perceptions. This is also influenced by the nature of the technologies used to capture and share such information. On-going work will consider these issues.

ACKNOWLEDGEMENTS

Thanks to Steffen Fritz and his team at IIASA for the use of the Geo-Wiki data. The data and the code will be available to researchers on request.

REFERENCES

Cohn, J.P. (2008). Citizen science: can volunteers do real research? *BioScience* 58(3):192–197. doi:10.1641/B580303 .

Coleman, D. (2010). The potential and early limitations of volunteered geographic information. *Geomatica* 64(2): 27–39.

Comber, A., See, L., Fritz, S., Van der Velde, M., Perger, C., Foody, G.M. (2013a). Using control data to determine the reliability of volunteered geographic information about land cover. *International Journal of Applied Earth Observation and Geoinformation*, 23: 37–48

Comber, A., Brunsdon, C., See, L., Fritz, S. and McCallum, I. (2013b). Comparing expert and non-expert conceptualisations of the land: an analysis of crowdsourced land cover data. *Lecture Notes in Computer Science: Spatial Information Theory*, 8116: 243-260, doi: 10.1007/978-3-319-01790-7_14

Comber, A.J., Fisher, P.F. and Wadsworth, R.A., (2008). Semantics, Metadata, Geographical Information and Users. Editorial *Transactions in GIS,* 12(3): 287–291

Comber, A.J., Fisher, P.F., Wadsworth, R.A., (2005). What is land cover? *Environment and Planning B: Planning and Design*, 32:199-209.

Derungs, C, Wartmann, F, Purves, R S, and Mark, D M, (2013). The meanings of the generic parts of toponyms: use and limitations of gazetteers in studies of landscape terms. *Spatial Information Theory* (pp. 261-278).

Foody, G. M., See, L., Fritz, S., Van der Velde, M., Perger, C., Schill, C., & Boyd, D. S. (2013). Assessing the accuracy of volunteered geographic information arising from multiple contributors to an internet based collaborative project. *Transactions in GIS, 17*(6), 847-860.

Foody, G. M., See, L., Fritz, S., Van der Velde, M., Perger, C., Schill, C., ... & Comber, A. (2014). Accurate attribute mapping from volunteered geographic information: issues of volunteer quantity and quality. *The Cartographic Journal*, 1743277413Y-0000000070.

Fritz, S., McCallum, I., Schill, C., Perger, C., Grillmayer, R., Achard, F., ... & Obersteiner, M. (2009). Geo-Wiki. Org: The use of crowdsourcing to improve global land cover. *Remote Sensing, 1*(3), 345-354.

Fritz, S., See, L., McCallum, I., Schill, C., Obersteiner, M., Van der Velde, M., ... & Achard, F. (2011). Highlighting continued uncertainty in global land cover maps for the user community. *Environmental Research Letters, 6*(4), 044005.

Goodchild, M.F. (2007). Citizens as sensors: the world of volunteered geography. *Geojournal* 69: 211-221.

Haklay, M. (2010). How good is volunteered geographical information? A comparative study of openstreetmap and ordnance survey datasets. *Environment Planning B*, 37(4):682–703

Hand, E. (2010). Citizen science: people power. *Nature* 466(7307):685–687.

Kinley, L. (2013). Towards the use of Citizen Sensor Information as an Ancillary Tool for the Thematic Classification of Ecological Phenomena. *Proceedings of the 2nd AGILE (Association of Geographic Information Laboratories for Europe) PhD School 2013.*

Mark, DM and Turk, AG (2003). Landscape categories in yindjibarndi: Ontology, environment, and language. In: Kuhn, W., Worboys, M.F., Timpf, S. (eds.) *COSIT 2003*. LNCS, vol. 2825,pp. 28–45. Springer, Heidelberg

Perger C, Fritz S, See L, Schill C, Van der Velde M, et al. (2012). A campaign to collect volunteered geographic Information on land cover and human impact. In: Jekel T, Car A, Strobl J, Griesebner G, editors. GI_Forum 2012: Geovisualisation, Society and Learning. Berlin / Offenbach: Herbert Wichmann Verlag. pp. 83–91.

Pistorius, T., & Poona, N. (2014). Accuracy assessment of game-based crowdsourced land-use/land cover image classification. In *Geoscience and Remote Sensing Symposium (IGARSS), 2014 IEEE International* (pp. 4780-4783). IEEE.

Raddick, M.J, Bracey G, Gay P L, Lintott C J, Murray P, Schawinski K, Szalay AS & Vandenberg, J. (2010). Galaxy zoo: Exploring the motivations of citizen science volunteers. *Astronomy Education Review*, *9*(1), 010103.

Schepaschenko, D., See, L., Lesiv, M., McCallum, I., Fritz, S., Salk, C., ... & Ontikov, P. (2015). Development of a global hybrid forest mask through the synergy of remote sensing, crowdsourcing and FAO statistics. *Remote Sensing of Environment*, *162*, 208-220.

See, L., Schepaschenko, D., Lesiv, M., McCallum, I., Fritz, S., Comber, A. Perger C, Schill C, Zhao Y, Maus V, Siraj MA, Albrecht F, Cipriani A, Vakolyuk M, Garcia A, Rabia AH, Singha K, Marcarini AA, Kattenborn T, Hazarika R, Schepaschenko M, van der Velde M, Kraxner F, Obersteiner, M (2015). Building a hybrid land cover map with crowdsourcing and geographically weighted regression. *ISPRS Journal of Photogrammetry and Remote Sensing*, *103*, 48-56.

Silvertown, J. (2009). A new dawn for citizen science. *Trends in ecology & evolution*, *24*(9), 467-471.

Smith, B. and Mark, D.M. (1998). Ontology and Geographic Kinds. In *Proceedings of 8th International Symposium on Spatial Data Handling*, editors T. K. Poiker and N. Chrisman, (International Geographical Union, Vancouver) pp308-320

Vaz, E., & Jokar Arsanjani, J. (2015). Crowdsourced mapping of land use in urban dense environments: An assessment of Toronto. *The Canadian Geographer/Le Géographe Canadien*, DOI: 10.1111/cag.12170

18

THE ENMAP CONTEST: DEVELOPING AND COMPARING CLASSIFICATION APPROACHES FOR THE ENVIRONMENTAL MAPPING AND ANALYSIS PROGRAMME – DATASET AND FIRST RESULTS

A. Ch. Braun[a], M. Weinmann[b*], S. Keller[b], R. Müller[c], P. Reinartz[c], S. Hinz[b]

[a] Institute of Regional Science, Karlsruhe Institute of Technology (KIT)
Reinhard-Baumeister-Platz 1, 76131 Karlsruhe, Germany - andreas.ch.braun@kit.edu
[b] Institute of Photogrammetry and Remote Sensing, Karlsruhe Institute of Technology (KIT)
Englerstr. 7, 76131 Karlsruhe, Germany - {martin.weinmann, sina.keller, stefan.hinz}@kit.edu
[c] Remote Sensing Technology Institute (IMF), German Aerospace Center (DLR)
82234 Wessling, Germany - {rupert.mueller, peter.reinartz}@dlr.de, [*] Corresponding Author

Commission III, ICWG III/VII

KEY WORDS: Hyperspectral Imaging, EnMAP, Classification, Benchmark Datasets, Evaluation Methodology

ABSTRACT:

The Environmental Mapping and Analysis Programme EnMAP is a hyperspectral satellite mission, supposed to be launched into space in the near future. EnMAP is designed to be revolutionary in terms of spectral resolution and signal-to-noise ratio. Nevertheless, it will provide a relatively high spatial resolution also. In order to exploit the capacities of this future mission, its data have been simulated by other authors in previous work. EnMAP will differ from other spaceborne and airborne hyperspectral sensors. Thus, the assumption that the standard classification algorithms from other sensors will perform best for EnMAP as well cannot by upheld since proof. Unfortunately, until today, relatively few studies have been published to investigate classification algorithms for EnMAP. Thus, the authors of *this* study, who have provided some insights into classifying simulated EnMAP data before, aim to encourage future studies by opening the EnMAP contest. The EnMAP contest consists in a benchmark dataset provided for algorithm development, which is presented herein. For demonstrative purposes, this report also represents two classification results which have already been realized. It furthermore provides a roadmap for other scientists interested in taking part in the EnMAP contest.

1 INTRODUCTION

The Environmental Mapping and Analysis Programme, acronym EnMAP, is a spaceborne hyperspectral sensor to be launched during the forthcoming years (Kaufmann et al., 2008; Stuffler et al., 2009). EnMAP is designed as an imaging pushbroom hyperspectral sensor mainly based on modified existing or pre-developed technology. EnMAP offers a spectral range provided by two instruments from 420 nm to 1000 nm (VNIR) and from 900 nm to 2450 nm (SWIR). One important property is the high radiometric resolution and stability in both instruments. Its swath width is 30 km at a spatial resolution of 30 m × 30 m, which of course is high for a spaceborne hyperspectral instrument but low in comparison to airborne instruments. EnMAP will have a fast target revisit of only 4 days (Stuffler et al., 2007; Kaufmann et al., 2006).

Since the launch of EnMAP has not been realized yet, data similar to those expected to be produced by EnMAP have to be simulated. Sophisticated approaches to simulate EnMAP data have been published by Guanter et al. (2009) and Segl et al. (2010, 2012). One dataset simulated by these approaches is the EnMAP Alpine Foreland dataset, showing the Ammersee region in Bavaria, Germany.

Due to these properties, EnMAP data will differ from other hyperspectral data available to users of remote sensing datasets. EnMAP will be the first instrument to provide a radiometric quality largely comparable to airborne instruments (especially in terms of signal-to-noise values) but with a spatial resolution comparable to Landsat data. Since EnMAP will neither be similar to airborne hyperspectral sensors (like HyMap, for instance) nor to other spaceborne instruments (like Hyperion, for instance), it cannot be generally assumed that classification methods appropriate for such instruments will work well for EnMAP too (Braun et al.,

2012). Hence, research is needed to develop appropriate classification techniques even before the launch of EnMAP.

In order to stimulate research on high performance classification, this paper provides a simulated EnMAP benchmark dataset – derived from the EnMAP Alpine Foreland data, which cover 900 km^2 in a 1000 × 1000 pixel image. The dataset comprises 20 land use classes which are spectrally relatively similar and intricate to classify. Evaluation is performed on the basis of overall accuracy, completeness, correctness and quality. Besides describing the benchmark dataset, two results using state-of-the-art classifiers are presented. The results are based on the use of a Support Vector Machine (Cortes and Vapnik, 1995) and a Random Forest (Breiman, 2001).

After presenting the data and results on the conference, the benchmark dataset will be made available on the homepage of the authors. From there, it can be downloaded by further researchers, who will develop their approaches, compare them among one another and publish the results in future publications. This EnMAP contest will provide insight into best practices for EnMAP data classification. It will be helpful to the sensors developers and operators, because the datasets can be delivered with helpful information on data exploitation to interested users. It will furthermore be helpful to users in order to extract better results from their data. Finally, the EnMAP contest will be scientifically interesting by showing common points and differences between traditional hyperspectral classification and classification specifically designed for EnMAP data.

2 RELATED WORK

Few approaches have been published which provide classification results on simulated EnMAP data. This research gap is especially deplorable, since intense efforts are being undertaken to develop an EnMAP box which could integrate such approaches and make them readily available to future users (Held et al., 2012). Braun et al. (2012) compare three different state-of-the-art kernel-based classifiers (Support Vector Machine, Import Vector Machine, Relevance Vector Machine) on simulated EnMAP data, concluding that Import Vector Machine and Relevance Vector Machine outperform the Support Vector Machine. They furthermore outline the particular differences of the three methods and point out, why a combination of the three methods could even enhance performance[1]. Dörnhöfer and Oppelt (2015) produce a bio-optical model to analyse profundity and benthos of the coast of Helgoland, Germany. They use simulated EnMAP scenes to produce and test the model, concluding that EnMAP reliably detects micro-variations of structures and that water remote sensing will benefit from EnMAP's properties. Schwieder et al. (2014) use simulated EnMAP data for mapping shrub cover fraction in Southern Portugal, comparing three machine learning techniques (Support Vector Regression, Random Forest, Partial Least Squares Regression). Support Vector Regression performed best and thus, EnMAP and Support Vector Regression is attributed a great potential in quantifying fractional vegetation cover and monitoring gradual land use change processes. Bracken (2014) used EnMAP data to estimate soil erosion in semi arid Mediterranean environments. The Ph.D. thesis is one of the first to fully document the benefits of the new mission for a relevant environmental problem. Faßnacht et al. (2011) present a method to automatically extract tree covered areas from several hyperspectral datasets, including EnMAP. This method is based on the extended Normalized Difference Vegetation Index (NDVI).

In order to stimulate further research on classifying EnMAP data, a benchmark dataset will be introduced in the following. The dataset will be made available to the research community and published results will be regularly compared.

3   DATASET

This section introduces the EnMAP contest dataset. The dataset is based on the simulated EnMAP Alpine Foreland image, provided by Guanter et al. (2009). The colleagues produce a 1000 × 1000 pixel datasets, covering 30 × 30 km regions. Hence, the ground sampling distance is 30 m. The datasets cover the 420 to 2450 nm spectral range at a varying spectral sampling of 6.5-10 nm. The images consist of 244 simulated spectral channels. Figure 1 shows a near natural color visualization of the 244 channel simulated EnMAP dataset Alpine Foreland. The image depicts the area around the Ammersee in Bavaria, Germany. Note the diversity of different agricultural, vegetation, urban and industrial, and water classes (cf. Guanter et al. (2009)).

This diversity is represented in the EnMAP contest dataset. The EnMAP contest dataset comprises 20 different land use classes, and it represents a typical scenario for modern remote sensing: high accuracy is aimed for, whereas only small training data sets are provided. The $N_C = 20$ classes are defined to be spectrally very similar and thus, provide a dataset which is difficult to classify, cf. Figure 2. The mean spectra are shown in Figure 3 to visualize spectral similarity. More specifically, classes have been defined on the screen by focussing on visual differences in the images (considering several channel combinations) but also by checking pixels' individual spectra. Then, typical areas have been assigned a class label $l \in \{l^1, \ldots, l^{N_C}\}$, where N_C represents the number of classes and $N_C = 20$ for this dataset. From

these areas, a random subsample has been drawn for each class, since the number of pixels within the entire areas was too large to be exploited conveniently with state-of-the-art classifiers (which, while in the developing phase, tend to be time consuming given larger training numbers). The pixels from this subsample within these areas have afterwards been randomly split a *second time*, this time into a training set $\mathcal{X}$ and a test set $\mathcal{Y}$ (see also, Section 5). The splitting for each class is similar, 70% of the pixels are in $\mathcal{X}$ and 30% are in $\mathcal{Y}$. In total, $\mathcal{X}$ comprises 2617 pixels and $\mathcal{Y}$ 1124 pixels.

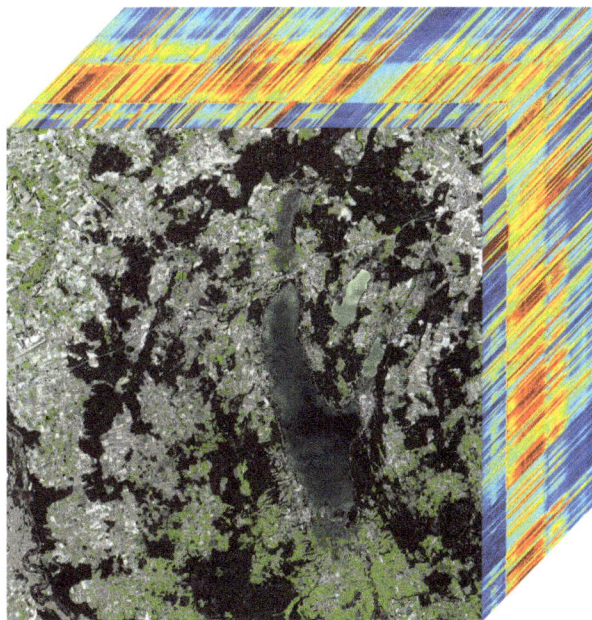

Figure 1. Near natural color visualization of the 244 channel simulated EnMAP dataset Alpine Foreland, showing the area around the Ammersee in Bavaria, Germany. Dataset produced by Guanter et al. (2009), courtesy of Dr. K. Segl. Larger image available at: www.ipf.kit.edu/code.php

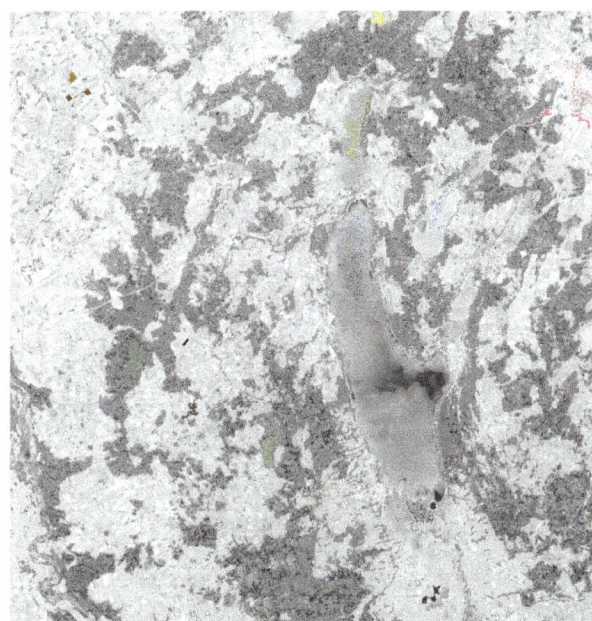

Figure 2. Location of labelled data within the simulated EnMAP dataset from Figure 1. Larger image available at: www.ipf.kit.edu/code.php

---

[1] In Braun et al. (2014), such a combination is presented, albeit on other data than on EnMAP.

(a) Water classes

(b) Forest/meadow classes

(c) Agriculture classes

(d) Other classes

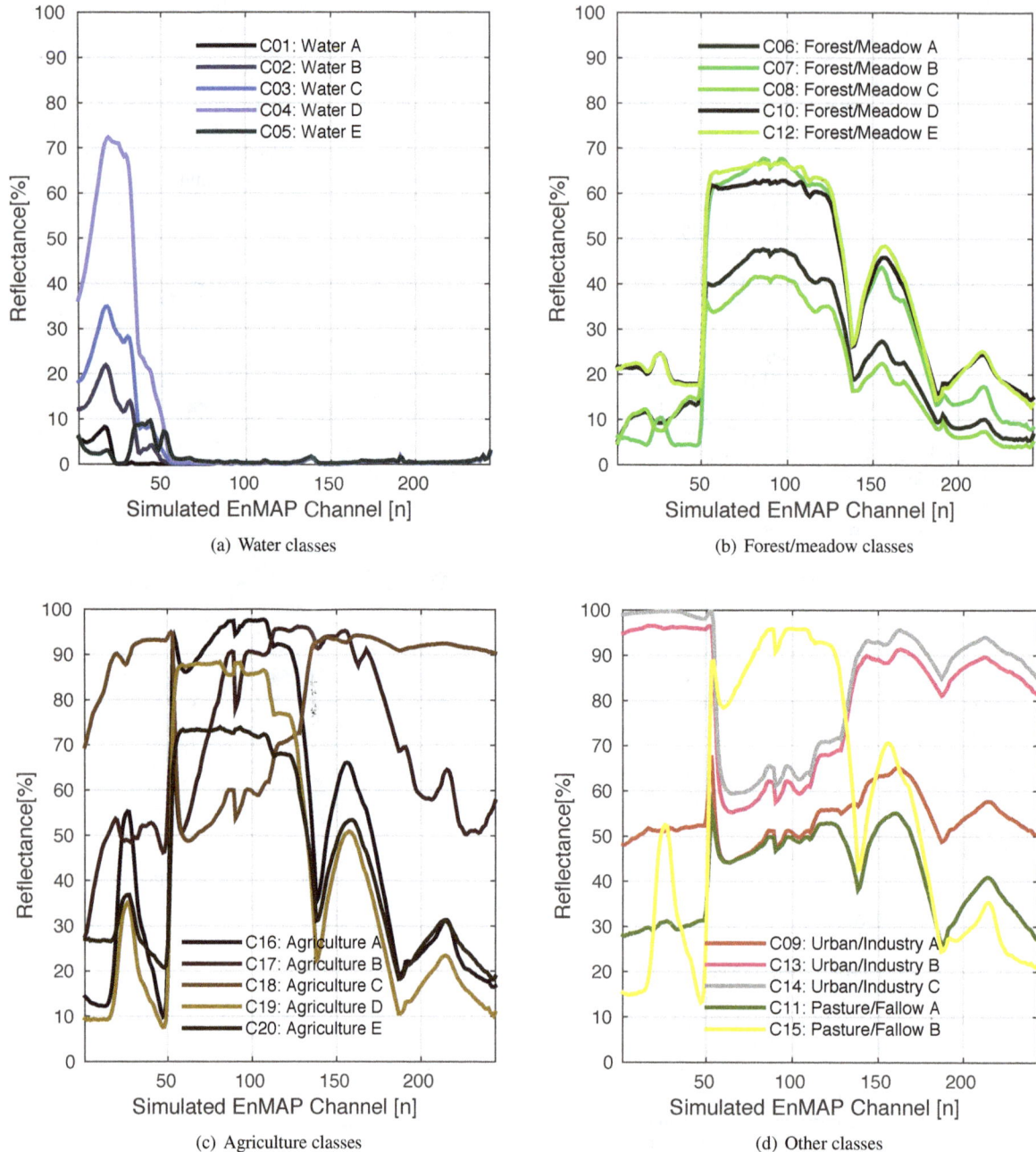

Figure 3. Average spectra for the 20 land use classes of the EnMAP contest dataset: (a) water classes, (b) forest/meadow classes, (c) agriculture classes, (d) other classes.

4   EVALUATION

Our evaluation focuses on a comparison of the performance of different approaches for classifying the provided dataset. For each approach, the test set is classified, and the resulting labels are compared to the reference labels on a per-pixel basis. We determine the respective confusion matrices and derive a measure indicating the overall performance as well as per-class measures indicating the class-wise performance. More specifically, the confusion matrix $\mathbf{C} = [c_{ij}]$ is defined in a way that the reference is given in row direction, while the prediction is given in column direction. Based on the confusion matrix, we consider the *overall accuracy*

$$\text{overall accuracy} = \frac{\sum_i c_{ii}}{\sum_i \sum_j c_{ij}} \quad (1)$$

in order to argue about the overall effectiveness of a specific approach. For the class-wise considerations, we assign the i-th class the following measures:

- True Positive (TP):
$$\text{TP}_i = c_{ii} \quad (2)$$

- False Positive (FP):
$$\text{FP}_i = \sum_{j, j \neq i} c_{ij} \quad (3)$$

- False Negative (FN):
$$\text{FN}_i = \sum_{j, j \neq i} c_{ji} \quad (4)$$

- True Negative (TN):

$$TN_i = \sum_{i,j} c_{ij} - TP_i - FP_i - FN_i \qquad (5)$$

Based on these measures, we derive the *completeness* (*recall*) given by

$$completeness_i = \frac{TP_i}{TP_i + FN_i} \qquad (6)$$

as well as *correctness* (*precision*) given by

$$correctness_i = \frac{TP_i}{TP_i + FP_i} \qquad (7)$$

and we furthermore derive the measure of *quality* which represents a compound metric indicating a good trade-off between omission and commission errors (Heipke et al., 1997):

$$quality_i = \frac{TP_i}{TP_i + FP_i + FN_i} \qquad (8)$$

$$= \frac{1}{completeness_i^{-1} + correctness_i^{-1} - 1} \qquad (9)$$

Further evaluation measures, that better capture the topological properties of the classified regions (see e.g. (Weidner, 2008)), could be introduced at a later stage.

5   CLASSIFICATION

In the scope of this paper, we focus on a supervised classification of individual pixels by using the training data in order to train a classifier which should afterwards be able to generalize to new, unseen data. Introducing a formal description, the training set $\mathcal{X} = \{(\mathbf{x}_i, l_i)\}$ with $i = 1, \ldots, N_{\mathcal{X}}$ consists of $N_{\mathcal{X}}$ training examples. Each training example encapsulates a feature vector $\mathbf{x}_i \in \mathbb{R}^d$ in a d-dimensional feature space and the respective class label $l_i \in \{l^1, \ldots, l^{N_C}\}$, where N_C represents the number of classes. In contrast, the test set $\mathcal{Y} = \{\mathbf{x}_j\}$ with $j = 1, \ldots, N_{\mathcal{Y}}$ only consists of $N_{\mathcal{Y}}$ feature vectors $\mathbf{x}_j \in \mathbb{R}^d$. If available, the respective class labels $l_j \in \{l^1, \ldots, l^{N_C}\}$ may be used for evaluation (this is the case for the test set of the EnMAP contest dataset, see Section 3). For multi-class classification, we involve two classifiers represented by a Support Vector Machine (Cortes and Vapnik, 1995) and a Random Forest (Breiman, 2001).

5.1   Classification Based on a Support Vector Machine

Given the training set $\mathcal{X}$, the Support Vector Machine – just as comparable kernel-based models like the Import Vector Machine or Relevance Vector Machine – optimizes a linearly solvable classification problem depending only on the input features $\mathbf{x}_i$, a weight vector w and a bias b. Kernel-based methods introduce non-linear functions $\phi(\mathbf{x}_i)$ to easily find a linear solution in a Reproducing Kernel Hilbert Space (RKHS), these higher dimensional feature spaces are induced implicitly by kernel functions $K(\mathbf{x}_i, \mathbf{x}_j) = \langle \phi(\mathbf{x}_i), \phi(\mathbf{x}_j) \rangle$. Finally, all methods look for a subset $\mathcal{V} \subset \mathcal{X}$ of training samples to sparsely induce these spaces.

The Support Vector Machine was designed to solve large margin classification problems as an implementation of statistical learning theory. It establishes a separating hyperplane and a maximal margin free of training data by choosing a subset $\mathcal{SV} \subset \mathcal{X}$ called support vectors (SVs). The optimization problem is given by Equations 10 and 11. The SVs are used to calculate the normal vector w on the hyperplane and the bias b to fulfil the constraint on the optimization problem.

$$\min \quad \frac{||w||^2}{2} + C \sum_{i=1}^{n} \xi_i \qquad (10)$$

$$\text{subject to:} \quad l_i(w \cdot \mathbf{x}_i + b) \geq 1 - \xi_i \qquad (11)$$

It can be shown that minimizing Equation 10 is equal to maximizing the *margin*. The slack variables ξ_i allow for falsely assigned training data in favour of generalization.

5.2   Classification Based on a Random Forest

A Random Forest (Breiman, 2001) is an ensemble of randomly trained decision trees. In the training phase, a pre-defined number N_T of individual decision trees are trained on different subsets of the given training data, where the subsets are randomly drawn with replacement. Thus, the decision trees are all randomly different from one another which results in a de-correlation between individual tree predictions. In the classification phase, the feature vectors $\mathbf{x}_i$ are classified by each tree, i.e. each tree casts a vote for one of the class labels l^k with $k = 1, \ldots, N_C$. Thus, the posterior probability $p\left(l_i = l^k | \mathbf{x}_i\right)$ of a class label l_i belonging to class l^k given the feature vector $\mathbf{x}_i$ may be expressed as the ratio of the number N_k of votes cast for class l^k across all decision trees and the number N_T of involved decision trees:

$$p\left(l_i = l^k | \mathbf{x}_i\right) = \frac{N_k}{N_T} \qquad (12)$$

Instead of a probabilistic consideration, the assignment of a respective class label l_i to an observed feature vector $\mathbf{x}_i$ is typically based on the majority vote across all decision trees which results in an improved generalization and robustness (Criminisi and Shotton, 2013).

6   RESULTS

The data described in Section 3 have been classified by a Support Vector Machine and a Random Forest. For the Support Vector Machine, a one-against-one approach and a Radial Basis Function (RBF) kernel have been used. The kernel parameter γ has been optimized in the range between $\gamma = 10^{-15}, \ldots, 10^5$ and the cost parameter C in the range between $C = 10^{-5}, \ldots, 10^{15}$ by cross validation with five fold grid search and exponent 5 increments in each parameter, for fully comprehensive instructions cf. Braun et al. (2010, 2012). The Random Forest has been trained using a maximum of 500 trees. For each classifier, the entire dataset $\mathcal{X} = \{(\mathbf{x}_i, l_i)\}$ with all 244 channels has been used.

The classification results of the Support Vector Machine and the Random Forest are presented in Figure 4, a near-natural color visualisation is also given in order to facilitate visual comparison. Obviously, both classifiers produce visually rather similar results. It should be kept in mind though, that an area of 900 square kilometres is observed and that both classifiers are state-of-the-art methods not expected to fail on large numbers of pixels. Hence, such similar results had to be expected beforehand. The largest failure of both classifiers is that they over-estimate the appearance of class 9, which relates to urban and industrial areas and confuse them with agricultural areas.

When preparing this paper, the authors have evaluated several other classifiers, like Import Vector Machines, Relevance Vector Machines, AdaBoost, Neural Networks, Gaussian Mixture Models, but also more traditional techniques like Spectral Angle Mapper and Maximum Likelihood (see webpage for details). Some

(a) SVM Result   (b) View   (c) RF Result

Figure 4. Results of the classified EnMAP contest dataset with two state-of-the-are classifiers. Left: Support Vector Machine (SVM); Centre: view for visual evaluation; Right: Random Forest (RF). Larger image available at: www.ipf.kit.edu/code.php

Table 1. Confusion matrix values for Support Vector Machine and Random Forest. Average, minimum and maximum values of class-specific measures represented by quality (QLT), correctness (COR) and completeness (CMP).

Classifier	avrg.QLT	avrg.COR	avrg.CMP
Support Vector Machine	75.23%	84.54%	85.02%
Random Forest	75.11%	83.44%	84.10%

Classifier	min.QLT	min.COR	min.CMP
Support Vector Machine	38.35%	66.66%	46.66%
Random Forest	37.72%	52.72%	48.33%

Classifier	max.QLT	max.COR	max.CMP
Support Vector Machine	100.00%	100.00%	100.00%
Random Forest	100.00%	100.00%	100.00%

of those produced obvious visual differences. However, the main goal of this report is to concentrate on quantitative figures. Thus, two of the most high ranking techniques in terms of overall accuracy have been selected for this report.

On the basis of the results visible in Figure 4, confusion matrices have been computed for both results, they are found in Figure 5. With an overall accuracy of 84.6% for Support Vector Machine, and 83.2% for Random Forest, both classifiers performed particularly well on the EnMAP contest dataset. For comparison, values of class-specific quality figures are provided in Table 1. As can be seen, the Support Vector Machine outperforms the Random Forest approach for the average and minimum class-specific quality figures also. The slight exception for minimum completeness does not neglect the general trend. Of course, there are some individual classes for which Random Forest values were higher than Support Vector Machine values, however, since there were no interpretable trends, comparing classes individually is omitted here.

Hence, in total, it can be concluded that both classifiers are well suited for hyperspectral EnMAP data, with Support Vector Machine being slightly superior to Random Forest, a finding confirmed by Pal (2006), Waske et al. (2010) and Chi et al. (2008) for other hyperspectral data.

7   THE ENMAP CONTEST

This contribution has described the dataset for the EnMAP contest and some first results. Now, it encourages the scientific com-

munity to take part in the EnMAP contest to promote scientific cooperation on producing high performance algorithms even before the launch of EnMAP. Therefore, it is required to provide some details about how scientist can take part in the contest. Fully instructive information will be given in a Portable Document File at www.ipf.kit.edu/code.php. Interested scientists will have to realize the following steps.

1. Go to www.ipf.kit.edu/code.php and read the instructions (PDF)

2. Download the training data $\mathcal{X}$ and the test data $\mathcal{Y}$ (provided as *.mat and *.txt (ASCII) data)

3. Download the entire image $\mathcal{I}$ (provided as *.mat and *.txt (ASCII) data)

4. Develop a classification algorithm $\mathcal{A} : f(\mathbf{x}_i) \rightarrow l_i$

5. Train the algorithm $\mathcal{A}$ on $\mathcal{X} = \{(\mathbf{x}_i, l_i)\}$

6. Apply the algorithm $\mathcal{A}$ to $\mathcal{Y}$

7. Calculate the quality figures overall accuracy, $completeness_i$, $correctness_i$ and $quality_i$ based on $\mathcal{Y}$

8. Apply the algorithm $\mathcal{A}$ to $\mathcal{I}$

9. Report the quality figures to the corresponding author and provide the classified image $\mathcal{I}$ (for control)

The authors of this paper will evaluate the quality figures reported for $\mathcal{Y}$ by checking them on the provided results for $\mathcal{I}^2$. The first EnMAP contest will go until 31st of December 2015. Then, the results will be submitted to the ISPRS Journal of Photogrammetry and Remote Sensing in a condensed manner. Scientist whose results are among the ten highest overall accuracy values confirmed will be invited as co-authors in the ranking of their results.

---

[2]The authors posses an image with the $\mathcal{X}$ and $\mathcal{Y}$ pixels' positions in the image. These positions are not available to contestants and cannot be made available. Thus, falsification of results is avoided.

8   DISCUSSION AND CONCLUSION

This contribution has elaborated a benchmark dataset for hyperspectral simulated EnMAP data. The benchmark is made available at www.ipf.kit.edu/code.php, where results will also be available at higher resolution for a more detailed comparison. Thus, the paper aims to promote research on classifying EnMAP data before its launch. Given a set of comprehensive studies comparing classification approaches, the authors believe that the EnMAP mission will be an even greater success, since confusion of future users about which algorithm to use is reduced. Similar benchmarks have been provided for other hyperspectral datasets, for instance, the ROSIS Pavia dataset, the AVIRIS Indian Pines and Salinas datasets, or the HYDICE Washington D.C. dataset. These datasets have provided some objective insights into the performance expected from individual algorithms.

Although its main goal is to introduce the dataset, this paper has also presented two results of state-of-the-art classifiers for the simulated EnMAP data. It has shown that classifiers know to perform well on other hyperspectral data, i.e. kernel-based and ensemble techniques are also applicable to EnMAP. A finding which is not necessarily expected in the first place as argued above. The Support Vector Machine performed slightly better than the Random Forest in both global and (generally) class-specific figures. More classifiers have been evaluated by the authors and the respective results will be published in future work.

9   ACKNOWLEDGEMENT

The authors acknowledge the simulation of the EnMAP dataset and, perhaps more importantly, the kind support provided by Guanter et al. (2009) and especially Dr. Karl Segl of the GFZ German Research Centre for Geosciences who have helped in understanding the properties of and using the dataset.

References

Bracken, A. H., 2014. Detecting soil erosion in semi-arid Mediterranean environments using simulated EnMAP data. Thesis (M.Sc.), Department of Geography, University of Lethbridge, Lethbridge, Canada.

Braun, A. C., Rojas, C., Echeverria, C., Rottensteiner, F., Bähr, H.-P., Niemeyer, J., Aguayo Arias, M., Kosov, S., Hinz, S. and Weidner, U., 2014. Design of a spectral-spatial pattern recognition framework for risk assessments using Landsat data – A case study in Chile. *IEEE Journal of Selected Topics in Applied Earth Observations and Remote Sensing* 7(3), pp. 917–928.

Braun, A. C., Weidner, U. and Hinz, S., 2010. Support vector machines for vegetation classification – A revision. *Photogrammetrie – Fernerkundung – Geoinformation* 2010(4), pp. 273–281.

Braun, A. C., Weidner, U. and Hinz, S., 2012. Classification in high-dimensional feature spaces – Assessment using SVM, IVM and RVM with focus on simulated EnMAP data. *IEEE Journal of Selected Topics in Applied Earth Observations and Remote Sensing* 5(2), pp. 436–443.

Breiman, L., 2001. Random forests. *Machine Learning* 45(1), pp. 5–32.

Chi, M., Feng, R. and Bruzzone, L., 2008. Classification of hyperspectral remote-sensing data with primal SVM for small-sized training dataset problem. *Advances in Space Research* 41(11), pp. 1793–1799.

Cortes, C. and Vapnik, V., 1995. Support-vector networks. *Machine Learning* 20(3), pp. 273–297.

Criminisi, A. and Shotton, J., 2013. *Decision forests for computer vision and medical image analysis*. Advances in Computer Vision and Pattern Recognition, Springer, London, UK.

Dörnhöfer, K. and Oppelt, N., 2015. Anwendung eines bio-optischen Modells zur Erfassung von Benthos und Wassertiefen in Küstengewässern – ein Test mit simulierten EnMAP Daten. *DGPF Tagungsband* 24, pp. 384–391.

Faßnacht, F., Weinacker, H. and Koch, B., 2011. Automatic forest area extraction from imaging spectroscopy data using an extended NDVI. *Proceedings of the 7th EARSeL Workshop on Imaging Spectroscopy.*

Guanter, L., Segl, K. and Kaufmann, H., 2009. Simulation of optical remote-sensing scenes with application to the EnMAP hyperspectral mission. *IEEE Transactions on Geoscience and Remote Sensing* 47(7), pp. 2340–2351.

Heipke, C., Mayer, H., Wiedemann, C. and Jamet, O., 1997. Evaluation of automatic road extraction. *The International Archives of the Photogrammetry, Remote Sensing and Spatial Information Sciences*, Vol. XXXII/3-4W2, pp. 151–160.

Held, M., Jakimow, B., Rabe, A., van der Linden, S., Wirth, F. and Hostert, P., 2012. EnMAP-Box Manual: Version 1.4. Technical Report, Humboldt-Universität zu Berlin, Berlin, Germany.

Kaufmann, H., Segl, K., Chabrillat, S., Hofer, S., Stuffler, T., Mueller, A., Richter, R., Schreier, G., Haydn, R. and Bach, H., 2006. EnMAP – A hyperspectral sensor for environmental mapping and analysis. *Proceedings of the IEEE International Geoscience and Remote Sensing Symposium (IGARSS)*, pp. 1617–1619.

Kaufmann, H., Segl, K., Guanter, L., Hofer, S., Foerster, K.-P., Stuffler, T., Mueller, A., Richter, R., Bach, H., Hostert, P. and Chlebek, C., 2008. Environmental Mapping and Analysis Program (EnMAP) – Recent advances and status. *Proceedings of the IEEE International Geoscience and Remote Sensing Symposium (IGARSS)*, Vol. 4, pp. 109–112.

Pal, M., 2006. Support vector machine-based feature selection for land cover classification: a case study with DAIS hyperspectral data. *International Journal of Remote Sensing* 27(14), pp. 2877–2894.

Schwieder, M., Leitão, P. J., Suess, S., Senf, C. and Hostert, P., 2014. Estimating fractional shrub cover using simulated EnMAP data: a comparison of three machine learning regression techniques. *Remote Sensing* 6(4), pp. 3427–3445.

Segl, K., Guanter, L., Kaufmann, H., Schubert, J., Kaiser, S., Sang, B. and Hofer, S., 2010. Simulation of spatial sensor characteristics in the context of the EnMAP hyperspectral mission. *IEEE Transactions on Geoscience and Remote Sensing* 48(7), pp. 3046–3054.

Segl, K., Guanter, L., Rogass, C., Kuester, T., Roessner, S., Kaufmann, H., Sang, B., Mogulsky, V. and Hofer, S., 2012. EeteS – The EnMAP end-to-end simulation tool. *IEEE Journal of Selected Topics in Applied Earth Observations and Remote Sensing* 5(2), pp. 522–530.

Stuffler, T., Förster, K., Hofer, S., Leipold, M., Sang, B., Kaufmann, H., Penné, B., Mueller, A. and Chlebek, C., 2009. Hyperspectral imaging – An advanced instrument concept for the EnMAP mission (Environmental Mapping and Analysis Programme). *Acta Astronautica* 65(7-8), pp. 1107–1112.

Stuffler, T., Kaufmann, C., Hofer, S., Förster, K. P., Schreier, G., Mueller, A., Eckardt, A., Bache, H., Penné, B., Benz, U. and Haydn, R., 2007. The EnMAP hyperspectral imager – An advanced optical payload for future applications in Earth observation programmes. *Acta Astronautica* 61(1-6), pp. 115–120.

Waske, B., van der Linden, S., Benediktsson, J. A., Rabe, A. and Hostert, P., 2010. Sensitivity of support vector machines to random feature selection in classification of hyperspectral data. *IEEE Transactions on Geoscience and Remote Sensing* 48(7), pp. 2880–2889.

Weidner, U., 2008. Contribution to the assessment of segmentation quality for remote sensing applications. *The International Archives of the Photogrammetry, Remote Sensing and Spatial Information Sciences*, Vol. XXXVII-B7, pp. 479–484.

SVM	C01	C02	C03	C04	C05	C06	C07	C08	C09	C10	C11	C12	C13	C14	C15	C16	C17	C18	C19	C20	QLT	COR	CMP
C01	60	0	0	0	0	0	0	0	0	0	0	0	0	0	0	0	0	0	0	0	100,00	100,00	100,00
C02	0	55	5	0	0	0	0	0	0	0	0	0	0	0	0	0	0	0	0	0	91,67	100,00	91,67
C03	0	0	60	0	0	0	0	0	0	0	0	0	0	0	0	0	0	0	0	0	92,31	92,31	100,00
C04	0	0	0	60	0	0	0	0	0	0	0	0	0	0	0	0	0	0	0	0	100,00	100,00	100,00
C05	0	0	0	0	60	0	0	0	0	0	0	0	0	0	0	0	0	0	0	0	100,00	100,00	100,00
C06	0	0	0	0	0	33	5	18	0	0	3	0	0	0	0	1	0	0	0	0	43,42	67,35	55,00
C07	0	0	0	0	0	5	55	0	0	0	0	0	0	0	0	0	0	0	0	0	84,62	91,67	91,67
C08	0	0	0	0	0	10	0	50	0	0	0	0	0	0	0	0	0	0	0	0	64,10	73,53	83,33
C09	0	0	0	0	0	1	0	0	56	0	2	0	1	0	0	0	0	0	0	0	80,00	84,85	93,33
C10	0	0	0	0	0	0	0	0	0	54	1	4	0	0	0	0	0	0	0	1	70,13	76,06	90,00
C11	0	0	0	0	0	0	0	0	2	2	56	0	0	0	0	0	0	0	0	0	82,35	87,50	93,33
C12	0	0	0	0	0	0	0	0	0	13	2	40	0	0	1	0	0	0	2	2	62,50	90,91	66,67
C13	0	0	0	0	0	0	0	0	7	0	0	0	28	18	0	0	1	6	0	0	38,36	68,29	46,67
C14	0	0	0	0	0	0	0	0	1	0	0	0	9	49	0	0	0	1	0	0	61,25	71,01	81,67
C15	0	0	0	0	0	0	0	0	0	0	0	0	0	0	40	19	0	0	1	0	50,00	66,67	66,67
C16	0	0	0	0	0	0	0	0	0	0	0	0	0	0	17	42	0	0	1	0	51,85	66,67	70,00
C17	0	0	0	0	0	0	0	0	0	0	0	0	0	0	0	0	60	0	0	0	98,36	98,36	100,00
C18	0	0	0	0	0	0	0	0	0	0	0	0	3	2	0	0	0	43	0	0	76,79	84,31	89,58
C19	0	0	0	0	0	0	0	0	0	2	0	0	0	0	2	0	0	1	37	0	80,43	90,24	88,10
C20	0	0	0	0	0	0	0	0	0	0	0	0	0	0	0	1	0	0	0	13	76,47	81,25	92,86

(a) SVM Result

RF	C01	C02	C03	C04	C05	C06	C07	C08	C09	C10	C11	C12	C13	C14	C15	C16	C17	C18	C19	C20	QLT	COR	CMP
C01	60	0	0	0	0	0	0	0	0	0	0	0	0	0	0	0	0	0	0	0	100,00	100,00	100,00
C02	0	56	4	0	0	0	0	0	0	0	0	0	0	0	0	0	0	0	0	0	93,33	100,00	93,33
C03	0	0	60	0	0	0	0	0	0	0	0	0	0	0	0	0	0	0	0	0	93,75	93,75	100,00
C04	0	0	0	60	0	0	0	0	0	0	0	0	0	0	0	0	0	0	0	0	100,00	100,00	100,00
C05	0	0	0	0	60	0	0	0	0	0	0	0	0	0	0	0	0	0	0	0	100,00	100,00	100,00
C06	0	0	0	0	0	29	6	21	0	0	2	0	0	0	0	1	0	0	0	0	33,72	52,73	48,33
C07	0	0	0	0	0	7	53	0	0	0	0	0	0	0	0	0	0	0	0	0	80,30	89,83	88,33
C08	0	0	0	0	0	19	0	41	0	0	0	0	0	0	0	0	0	0	0	0	50,62	66,13	68,33
C09	0	0	0	0	0	0	0	0	54	0	4	0	1	0	0	0	0	0	0	0	81,82	90,00	90,00
C10	0	0	0	0	0	0	0	0	0	52	2	4	0	0	0	0	0	0	0	1	66,67	74,29	86,67
C11	0	0	0	0	0	0	0	0	0	2	55	3	0	0	0	0	0	0	0	0	78,57	84,62	91,67
C12	0	0	0	0	0	0	0	0	0	14	2	39	0	0	1	0	0	0	2	2	58,21	84,78	65,00
C13	0	0	0	0	0	0	0	0	4	0	0	0	31	18	0	0	1	6	0	0	44,29	75,61	51,67
C14	0	0	0	0	0	0	0	0	1	0	0	0	9	49	0	0	0	1	0	0	58,54	68,57	80,00
C15	0	0	0	0	0	0	0	0	0	0	0	0	0	0	40	19	0	0	1	0	52,38	64,71	73,33
C16	0	0	0	0	0	0	0	0	0	0	0	0	0	0	17	42	0	0	1	0	49,35	69,09	63,33
C17	0	0	0	0	0	0	0	0	0	0	0	0	0	0	0	0	60	0	0	0	96,77	96,77	100,00
C18	0	0	0	0	0	0	0	0	1	0	0	0	0	2	0	0	0	43	0	0	77,19	83,02	91,67
C19	0	0	0	0	0	0	0	0	0	2	0	0	0	0	2	0	0	1	37	0	84,44	92,68	90,48
C20	0	0	0	0	0	0	0	0	0	0	0	0	0	0	0	1	0	0	0	13	82,35	82,35	100,00

(b) RF Result

Figure 5. Confusion matrices of the Support Vector Machine (above) and Random Forest (below) result. Rows: known classes, Columns: predicted classes. Overall accuracy and class-specific quality measures included. Larger image available at: www.ipf.kit.edu/code.php

19

A SEMI-AUTOMATIC PROCEDURE FOR TEXTURING OF LASER SCANNING POINT CLOUDS WITH GOOGLE STREETVIEW IMAGES

J. F. Lichtenauer[a],*, B. Sirmacek[b]

[a] Laan der Vrijheid 92, 2661HM Bergschenhoek, The Netherlands - jeroenlichtenauer@gmail.com
[b] Department of Geoscience and Remote Sensing, Delft University of Technology,
Stevinweg 1, 2628CN Delft, The Netherlands - www.berilsirmacek.com

Commission VI, WG VI/4

KEY WORDS: Point Clouds, Terrestrial Laser Scanning, Structure from Motion (SfM), Texturing, Google Streetview

ABSTRACT:

We introduce a method to texture 3D urban models with photographs that even works for Google Streetview images and can be done with currently available free software. This allows realistic texturing, even when it is not possible or cost-effective to (re)visit a scanned site to take textured scans or photographs. Mapping a photograph onto a 3D model requires knowledge of the intrinsic and extrinsic camera parameters. The common way to obtain intrinsic parameters of a camera is by taking several photographs of a calibration object with a priori known structure. The extra challenge of using images from a database such as Google Streetview, rather than your own photographs, is that it does not allow for any controlled calibration. To overcome this limitation, we propose to calibrate the panoramic viewer of Google Streetview using Structure from Motion (SfM) on any structure of which Google Streetview offers views from multiple angles. After this, the extrinsic parameters for any other view can be calculated from 3 or more tie points between the image from Google Streetview and a 3D model of the scene. These point correspondences can either be obtained automatically or selected by manual annotation. We demonstrate how this procedure provides realistic 3D urban models in an easy and effective way, by using it to texture a publicly available point cloud from a terrestrial laser scan made in Bremen, Germany, with a screenshot from Google Streetview, after estimating the focal length from views from Paris, France.

1. INTRODUCTION

In recent years, a lot of progress has been made with the development of three dimensional (3D) sensors, 3D visualisation technologies and data storage-, processing- and distributions possibilities. This seems good news for one of its most useful applications: 3D mapping of our urban environments. However, the main limitation still left for large-scale realistic digitization of urban environments is the collection of data itself. Although flying- and Earth-orbiting cameras and 3D sensors allow coverage of the entire globe at relatively low cost, a view from above is severely restricted by occlusions. Unfortunately, the most relevant urban areas for daily use are mostly at ground level, rather than on rooftops or above tree canopies.

This means that terrestrial short-range measurements still need to be made to map urban environments realistically in 3D. While state-of-the-art mobile laser scanners can capture high quality 3D data together with coloured texture, their prices are high and the acquisition of texture data often requires significant extra scanning time over the capture of 3D data alone. It is mainly because of the cost of these sensors and the time it takes to use them to acquire measurements from all the required locations, that realistic 3D urban maps are still not widely available today.

Therefore, in order to make full use of the available 3D data technologies for realistic urban mapping, we also need to be able to include data from fast, low-cost scanning methods, as well as from previously recorded data, which do not always include colour texture information together with the 3D measurements. And even with 3D scans that include colour texture, unsatisfactory lighting conditions during a scan might still require new texture to be added afterwards.

To overcome this limitation, we propose a simple semi-automatic method to accurately texture 3D data with photographs, which consists of a novel combination of currently available free software implementations of pre-existing algorithms for Structure from Motion (SfM), camera pose estimation and texture mapping. Only three or more tie points per photo are needed to accurately align it to the corresponding 3D data. These tie points can either be obtained from a simple manual procedure that can be done by anyone without special training, or from an automatic 2D to 3D feature matching algorithm. To solve the problem of calibrating a panoramic viewer without being able to use an a priori known calibration pattern, we propose to obtain intrinsic camera parameters from a Structure from Motion algorithm applied to any suitable data that is already available from the viewer.

To demonstrate that our methods works effectively, we applied it to texture a laser scan of building faades at the Bremer Market Square in Germany with a screenshot from Google Streetview, using a focal length estimation made from views of the 'Arc de Triomphe' in Paris, France. The results indicate that our method allows to easily make realistic-looking digital urban 3D models by combining pre-recorded data from completely unrelated sources.

2. PREVIOUS WORK

A promising way to easily generate photo-realistic urban 3D models is the use of Structure from Motion (SfM). Snavely et al. (2008) have shown that photos from the internet can be used with SfM to automatically reconstruct buildings and small parts of cities. This produces fully textured 3D models without even having to visit a site. Unfortunately, it requires having at least 3 good quality photos available on the internet of every building that needs to be reconstructed. This limits the application of

---

*Corresponding author

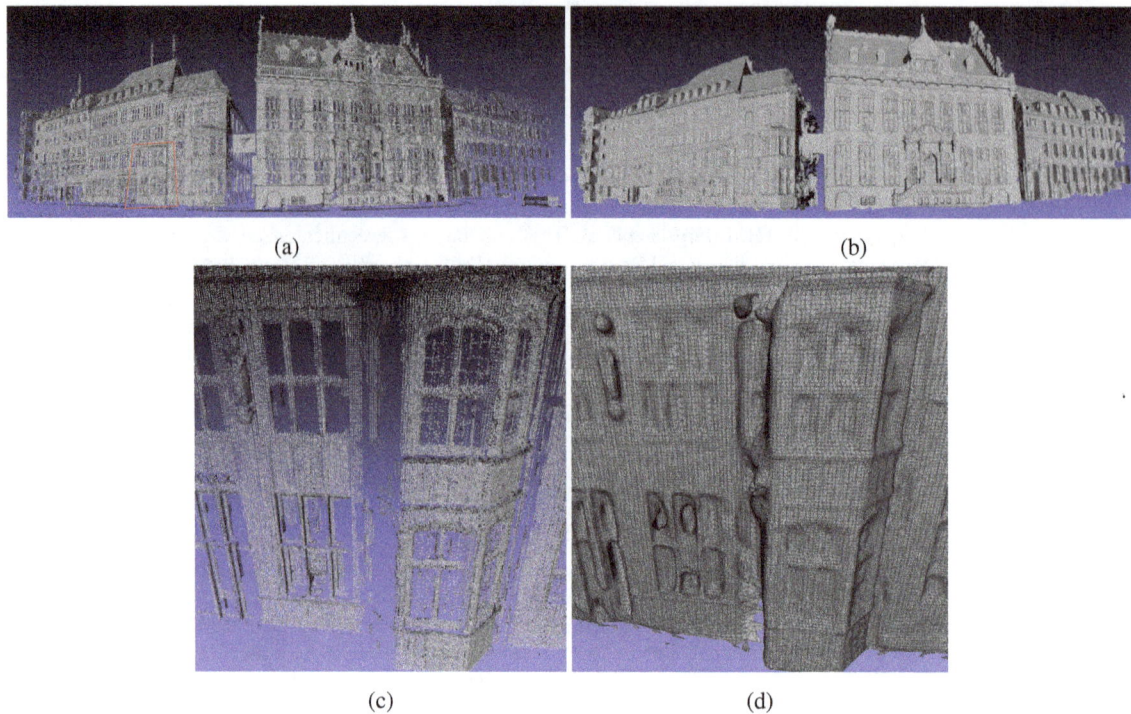

Figure 1: 3D Surface reconstruction from point clouds. (a) Point cloud extracted from a laser scan data set taken in Bremen, Germany. (b) Mesh surface obtained using Poisson surface reconstruction on the individual building faades. (c) Close-up view of the point cloud for the selected area in subfigure (a). (d) Close-up view of the triangulation.

this method to the reconstruction of popular tourist attractions. Furthermore, the 3D data from such reconstructions usually contain many missing parts and inaccuracies. Combining SfM results with more accurate 3D data from other sensors, such as laser scanners, still requires additional data alignment solutions.

Several methods have been suggested to automatically align a 2D image to a 3D model, by computing some sort of similarity measure between 2D image contents and structure of the 3D model. Lensch et al. (2000) proposed a silhouette-based alignment method that includes an automatic initialisation. However, the procedure requires an isolated object that is entirely visible in the photo, which is often not the case in 3D scans of urban environments. Viola and Wells III (1997) proposed to use surface normals to compute mutual information between an image and a 3D model. Corsini et al. (2009) improved on this by including more kinds of data to calculate mutual information. Although the mutual information registration methods can be applied to urban models, a sufficiently close pose and focal length initialisation has to be supplied manually for each photo, which comprises of the simultaneous adjustment of 7 parameters (3D location + 3D orientation + focal length). This might easily require more time and skill than annotating 3 or more tie points.

Morago et al. (2014) have chosen to automate the alignment of new photos with LIDAR scans by 2D feature point matching with photos that were taken together with the LIDAR scans, making use of their available 3D registration. However, this approach does not solve the problem of how to align photos with 3D scans for which no photos have been registered yet.

Using tie points to align 2D images to 3D models is a well-studied problem, for which many effective solutions already have been found such as by Hesch and Roumeliotis (2011) and Zheng et al. (2013). Accurate 2D-3D alignment can already be achieved reliably with only three or more well-spread tie-points (or at least four to prevent multiple ambiguous solutions). These methods

make use of the perspective projection (pinhole camera) model and therefore require calibration of a camera's intrinsic parameters. A requirement which also holds for projecting a photo onto a 3D model once its correct pose is estimated. For self-made photographs, intrinsic camera calibration may be done by taking several photographs of a calibration pattern with known structure. However, this cannot be done for photos from an image database such as Google Streetview.

3. MAIN CONTRIBUTIONS

The main contributions of our proposed procedure that set it apart from previously proposed methods are as follows: - Our method does not only work with self-made photographs, but also with screen shots from panoramic viewers such as Google Streetview, which already contain a large coverage of terrestrial urban photographs. - Since it doesn't depend on a 3D reconstruction from photogrammetry other than for camera calibration, it works with any source of 3D data and requires only one photo to texture any surface area visible in that photo. - Three or more tie points per camera are sufficient, without the need for a manual initialisation of camera pose. - Since the method is fully tie-point-based, as long as 3 or more tie points can be provided accurately, it works on any type of scene, any photo quality and doesn't require segmentation of individual buildings or objects. - Free software implementations are currently available for all of the steps that our procedure requires to convert a 3D point cloud into a textured 3D mesh. To our knowledge, we are the first to propose a procedure that covers a complete work flow with all of these advantages that are crucial for obtaining large-scale urban 3D maps.

4. TEXTURING A 3D MESH SURFACE WITH A PHOTO

In principle, texturing a 3D model from a photo is the inverse of how the light from the scene was projected onto the photo.

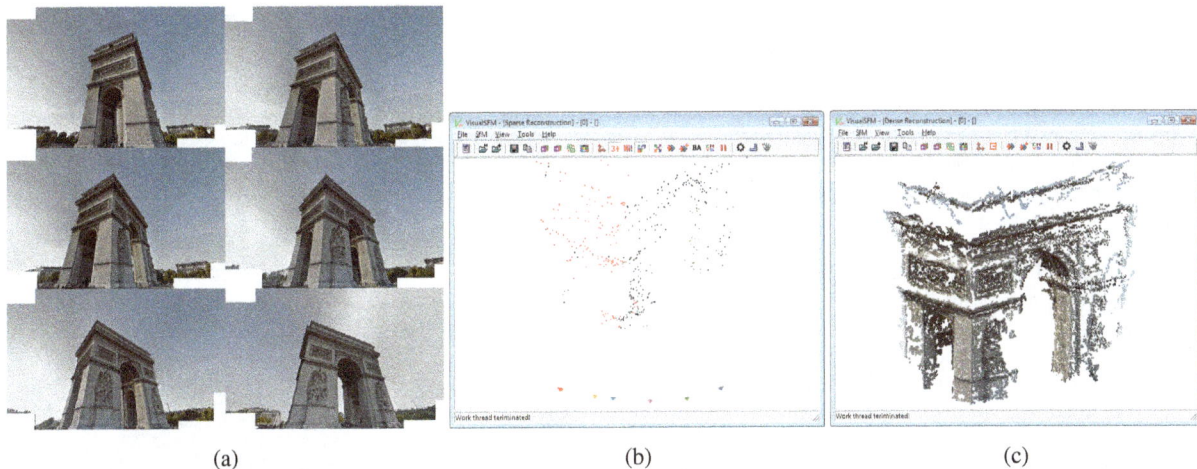

(a)   (b)   (c)

Figure 2: Structure from Motion for calibrating the focal length of Google Streetview. (a) Google Streetview screen shots with the on-screen information covered with white rectangles. (b) Reconstruction of the 6 camera poses and 3D feature point locations using VisualSFM. (d) Denser surface reconstruction using the CMVS plugin of VisualSFM.

By projecting the image points back into the 3D scene from the same relative position and orientation, the visible surface of the 3D model can be coloured correctly from the photo. This can be done using the freely available Meshlab software by Cignoni et al. (2008). The procedures described below are demonstrated in more detail in our three video tutorials: Lichtenauer and Sirmacek (2015a,b,c).

4.1 Generating a Surface Model

Texturing requires a surface, so a point cloud must be converted into a 3D mesh first, as shown in figure 1. This can be done with the Poisson surface reconstruction method in Meshlab. We assume that the point cloud is clean and only includes points which are representing the faade surface samples (fig. 1(a)). If the point cloud is too noisy to represent the faade surface correctly, or else if there are occluding objects, the point cloud must be pre-processed before mesh generation. This pre-process must remove the points coming from noise and occlusion effects. The best results are achieved by reconstructing relatively straight parts individually. Therefore, it will help to segment building point clouds into their separate sides, reconstruct their surfaces separately and then combine the surface meshes of all the sides again afterwards (fig. 1(b)). Note that the Poisson reconstruction method in Meshlab requires surface normals of the point cloud to be estimated first.

4.2 Obtaining intrinsic camera parameters from SFM

To correctly estimate camera pose and to project an image accurately onto a 3D surface, we need to know the intrinsic parameters of the camera that was used to generate the photo. The intrinsic camera parameters are coefficients of a mathematical approximation that relates the pinhole camera model to the actual pixel locations in a photograph. In a pinhole camera model, a point in a 3D scene is represented by the 2D location of the intersection of the straight line between the scene point and the camera's focal point (the pin hole) with a flat surface at focal distance f from the focal point.

The most common way to obtain intrinsic camera parameters is by performing a separate calibration procedure in which additional photographs are taken from a known calibration pattern or -structure. However, such procedures require extra time, equipment and skills. Furthermore, in some cases a separate calibration

might not even be possible at all. For instance, when the camera that has been used for the texture photographs cannot be accessed, or when a camera is used that has a variable focal length (zoom function). An example of such a situation might be when someone wants to texture a 3D mesh with a photograph from a public database such as Google Streetview Google Streetview - Google Maps (n.d.), which doesn't include details about the viewer's intrinsics.

A panoramic viewer such as the one for Google Streetview uses a single 360-degree spherical per location. Since all source photos have already been converted to the same spherical camera model that the panoramic viewer requires, it doesn't matter where the photos have been taken and what equipment has been used. All we need to do is to calibrate the viewer's own intrinsic parameters. Note that the view that is being shown to the user is only a part of the whole spherical image, converted to a perspective view centred around the viewing direction chosen by the user. Since Google Streetview currently has no radial distortion added, The only intrinsic parameter we need to estimate is the focal length. Because focal length is measured in screen pixels, it depends on the user's screen resolution. This means that the focal length has to be estimated separately for each screen resolution at which screen shots are taken from Google Streetview for texturing.

To estimate the intrinsic parameters for a public database such as Google Streetview, without being able to perform a regular camera calibration procedure, we propose to use Structure from Motion (SfM). Besides generating a 3D structure of the photographed scene, SfM software also provides the estimates of the intrinsic and extrinsic camera parameters for each of the source photographs. And since the focal length of Google Streetview does not depend on the location, we can choose a suitable scene anywhere in the world. For our example, we have chosen the Arc de Triomphe in Paris, France. This structure has views available from all around at small steps, which is exactly what is required for SfM.

Figure 2(a) shows the 6 screen shots we have taken from Google Streetview, with the on-screen information boxes removed to prevent false feature point matches between the images. Figure 2(b) shows the result of estimating structure (3D feature point locations) from motion from these images with the free VisualSFM software created by Wu (2011); Wu et al. (2011). We have disabled radial distortion in this VisualSFM, since the images were already distortion-free. By forcing the different camera views to have the same intrinsic parameters estimated, accuracy can be increased (since all images of come from the same panoramic viewer). However, performing the initial sparse SfM reconstruction in Visual SFM with 'shared calibration' enabled prevents the SfM algorithm from converging to the right solution. Instead, the initial SfM result without shared calibration can be refined afterwards using the Bundle Adjustment (BA) step with the 'shared-calibration' option enabled.

After the sparse SfM reconstruction, VisualSFM can also make a denser reconstruction, by using the CMVS option. For the purpose of texturing a 3D mesh with a photograph, we don't need a 3D reconstruction from SfM for any other purpose than to estimate the intrinsic camera parameters. However, the dense reconstruction is useful to visually inspect whether SfM has succeed. If some angles that should be straight have not been reconstructed perfectly straight, it means that the intrinsic camera parameters also have not been estimated correctly. Figure 2(c) shows the dense reconstruction for our example of the Arc de Triomphe. Furthermore, executing the CMVS procedure in VisualSFM also generates some output files that are useful for loading aligned texture photos into Meshlab.

The file 'bundle.rd.out' can be opened with Meshlab to load aligned rasters for texturing. For a description of the Bundler file format, see Snavely (2008-2009). It contains a list of intrinsic parameters for each camera view used in SfM, followed by a list of all structure points. When loading a Bundler file into Meshlab, it also asked for a text file containing a list of images corresponding to the cameras defined in the 'rd.out' file. The focal length of Google can also be found in the 'bundle.rd.out' file as the first value for each camera. The focal lengths for all views should be almost the same after having used Bundle Adjustment with the 'shared-calibration' option.

When using self-made photos for texturing which might include radial distortion, another possibility is to use VisualSFM on several photos taken from the scene that needs to be textured. VisualSFM will then not only calculate the intrinsic parameters for your camera automatically, but also rectify any radial distortion that might be present. After performing the CMVS procedure in VisualSFM, the rectified versions of your photos can be found in the 'visualise' directory of the output from VisualSFM, which correspond to the camera parameters in the 'bundle.rd.out' file in the order specified in the list.txt file. These rectified photos are suitable for texturing 3D lasers scans.

VisualSFM uses the following camera model that relates pinhole model coordinates $(n_x, n_y)^\top$ to image coordinates $(m_x, m_y)^\top$ (which are relative to the image center):

$$\begin{bmatrix} n_x \\ n_y \end{bmatrix} = f \begin{bmatrix} X_c/Z_c \\ Y_c/Z_c \end{bmatrix} \tag{1}$$

$$= \left(1 + r(m_x^2 + m_y^2)\right) \begin{bmatrix} m_x \\ m_y \end{bmatrix}, \tag{2}$$

where $(X_c, Y_c, Z_c)^\top$ are the 3D coordinates of the visible point with respect to the camera frame. The above camera model has only two intrinsic camera parameters: the focal length f and a radial distortion coefficient r. This model assumes that the image center coincides with the optical center, that the camera pixels are perfectly square and that the lens distortion is quadratic, without higher order components.

By using the f and r estimated by the SfM software, equations 1 and 2 can also be used to obtain the texture colour of a 3D point that is visible in the image at pixel coordinates $(m_x, m_y)^\top$. For

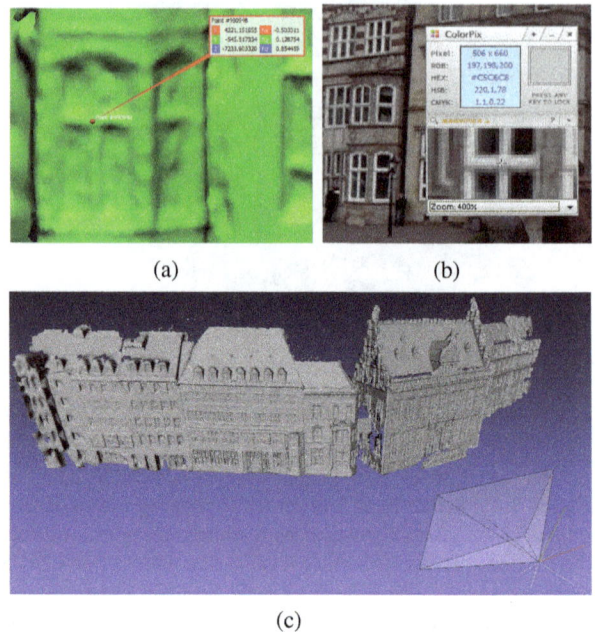

(a)   (b)

(c)

Figure 3: Pose estimation from tie point annotation. (a) Mesh vertex selected to see 3D point location and surface normal information. (b) Retrieving corresponding image coordinates of the same point. (c) Camera pose estimated from multiple tie points.

faster texturing of many 3D points, a rectified image can be generated using equation 2. When the SfM procedure includes the image that is to be used for texturing, a rectified image is already generated by VisualSFM itself. In the rectified image, the pixel coordinates are corresponding to the pinhole model coordinates $(n_x, n_y)^\top$. Therefore, texturing of 3D points can be done with a rectified image by only applying equation 1. Furthermore, this also allows to use third-party software, such as the texture function of Meshlab, which we use for our example.

4.3   Obtaining extrinsic camera parameters

Equation 1 requires the 3D coordinates of points to be defined in the reference frame of the camera. The coordinates of a point in the scanned 3D structure $(X_s, Y_s, Z_s)^\top$ are converted to the reference frame of the texture camera by:

$$\begin{bmatrix} X_c \\ Y_c \\ Z_c \end{bmatrix} = \mathbf{R} \begin{bmatrix} X_s \\ Y_s \\ Z_s \end{bmatrix} + \mathbf{T}, \tag{3}$$

$$\text{where} \quad \mathbf{R} = \begin{bmatrix} R_{11} & R_{12} & R_{13} \\ R_{21} & R_{22} & R_{23} \\ R_{31} & R_{32} & R_{33} \end{bmatrix}, \mathbf{T} = \begin{bmatrix} T_1 \\ T_2 \\ T_3 \end{bmatrix}.$$

The rotation matrix $\mathbf{R}$ and the translation vector $\mathbf{T}$ are the extrinsic camera parameters (or 'camera pose') of the texturing camera with respect to the 3D mesh that has to be textured.

The SfM procedure described above also computes extrinsic camera parameters $\mathbf{R}$ and $\mathbf{T}$. However, these are only suitable for alignment with the 3D reconstruction made by SfM. They cannot be used for texturing another 3D model, such as a laser scan, which will have an entirely different scale and alignment. Fortunately, the camera pose with respect to any 3D structure can be accurately estimated from as little as three point correspondences between the photo and the surface of the 3D structure. To prevent multiple solutions from which the correct one has to be validated by visual verification, at least four tie points are needed. Hesch

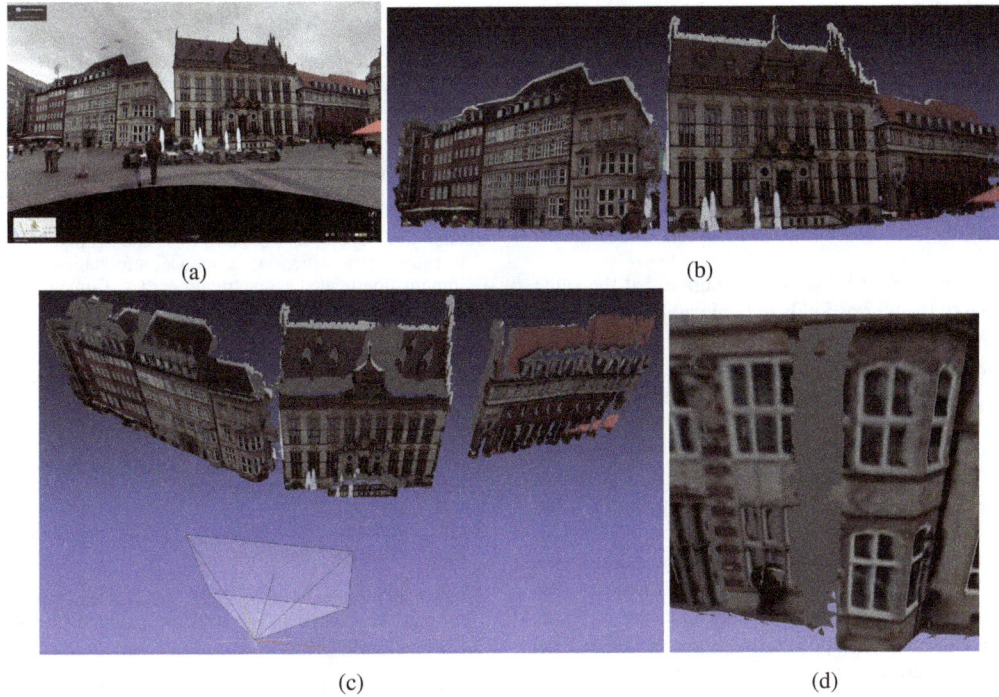

Figure 4: Texturing a laser scan with a Google Streetview screen shot (a) The screen shot from Google Streetview. (b) The textured 3D mesh seen from the view of the Google Streetview screen shot. (c) The textured 3D mesh from a different angle, with parts occluded from the camera view coloured in gray. (d) Close up of the textured 3D mesh from a different angle.

and Roumeliotis (2011) presented a Direct Least-Squares (DLS) method for computing camera pose which, unlike some other methods, doesn't require any initial (rough) pose estimate. The Matlab code can be downloaded from their website. Although the authors propose to refine the result of their DLS method with an additional Bundle Adjustment (BA) step, the accuracy of the DLS method is already very accurate by itself. For the purpose of texturing a 3D surface, the improvement of the additional BA refinement will likely be unnoticeable.

The tie points can either be determined by manual annotation, or by an automatic procedure that matches corner points in the image with corner points of the 3D structure. Figure 3(a) and (b) show how a tie point can be found through manual annotation. The 3D location of a vertex on a 3D mesh surface can be found using the point picker option of the free software CloudCompare *EDF R&D* Telecom ParisTech, CloudCompare (version 2.6.1).

To obtain the most accurate texture mapping, it is best to choose the points as far apart as possible, unless a significant lens distortion is present, in which case points close to the photo border might not be accurately approximated by the radial distortion model. For ease of computation, use the rectified version of the texture photograph to determine the image coordinates of the points (so that the radial distortion represented by equation 2 can be ignored). Note that the DLS method requires normalised image coordinates with unit focal length: $\frac{1}{f}(n_x, n_y)^\top$.

The Matlab code from Hesch and Roumeliotis (2011) to estimate camera pose from tie points can be used with the free software Octave Eaton et al. (2015). To convert the estimated pose to the format of Bundler, it has to be flipped over, using:

$$\mathbf{R}_b = \mathbf{FR} \tag{4}$$

$$\mathbf{T}_b = \mathbf{FT} \tag{5}$$

$$\mathbf{F} = \begin{bmatrix} 1 & 0 & 0 \\ 0 & -1 & 0 \\ 0 & 0 & -1 \end{bmatrix} \tag{6}$$

4.4 Projecting a 2D Image on a 3D Mesh Surface

Texturing from a photo can be done with the Meshlab software, using the option 'parameterization + texturing from registered rasters'. To load the (rectified) texture photos and their corresponding focal length and extrinsic calibration into Meshlab, the Bundler file format can be used, as mentioned above. This allows to load multiple photos at once, for combined texturing. The texturing procedure of Meshlab has several options of how to combine overlapping photos with possibly different illumination conditions. Figure 4(a) shows the screenshot from Google Streetview that we used for our example. Figure 4(b) shows the result from texturing from the same viewing angle as the Google Streetview screen shot. Figure 4(c) shows the texture result under a different angle, which includes parts that were occluded from the view and thus have no texture information. These occluded parts have been coloured gray. Figure 4(d) shows a close up of the texture result.

5. SHOWCASE EXAMPLE AND DISCUSSION

Our 3D texturing procedure consists of a novel combination of free software implementations of pre-existing algorithms for Structure from Motion, pose estimation and texture mapping. For an evaluation of the performances of each of the selected methods, please refer to our references to their respective literature in the previous sections.

To show the effectiveness of our complete procedure and give an impression of the quality that can be obtained, we have demonstrated the required steps above using a challenging example that we find representative of a common practical situation for which our procedure is intended to work. The 3D point cloud in our example was derived from terrestrial laser scans made with the Riegl VZ-400 scanner of a row of buildings at the Market Square in Bremen, Germany. It was part of a larger dataset collected by Borrmann and Nüchter (n.d.) and available for download. The show case example shows that the method is easy to use and that the accuracy of the method can be very high depending on the tie point selection.

The accuracy of the final texture alignment depends on several factors. First of all, the accuracy of the 3D model. When the photo for texturing is not made from the same location as the 3D scan, an error in the estimated distance from the 3D scanner will result in a shift in the projection from the photo.

Secondly, the tie points must be selected accurately at the same location in the photo that corresponds to the location selected on the 3D model. A tie point misalignment results in an inaccuracy of the camera pose estimation. Random errors in tie point alignment can be compensated by taking a larger number of tie points.

Thirdly, the intrinsic camera parameters must be estimated accurately. Lens distortion must be removed completely. This is not possible if the lens that was used has a more complex distortion than what is modelled by the (radial) lens distortion model. Unfortunately, Google Streetview contains a lot of misalignments in the multiple views that have been combined to generate the spherical images. This will cause inaccuracies in the tie point alignment (leading to inaccurate pose estimation), as well as misalignment of the mapped texture. The effects of this can be mitigated by carefully selecting the tie points at places that seem undistorted and by combining multiple views from different positions, to not use the edges of the photos, which will be effected the worst. An error in the focal length estimation from Structure from Motion will also cause texture misalignment, but by using the same inaccurate focal length for pose estimation as for texturing, the effect of an error in focal length is relatively small and distributed over the middle and outer parts of the photo.

Lastly, high quality texturing obviously depends on the quality of the photos that are used. A complete texturing of all parts of a 3D model requires well-placed views that do not leave any parts of the 3D surface unseen. Furthermore, differences between lighting conditions and exposures of the different photos can cause edges to be visible where the texture from one photo transitions into that of another.

6. CONCLUSION

Generation of realistic 3D urban structure models is valuable for urban monitoring, planning, safety, entertainment, commercial and many other application fields. After having the 3D model of the structure by laser scanning, photogrammetry or using CAD design software tools, the most realistic views are obtained after texturing the 3D model with real photos. Google streetview provides opportunity to access photos of urban structures almost all around the world. In this study, we introduce a semi-automatic method for obtaining realistic 3D building models by texturing them with Google streetview images. We discuss advantages and also the limitations of the proposed texturing method.

ACKNOWLEDGEMENTS

This research is funded by the FP7 project IQmulus (FP7-ICT-2011-318787) a high volume fusion and analysis platform for geospatial point clouds, coverages and volumetric data set.

References

Borrmann, D. and Nüchter, A., n.d. 3d scan data set recorded in city center of bremen, germany as part of the thermalmapper project. http://kos.informatik.uni-osnabrueck.de/3Dscans/.

Cignoni, P., Callieri, M., Corsini, M., Dellepiane, M., Ganovelli, F. and Ranzuglia, G., 2008. Meshlab: an open-source mesh processing tool. In: Sixth Eurographics Italian Chapter Conference, pp. 129–136.

Corsini, M., Dellepiane, M., Ponchio, F. and Scopigno, R., 2009. Image-to-geometry registration: a mutual information method exploiting illumination-related geometric properties. Computer Graphics Forum 28(7), pp. 1755–1764.

Eaton, J. W., Bateman, D., Hauberg, S. and Wehbring, R., 2015. Gnu octave version 4.0.0 manual: a high-level interactive language for numerical computations. http://www.gnu.org/software/octave/doc/interpreter.

EDF R&D Telecom ParisTech, CloudCompare (version 2.6.1), n.d. http://www.cloudcompare.org/.

Google Streetview - Google Maps, n.d. https://www.google.com/maps/views/streetview.

Hesch, J. and Roumeliotis, S., 2011. A direct least-squares (dls) method for pnp. In: Computer Vision (ICCV), 2011 IEEE International Conference on, pp. 383–390.

Lensch, H. P. A., Heidrich, W. and peter Seidel, H., 2000. Automated texture registration and stitching for real world models. In: in Pacific Graphics, pp. 317–326.

Lichtenauer, Jeroen, F. and Sirmacek, B., 2015a. Texturing a 3d mesh with google streetview 1 of 3: Manual alignment. http://youtu.be/Nu3VaeBxGHc.

Lichtenauer, Jeroen, F. and Sirmacek, B., 2015b. Texturing a 3d mesh with google streetview 2 of 3: Estimate focal length with visualsfm. http://youtu.be/OHZDuI48IHc.

Lichtenauer, Jeroen, F. and Sirmacek, B., 2015c. Texturing a 3d mesh with google streetview 3 of 3: Calculate pose from tie points. http://youtu.be/8i86ys9Boqc.

Morago, B., Bui, G. and Duan, Y., 2014. Integrating lidar range scans and photographs with temporal changes. In: Computer Vision and Pattern Recognition Workshops (CVPRW), 2014 IEEE Conference on, pp. 732–737.

Snavely, N., 2008-2009. Bundler v0.4 user's manual. http://www.cs.cornell.edu/~snavely/bundler/.

Snavely, N., Seitz, S. and Szeliski, R., 2008. Modeling the world from internet photo collections. International Journal of Computer Vision 80 (2), pp. 189–210.

Viola, P. and Wells III, W. M., 1997. Alignment by maximization of mutual information. Int. J. Comput. Vision 24(2), pp. 137–154.

Wu, C., 2011. Visualsfm: A visual structure from motion system. http://ccwu.me/vsfm/.

Wu, C., Agarwal, S., Curless, B. and Seitz, S. M., 2011. Multicore bundle adjustment. IEEE International Conference on Computer Vision and Pattern Recognition (CVPR).

Zheng, Y., Kuang, Y., Sugimoto, S., Astrom, K. and Okutomi, M., 2013. Revisiting the pnp problem: A fast, general and optimal solution. In: Computer Vision (ICCV), 2013 IEEE International Conference on, pp. 2344–2351.

20

AUTOMATIC EXTRACTION AND TOPOLOGY RECONSTRUCTION OF URBAN VIADUCTS FROM LIDAR DATA

Yan Wang [a]*, Xiangyun Hu [a]

[a] School of Remote Sensing and Information Engineering, Wuhan University, Luoyu Road 129, Wuhan 430079, China - (wang_yan, huxy)@whu.edu.cn

Commission III, WG III/4

KEY WORDS: Airborne LiDAR, Point Cloud, Overpass, Urban Viaducts Recognition, Topology Reconstruction

ABSTRACT:

Urban viaducts are important infrastructures for the transportation system of a city. In this paper, an original method is proposed to automatically extract urban viaducts and reconstruct topology of the viaduct network just with airborne LiDAR point cloud data. It will greatly simplify the effort-taking procedure of viaducts extraction and reconstruction. In our method, the point cloud first is filtered to divide all the points into ground points and none-ground points. Region growth algorithm is adopted to find the viaduct points from the none-ground points by the features generated from its general prescriptive designation rules. Then, the viaduct points are projected into 2D images to extract the centerline of every viaduct and generate cubic functions to represent passages of viaducts by least square fitting, with which the topology of the viaduct network can be rebuilt by combining the height information. Finally, a topological graph of the viaducts network is produced. The full-automatic method can potentially benefit the application of urban navigation and city model reconstruction.

1. INSTRUCTIONS

1.1 Background

Urban viaducts are usually transportation junctions of a city and play a key role in the whole transportation system. We have developed an original method that can automatically extract urban viaducts and reconstruct topology of the network just with airborne LiDAR point cloud data. To the best of our knowledge, there are few studies in this field. Our method provides a complete solution to recognize and extract urban viaducts full-automatically. It can be of great help in city model reconstruction, city road network extraction, navigation and so on.

In fact, viaducts are roads. A few researches have been conducted to extract road from LiDAR point cloud data and have achieved quite an accomplishment in recent years (Mena, 2013).

Combining the multi-spectral remote sensing images and point cloud data to extract road network is a popular study field (Akel and et al., 2005. Hu and Tao, 2007. Rottensteiner and Clode, 2008. Samadzadegan and et al. 2009. Hu and et al., 2004. Shao and et al. 2011,). The remote sensing images contains distinguishable multi-spectral information, in the meanwhile, point cloud data has advantages of containing information of geometry and height, which can hardly be effected by occlusion and luminance variation. It turns out to be effective to take the advantages from different data sources. To detect road just by the point cloud data, which relies on a single kind of sensors and can avoid the complexity of merging different data sources, is also proved acceptable and has achieved significant improvement(Choi and et al., 2008. Hu and et al., 2014. Clode

and et al., 2004. Boyko and Funkhouser, 2011). However, the main subject of the studies for road extraction focuses on the roads lying on the ground and let alone the complex viaducts network. The problem of automatically extracting the complex viaducts network remains unsolved.

Building a model of viaducts needs not only a vivid appearance, but also detailed topology information for many applications. Point clouds of mobile laser scanning can generate viaducts model semi-automatically (Yang and et al., 2013). It builds the model with details, but it is time-consuming and has difficulties in covering large area quickly.

There are lots of algorithms developed to reconstruct the complex viaducts or overpass by combining road maps in 2D and 3D airborne point cloud (Harvey and McKeown, 2008. Chen and Lo, 2009. Wu and et al. 2014. González-Jorge and et al., 2013. Schpok, 2011.). It avoids the troubles to reconstruct topology. As the 2D road maps are manually acquired or generated from other data resources, these methods are also labour-consuming work and have its limitation in many applications.

With the increasing amounts of vehicles and the urge to avoid traffic jam, more and more viaducts are constructed. Our algorithm aims at extracting the complex viaducts network and reconstructing its topology quickly and full-automatically from the airborne LiDAR point cloud of an area.

1.2 An Introduction to Our Algorithm

Our study contains two main parts: urban viaducts recognition and topology reconstruction.

---

* Yan Wang is with School of Remote Sensing and Information Engineering, Wuhan University.
  Email: wang_yan@whu.edu.cn

The objective of urban viaducts recognition is to find the viaducts points in the point cloud. It takes 3 steps to distinguish whether a point belongs to a viaduct or not.

1) Filter the point cloud to divide all the points into ground points and none-ground points (Axelsson, 2000). The viaducts points will be in the none-ground points.

2) Find the points that locate in the boundaries between ground points and none-ground points and treat them as the seed points.

3) Viaducts are artificial structures that are built according to some specific general prescriptive designation rules. A series of features are concluded from these rules as a region growth strategy. Conduct a region growing algorithm (Garcia and et al. 2009) from the seed points according to the region growth strategy, and then all the points that belong to the viaducts can be detected.

As most of the viaducts are complex networks where a passage may has multiple forks and viaducts can be overlapped by each other. By finding the viaduct points is not enough to tell how many viaducts are there and whether they are connected or not. Our algorithm tries to reconstruct the topology of the complex viaducts networks and give answers to the questions. Below is how we rebuild the topology of the complex viaducts networks.

1) The viaduct points are projected into 2D images.

2) To remove the noise in the images, flood fill algorithm (Ho and Marshall, 1990) and morphology operations are adopted.

3) The centrelines of the viaducts are extracted and every passage of a viaduct is represented by a corresponding cubic function by least square fitting.

4) With the best fitted functions, combining the height information, we can rebuild topology relationship of the viaduct network.

2. URBAN VIADUCTS RECOGNITION

2.1 Study Region

We take Zhuodaoquan viaduct in Wuhan China as our test subject. Zhuodaoquan viaduct is one of busiest transportation junction in Wuhan and plays a key role in relieving the traffic pressure of this city.

As shown in Figure 1 and 2, there are roads, multiple layers of passes, overlapped viaducts, trees, tall and low buildings, park, lake and so on within this region. The point cloud data contains 4.3 million points with the density of 15 points/m2. What we will do next is extracting the viaducts points from the whole point cloud.

Figure 1. Overview of the Zhuodaoquan Viaduct

Figure 2. Point cloud of Zhuodaoquan Viaduct

2.2 Viaducts Extraction Algorithm

Viaducts are artificial structures that are built according to the specific general prescriptive designation rules. A series of features are concluded from these rules as a region growth strategy.

1) It's obvious that all the viaducts are above the ground and connected to the ground.

2) The curves of viaducts are generally smooth and continuous, some of which are merged, intercepted or overlapped.

3) Surfaces of the viaducts are smooth.

4) The slop of overpasses/bridges <=9%, according to 'code for design of urban road engineering' from China.

5) The width of a passage is 3.75 m according to 'General code for design of high way bridges and culverts' from China.

With a strategy composed of the above features, an iterative region growth algorithm from the seed points to the none-ground is conducted. The detail flow chart of the algorithm is shown in Figure 3.

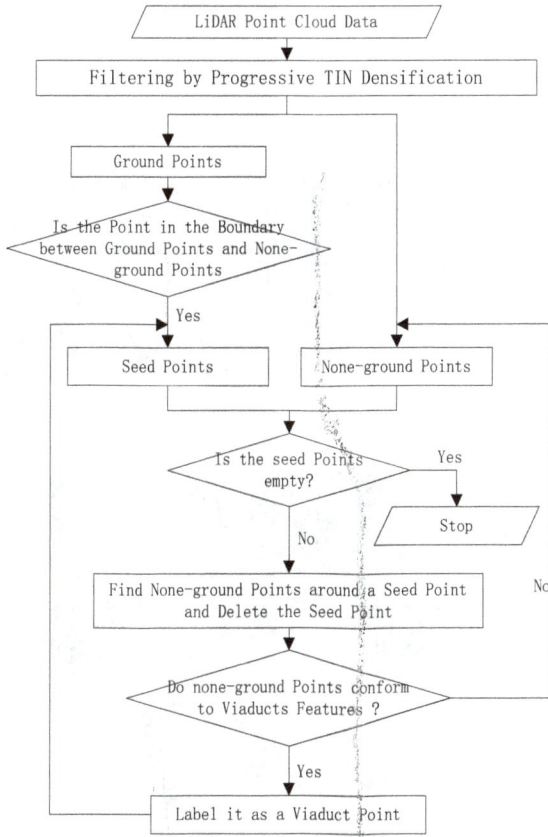

Figure 3. Algorithm Flow Chart of Extracting Viaducts Points

2.2.1 Filtering: The point cloud is filtered by progressive TIN densification (Axelsson, 2000) to divide all the points into ground points and none-ground points as in Figure 4. Purple points are ground points and yellow points are none-ground points. The viaducts points are in the none-ground points.

Figure 4. Filtered Point Cloud.

2.2.2 Seed Points Selection and Region Growth: The problems of a region growth algorithm focus on two aspects, 1how to select the seed points and what's the region growth strategy.

Selecting the original seed points is simply taking the points lying in the boundary between ground points and none-ground points.

The region growth strategy is composed of the features derived from the specific designation rules as shown in Table 1.

Criterion	Threshold
Is the point a none-ground point?	*
Is the height difference between the point and seed point $< T_{height}$?	$T_{height} = 0.03m$
Is the slop difference between the point and seed point $< T_{slop}$?	$T_{slop} = 3\%$
Is the slop of the surface fitted by the points $< T_{Sslop}$?	$T_{Sslop} = 9\%$
Is the intensity of the point $> T_{i\_min}$ and $< T_{i\_max}$?	$T_{i\_min} = 0$, $T_{i\_max} = 60$

Table 1. Margin settings for A4 size paper

The T_{height} may be different with different point densities. If a point conform all of the features, as said in Figure 3, we take this points as a new seed points and label is as a viaduct point. The result is shown in Figure 5.

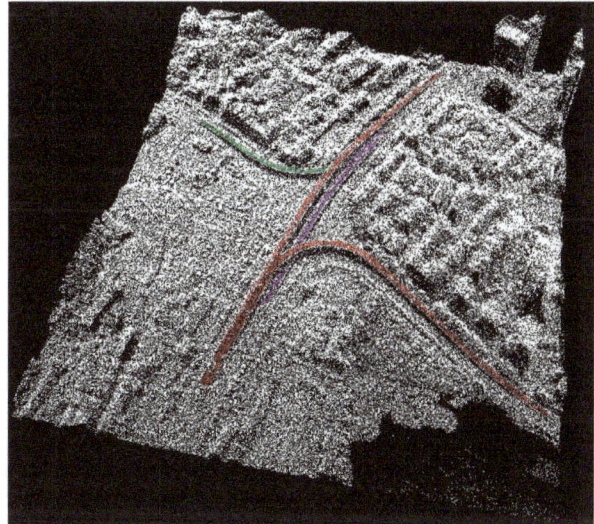

Figure 5. Viaducts Extraction Result by Seed Growing

The purple passage and green passage in Figure 5 should be connected. Because of occlusion, they break. What we will do next is bridging the gap caused by overlap and reconstructing their topology.

3. TOPOLOGY RECONSTRUCTION

We can extract the viaduct points from the point cloud now. But how many viaducts are there? Which two viaducts are connected? How many forks do the viaducts have? Which viaduct is higher? Only with correct topological structure information can we answer these questions. In this chapter, our method of reconstructing topology is introduced in detail.

It's a challenging task to rebuild a topology because of the difficulties in representing and operating a 3D curve. So the viaduct points are reversibly projected into 2D binary images like in Figure 6. The viaducts topology is analysed in 2D images and obtain 2D connectivity. And then the viaducts are projected back to 3D space in order to use the height information for generating 3D connectivity. Topological graphs of the viaducts network are finally produced.

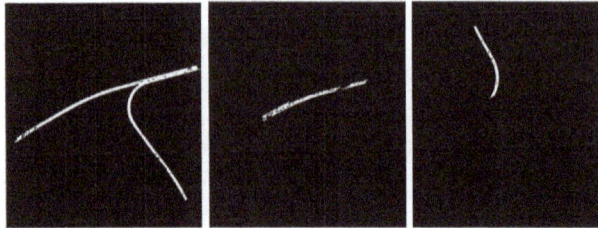

Figure 6. Viaducts in 2D Images

A whole process of reconstructing topology is shown in Figure 7.

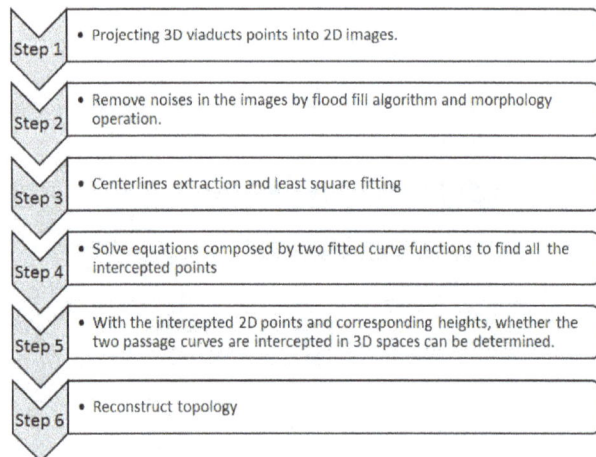

Step 1	• Projecting 3D viaducts points into 2D images.
Step 2	• Remove noises in the images by flood fill algorithm and morphology operation.
Step 3	• Centerlines extraction and least square fitting
Step 4	• Solve equations composed by two fitted curve functions to find all the intercepted points
Step 5	• With the intercepted 2D points and corresponding heights, whether the two passage curves are intercepted in 3D spaces can be determined.
Step 6	• Reconstruct topology

Figure 7. Reconstruction Flow

3.1 Remove the Noise

Because of all kinds object on the viaducts, such as cars, trees, lighting facility and so on, it leads to two kinds noise in the images when projecting the 3D points to binary images. The holes caused by the objects on the viaducts and gaps caused by occlusion.

Firstly, flood fill algorithm (Ho and Marshall, 1990) is adopted to fill all the holes in the images. Secondly, morphology operations are conducted to bridge the gaps in the images. The two de-noise steps results in inputs in good condition for the further extraction of the centrelines.

3.2 Centrelines Extraction and Least Square Fitting

Extract the centrelines of all the viaducts by Zhang-Suen algorithm(Zhang and Suen, 1984) and fit every passage with a cubic function by least square fitting, There are several reasons why to represent every passage of a viaduct with a corresponding cubic function. First of all, it is appropriate for math analysis. All the interception points can be found by solving two curve equations which is simple and reliable.

Secondly, because of the annoying noises, the results of Zhang-Suen algorithm don't show in good appearances, expressing every passage of a viaduct with a beautiful polynomials function, where cubic function is enough according to our experiments, will lead to a good visual perception (see Figure 8).

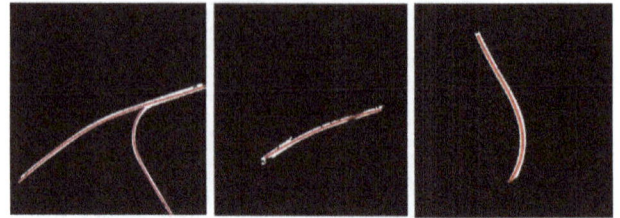

Figure 8. Cubic Function Expressions of the passages

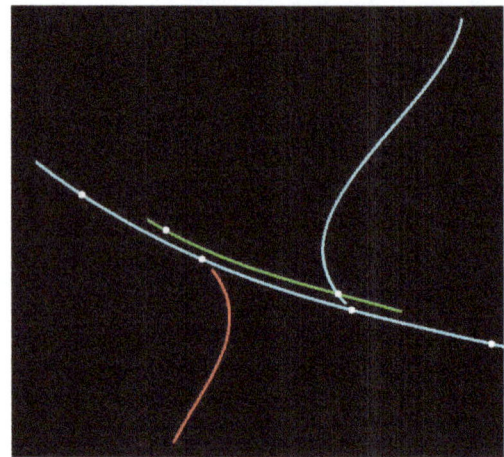

Figure 9. All the Potential Intersection Points

3.3 Topology Reconstruction

By solving the equations, intersection points can be calculated. Inversely project the intersection points into 3D space and get the height information of every intersection points, whether two passages are intersected in 3D space or not is obvious according to the height difference. As in Figure 10, topology relationship of the viaducts are reconstructed.

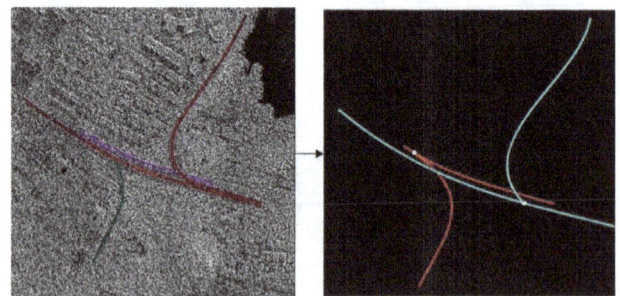

Figure 10. Topology Reconstruction Result

4. CONCLUDSION

In this paper, we have introduced a method that can automatically extract urban viaducts and reconstruct topology of the viaduct network with airborne LiDAR point cloud data. A topological graph of the viaducts network is finally produced. It is useful in urban navigation and city model reconstruction.

The future work will focus on improving the algorithm and constructing 3D digital viaducts models full-automatically.

ACKNOWLEDGEMENTS

Thanks school of remote sensing and information engineering of wuhan university for providing the airborne LiDAR point cloud data for our experiment.

REFERENCES

Mena J, B., 2003. State of the art on automatic road extraction for GIS update: a novel classification. *Journal Pattern Recognition Letters Archive*, 24(16), pp. 3037–3058.

Akel A. N., Kremeike K., Filin S., Sester, M., Doytsher, Y., 2005. Dense DTM generalization aided by roads extracted from LiDAR data. *The international Archives of the Photogrammetry, Remote Sensing and Spatial Information Sciences*, Volume XXXVI-3/W19, pp. 54-59.

Hu X, Tao V., 2007 Automatic extraction of main road centerlines from high resolution satellite imagery using hierarchical grouping. *Photogrammetric Engineering and Remote Sensing*, 73(9), pp. 1049-1056.

Rottensteiner F, Clode S., 2008. Building and road extraction by LiDAR and imagery. Topographic Laser Ranging and Scanning: Principles and Processing, Edited by Jie Shan, Charles K. Toth, CRC Press Tayloar & Francis Group, pp. 445-478.

Samadzadegan F, Hahn M, Bigdeli B., 2009. Automatic road extraction from LIDAR data based on classifier fusion. *Joint Urban Remote Sensing Event*, Shanghai, China, pp. 1-6.

Hu X, Tao C V, Hu Y., 2004. Automatic road extraction from dense urban area by integrated processing of high resolution imagery and lidar data. *International Archives of Photogrammetry, Remote Sensing and Spatial Information Sciences*, Vol. XXXV-B3, pp. 35-40.

Shao Y, Guo B, Hu X, et al., 2011. Application of a fast linear feature detector to road extraction from remotely sensed imagery. *IEEE Journal of Selected Topics in Applied Earth Observations and Remote Sensing*, 4(3), pp. 626-631.

Choi Y W, Jang Y W, Lee H J, et al., 2008. Three-dimensional LiDAR data classifying to extract road point in urban area. *Geoscience and Remote Sensing Letters, IEEE*, 5(4), pp.725-729.

Hu X, Li Y, Shan J, et al., 2014. Road Centerline Extraction in Complex Urban Scenes From LiDAR Data Based on Multiple Features. *IEEE Transactions on Geoscience and Remote Sensing*, 52(11), pp. 7448-7456.

Clode S, Kootsookos P J, Rottensteiner F., 2004. The automatic extraction of roads from LIDAR data. *International Archives of Photogrammetry, Remote Sensing and Spatial Information Sciences*, Vol. XXXV-B3, pp. 231-236.

Boyko A, Funkhouser T., 2011. Extracting roads from dense point clouds in large scale urban environment. *ISPRS Journal of Photogrammetry and Remote Sensing*, 66(6) pp. S2-S12.

Yang B, Fang L, Li J., 2013. Semi-automated extraction and delineation of 3D roads of street scene from mobile laser scanning point clouds. *ISPRS Journal of Photogrammetry and Remote Sensing*, 79, pp. 80-93.

Harvey W A, McKeown Jr D M., 2008. Automatic Compilation of 3D Road Features Using LIDAR and Multi-spectral Source Data. Portland, OR: ASPRS. Accessed January 1, 2014. http://www.terrasim.com/brochures/events/2008/2008ASPRS_P aper.pdf.

Chen L C, Lo C Y., 2009. 3D road modeling via the integration of large-scale topomaps and airborne LIDAR data. *Journal of the Chinese Institute of Engineers*, 32(6), pp. 811-823.

Wu Y, Cheng L, Wang Y, et al., 2014. Process of airborne lidar data for detection of complex overpass *Geoinformatics (GeoInformatics), 2014 22nd International Conference on. IEEE*, 2014, pp. 1-5.

González-Jorge H, Puente I, Riveiro B, et al., 2013. Automatic segmentation of road overpasses and detection of mortar efflorescence using mobile LiDAR data. *Optics & Laser Technology*, 54, pp. 353-361.

Schpok J., 2011. Geometric overpass extraction from vector road data and dsms. Proceedings of the 19th ACM SIGSPATIAL International Conference on Advances in Geographic Information Systems. ACM, 2011, pp. 3-8.

Axelsson P., 2000. DEM generation from laser scanner data using adaptive TIN models. *International Archives of Photogrammetry and Remote Sensing*, 33(B4/1; PART 4) pp. 111-118.

Garcia Ugarriza L., Saber E., Vantaram S. R., Amuso, V., Shaw M. and Bhaskar R., 2009. Automatic image segmentation by dynamic region growth and multiresolution merging. *IEEE Transactions on Image Processing*, 18(10), pp. 2275-2288.

Ho C M W, Marshall G R., 1990. Cavity search: an algorithm for the isolation and display of cavity-like binding regions. *Journal of computer-aided molecular design*, 4(4), pp. 337-354.

Zhang T Y, Suen C Y., 1984. A fast parallel algorithm for thinning digital patterns. *Communications of the ACM*, 27(3), pp. 236-239.

21

SEMANTIC INTERPRETATION OF INSAR ESTIMATES USING OPTICAL IMAGES WITH APPLICATION TO URBAN INFRASTRUCTURE MONITORING

Yuanyuan Wang [a], Xiao Xiang Zhu [a, b, *]

[a] Helmholtz Young Investigators Group "SiPEO", Technische Universität München, Arcisstraße 21, 80333 Munich, Germany.
wang@bv.tum.de
[b] Remote Sensing Technology Institute (IMF), German Aerospace Center (DLR), Oberpfaffenhofen, 82234 Weßling, Germany.
xiao.zhu@dlr.de

Commission III, WG III/4

KEY WORDS: optical InSAR fusion, semantic classification, InSAR, SAR, railway monitoring, bridge monitoring

ABSTRACT:

Synthetic aperture radar interferometry (InSAR) has been an established method for long term large area monitoring. Since the launch of meter-resolution spaceborne SAR sensors, the InSAR community has shown that even individual buildings can be monitored in high level of detail. However, the current deformation analysis still remains at a primitive stage of pixel-wise motion parameter inversion and manual identification of the regions of interest. We are aiming at developing an automatic urban infrastructure monitoring approach by combining InSAR and the semantics derived from optical images, so that the deformation analysis can be done systematically in the semantic/object level. This paper explains how we transfer the semantic meaning derived from optical image to the InSAR point clouds, and hence different semantic classes in the InSAR point cloud can be automatically extracted and monitored. Examples on bridges and railway monitoring are demonstrated.

1. INTRODUCTION

1.1 Deformation Monitoring by SAR Interferometry

Long term deformation monitoring over large area is so far only achievable through differential SAR interferometry (InSAR) techniques such as persistent scatterer interferometry (PSI) (Adam et al., 2003; Ferretti, Prati & Rocca, 2001; Gernhardt & Bamler, 2012; Kampes, 2006) and differential SAR tomography (TomoSAR) (Fornaro et al., 2015; Fornaro, Reale & Serafino, 2009; Lombardini, 2005; Zhu & Bamler, 2010a; Zhu & Bamler, 2010b). Through modelling the interferometric phase of the scatterers in the SAR image, we are able to reconstruct their 3-D positions and the deformation histories.

The focus of development in differential InSAR techniques has always been on the estimation of the phase history parameters (elevation, motion parameters, etc.) under different scattering models including single deterministic scattering (persistent scatterer), distributed scattering (distributed scatterer), and layover of multiple scatterings (TomoSAR).

In regard of large area deformation monitoring, PSI is the workhorse among the InSAR methods. Distributed scatterer (DS)-based methods such as SqueeSAR and its alternatives (Ferretti et al., 2011; Jiang et al., 2015; Wang, Zhu & Bamler, 2012) have emerged in the last few years to complement the drawback of few PS in nonurban area, while TomoSAR has become the most competent method for urban area monitoring because of its capability of separating multiple scatterers in a resolution cell. With meter-resolution SAR data, it is also demonstrated that even individual building could be monitored in very high level of detail from space.

In summary, great progress has been made in inversion problems of the coherent signals from SAR data stacks.

Millimetre-precision in the linear deformation rate can be achieved (Ferretti, Prati & Rocca, 2001; Bamler et al., 2009).

1.2 Motivation

The current D-InSAR methods are able to produce excellent deformation estimates. However, they are based on pixel-wise parameters inversion and manual identification of the region of interest. It lacks a systematic way to monitor the region of interest, for example, the railway or the road network in a city. Therefore, the next generation InSAR techniques in urban area should be aimed towards exploiting and understanding the regularities and semantics of the manmade world that is imaged.

With such vision in mind, we aim to bridge the InSAR and optical field by complementing InSAR's high precision deformation measurement with the high interpretability of optical images.

1.3 Methodology

This work is aimed towards a future generation of InSAR techniques that are contextually aware of the semantics in a SAR image. It enables the object-level deformation reconstruction and analysis from SAR images. The proposed approach brings the first such analysis via a semantic classification in the InSAR point cloud.

The general framework of the proposed approach is shown in Figure 1. The semantic classification of the InSAR point cloud is achieved by co-registering the InSAR point cloud and an optical image to a common reference 3-D model, so that the semantic classification in the optical image can be transferred to the InSAR point cloud. The general procedures are as follows.

---

* Corresponding author

a. Retrieve the 3-D positions of the scatterers from SAR image stacks. Since urban area is of our main interest, tomographic SAR inversion should be employed in order to resolve a substantial amount of layovered scatterers.

b. Absolute geo-reference the 3-D InSAR point cloud, due to the relative position of the InSAR point cloud w.r.t. a reference point. This step is achieved by co-registering the InSAR point cloud with a reference 3-D model.

c. Texturing the reference 3-D model with high resolution optical images, so that each SAR scatterer can be traced in the optical image.

d. Classify the optical image pixels based on its semantic meaning, e.g. geometry, material, and so on.

e. Perform further analysis on object-level in the InSAR point cloud based on their semantic class.

Since the fusion of the InSAR point cloud and the optical image is done in pure 3-D, which required strict 3-D reconstruction from both SAR images and the optical images, the work described in this paper is also different from many early research on SAR and optical image fusion, such as (Gamba & Houshmand, 2002; Wegner, Ziehn & Soergel, 2014; Wegner, Thiele & Soergel, 2009).

To summarize, the proposed method requires in addition only a stereo pair of optical images of the same area. The InSAR point cloud is co-registered with the 3-D point cloud derived from the optical image pair, and is therefore, co-registered to the classification derived from the optical images. Additionally, the user could also use reference model from different sources, e.g., LiDAR.

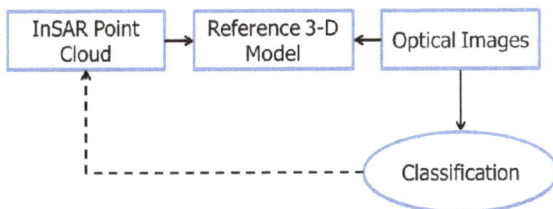

Figure 1. Flowchart of the proposed method. The semantic classification of the InSAR point cloud is achieved by co-registering the InSAR point cloud and the optical image to a reference model.

2. SAR TOMOGRAPHIC INVERSION

TomoSAR aims at separating multiple scatterers possibly layovered in the same pixel, and retrieving their third coordinates *elevation* in the SAR native coordinate system. Displacement of the scatterers can also be modeled and estimated, using stack of images acquired at different times. The technique is commonly known as differential SAR tomography (D-TomoSAR)

The D-TomoSAR processing was done by DLR's D-TomoSAR software Tomo-GENESIS. For an input data stack, Tomo-GENESIS retrieves the following information:

- the number of scatterers inside each pixel,
- the scattering amplitude and phase of each scatterer,
- and their 3D positions and motion parameters, e.g. linear deformation rate and amplitude of seasonal motion.

The scatterers' 3D positions in SAR coordinates are converted into a local Cartesian coordinate system, such as Universal Transverse Mercator (UTM), so that the results from multiple data stacks with different viewing angles can be combined. For our test area Berlin, two TerraSAR-X high resolution image stacks – one ascending orbit, the other descending orbit – are processed. These two point clouds are fused to a single one, following a feature-based matching algorithm which estimates and matches common building edges in the two point clouds (Wang & Zhu, 2015). Figure 2 is the fused point cloud which provides a complete monitoring over the whole city of Berlin.

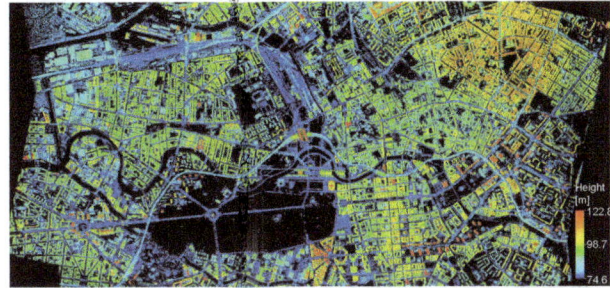

Figure 2. The fused TomoSAR point cloud of Berlin, which combines the result from an ascending stack and a descending stack. The height is color-coded.

3. GEOREFERENCE TOMOSAR POINT CLOUDS

Due to the relative position of the TomoSAR point cloud w.r.t. a reference point selected during the processing, it must be co-registered to a reference 3-D model in order to align with an optical image.

Our reference model is the 3-D point cloud derived from a pair of optical images by means of stereo matching. As a by-product, the optical images are also geo-localized in the optical point cloud. Other precise reference 3-D model such as LiDAR point cloud can also be used, with the additional effort of aligning it with the optical image.

3.1 Co-registration workflow

Since both the TomoSAR and optical point clouds are geo-coded into local coordinate systems, the co-registration problem is the estimation of translation between two rigid point clouds, subject to a certain tolerance on rotation and scaling. However, the optical point cloud is nadir-looking, in contrast to the side-looking geometry of SAR. In another word, façade point barely appears in optical point cloud while it is prominent in TomoSAR point cloud. This difference is exemplified in Figure 3, where the left and the right subfigures correspond to the TomoSAR and optical point clouds of the same area. The same conclusion should also apply to nadir-looking LiDAR point cloud. These unique modalities have driven our algorithm to be developed in the following way:

1   Edge extraction
   a. The optical point cloud is rasterized into a 2D height image.
   b. The point density of TomoSAR point cloud is estimated on the rasterized 2D grid.
   c. The edges in the optical height image and the TomoSAR point density image are detected.
2   Initial alignment
   a. Horizontally by cross-correlating the two edge images.

    b. Vertically by cross-correlating the height histogram of the two point clouds.

3   Refined solution

    a. The façade points in both point clouds are removed.

    b. The final solution is obtained using iterative closest point (ICP) applied on the two reduced point clouds.

(a)

(b)

Figure 3. (a) TomoSAR point cloud of high-rise buildings, and (b) the optical point cloud of the same area. Building façades are almost invisible in the optical point cloud, while it is prominent in the TomoSAR point cloud.

3.2 2-D Edge Extraction

The 2-D edge images of the TomoSAR and optical point cloud are extracted from their rasterized height image, and point density image, respectively. Here we use 2×2 m for our dataset. For the optical point cloud, the mean height in each grid cell is computed, while for the TomoSAR point cloud, the number of points inside the grid cell is counted. The edges can be extracted from these two images using an edge detector, such as Sobel filter (Sobel, 1968). The thresholds in the edge detector are decided adaptively, so that the numbers of edge pixels in the two edge images are on the same order. Figure 4 is a close up view of the two edge images near downtown Berlin.

3.3 Initial Alignment

The initial alignment provides an initial solution to the iterative closest point (ICP) algorithm which is known to suffer from finding possibly a local minimum. The initial alignment consists of independently finding the horizontal and the vertical shifts. The horizontal shift is found by cross-correlating the edge images of the two point clouds. In most of the cases, a unique peak can be found, due to the complex, hence pseudorandom, structures of a city. Please see Figure 5 for the 2D correlation of two edge images, where a single prominent peak is found. The vertical shift is found by cross-correlating the height histogram of the two point clouds, which is shown in Figure 6. We also set the bin spacing of the height histograms to be 2m in our experiment. The accuracy of the shift estimates are

of course limited by the discretization in the three directions. However, this is sufficient for the final estimation.

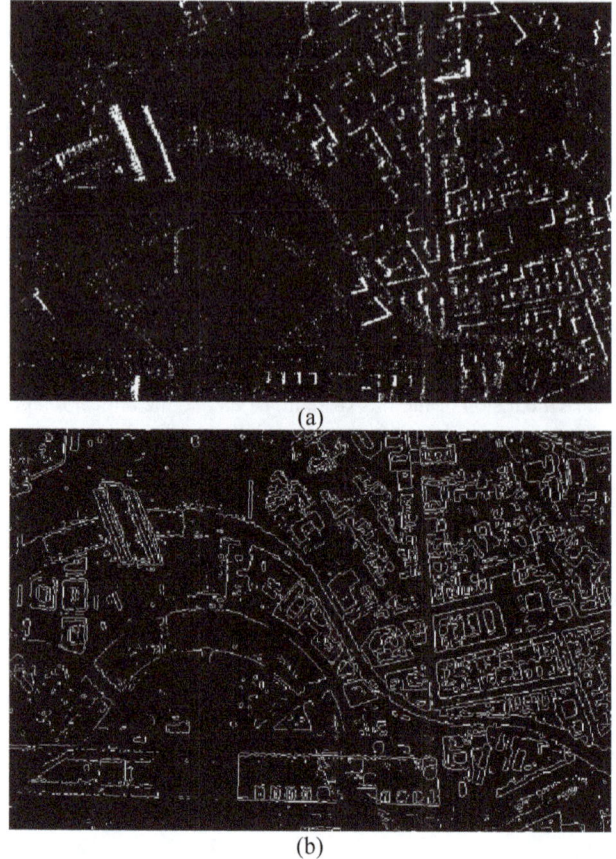

(a)

(b)

Figure 4. (a) A crop of the edge image of the TomoSAR point cloud in downtown Berlin, and (b) the edge image of the reference optical point cloud roughly at the same area.

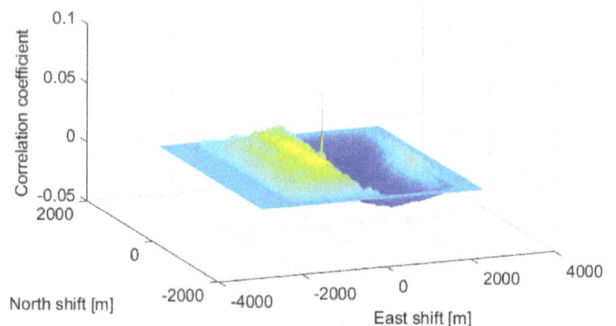

Figure 5. 2D cross-correlation of the edge images of TomoSAR and optical point clouds. Due to the pseudorandom nature of the urban infrastructure, a single prominent peak can always be found.

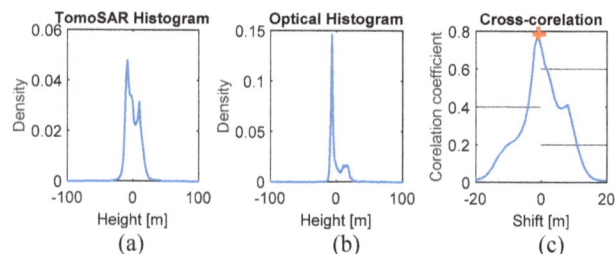

(a)   (b)   (c)

Figure 6. (a) The height histogram of TomoSAR point cloud, (b) the height histogram of LiDAR point cloud, and (c) the correlation of (a) and (b), where the red cross marks the peak position which is at -2 m.

3.4 Final Solution

The final solution is obtained using a normal ICP algorithm based on the initial solution calculated from the previous step. ICP solves the following equation

$$\left\{\hat{\mathbf{R}}, \hat{\mathbf{t}}\right\} = \min \sum_i \left\| \mathbf{x}_i - \mathbf{R}\mathbf{p}_i - \mathbf{t}\right\|_2^2 \qquad (1)$$

where $\mathbf{R}$ and $\mathbf{t}$ are the rotation matrix and the translation vector, and $\mathbf{x}_i$ and $\mathbf{p}_i$ are the corresponding point pair. Given correct point pairs of the two point clouds, (1) can be easily solved. Assuming the closest points being the corresponding point pair, ICP iteratively improves the co-registration results. However, it suffers from finding local minimum. Therefore, giving a good initial estimate is the key to the success of ICP.

In our implementation, the façade points in the TomoSAR point clouds are removed to prevent ICP from finding a wrong solution. Figure 7(a) demonstrates the co-registered point cloud combining the optical images-derived one and two TomoSAR point clouds from ascending and descending viewing angles with color representing the height. Successful co-registration can be confirmed by seeing the correct location of the façade points in Figure 7(b) which shows the top view of fused point cloud with different colors representing different point clouds.

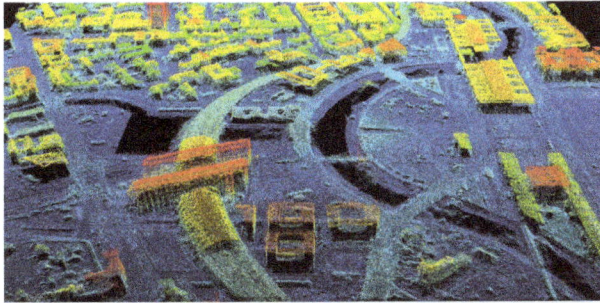

(a)

(b)

Figure 7. (a) the fused point cloud combining the optical images-derived one and two TomoSAR point clouds from ascending and descending viewing angle with color representing the height, and (b) the top view of co-registered point cloud where red, blue, green representing the points from descending TomoSAR, optical, and ascending TomoSAR, respectively.

4. SEMANTIC CLASSIFICATION IN OPTICAL IMAGE

The semantic classification is done in a sliding patch manner. Each patch is described using a dictionary, to be specific, the occurrence of the atoms in the dictionary. Such model is known

as the Bag of Words (BoW) (Csurka et al., 2004). The final patch classification is achieved using support vector machine (SVM). The detailed workflow is as follows.

4.1 BoW Model

BoW originates from text classification, where a text is modeled as the occurrence of the words in a dictionary, disregarding the grammar as well as the order. This is also recently employed in computer vision, especially in image classification. Analogous to text, the BoW descriptor $\mathbf{w}$ of an image (in our case an image patch) $\mathbf{Y}$ is modeled as the occurrence of the "visual" words in a predefined dictionary $\mathbf{D}$, i.e.:

$$\mathbf{w} = h_{\mathbf{D}}\left(\psi\left(\mathbf{Y}\right)\right) \qquad (2)$$

where $h(\cdot)$ is the histogram operator, and $\psi(\cdot)$ is the transformation function from the image space to the feature space. Hence the visual words refer to the representative features in the image, whose ensemble constructs the dictionary.

4.2 Dictionary Learning

Define the dictionary matrix as $\mathbf{D} \in \mathbb{R}^{N \times k}$, where N is the dimension of the word, i.e. feature vector/atom, and k is the number of words. The k feature vector should include representative features appearing in the whole image, so that each patch can be well described. Therefore, the dictionary is usually overcomplete.

We adopt a dictionary learning approach commonly used in the computer vision community:

1.   Sample the whole image sufficiently dense, and computing the feature at each sampled location. To this end, a large number of feature vectors are collected.

2.   Reduce the number of feature vectors by quantization in the feature space. Here, we perform an unsupervised clustering, e.g. k-means. The cluster centroids are extracted as the final dictionary. Figure 8 exemplify the quantization in a 2-D feature space. The colored crosses are the features extracted from the whole image. The cluster centroids are the final words in the dictionary.

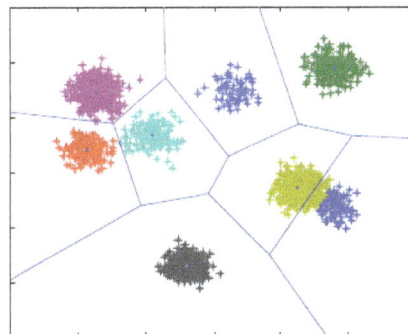

Figure 8. Demonstration of dictionary learning in two dimensional feature space. The colored crosses are the features collected from all the patches in the image. A k-means clustering is performed to get k cluster centers, i.e. the dictionary atoms. Image modified from (Cui, 2014).

4.3 Patch Description

Each patch is described using Equation (2). Similar to the feature extraction in the dictionary learning step, we calculate the dense local features of each patch, i.e. the feature is computed in a sliding window through the patch. This is described in Figure 9 where the red window traverses the patch, and computes one local feature vector at each position.

The descriptor of a patch is the occurrence (histogram) of the collected features in the dictionary. This is calculated by assigning the features to their nearest neighbours in the dictionary words. To this end, the patch descriptor is a vector $\mathbf{v} \in \mathbb{R}^k$.

Several commonly used features have been tested, which includes the most popular scale-invariant feature transform (SIFT) suggested by many literatures. In our problem, the simple vectorized RGB pixel values in a 3×3 sliding window turned out to be the most appropriate feature, and hence are selected for further classification.

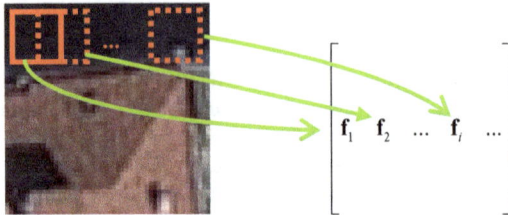

Figure 9. Demonstration of dense local feature computed in an image patch. The feature is computed in the red sliding window, and the local features are denoted as $\mathbf{f}_i$ in the

4.4 Classification

The classification is done using a linear SVM (Cortes & Vapnik, 1995) implemented in an open source library VLFeat (Andrea Vedaldi, 2010). The SVM classifier finds a hyperplane which separates two classes of training samples with maximal margin. Giving the patch descriptor $\mathbf{v}$, its SVM classification is:

$$f(\mathbf{v}) = sign(\mathbf{w}^T \mathbf{v} + b) \tag{3}$$

where $\mathbf{w} \in \mathbb{R}^k$ and b are the parameters of the hyperplane, and $sign(\bullet)$ is the sign operator which outputs ±1.

For an m-class ($m>2$) problem, we follow the one-against-rest approach. Different SVM is trained for each class. The final classification of a patch $\mathbf{v}$ is assigned to the one with the largest SVM score, i.e.:

$$f(\mathbf{v}) = \max(\mathbf{W}^T \mathbf{v} + \mathbf{b}) \tag{4}$$

where $\mathbf{W} \in \mathbb{R}^{k \times m}$ and $\mathbf{b} \in \mathbb{R}^m$ are the concatenated parameters of m hyperplanes.

We classify every 4×4 pixel in our test image (5000×5000) taking into account the 50×50 pixel patch around it. We manually selected 570 50×50 pixel patches as training samples. Four classes are preliminarily defined: building, roads/rail, river, and vegetation. Each of them has 240, 159, 39, and 132 training patches, respectively. The feature in our experiment is simply the vectorized RGB pixel values in a 3×3 sliding window, which results in a feature space of 27 dimensions.

Figure 10 shows the classification result of a region in the entire image, where the left image is the optical image, and in the right image, classified building, road, river, and vegetation depicted in red, blue, green, and blank respectively. Despite the extremely simple feature we used, the four classes are very well distinguished.

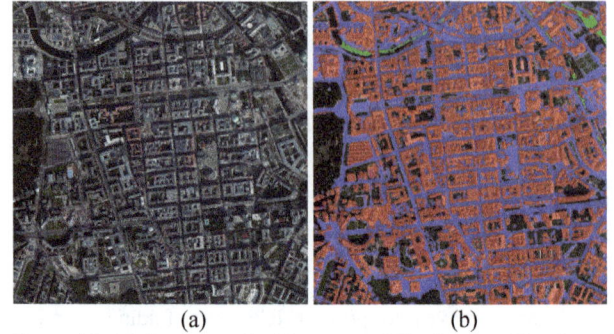

Figure 10. (a) the test optical image, and (b) the classification of building, road, river, and vegetation, where they are colored in red, blue, green, and blank.

Since we are particularly interested in building, its classification performance is evaluated by classifying half of training samples using the SVM trained with the other half of the samples. The average precision of the current algorithm is 98%. The full precision and recall curve is plotted in Figure 11(a). The equivalent receiver operating characteristic curve is also shown in Figure 11(b), for the readers who are more familiar with it. The red cross marks our decision threshold which gives a detection rate of 90%, and false alarm rate of 3%.

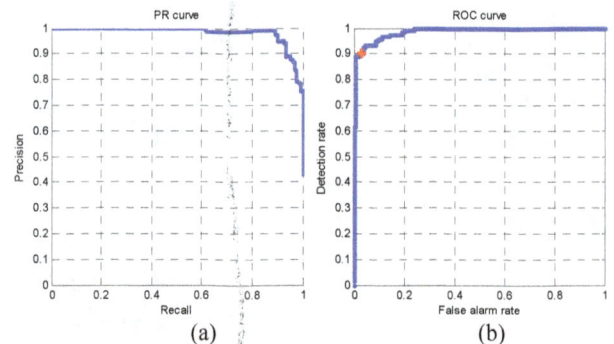

Figure 11. (a) precision and recall curve of the building classification with an average precision is 98%, and (b) the ROC curve of the classification. The red cross marks our decision point which gives a detection rate of 90%, and false alarm rate of 3%.

5. SEMANTIC-LEVEL ANALYSIS

5.1 Automatic Railway Monitoring

We applied the semantic classification scheme on an orthorectified optical image centered at the Berlin central station. For this analysis, we particularly classified the railway and river class. Figure 12 shows the classification map where the railway class and river class are labelled in green and red, respectively. The classification performance is consistent with the evaluation shown in Figure 11. Some false alarm appeared as small clusters, but they can be removed by post-processing.

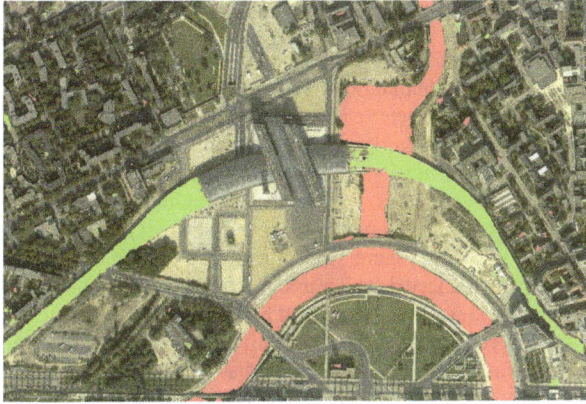

Figure 12. River (red) and railway (green) classified using the BoW method. The classification performance is consistent as the evaluation in Figure 11 shows.

5.1.1   Railway classification refinement using smooth spline

Based on the classification, the corresponding points in the TomoSAR point cloud can be extracted. Assuming the railway is smooth and continuous, a smooth spline function was fitted to the x and y (east and north) coordinates of the railway points to connect separated segments, i.e.:

$$\hat{\mathbf{s}} = \arg\min_{\mathbf{s}} \left\{ \lambda \|\mathbf{y} - \mathbf{s}\|_2^2 + (1 - \lambda) \|\Delta\mathbf{s}\|_2^2 \right\} \tag{5}$$

where $\mathbf{y}$ is the y (north) coordinates of the railway points, $\mathbf{s}$ is the spline function (quadratic or cubic) w.r.t. the x (east) coordinates of the railway points, Δ is the Laplace operator, and $\lambda \in [0,1]$ the smoothing parameter. The regularization on the L2 norm of the second order derivative grant the smoothness of the spline function. The smooth spline is centered in the railway, and the width of the railway is adaptively estimated at each position. Therefore, we are able to interpolate the gap of the railway due to miss classification (here due to the presence of the Berlin central station). Figure 13(a) shows the extracted continuous railway points overlaid on the optical image. The color shows the amplitude of seasonal motion caused by thermal dilation.

5.1.2   Total variation denoising

Due to the high dynamics of the motion in high resolution SAR data, we introduce an additional step of denoising in order to retrieve some higher level information such as the joints of the railway sections.

Because the thermal dilation of the steel railway beam is mostly proportional to its length (Kerr, 1978), it is plausible to assume that the scatterers on the same cross-section of the railway undergo identical deformation. Therefore, the scatterers' deformation along the railway can be transformed into one dimension, i.e. the railway distance. The original deformation estimates in the radar's line-of-sight direction must also be projected to the railway direction.

In order to preserve the edge and the piecewise linear structure of the railway deformation parameters which can be observed in Figure 13(a), we employ a minimization of total variation of the second order derivative of the deformation function along the railway:

$$\hat{\mathbf{g}} = \arg\min_{\mathbf{g}} \left\{ \frac{1}{2} \|\mathbf{g} - \mathbf{v}\|_2^2 + \lambda \|\Delta\mathbf{g}\|_1 \right\} \tag{6}$$

where $\mathbf{v}$ is the deformation estimates along the railway direction, $\mathbf{g}$ is its denoised version. As shown in literatures of total generalized variation (Bredies, Kunisch & Pock, 2010; Knoll et al., 2011), the L1 norm of second order derivative is convex and lower semi-continuous, one can solve it using convex optimization solvers. Figure 14 (a) shows the original and denoised deformation estimates as a function of the railway direction. The edges due to different railway segments are clearly preserved. Figure 13(b) shows the denoised deformation estimates re-projected back into the world coordinate, in order to have a visual comparison with Figure 13(a).

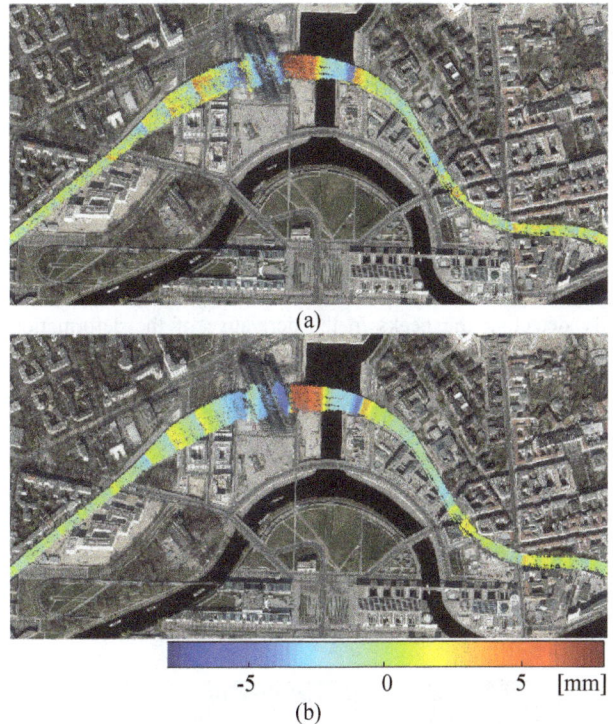

(a)

(b)

Figure 13. (a) Railway points extracted from the TomoSAR point cloud. The color shows the amplitude of seasonal motion due to the thermal expansion of the steel, and (b) denoised deformation estimates using minimization of total variation.

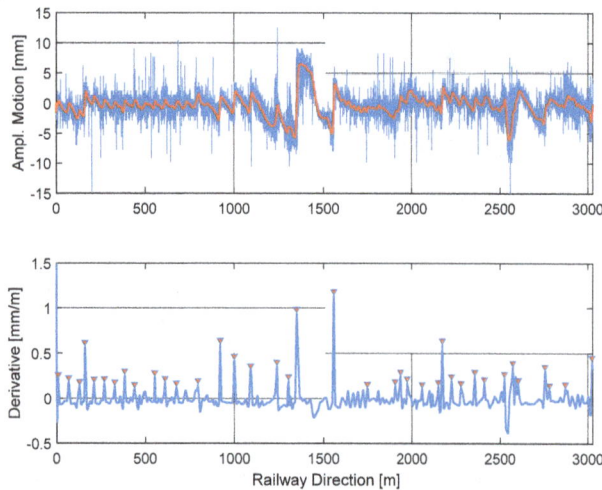

Figure 14. (a) Original estimates of the amplitude of seasonal deformation in blue, and the denoised deformation estimates in red estimated by regularizing on the total variation of its second order derivative, and (b) peaks detected in the derivative of the denoised deformation function along the railway direction. They are the locations of the railway segment joints.

5.1.3   Railway joints detection

By detecting the peaks in the derivative of the deformation function along the railway direction, the joints of railways can be detected. Constraint is put on the minimum distance between two peaks representing the minimum length of a railway segment. The detected peaks in the deformation's derivative can be seen in (b). The positions of the peaks on the railway are shown as the green dots in Figure 15(a). Each green dot represents the midpoint of the joint cross-section. In Figure 15(c), we provide the close up view of the two joints in the optical image with 20cm pixel size. It can be clearly observed that the railway joint shown up as dark lines in the optical image.

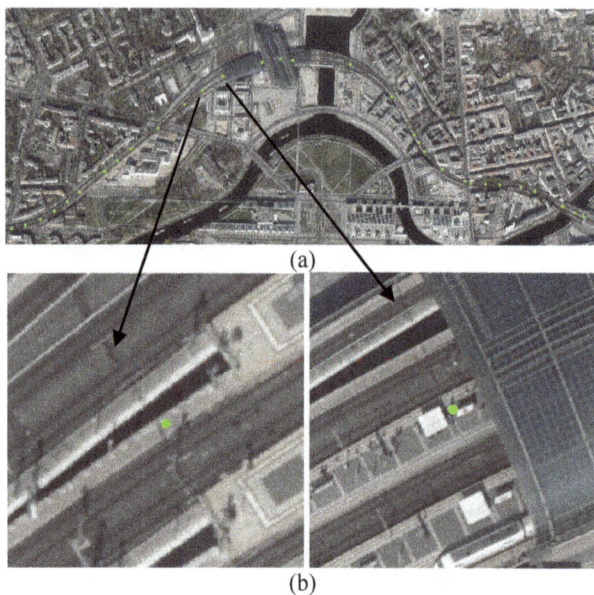

Figure 15. (a) the midpoint of the detected railway joint cross-section marked in green, and (b) close up view of the railway joint. The background optical image has a ground spacing of 20cm.

5.2   Automatic Bridge Monitoring

By analysing the discontinuity of the river segmentation and assuming the discontinuities are caused by bridges, the bridges' positions can be detected automatically. The workflow of identifying all the bridge is fairly simple. Starting with the initial river classification, small clusters are removed by setting threshold on the area of connected region. Followed by morphological closing operation, the gaps caused by bridges are closed as shown in Figure 16(b). Lastly, by finding the convex hull of each connected region in the difference image of Figure 16(b) and (a), the bridge mask can be identified. As we can see, all the bridges in Figure 12 are identified, except a very narrow one that was originally false classified as river.

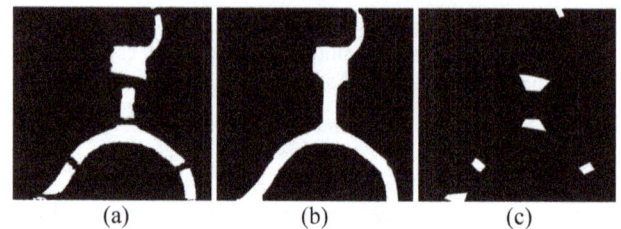

Figure 16. (a) initial river mask with false detected small cluster removed, (b) connected river mask by applying morphological closing, and (c) the final bridge mask by finding the convex hull of each region in the difference between (b) and (a).

The corresponding bridge points are extracted from the TomoSAR point cloud, and projected to the optical image. The projected bridge points are shown in Figure 17 where the color represents the amplitude of seasonal deformation. The upper most bridge belongs to a segment of the railway which is known to have thermal expansion. The middle bridge undergoes a 5mm seasonal motion in its west end and 2mm at the east end. This suggests a more rigid connection of the bridge with the foundation at its east end. The two lower bridges are stable according to the motion estimates.

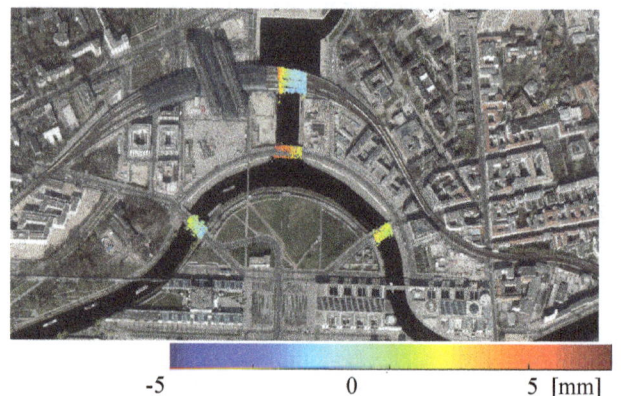

Figure 17. Overlay of the amplitude of seasonal motion of brides extracted from the TomoSAR point cloud on the optical image. The bridges are automatically detected from the classification map shown in Figure 12 using discontinuity analysis.

CONCLUSION

This paper presents the first systematic semantic analysis of very high resolution InSAR point cloud in urban area. Through co-registering optical image and InSAR point cloud to a common reference 3-D model, we are able to relate the semantic meaning extracted from the optical image to the

InSAR point cloud. The complementary information provided by the two data types enables an object-level InSAR deformation and 3-D analysis.

In the future, we will include more semantic classes, such as high-rise buildings, residential area, or even specific landmarks, and so on. To reduce the human interaction, we aim at a completely unsupervised semantic classification.

REFERENCES

Adam, N., Kampes, B., Eineder, M., Worawattanamateekul, J. & Kircher, M. 2003. The development of a scientific permanent scatterer system. In *ISPRS Workshop High Resolution Mapping from Space, Hannover, Germany*, pp. 6.

Andrea Vedaldi, B.F. 2010. VLFeat: an open and portable library of computer vision algorithms. In *Proceedings of the 18th International Conference on Multimedea 2010*, pp. 1469–1472. , Firenze, Italy.

Bamler, R., Eineder, M., Adam, N., Zhu, X. & Gernhardt, S. 2009. Interferometric Potential of High Resolution Spaceborne SAR. *Photogrammetrie - Fernerkundung - Geoinformation* 2009(5), 407–419.

Bredies, K., Kunisch, K. & Pock, T. 2010. Total Generalized Variation. *SIAM Journal on Imaging Sciences* 3(3), 492–526.

Cortes, C. & Vapnik, V. 1995. Support-vector networks. *Machine Learning* 20(3), 273–297.

Csurka, G., Dance, C., Fan, L., Willamowski, J. & Bray, C. 2004. Visual categorization with bags of keypoints. In *Workshop on statistical learning in computer vision, ECCV*, pp. 1–2.

Cui, S. 2014. Spatial and temporal SAR image information mining. Ph.D. thesis. Universität Siegen.

Ferretti, A., Prati, C. & Rocca, F. 2001. Permanent scatterers in SAR interferometry. *IEEE Transactions on Geoscience and Remote Sensing* 39(1), 8–20.

Ferretti, A., Fumagalli, A., Novali, F., Prati, C., Rocca, F. & Rucci, A. 2011. A New Algorithm for Processing Interferometric Data-Stacks: SqueeSAR. *IEEE Transactions on Geoscience and Remote Sensing* 49(9), 3460–3470.

Fornaro, G., Reale, D. & Serafino, F. 2009. Four-Dimensional SAR Imaging for Height Estimation and Monitoring of Single and Double Scatterers. *IEEE Transactions on Geoscience and Remote Sensing* 47(1), 224–237.

Fornaro, G., Verde, S., Reale, D. & Pauciullo, A. 2015. CAESAR: An Approach Based on Covariance Matrix Decomposition to Improve Multibaseline-Multitemporal Interferometric SAR Processing. *IEEE Transactions on Geoscience and Remote Sensing* 53(4), 2050–2065.

Gamba, P. & Houshmand, B. 2002. Joint analysis of SAR, LIDAR and aerial imagery for simultaneous extraction of land cover, DTM and 3D shape of buildings. *International Journal of Remote Sensing* 23(20), 4439–4450.

Gernhardt, S. & Bamler, R. 2012. Deformation monitoring of single buildings using meter-resolution SAR data in PSI. *ISPRS Journal of Photogrammetry and Remote Sensing* 73, 68–79.

Jiang, M., Ding, X., Hanssen, R.F., Malhotra, R. & Chang, L. 2015. Fast Statistically Homogeneous Pixel Selection for Covariance Matrix Estimation for Multitemporal InSAR. *IEEE Transactions on Geoscience and Remote Sensing* 53(3), 1213–1224.

Kampes, B.M. 2006. Radar *Interferometry - Persistent Scatterer Technique*, Dordrecht, The Netherlands: Springer, 211p.

Kerr, A.D. 1978. Analysis of thermal track buckling in the lateral plane. *Acta Mechanica* 30(1-2), 17–50.

Knoll, F., Bredies, K., Pock, T. & Stollberger, R. 2011. Second order total generalized variation (TGV) for MRI. *Magnetic resonance in medicine* 65(2), 480–491.

Lombardini, F. 2005. Differential tomography: a new framework for SAR interferometry. *IEEE Transactions on Geoscience and Remote Sensing* 43(1), 37–44.

Sobel, I. 1968. An Isotropic 3x3 Image Gradient Operator. *Presentation at Stanford A.I. Project 1968.*

Wang, Y. & Zhu, X. 2015. Automatic Feature-based Geometric Fusion of Multi-view TomoSAR Point Clouds in Urban Area. *IEEE Journal of Selected Topics in Applied Earth Observation and Remote Sensing* 8(3), 953 – 965.

Wang, Y., Zhu, X. & Bamler, R. 2012. Retrieval of Phase History Parameters from Distributed Scatterers in Urban Areas Using Very High Resolution SAR Data. *ISPRS Journal of Photogrammetry and Remote Sensing* 73, 89–99.

Wegner, J.D., Thiele, A. & Soergel, U. 2009. Fusion of optical and InSAR features for building recognition in urban areas. *International Archives of the Photogrammetry, Remote Sensing and Spatial Information Sciences* 38(Part 3), W4.

Wegner, J.D., Ziehn, J.R. & Soergel, U. 2014. Combining High-Resolution Optical and InSAR Features for Height Estimation of Buildings With Flat Roofs. *IEEE Transactions on Geoscience and Remote Sensing* 52(9), 5840–5854.

Zhu, X. & Bamler, R. 2010a. Tomographic SAR Inversion by L1-Norm Regularization -- The Compressive Sensing Approach. *IEEE Transactions on Geoscience and Remote Sensing* 48(10), 3839–3846.

Zhu, X., & Bamler, R. 2010b. Very High Resolution Spaceborne SAR Tomography in Urban Environment. *IEEE Transactions on Geoscience and Remote Sensing* 48 (12), 4296–4308.

22

INTEGRATION OF REMOTE SENSING DATA AND BASIC GEODATA AT DIFFERENT SCALE LEVELS FOR IMPROVED LAND USE ANALYSES

G. Waldhoff [a, *], S. Eichfuss [a], G. Bareth [a]

[a] Institute of Geography, University of Cologne, Albertus-Magnus-Platz, 50923 Cologne, Germany - (guido.waldhoff, seichfus, g.bareth)@uni-koeln.de

Commission VII, WG VII/5

KEY WORDS: GIS, Land Use, Digital Landscape Model, Cadastral Data, WorldView-2, ATKIS, ALK, Multi Data Approach

ABSTRACT:

The classification of remote sensing data is a standard method to retrieve up-to-date land use data at various scales. However, through the incorporation of additional data using geographical information systems (GIS) land use analyses can be enriched significantly. In this regard, the Multi-Data Approach (MDA) for the integration of remote sensing classifications and official basic geodata for a regional scale as well as the achievable results are summarised. On this methodological basis, we investigate the enhancement of land use analyses at a very high spatial resolution by combining WorldView-2 remote sensing data and official cadastral data for Germany (the Automated Real Estate Map, ALK). Our first results show that manifold thematic information and the improved geometric delineation of land use classes can be gained even at a high spatial resolution.

1. INTRODUCTION

Satellite remote sensing is a standard method for the generation of land use and land cover data for global or regional investigations. Also at local scales, the method is still competitive using high spatial resolution data, where up-to-date information is needed for wider areas. For the global scale, remote sensing is often the only way to acquire coherent land cover information (Mora et al., 2014). However, for the regional and the local scale several additional sources of spatial information can be incorporated in a GIS to enrich the land use analysis. Although initiatives for the integration of remote sensing and GIS date back to the late 1970s (Hutchinson, 1982; Merchant and Narumalani, 2009), today a strong integration of GIS methods in remote sensing-based land use analyses is still rather the exception than the rule. Often additional spatial information is used to focus on a single subject. For example agricultural parcel boundaries are incorporated to reduce misclassification (Aplin and Smith, 2008; De Wit and Clevers, 2004; Esch et al., 2014; Lucas et al., 2007; Smith and Fuller, 2001; Turker and Arikan, 2005).

Nowadays, various official basic geospatial datasets like digital landscape models (DLM) or digital cadastral information systems are available at different scale levels for many countries. These datasets usually contain manifold high-quality information on various land use categories. Nevertheless, approaches that incorporate DLM data more comprehensively are rare (e.g. Hazeu et al., 2014) and the intensive incorporation of cadastral data is even rarer.

Hence, at first we summarise a methodology developed for the integration of remote sensing-based land use information and official DLM data for Germany for a regional scale land use analysis. Afterwards, the potential of the methodology and the usability of official German cadastral data are investigated for land use analyses at a significant higher spatial resolution (local scale).

2. MULTI-DATA APPROACH

The methodological background of the presented studies is the Multi-Data Approach (MDA) by Bareth (2008), which was adapted and further developed in the raster data model (Waldhoff, 2014). The key aspect of the MDA is the emphasis on the integration of remote sensing-based classification results and officially available basic geodata in a GIS, to generate land use data of enhanced information content.

In the initial remote sensing part of the MDA, (multi temporal) satellite images are analysed individually using supervised classification methods like support vector machines (SVM) or maximum likelihood (MLC). With regard to the information content of the additionally available datasets, the classification of each remote sensing image focuses mostly on the vegetation information that is only insufficiently provided by other sources.

In the GIS part of the MDA, the remote sensing classification results and preselected information of the basic geodata are combined to a multi-layer dataset. By the means of expert knowledge-based production rules, the valuable information of each information layer is then combined in the final land use data product. In this way, either categorical information, geometrical information or both were at the centre of attention.

3. MDA IMPLEMENTATION ON A REGIONAL SCALE

The further development of the MDA for a regional scale land use analysis (ca. 1:50.000) was conducted for the Rur catchment, being the study area of the CRC/TR 32 "Patterns in Soil-Vegetation-Atmosphere-Systems: Monitoring, Modelling and Data Assimilation" (www.tr32.de). The Rur catchment (ca. 3000 km²) is mainly situated in western Germany, but it also reaches into the Netherlands and Belgium (cf. Figure 1). It is

---

* Corresponding author

characterised by intensive arable land use in the north and mainly grassland and forest in the south.

The main objectives of this land use study were therefore the adequate disaggregated differentiation of the various land use types for a regional scale and the differentiation of the major crop types on parcel-level. For this purpose, a cell size of 15 m was chosen.

The handling of agricultural land and the differentiation of the crop types through the multi temporal analysis of moderate spatial resolution remote sensing data (e.g. ASTER, Landsat-7 & -8) is treated in more detail in Waldhoff et al. (2012) or in Waldhoff (2014).

Besides the crop classification, also the basic differentiation of the other major land use classes, in particular concerning non-agricultural vegetation and impervious surfaces, was obtained from the remote sensing analysis (cf. Figure 2a).

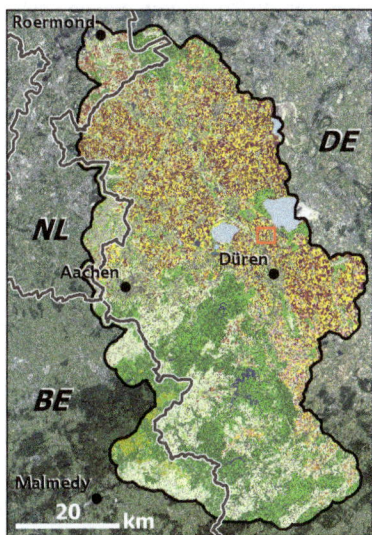

Figure 1. Land use overview of the Rur catchment.

To enhance the remote sensing results outside of agricultural land, selected land use classes of the ATKIS Basis DLM (ATKIS = Authorative Topographic-Cartographic Information System) were used for the German part of the Rur catchment (Waldhoff and Bareth, 2009). The ATKIS is provided by the official state survey and mapping agencies at a scale of 1:25,000 and is currently updated in total every three years (BRK.NRW, 2012). It includes diverse information for example on the transport system, but it also distinguishes built-up areas, different residential and industrial land uses, forest areas, arable land or grassland. The data are provided in vector format with a spatial accuracy of ±3 m (AdV, 2008b). For the study, selected ATKIS layers were rasterised and resampled to a spatial resolution of 15 m.

Figure 2b shows selected rasterised ATKIS layers for the same area as in Figure 2a. Red and orange tones indicate different residential and commercial land uses, while green colours (from dark to light) denote forest areas, non-agricultural vegetation areas and grassland. Grey colours represent different road categories (e.g. carriage way or municipal road).

Compared to Figure 2a, even after the resampling to 15 m pixel size, the delimitation of the individual land use areas appears much sharper. Also, most areas are differentiated into more land use classes than in the remote sensing result. However, the information content of some land use classes is highly aggregated. For example, there is no differentiation of arable land into different crop types. Additionally, for residential areas only sparse information is provided on the vegetation cover or

impervious surfaces. Thus, one step to improve the land use/land cover differentiation within the major land use categories (e.g. arable land, built-up, forest, other vegetation areas) was to combine either thematically or geometrically more precise ATKIS land use classes with the corresponding remote sensing classification results. In addition, land use classes that were more adequately represented in the ATKIS were fully transferred to the final land use dataset via overlay analysis.

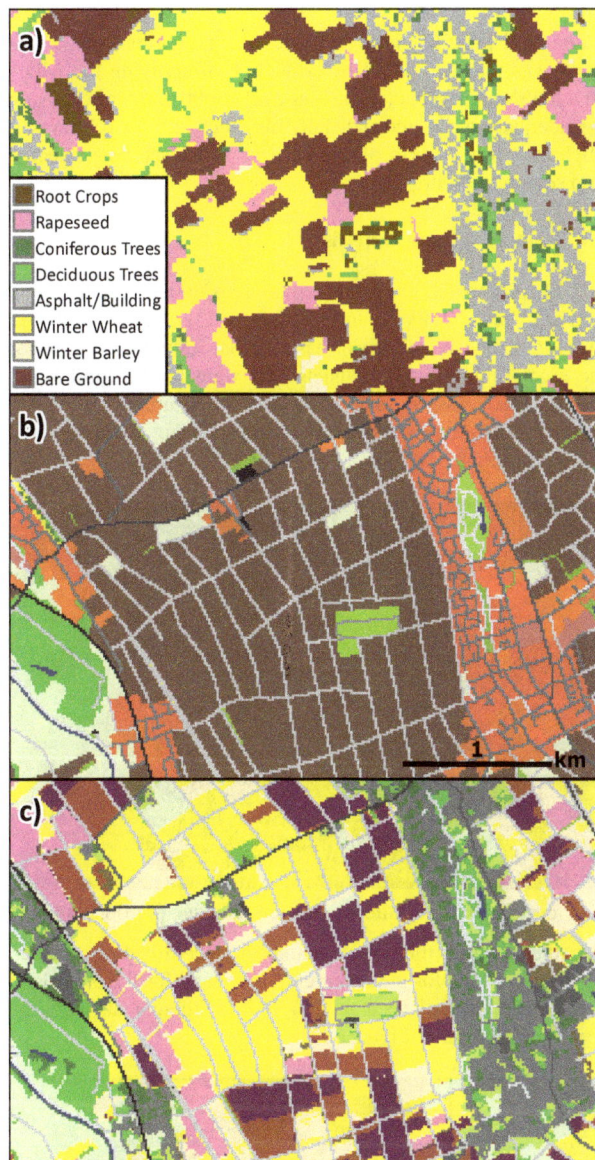

Figure 2. (a) Subset of the MLC result of a Landsat-8 image (05.05.2014). (b) Selected ATKIS classes (rasterised). Dark red to orange tones indicate different residential and commercial land uses, dark to light green colours denote forest, non-agricultural vegetation and grassland whereas grey colours differentiate road categories. (c) MDA classification of 2014 (Lussem and Waldhoff, 2014).

Figure 2c depicts the final MDA land use classification for the subset area. Compared to Figure 2a, much more land use/land cover classes are differentiated. For example, now areas of different settlement-related land uses are distinguished (grey colours). Likewise, arable land is now differentiated into the major crops, based on the remote sensing results. Moreover, the misclassification of crops outside of arable land areas (e.g.

winter wheat in settlement areas, cf. Figure 2b), was reduced dramatically with the ATKIS data. Yet, the classification of vegetation in settlement areas per se was maintained to differentiate vegetated and impervious surface areas within the corresponding ATKIS classes. Finally, the final MDA result now contains areas that are occupied by the transportation network. Such areas are usually not adequately represented in remote sensing classifications.

4. LOCAL SCALE INVESTIGATION

Based on the results that were achieved for a regional scale, this study was intended as a first evaluation of applying the MDA on a high spatial scale (in the range of 1:5000 and higher) using the adequate basic geodata. For this study, a part of the Rur catchment in the north of Düren, near Selhausen was selected (red square in Figure 1).

4.1 Remote sensing analysis

For the high spatial resolution remote sensing land use mapping a WorldView-2 (WV-2) image of June 11, 2014 was chosen (Figure 3). The WV-2 sensor has a spatial resolution of 2 m for the eight multispectral channels that were used (Updike and Comp, 2010). In accordance with the procedure of the MDA, the WV-2 image was separately classified using the supervised MLC algorithm.

Figure 3. Subset of the WorldView-2 image (11.06.2014).

For the aim of the investigation the classification yielded satisfying results for the major crop types and for the other land use classes (overall accuracy: 91.30 %). Figure 4 displays the result for a subset of Figure 3.

Figure 4. Subset of the MLC result of the WordView-2 scene.

Owing to the acquisition date of the WV-2 image, many agricultural parcels which comprise summer crops either still show more or less bare ground (brown and olive colours and in Figure 4) or the development stage of the crops was inadequate for the identification of the crop. As a result, no specific crop types could be classified on the corresponding parcels whereas winter crops were adequately differentiated.

4.2 Integration of additional data

Concerning the integration of the remote sensing results with additional data at this spatial resolution, cadastral information systems are the first choice in terms of geometric accuracy and thematic resolution. Thus, in this study the usage of the Automated Real Estate Map (Automatisierte Liegenschaftskarte, ALK) was investigated. This data product is available in vector format for whole Germany and is produced by the official cadastre offices (Ehrmanntraut and Nerkamp, 2011). As a result, it is continuously updated and of high quality. However, for this study only polygon data of 2009 was available, so that there is quite a large time gap between the acquisition date of the remote sensing data and the production date of the ALK. Yet, for the aim of the study the data was considered suitable. The ALK is a spatially comprehensive data product and describes the actual use of land parcels at a resolution of about 1:1000. It contains diverse and detailed information on the use of each land parcel, like specific types of residential or commercial use, dominant vegetation cover (e.g. deciduous, coniferous or mixed forest, copse or pasture) or the spatial extent of the different transportation route categories. Moreover, it also informs about the existence and the geometry of small vegetation patches which often accompany roads (turquoise colour in Figure 5). Furthermore, the surface areas occupied by buildings are included.

4.3 Results

For this study, an area of roughly 45 km² was available, where the ALK data and the WV-2 scene overlapped. This area includes 17,776 single land parcel objects that are organised into 131 different classes. To handle the vast amount of thematic information, associated classes where aggregated to 29 superordinate categories that are relevant for a common land use analysis. In this regard, for example land use classes like health, public administration, research and education were combined to the class public facilities.

Figure 5. Example of the selected and aggregated ALK classes for the land use analysis (class colours correspond with Figure 2b).

Figure 5 illustrates the thematically aggregated ALK layer for the selected subset. Still manifold land use categories are differentiated. The grey lines delimit the individual land parcels. Following the general colouring of the ATKIS data for the regional scale analysis (cf. Figure 2b), for example green colours indicate forest areas, copse or pasture (from dark to light colours). Dark brown areas signify arable land. Red colours differentiate various residential and commercial land use types. Additionally, the building layer (black colour) is overlaid.

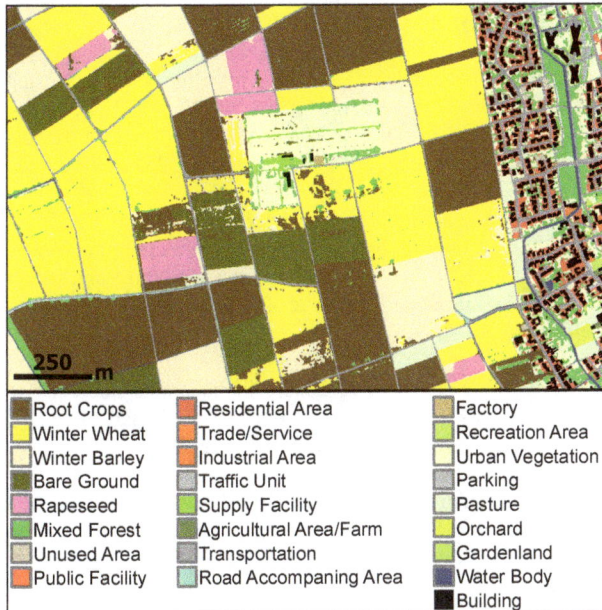

Figure 6. Integration result of selected remote sensing derived class allocations and ALK information.

In accordance with the MDA procedure presented before, at first, the land parcel layer and the layer of the building geometries were converted to raster layers with the same cell size as the WV-2 data. In the MDA-GIS-part, the classification results and the ALK raster were then combined to a layer stack. Afterwards, selected information of each layer was subsequently integrated into the final land use layer to enrich its information content (cf. Figure 6).

On the one hand, arable land, as indicated by the ALK, was disaggregated with the crop type distribution information of the remote sensing classification for the corresponding land parcels. Additionally, the misclassification of impervious surfaces within arable land (caused by the high spectral similarity e.g. of asphalt and dry soil) could be reclassified to bare soil through the utilisation of the ALK data (compare centres of Figures 4 & 6). On the other hand, for non-agricultural land use classes, the ALK land use information was combined with the remote sensing-based land use/land cover information. In doing so, the classification results on vegetation coverage, impervious surfaces or bare ground were used to disaggregate the land use information within the cadastral parcels of settlement areas.

Furthermore, various land use classes of the ALK were considered as more spatially precise or thematically detailed than the class allocations of the remote sensing classification. As a result, ALK classes like the road network, road accompanying area, water body or pasture completely substituted the corresponding remote sensing class allocations in the final dataset. Also, areas occupied by buildings were transferred from the ALK.

5. DISCUSSION & CONCLUSION

In this contribution, at first the MDA for the integration of DLM data and remote sensing-derived land cover/land use

information for the regional scale was recapped. It was shown that by using multiple data sources in a knowledge-based approach, manifold additional information can be derived that are not obtainable from one of the incorporated sources alone. On this basis, the second study was designated to investigate the implementation of this methodological approach for a considerably higher spatial scale by incorporating geometrical and categorical information of a cadastral information system.

Regarding the remote sensing data analysis, the high spatial resolution WV-2 data obviously provided much more geometrical detail, than the moderate spatial resolution remote sensing data that was used for the regional scale analysis. However, the degree of the categorical information that was obtained from the WV-2 classification was principally not very different to the results, which were obtained for the regional scale analysis (cf. Figure 2a & 4). In addition, even object-based classification approaches are not expected to lead to significant improvements without meticulous and time-consuming iterative algorithm refinement.

Concerning the enrichment of the categorical information content, already the conducted steps of integrating classification results and the cadastral ALK data revealed the high potential of this approach for large-scale land use/land cover analyses. In this study, the first focus was on the disaggregation of arable land using the remote sensing crop class allocations in conjunction with the ALK geometric information on arable land parcels. Thus, the classification of crops was restricted to arable land. For the sake of simplicity, adjacent arable land parcels were regarded as one coherent area. In future analyses, the usage of the single land parcel information will be investigated to further reduce classification errors (Smith and Fuller, 2001).

For non-agricultural land use classes, the ALK land use information was combined with the remote sensing-based land use/land cover information on vegetated, bare or impervious surfaces to disaggregate the parcel-based land use information. In this regard, also misclassifications of crops in built-up areas were considered as valuable information for the identification of vegetation areas. In this way, also a higher geometrical accurateness concerning the delimitation of land cover patches was achieved.

In addition, the representation of the transportation network areas is still improved through the data integration. Especially the relatively small accompanying areas of streets are now better distinguishable from other features. Corresponding vegetation areas like small patches of bushes that accompany arable land are otherwise easily ignored or confused with the adjacent agricultural crop. The totality of such features should sum up to a significant areal extent, with regard to total areas of vegetation classes.

For this study only a rather outdated ALK dataset was available. However, for the purpose of this investigation, this aspect was negligible. The fact that cadastral data is usually continuously updated, is one of the main reasons why this data is especially attractive for the presented approach. While no dataset is without errors, there is still a high certainty of correctness in current ALK data.

In terms of the data product of the cadastral data of Germany, the ALK is currently successively replaced by the ALKIS (Official Real Estate Cadastre Information System). However, the general findings of this investigation will still apply to the ALKIS. Moreover, compared to the ALK, the ALKIS conveys even more valuable information. For example, the ALKIS also contains multiple point or line features, with attributes concerning their areal extent (AdV, 2008a). The integration of such information will be investigated in the future.

ACKNOWLEDGEMENTS

We thank Geobasis.NRW for the provision of the ATKIS Base DLM and the ALK data. The Landsat data and the ASTER data was obtained from the U.S. Geological Survey.

REFERENCES

AdV, 2008a. Dokumentation zur Modellierung der Geoinformationen des amtlichen Vermessungswesens (GeoInfoDok) - ALKIS-Objektartenkatalog, Version 6.0, Stand: 11.04.2008. Arbeitsgemeinschaft der Vermessungs-verwaltungen der Länder der Bundesrepublik Deutschland.

AdV, 2008b. Dokumentation zur Modellierung der Geoinformationen des amtlichen Vermessungswesens (GeoInfoDok) - ATKIS-Katalogwerke - ATKIS-Objektarten-katalog Basis-DLM - NRW-Erfassung, Version 6.0, Stand: 11.04.2008. Arbeitsgemeinschaft der Vermessungs-verwaltungen der Länder der Bundesrepublik Deutschland, p. 192.

Aplin, P., Smith, G.M., 2008. Advances in Object-Based Image Classification International Society of the Photogrammetry and Remote Sensing, Beijing, pp. 725-728.

Bareth, G., 2008. Multi-Data Approach (MDA) for Enhanced Land Use/Land Cover Mapping, The International Archives of the Photogrammetry, Remote Sensing and Spatial Information Sciences, Vol. XXXVII. Part B8. Beijing 2008. International Society of the Photogrammetry and Remote Sensing Beijing, pp. 1059-1066.

BRK.NRW, 2012. Amtliches Topographisch-Kartographisches Informationssystem (ATKIS®) - Digitale Landschaftsmodelle. Bezirksregierung Köln - Abteilung Geobasis NRW, Bonn, p. 8.

De Wit, A.J.W., Clevers, J.G.P.W., 2004. Efficiency and accuracy of per-field classification for operational crop mapping. *International Journal of Remote Sensing* 25, 4091-4112.

Ehrmanntraut, E., Nerkamp, K.-H., 2011. Konzeption und Nutzung des automatisiert geführten Liegenschaftskatasters, In: Kummer, K., Frankenberger, J. (Eds.), Das deutsche Vermessungs- und Geoinformationswesen 2012. Wichmann, Berlin, pp. 137-164.

Esch, T., Metz, A., Marconcini, M., Keil, M., 2014. Differentiation of Crop Types and Grassland by Multi-scale Analysis of Seasonal Satellite Data, In: Manakos, I., Braun, M. (Eds.), Land Use and Land Cover Mapping in Europe: Practices & Trends. Springer, Dordrecht Heidelberg, pp. 329-339.

Hazeu, G.W., Schuiling, C., Dorland, v.G.J., Roerink, G.J., Naeff, H.S.D., Smidt, R.A., 2014. Landelijk Grondgebruiksbestand Nederland versie 7 (LGN7): vervaardiging, nauwkeurigheid en gebruik. Alterra Wageningen UR, Wageningen.

Hutchinson, C.F., 1982. Techniques for Combining Landsat and Ancillary Data for Digital Classification Improvement. *Photogrammetric Engineering and Remote Sensing* 48, 123-130.

Lucas, R., Rowlands, A., Brown, A., Keyworth, S., Bunting, P., 2007. Rule-based classification of multi-temporal satellite imagery for habitat and agricultural land cover mapping. *ISPRS Journal of Photogrammetry and Remote Sensing* 62, 165-185.

Lussem, U., Waldhoff, G., 2014. Enhanced land use classification 2014 of the Rur catchment. CRC/TR32 Database (TR32DB), DOI: 10.5880/TR32DB.12.

Merchant, J., Narumalani, S., 2009. Integrating Remote Sensing and Geographic Information Systems, In: Warner, T.A., Nellis, M.D., Foody, G.M. (Eds.), The SAGE handbook of remote sensing. SAGE Publications Ltd., London, pp. 257-269.

Mora, M., Tsendbazar, N.-E., Herold, M., Arino, O., 2014. Global Land Cover Mapping: Current Status and Future Trends, In: Manakos, I., Braun, M. (Eds.), Land Use and Land Cover Mapping in Europe: Practices & Trends. Springer, Dordrecht Heidelberg, pp. 11-30.

Smith, G.M., Fuller, R.M., 2001. An integrated approach to land cover classification: An example in the Island of Jersey. *International Journal of Remote Sensing* 22, 3123-3142.

Turker, M., Arikan, M., 2005. Sequential masking classification of multi-temporal Landsat7 ETM+ images for field-based crop mapping in Karacabey, Turkey. *International Journal of Remote Sensing* 26, 3813-3830.

Updike, T., Comp, C., 2010. Radiometric Use of WorldView-2 Imagery - Technical Note. DigitalGlobe, Inc.

Waldhoff, G., 2014. Multidaten-Ansatz zur fernerkundungs- und GIS-basierten Erzeugung multitemporaler, disaggregierter Landnutzungsdaten. Methodenentwicklung und Fruchtfolgen-ableitung am Beispiel des Rureinzugsgebiets. Universität zu Köln, Köln, p. 334.

Waldhoff, G., Bareth, G., 2009. GIS- and RS-based land use and land cover analysis: case study Rur-Watershed, Germany, Geoinformatics 2008 and Joint Conference on GIS and Built Environment: Advanced Spatial Data Models and Analyses, Proc. SPIE 7146. SPIE, pp. 714626-714628.

Waldhoff, G., Curdt, C., Hoffmeister, D., Bareth, G., 2012. Analysis of multitemporal and multisensor remote sensing data for crop rotation mapping. *ISPRS Annals of the Photogrammetry, Remote Sensing and Spatial Information Sciences* I-7, 177-182.

23

A TASK-DRIVEN DISASTER DATA LINK APPROACH

L.Y. Qiu [a], Q. Zhu [b, c], J.Y. Gu [a], Z.Q. Du [a, d], *

[a] State Key Laboratory of Information Engineering in Surveying, Mapping and Remote Sensing, Wuhan University, 430079 Wuhan, China – qiu_linyao@163.com, duzhiqiang@whu.edu.cn, gujieye@126.com
[b] State-province Joint Engineering Laboratory of Spatial Information Technology for High-Speed Railway Safety, Southwest Jiaotong University, 610000, Chengdu, China – zhuq66@263.net
[c] Faculty of Geosciences and Environmental Engineering, Southwest Jiaotong University, 610000, Chengdu, China – zhuq66@263.net
[d] Collaborative Innovation Center of Geospatial Technology, 430079 Wuhan, China –duzhiqiang@whu.edu.cn

KEY WORDS: Emergency Task, Ontology, Disaster Data Management, Semantic Mapping, Spatial-temporal Correlation, Data Link

ABSTRACT:

With the rapid development of sensor networks and Earth observation technology, a large quantity of disaster-related data is available, such as remotely sensed data, historic data, cases data, simulation data, disaster products and so on. However, the efficiency of current data management and service systems has become increasingly serious due to the task variety and heterogeneous data. For emergency task-oriented applications, data searching mainly relies on artificial experience based on simple metadata index, whose high time-consuming and low accuracy cannot satisfy the requirements of disaster products on velocity and veracity. In this paper, a task-oriented linking method is proposed for efficient disaster data management and intelligent service, with the objectives of 1) putting forward ontologies of disaster task and data to unify the different semantics of multi-source information, 2) identifying the semantic mapping from emergency tasks to multiple sources on the basis of uniform description in 1), 3) linking task-related data automatically and calculating the degree of correlation between each data and a target task. The method breaks through traditional static management of disaster data and establishes a base for intelligent retrieval and active push of disaster information. The case study presented in this paper illustrates the use of the method with a flood emergency relief task.

1. INTRODUCTION

The number of incidents and magnitude of natural disasters worldwide have increased significantly due to climate changes in recent years (Ding et al., 2014; Iwata et al., 2014; Neumayer et al., 2014). A number of natural disasters (e.g., South Asia Tsunami, China Earthquake, Haiti Earthquake and Tohoku Earthquake) stroke across the globe, killing hundreds and causing billions of dollars in property and infrastructure damage (Grolinger et al., 2013).

Facing the urgent need for disaster mitigation, understanding how to enhance the capacity of effective monitoring, early warning and emergency response have become major challenges all around the world. On one hand, the amount of information and types of data related to disaster has increased greatly. Disaster data, including remote sensing images, history data, previous incidents records, simulation data, basic geographic data and disaster assessment products possess velocity, variety and veracity features converted from singleness and small amount. They put forward a higher requirement for integration, processing and analysis (Grolinger et al., 2013). On the other hand, government agencies in different levels and individual organizations master various data resources and take different disaster relief functions. In order to achieve good cooperation and collaboration in disaster management, the most effective data should be sent promptly to the most needed actors (Borkulo et al., 2006). In recent years, various types of sensors widely deployed in disaster monitoring network make it possible to continuously access disaster big data with high

spatial-temporal resolution and increasingly rich attribute information, which provides important support for enhancing capabilities of disaster emergency response. However, fast and easy acquisition and generation of heterogeneous data has exceeded existing ability of data management. The main reasons are as follows: 1) most existing disaster management systems operate in a typical passive data-centric mode (Ding et al., 2014). The functions and purpose of disaster information service are typically singular and direct, which could rapidly satisfy specific user needs, but will not fit the needs of the actual disaster management tasks of other user communities. The functions also do not generate products with high accuracy and veracity when the needed data source is limited or not accessible. 2) Current efforts to integrate geographic information data have been restricted to keyword-based-matching Spatial Information Infrastructure (SII) (Li et al., 2007). SII supports the discovery and retrieval of distributed geospatial data sources and geographic information services by providing catalogue services and syntactic interoperability standards (Lutz, 2007), but spatial-temporal characteristics of data (e.g., the spatial distribution of the cloud in a multispectral remote sensing image) are hidden inside the data file. Moreover, lack of a semantic association among multi-source heterogeneous data brings a passive result that the knowledge and discipline of the disaster are hardly found automatically. 3) Recent disaster data retrieval mainly relies on querying with keywords of metadata passively. The artificial experience plays an important role in finding available data because there is short

---

* Corresponding author: DU Zhiqiang. E-mail: duzhiqian@whu.edu.cn.

of mechanism of automatically discovering related data and disaster knowledge for computer reasoning (Fan and Zlatanova, 2011). In practice, trivial and time-consuming operations to integrate various resources have cost most the manual resources rather than improving the decision-making (Laniak et al., 2013). This is why most existing disaster management systems have been of limited use (Leskens et al., 2014) and resources cannot be fully utilized (Demir and Krajewski, 2013; Zhishan et al., 2012). This paper proposes a task-oriented disaster information link method, in which disaster emergency tasks are regarded as a key semantic factor to restrain, associate and aggregate spatial-temporal data.

Here, we discuss the challenge of managing disaster data to support various task processing in emergency response contexts. The paper is organized as follows: "Related Work" section presents related work on applying semantic-related technology and ontology for spatial data and emergency response. The section titled "Task and data ontologies for disaster management" firstly analyzes types and features of emergency tasks in disaster management and puts forward an ontology model describing them. Then it describes the semantic features of disaster data in regards to attribute, space-time and statistics. "Semantic mapping of task and data" discusses a map between characteristics of emergency tasks and disaster data in scale, attribute and spatial-temporal level and proposes task-oriented multi-dimensional data characteristics to analyse task preference to different data sets. The "Implementation" section introduces a case study illustrating how to aggregate data in a multilevel way to find the right data for a specific emergency task in storm-flood disaster chain. Finally, we conclude the article in "Conclusions and future work" section.

2. RELATED WORK

2.1 Related work on the semantic technology in disaster data management

As mentioned earlier, the existing disaster-related data is extremely heterogeneous and different vocabulary could be used in different sources. The reason why semantic-related technology is employed is that they can be used to identify and associate semantically-corresponding concepts with disaster-related information so heterogeneous data can be integrated and ingested (Hristidis et al., 2010). Many previous studies have discussed the importance of semantic-related technology for solving problems in geographical information systems (Cohn, 1997; Guarino, 1998). Cohn (1997) proposed that the human-computer interaction in GIS should be more concise and accurate than it is currently. Currently, aiming at resolving semantic diversity generating adverse effect on data management and achieving semantic interoperation among heterogeneous data, spatial semantic description has been used in disaster data management (Fan and Zlatanova, 2011; Li et al., 2007; Zhu et al., 2009; Schulz et al., 2012; Silva et al., 2013). Schulz (2012) and Silva (2013) established description of data by Linked Open Data (LOD). Based on semantic web knowledge, they adopted RDF (Resource Description Framework) to define standard and exchangeable data format for semantic annotation of disaster knowledge. LOD is considered an effective tool that could convert data relations to information computers could process, promoting automatic finding and reasoning of disaster knowledge (Foster and Grossman, 2003; Lausch et al., 2014). Michalowski (2004) also applied the Semantic Web technology to develop a Semantic Web-enabled management system. Such a system allows efficiently querying distributed information and effectively converting legacy data into more semantic representations (Michalowski et al., 2004). Zhu (2009) analysed challenges of intricate semantics in remote sensing information systems and proposed a hierarchical semantic restrain model as a uniform semantics description model. The connection between user semantics, data and processing services is established as basics of semantic reasoning in discovery, selection and composition of data and service.

2.2 Related work on ontology in disaster data management

Compared with the semantic methods mentioned, ontology has stronger semantic integrity and supports uniform description from data definition to operation. This is useful for automatic finding and mining of data (Babitski et al., 2009; Klien et al., 2006). Guarino (1998) analyzed the importance of the ontology concept in GIS system. However, he only proposed a possible ontology structure without attempting to implement it. Some researches provided conceptual structures of ontology in disaster management (Chatterjee and Matsuno, 2005; Li et al., 2009; Xu et al., 2009). Chatterjee and Matsuno (2005) discussed the necessity of using the ontology to solve the linguistic differences. Li (2009) proposed an ontology-based architecture for geo-objects in disaster systems. Xu (2009) also suggested building an ontology-based emergency response plan. Some researchers have studied specific ontology methods for semantic description (Huang and Yan, 2013; Wang et al., 2007; Yang et al., 2013). Huang (2013) proposed disaster domain ontology including hazard-affected body, disaster-inducing factors, inducing environment, disaster events and built connections among them by ontology. The model was experienced in disaster processing estimation and prediction. Babitski (2009) defined ontology of disaster damage, resource and the relations between them, so that available data resource could be quickly found while facing a certain assessment task. Wang (2007) put forward a spatial geographic ontology by analysing objects, relation and data in space. Such a description effectively presents hierarchical structure and semantic relation of spatial information. Yang (2013) developed a kind of task ontology, dividing task process from aspects of function, organization, spatial-temporal scale and complexity of calculation.

Although current semantic methods resolve problems of integration in disaster data management, most of them manage limited types of data and the semantic restraints or correlation on heterogeneous data are simple. Thus, a mature ontology-based data correlation method is required so that can both integrate heterogeneous data from different sensors and support automatic querying and reasoning functions.

3. TASK AND DATA ONTOLOGIES FOR DISASTER MANAGEMENT

3.1 A task ontology for emergency workflow

The need for up-to-date geospatial data in emergency situations is now widely recognized. Emergency responders may not be familiar with data standards or the appropriateness of certain data sets for a particular task. However, due to the critical nature of emergency response, responders rarely have time to sift through extensive query results and will not re-think what data sources and specific data characteristics are needed each time they face a task. Thus, it is worthwhile to formally delineate tasks and their relationships to types of data sources (Wiegand and García, 2007).

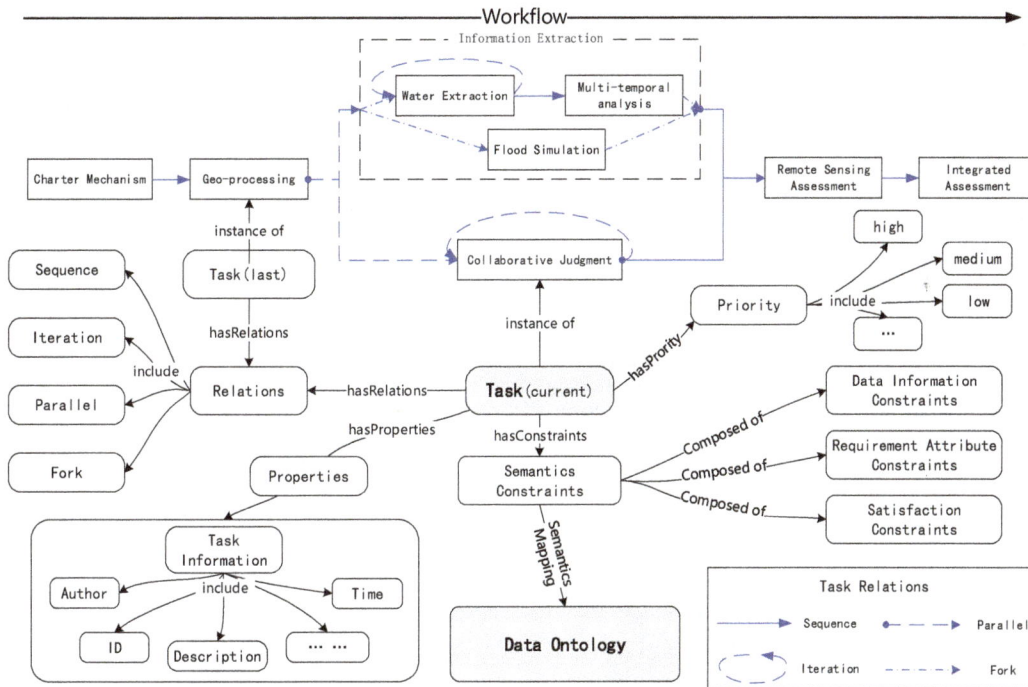

Figure 1. Structure of a task ontology in an emergency response workflow

Ontology is a description, in a formal, machine-readable format, that expresses concepts including the types of entities, attributes, relationships, and values found in a domain. Ontology can represent the semantics of emergency tasks and disaster data, thus helping create connections between them for emergency response processes. Therefore it contributes to the knowledge of workflow processing and task demand. In this paper, the task ontology in supporting emergency response is abstracted as a quintuple:

$$O_T = <C, R, S, P, I> \qquad (1)$$

C represents task basic information based on functional and non-functional properties; R represents the relationships between a task entity and others in an emergency workflow; S represents the semantic restraints of a task demand. P represents the priority of task execution and data retrieval in the emergency workflow; I represents the instances of tasks.

Taking a flood response workflow of the National Disaster Reduction Center in China (NDRCC) as an example, Figure 1 shows the ontological entities and relations of disaster tasks. A set of connections composes a directed graph that specifies how the task works. The whole emergency workflow contains a group of tasks such as charter mechanism based acquisition of satellite imageries, geo-processing of data, information extraction, collaborative judgment, remote sensing assessment and integrated assessment. Each task is an instance of an ontology entity. The relations including sequence, parallel, interaction and fork rules the logic order of task execution process. As the process is developing, the priority of a task is changing dynamically in real time, which further influences the queue of data retrieval and preparation. For instance, when the task of collaborative judgment following geo-processing proceeds, its execution priority is higher than its successors (such as remote sensing assessment) but the same as information extraction because they are parallel. The semantic

restraints describing the feature of task demand are composed by three parts: data information, requirement attributes and satisfaction. Data information confines the basic feature of input data including resolution, timeliness, types of sensors and spatial system. Requirement attributes describe preference and selection rules to data in the background of a certain disaster. Satisfaction represents the quality of task output influenced by data quality, environmental factors, response speed and overlap extent of data with the target area. The task ontology representing the function, attributes, process and need of tasks is a precondition to link task and data. It is presented as a RDF/OWL (Web Ontology Language) file (http://www.semanticweb.org/dell/ontologyies/tasks/task.owl). Some snippets of the file are listed in Table 1. Hereafter, ontologies are presented in protégé for clarity.

Table 1. Snippets of the task ontology file in Turtle

```
<!--http://www.semanticweb.org/dell/ontologies/task#Atomic-->
<owl:Class rdf:about="&task;Atomic">
    <rdfs:subClassOf rdf:resource="&task;TaskType"/>
</owl:Class>
<!-- http://www.semanticweb.org/dell/ontologies/task#AtomicTask
-->
<owl:Class rdf:about="&task;AtomicTask">
    <rdfs:subClassOf rdf:resource="&task;Task"/>
</owl:Class>
<!--
http://www.semanticweb.org/dell/ontologies/task#CompositTask --
>
<owl:Class rdf:about="&task;CompositTask">
    <rdfs:subClassOf rdf:resource="&task;Task"/>
</owl:Class>
<!-- http://www.semanticweb.org/dell/ontologies/task#Composite -
->
<owl:Class rdf:about="&task;Composite">
    <rdfs:subClassOf rdf:resource="&task;TaskType"/>
</owl:Class>
```

3.2　An ontology of disaster data

A success of disaster data management could be described as "getting the right resources to the right place at the right time; to provide the right information to the right people to make the right decisions at the right level at the right time (Xu and Zlatanova, 2007)." However, semantic heterogeneity of the spatial data remains one of the biggest challenges in disaster data management. Especially, as acquisition of multi-source data including remote sensing images, history data, case data, simulation data, basic geographic data and disaster assessment product has become increasingly easy and fast, metadata catalogs based data management can neither unify heterogeneous semantics nor explicitly represent correlation of various data. So data ontology is designed to solve the problem through the integration of disaster data and a triple is constructed for its description:

$$O_D =< T, F, I >　(2)$$

T represents the type classification of disaster data by defining a two-tuples composed of category and format. Category describes the conceptual classification, like observed data and history data, while format denotes specific file pattern, such as geotiff, img and shpfile (as shown in Figure 2). F represents the apparent and potential features of data from three aspects: attribute, space-time and statistics. Attribute contains inherent nature of data, which is obtained from data itself including spatial and temporal resolution, spatial reference and spectrums. Space-time describes spatial-temporal information including velocity of data acquisition and scope of the area covered by data. Such information is commonly obtained from record or calculation. Statistics show the rules and knowledge about data usage, such as the operating frequency of the data while facing a specific task. Then I represents the instances of data.

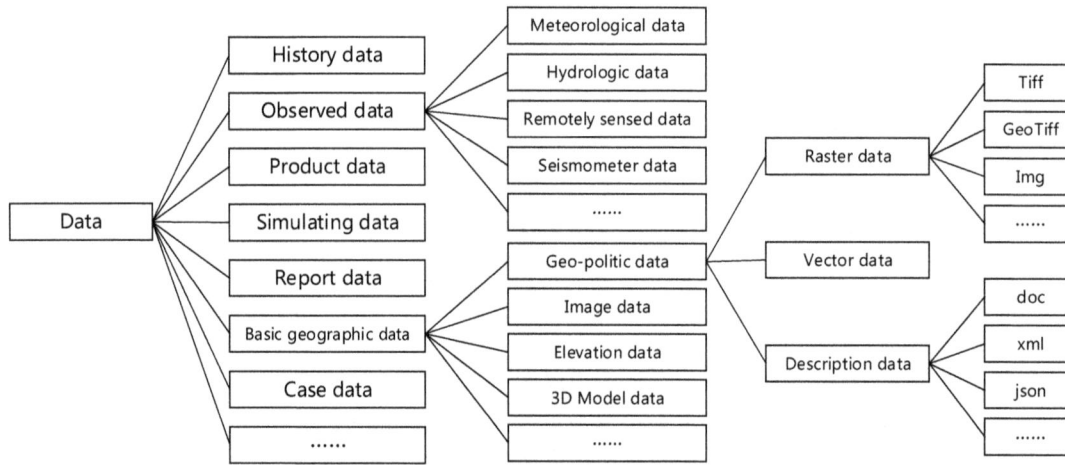

Figure 2. Disaster data classification

The relations of different data instances can be described in two aspects. For the data with disparate types, the correlation of them is described by statistics features in a common application environment, like the co-occurrence of heterogeneous data adopted in similar history cases. For data of the same type, the correlation is built by calculating the similarity of spatial and temporal features. The similarity is calculated by the following formula:

$$Sim_{case(i,j)} = w_t \times \alpha^{|D_j - D_i|} + w_s \times \beta^{-\ln\frac{Min(Area_i, Area_j)}{Max(Area_i, Area_j)}}　(3)$$

W_t is the weight of temporal similarity while W_s is the weight of spatial similarity. The sum of w_t and w_s equals 1, but their specific values rely on the task need. For example, w_t in temporal series analysis is higher than that in other tasks. α and β are two decay factors ranging from 0 to 1. $|D_j - D_i|$ represents the absolute interval value of two dates. Min and Max respectively means the overlap area of two sets of data and union area of minimum bounding box containing them. When the calculation is close to or equals 1, the degree of connection between two data sets are strong while if the value approximates 0, they have a weak connection. So the data ontology not only unifies the semantic description of heterogeneous data, but also

offers the correlation method to automatically find other related data resources in the searching process.

4.　SEMANTIC MAPPING OF TASK AND DATA

Due to a lack of semantic association between tasks and data in traditional disaster data management, the determination of which data source is the most appropriate for a specific task, as implicit knowledge, could not be commonly applied. However, a unified description containing task and data is complex and unnecessary because they belong to different domains and their own respective composition. Thus expressions with common semantic terms could neither highlight each feature characteristic nor help the system increase automatic understanding and analysis to disaster knowledge. Clearly, one should be able to connect tasks with data. Based on the ontologies, in order to build connections between task and data and further, convert task needs to specific data query filters, a mapping from semantics constraints of task ontology to that of data ontology is designed, which is expressed as:

$$O_T(S) \rightarrow O_D(F)　(4)$$

As shown in Figure 3, the mapping relation contains almost all the task needs and features of disaster data, and there are several mapping types, including one-to-one, one-to-many and many-to-many between them. A further classification including attribute, space-time and scale is built on the base of mapping relations. The attribute level describes some indicators showing which data set is more suitable for a specific task by analysing the statistics of data usage in similar historical cases. For instance, the high co-occurrence and adopting frequency represents the importance of a data source to a task. So mapping relations in the attribute level could describe what kind of data source is the most suitable to current emergency process and help the system analyse the feature for automatic retrieval. The spatial-temporal level contains the direct correlation of tasks

and data like task requirement on coordinate system, spatial reference and coverage area of data, which could filter the inappropriate data source and choose the potential source when the attribute-based retrieval does not find default suitable targets. The scale offers some flexible relevance factors like resolution of data. Such factors support analysing the correlation of task and potential data sets by calculating the degree of satisfaction and finally, a list of data based on quantitative estimation of correlation could be provided for task operators.

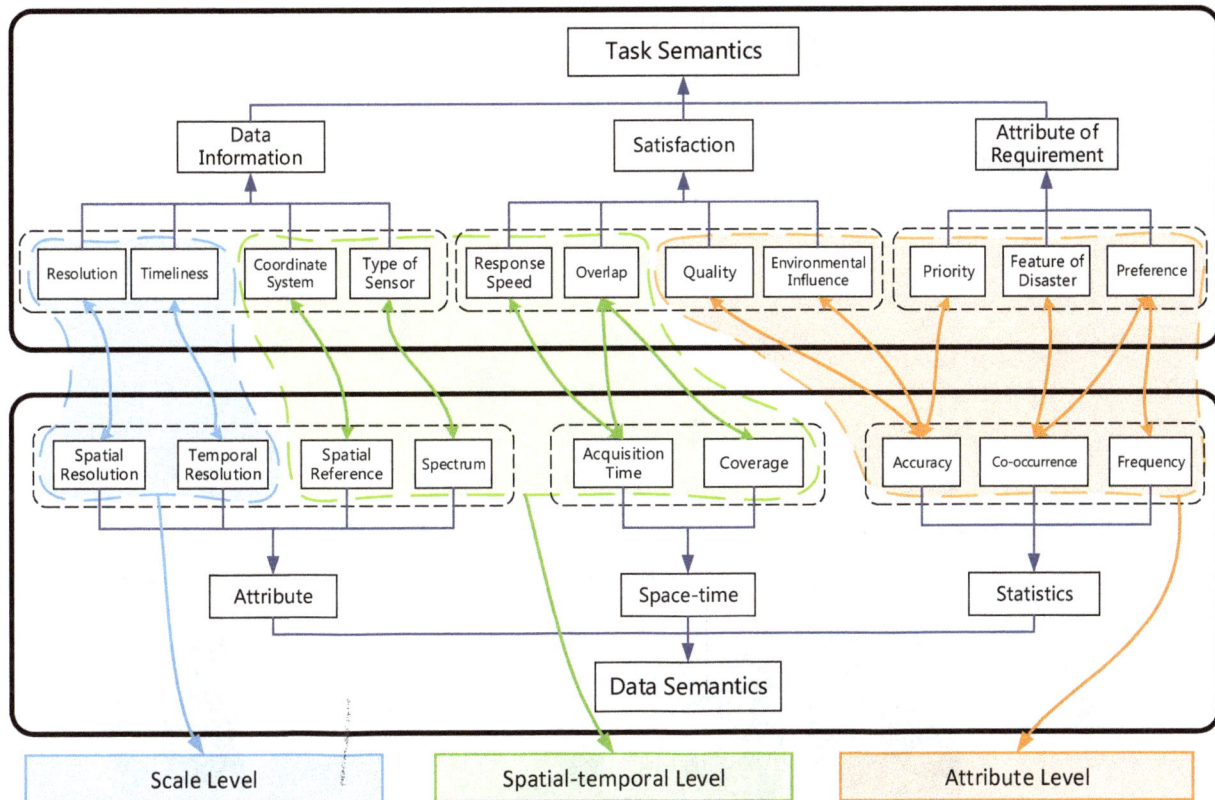

Figure 3. Semantic mapping between task and data

So far, the task-oriented disaster data link method proposed in this paper contains three parts. Firstly, the relation of tasks mentioned in section 3 builds the correlation of different tasks in a workflow, so the system could query and compose a workflow with needed tasks by itself. This sets a goal and order for automatic data preparation. Secondly, the similarity of data with the same category is calculated in spatial and temporal aspects. So while a data set well satisfying the running task is found, continuously some other related data could be searched by the system actively, accompanied with a quantitative analysis on relevance between potentially useful data and the target. Such operation helps users find more suitable resources for a specific task in advance. Thirdly, a classified mapping from task semantics to data semantics is set up for connecting emergency task and disaster data. Then the mapping drives a multi-level-based searching and filtering process to actively offer users the most suitable data satisfying target task quickly and precisely.

5. IMPLEMENTATION

A prototype system for building custom task flow and searching suitable data is developed and integrated in a SOA (Service-Oriented Architecture) based business-operation monitoring and management platform deployed in NDRCC, which is used for monitoring and managing disaster mitigation in whole emergency response period. Once users establish a new disaster task (as shown in Figure 4 (a)), by comparing factors like type, level and location of disaster with historical cases, the prototype system could set up a series of executable workflow and support man-machine manner (drag, drop and compose) to modify the process chain (as shown in Figure 4 (b)). We used the real-time emergency response task of the flood in Fushun, China in 2013 as an example. There are more than 40 typical flood events from the year of 2000 stored in the history database of NDRCC. Therefore, through selecting atomic tasks frequently used in similar history tasks, the system first composes a new workflow. Users could change it on the interface shown if necessary. Then the priority of each task is distributed relying on its location in the process chain. After that, data preparation including retrieval and selection process starts according to the priority rank.

Firstly, the system loads and parses RDF file of current task ontology using Jena and a list of related data type is created. Taking integrated assessment process for instance, the type list of its needed data includes raster (post-disaster urban image of Fushun, flood figure), vector (administrative map, flooding extraction figure, distribution diagram of damaged infrastructures and houses) and text (yearbook of population statistics and economy statistics in Fushun, reported data from disaster area). Then system starts to traverse the list to find the most suitable data for each type. Secondly, according to the correlation of task and data in attribute level, the data with the most frequency usage in history case will be searched. For example, the post-disaster raster image is used as a background

to show information of disaster area as rich as possible, the images of ZY-3 satellite was often chosen in this application, then the information will be obtain from RDF file directly. But if the ZY-3 images could not find it in the database, a further analysis to find potential right data proceeds. The system parses factors in spatial-temporal level to build a query condition, then images that can not satisfy Fushun flood in spatial reference, area and other conditions will be filtered and an available image set will be selected. Further, using formula (3), the system calculates the similarity of integrated assessment and each images in the data set in space, time and resolution and ranks them according to the correlation degree. Finally, a series of images labeled with relevance to the task will be arranged in data selection interface (shown as Figure 4 (c)) so users could choose the most suitable data for the assessment.

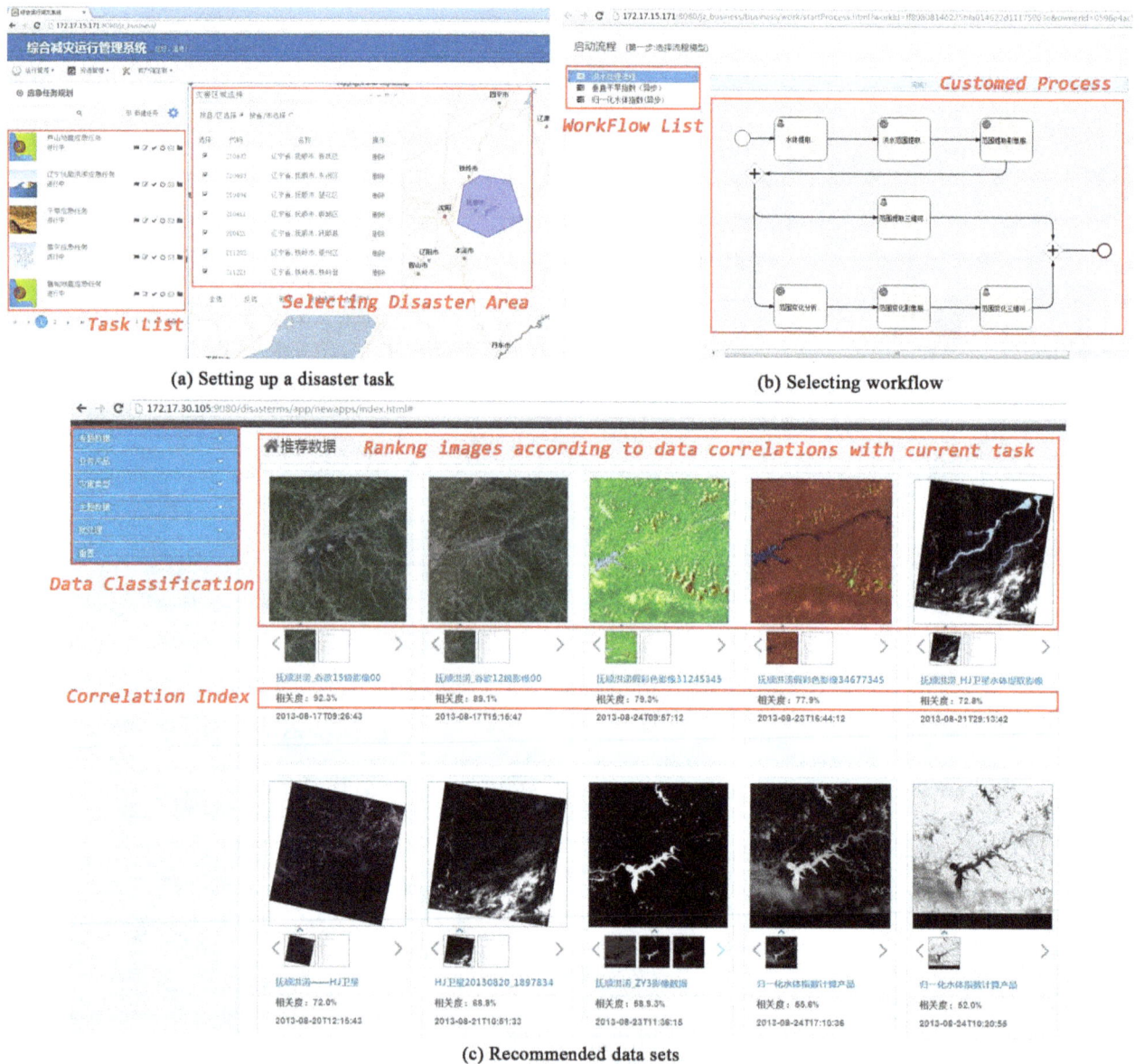

(a) Setting up a disaster task

(b) Selecting workflow

(c) Recommended data sets

Figure 4. The graphic interface of the prototype system

6. CONCLUSION AND FUTURE WORK

Compared to existing disaster data managing methods, there are several advantages to creating a task-oriented information link method using ontologies. Currently, searching for geospatial data can be overwhelming when one does know exactly which keywords to use. It can also be time-consuming to sift through undesirable results due to either poor keyword selection or bounding coordinate discrepancies within metadata. The method described here offers an innovative correlation method and lowers the complexity of man-machine interaction to find data.

In this study, ontologies for tasks and data sources are created independently and semantic mapping is set up between their features. The effort to create such a knowledge base is worthwhile because the independent ontologies and their association support semantic-related operations on spatial data, and helping users extract task-related information accurately. Then the analysis process of data searching is elaborated through introducing an emergency task scenario. The presented work is at an early stage and further research will focus on studying refining the statistical factors to take full advantage of historical cases and offering formulas to quantify the similarity between statistical factors and tasks.

ACKNOWLEDGEMENTS

This work was supported by the National Natural Science Foundation of China (Nos. 41171311, 41471320, 41471332), the National High Resolution Earth Observation System (the Civil Part) Technology Projects of China, and National High Technology Research and Development Program of China (2013AA122301).

REFERENCES

Chatterjee, R. and Matsuno, F., 2005. Robot description ontology and disaster scene description ontology: analysis of necessity and scope in rescue infrastructure context. Advanced Robotics, 19(8): 839-859.

Cohn, A.G., 1997. Qualitative spatial representation and reasoning techniques, KI-97: Advances in Artificial Intelligence. Springer, Berlin Heidelberg, pp. 1-30.

Demir, I. and Krajewski, W.F., 2013. Towards an integrated Flood Information System: Centralized data access, analysis, and visualization. Environmental Modelling & Software, 50: 77-84.

Ding, Y. et al., 2014. An integrated geospatial information service system for disaster management in China. International Journal of Digital Earth(ahead-of-print), pp. 1-28.

Fan, Z. and Zlatanova, S., 2011. Exploring ontologies for semantic interoperability of data in emergency response. Applied Geomatics, 3(2): 109-122.

Foster, I. and Grossman, R.L., 2003. Data integration in a bandwidth-rich world. Communications of the ACM, 46(11): 50-57.

Babitski, G., Probst, F., Hoffmann, J. and Oberle, D., 2009. Ontology Design for Information Integration in Disaster Management. GI Jahrestagung, 154: 3120-3134.

Grolinger K, Capretz M, Mezghani E, et al. Knowledge as a service framework for disaster data management. Enabling Technologies: Infrastructure for Collaborative Enterprises (WETICE), 2013 IEEE 22nd International Workshop on. IEEE, 2013: 313-318.

Guarino, N., 1998. Formal ontology in information systems: Proceedings of the first international conference (FOIS'98), 46. IOS press, Trento, Italy.

Hristidis, V., Chen, S., Li, T., Luis, S. and Deng, Y., 2010. Survey of data management and analysis in disaster situations. Journal of Systems and Software, 83(10): 1701-1714.

Huang, F. and Yan, L., 2013. Reasoning of ontology model for typhoon disasters domain based on Jena. Journal of Computer Applications, 3(33): 771-775, 779.

Iwata, K., Ito, Y. and Managi, S., 2014. Public and private mitigation for natural disasters in Japan. International journal of disaster risk reduction, 7: 39-50.

Klien, E., Lutz, M. and Kuhn, W., 2006. Ontology-based discovery of geographic information services—An application in disaster management. Computers, Environment and Urban Systems, 30(1): 102-123.

Laniak, G.F. et al., 2013. Integrated environmental modeling: A vision and roadmap for the future. Environmental Modelling & Software, 39: 3-23.

Lausch, A., Schmidt, A. and Tischendorf, L., 2015. Data mining and linked open data – New perspectives for data analysis in environmental research. Ecological Modelling, 295: 5-17.

Leskens, J.G., Brugnach, M., Hoekstra, A.Y. and Schuurmans, W., 2014. Why are decisions in flood disaster management so poorly supported by information from flood models? Environmental Modelling & Software, 53: 53-61.

Li, J., Zlatanova, S. and Fabbri, A.G., 2007. Geomatics solutions for disaster management. Berlin, Heidelberg, New York, Springer.

Li, B., Liu, J., Shi, L. and Wang, Z., 2009. A method of constructing geo-object ontology in disaster system for prevention and decrease, International Symposium on Spatial Analysis, Spatial-Temporal Data Modeling, and Data Mining. International Society for Optics and Photonics. 74923I-74923I.

Lutz, M., 2007. Ontology-based descriptions for semantic discovery and composition of geoprocessing services. Geoinformatica, 11(1): 1-36.

Michalowski, M. et al., 2004. Retrieving and semantically integrating heterogeneous data from the web. Intelligent Systems, IEEE, 19(3): 72-79.

Neumayer, E., Plümper, T. and Barthel, F., 2014. The political economy of natural disaster damage. Global Environmental Change, 24: 8-19.

Schulz, A., Döweling, S. and Probst, F., 2012. Integrating Process Modeling and Linked Open Data to Improve Decision Making in Disaster Management. Guest Editors, pp. 16.

Silva, T., Wuwongse, V. and Sharma, H.N., 2013. Disaster mitigation and preparedness using linked open data. Journal of Ambient Intelligence and Humanized Computing, 4(5): 591-602.

Borkulo, V.E., Barboza, V.S., Dilo, A., Zlatanova, S. and Scholten, H., 2006. Services for emergency response systems in the Netherlands, Proceedings of the Second Symposium on Gi4DM, Goa, India, pp. 6.

Wang, Y., Gong, J. and Dai, J., 2007. Spatial Data Semantic Query Based on Ontology. Journal of Geomatics, 2(32): 32-34.

Wiegand, N. and García, C., 2007. A Task - Based Ontology Approach to Automate Geospatial Data Retrieval. Transactions in GIS, 11(3): 355 - 376.

Xu, R., Dai, X., Yang, F. and Lin, P., 2009. Research on the construction method of emergency plan ontology based-on owl, The 2009 International Symposium on Web Information Systems and Applications, Nanchang, China, pp. 019-023.

Xu, W. and Zlatanova, S., 2007. Ontologies for disaster management response. Geomatics Solutions for Disaster Management. Springer, Berlin Heidelberg, pp. 185-200.

Yang, H., Lv, G. and Sheng, Y., 2013. Distributed Collaborative Geographic Modeling Task Decomposition Method Based on HTN Planning. Acta Geodaetica et Catrtographica Sinica, 42(3): 440-446.

Zhishan, Y., Run E, L., Yanjiang, W. and Xiaoling, S., 2012. The Research on Landslide Disaster Information Publishing System Based on WebGIS. Energy Procedia, 16: 1199-1205.

Zhu, Q., Li, H. and Yang, X., 2009. Hierarchical Semantic Constraint Model for Focused Remote Sensing Information Services. GEOMATICS AND INFORMATION SCIENCE OF WUHAN UNIVERSITY, 34(12): 1454-1457.

24

3D GIS BASED EVALUATION OF THE AVAILABLE SIGHT DISTANCE
TO ASSESS SAFETY OF URBAN ROADS

M. Bassani [a], N. Grasso [a], M. Piras [a],*

[a] Dept. of Environment, Land and Infrastructure Engineering, Politecnico di Torino, 24 corso Duca degli Abruzzi, Turin, 10024 Italy
(marco.bassani, nives.grasso, marco.piras)@polito.it

Commission III, WG III/4

KEY WORDS: 3D Sight Analysis, Sight Distance, Road Safety, Mobile Mapping, Digital Terrain Model, Sight Obstruction, Low-Cost Sensors, GNSS/IMU

ABSTRACT:

The available sight distance (ASD) in front of the driver to detect possible conflicts with unexpected obstacles is fundamental for traffic safety. In the last 20 years, road design software (RDS) has been continuously updated with dedicated modules to estimate ASD, thus assessing the quality of project from a safety point of view. Unfortunately, the evaluation of ASD still represents an issue in the case of existing road, and the object of discussion in the research community. To avoid problems related to the limitation associated with the use of digital terrain models typically employed in RDS, the Geographic Information Systems (GIS) software can use digital surface models (DSM) which are more flexible in the modelling of sight obstruction due to vegetation, street furniture, and vertical surfaces largely diffused in urbanized areas.

The paper deals with the evaluation of GIS in the estimation of ASD in a typical urban road where the density of sight obstruction along the roadside is relatively high. The work explores the case study of a collector road in the city of Turin (Italy). Results confirm the potentiality of GIS software in capturing the complex morphology of the urban environment, thus confirming that GIS could become an important analysis tool for road engineers in the field of road safety. The investigation here described is part of the Pro-VISION Project (funded in 2014 by the *Regione Piemonte*, Italy).

1. INTRODUCTION

About 75% of the accidents occur in the urban road network since this environment is affected by a high density of traffic made of different road users' (drivers, cyclists, and pedestrians) in terms of mass, size and speed. The number of conflicting points that occur where conflicting traffic flows intersect themselves is what distinguish the case of urban roads from the rural ones.

To reduce crashes and their worst consequences in terms of injuries and fatalities, road engineers are committed in protecting road users limiting vehicles' operating speeds and increasing the sight distances from the conflict points and/or potential obstacles along the driving path. According to standards and policies, the driver must have a visible space along his trajectory (called available sight distance, ASD) that she/he uses to control her/his vehicle, to avoid striking an unexpected object or other road users in the carriageway (Ministero delle Infrastrutture e Trasporti, 2001; AASHTO, 2011).

In the recent past, the evaluation of ASD has been implemented in road design software (RDS) that use digital terrain models (DTM), to design the horizontal and vertical alignments and to define the project surfaces (i.e., pavement, medians, margins and escarpments). The project model thus formed and the DTM are then analysed to derive the ASD from the most probable trajectories, which conventionally coincide with the centreline of each lane. With RDS, the evaluation of ASD is inaccurate since sight obstructions, such as vegetation and buildings, cannot be modelled and included in a DTM. In fact, the use of the digital surface models (DSM) will be better, because it includes the effective 3D model, where the buildings are included and some elements can be considered, but it depends on the DSM resolution.

RDS are more difficult to be used in the case of existing roads because it requires the regression of the real alignment, which is generally unknown, and the re-designed of project surfaces minimizing their distances from the DTM. Conversely, Geographic Information Systems (GIS) software can use DSM that are nowadays available for a large part of the urbanized areas. GIS can also incorporate every type of geospatial data as point cloud obtained by LiDAR, images, that may be used to include sight obstructions, and points or polygon (e.g they can be derived from Global Navigation Satellite System (GNSS) receivers), that can include vehicle trajectories defined as a set of 3D points (Khattak and Shamayleh, 2005; Castro et al., 2014). Moreover, GIS software has beneficiate from new tools in the field of sight analysis (Environmental Systems Research Institute, 2010). Unfortunately, GIS is not still diffused in the road designers community since it does not contains any tool to design road alignment and cross sections elements.

In 2013 the *Regione Piemonte* administration funded the Pro-VISION Project, whose participants were asked to cooperate in the tentative to fill the gap in knowledge and to promote the implementation of new tools in the field of road safety. The paper synthetized the results of the part of the

---

project dedicated to the estimation of the ASD along urban roadways. The authors explored the problems encountered in the formation of a 3D model of an urban road section here considered as a case study. In particular, shape and positions of sight obstructions have been collected and implemented in a DSM already available in GIS. Trajectories of vehicles and the positions of drivers' eyes have been first reconstructed and then explored to estimate the ASD from the driver's point of view. The information has been used to generate sight profiles which are fundamental in the safety assessment of existing infrastructures.

2. AVAILABLE SIGHT DISTANCE

2.1 Design standards

According to the Italian standard (Ministero delle Infrastrutture e Trasporti, 2001), the existence of appropriate ASD along roads is of primary condition for traffic safety. The ASD is the length of the longitudinal road section that the driver can see in front of her/him without considering the influence of traffic, weather and road lighting. The distance of unobstructed vision (Figure 1) have to be compared with the distances necessary to safely perform some basic manoeuvres like the emergency stopping in front of an unexpected obstacle (i.e., stopping sight distance, SSD), overtaking a slower vehicle (i.e., passing sight distance, PSD), lane change in the carriageway at singular points such as intersections and exits ramps (i.e., change lane distance, CLD). In US, the decision sight distance (DSD) to detect unexpected or otherwise difficult-to-perceive information source or condition in a roadway environment is also compared to ASD (AASHTO, 2011).

It is worth noting that current standards impose the execution of the sight analysis to new construction only (Ministero delle Infrastrutture e dei Trasporti, 2001), but in case of re-design of an existing road it could be extended to improve safety.

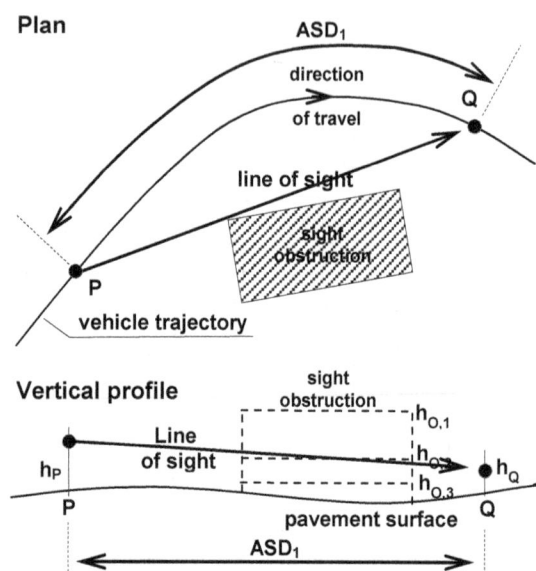

Figure 1. 3D estimation of ASD in a road sections

2.2 ASD estimation

Figure 1 shows the plan and the vertical profile of a typical driving situation along a horizontal curve where a driver at point P is able to see all the points along the trajectory to the point Q if the sight obstruction limits the field of vision. This is the case of sight obstructions tall enough (height equal to $h_{O,1}$) to obscure the part of trajectory over Q (i.e., ASD = ASD_1). Conversely, small sight obstructions with a height below the line of sight ($h_{O,3}$) do not create any visual limitations, so the driver can see over the point Q (i.e., ASD > ASD_1). In both cases, the evaluation of ASD is relatively simple and can be carried out referring to the plan view.

The case of obstruction with a height closed to the line of sight ($h_{O,2}$) cannot be analysed referring to planar representation (vertical or horizontal), and have to be investigated referring to the 3D analysis.

Such analysis is also necessary in case of vertical and horizontal curve combination, and when multiple sight obstructions are located along the roadside, like in the case of urban roads. In fact, the ASD is heavily affected by the road geometrics and by several sight obstructions located in the roadside and median like traffic barriers, vegetation, parked vehicles, street furniture, fences, buildings and other road elements. In night-time conditions on roads which are not provided of public lighting systems, the ASD is limited in vertical sag curves because the space illuminated by the vehicle headlights is shorter than the corresponding one along grades. According to Figure 1, ASD depends also on the height of the driver's eye above the road surface (h_P), and the object height above the pavement surface (h_Q) that road design standard establish in function of the driving manoeuvre considered (SSD, PSD, CLD, DSD, and others according to the national policy adopted). Finally, the ASD is estimated not considering the presence of other vehicles and the effects of adverse environmental conditions (i.e., fog).

As a result, in case of existing roads the estimation of ASD is possible through (a) direct field measurements, or (b) indirect measurements on a 3D model that represent the road scenario. The (a) case implies the total or partial closure of the road section to traffic for the time that is necessary to perform the survey, thus submitting the operators involved in the field to risky working conditions; furthermore, the evaluation is punctual and a lot of effort and time is necessary to get a sufficient quantity of ASD data. The (b) case imposes the acquisition and the processing of a consistent and robust quantity of geospatial data, which is nowadays possible recurring to existing database or re-creating it from field surveys. Geospatial data must include 3D elements like the pavement surface, the margins, and any potential sight obstructions (trees, buildings, street furniture, etc.). Afterwards, a relatively small effort is necessary to treat and process such data to form the 3D model. Finally, the vehicles trajectories are reconstructed and ASD can be evaluated in a greater number of points to achieve the desired level of accuracy.

3. SPATIAL DATA ACQUISITION AND TERRAIN MODELLING

3.1 ProVISION project

The main aim of the ProVISION project was to develop a set of tools to carry out visibility analysis on existing infrastructures and for any kind of road user. One of the objectives of this project was to implement a low-cost mobile mapping device to acquire images made of multiple sensors of mass market (e.g. GNSS, IMU, webcam and action-cam), and to provide georeferenced and spatially oriented images for photogrammetric purposes. Only the part of the investigation dedicated to vehicle users is commented in this paper.

To perform visibility analysis, a 3D model of the road environment was created integrating available data with new object surveyed in the field. As a case study, the urban road section of corso Castelfidardo in Turin (Italy) was considered as the tests site. The main steps of the Pro-VISION project are summarized as follows:

1. development of a low-cost mobile mapping system (hereafter MMS) for the acquisition of georeferenced spatial data (e.g. 3D images, trajectory, etc.);
2. integration between MMS data and aerial data (stereoimages and orthophotos) to generate an updated 3D model into the GIS environment;
3. road elements extraction (e.g. road markings, pedestrian crossing) to generate the trajectory (vehicle and obstacle) and to include the information about the road signals; and
4. definition of algorithms for the sight analysis.

In the following paragraphs describe the phases of creation of the 3D model, from initial data available, which were integrated with different techniques to generate the real model of the urban environment and the Digital Surface Model (DSM). The 3D model was built starting from the Turin's Municipal Technical Map at the scale of 1:1000 (edition 2014), the colored orthophotos at a resolution of 30 cm (edition 2012), and the digital terrain model (DTM) with a grid of 5 m of the *Regione Piemonte* (Italy).

3.2 Generation of the DSM model

From the Turin's Municipal Technical Map, a 2D model in a GIS system (coordinate system ETRF2000-UTM zone 32N) was created to enables the management of spatial data characterizing the road environment. In this kind of Technical Map, all the entities, which characterize the road infrastructure and the urban environment, are usually represented.

Subsequently, the model was integrated with the road markings by photogrammetric plotting using stereoscopic models of aerial images, by means the use of the ZMap software (Baz et al, 2008). In the first phase, all the road markings were defined by analysing the characteristics and differentiating dashed and continuous lines, punctual signals (stop and priority), pedestrian crossings and parking lanes. The final products were extracted in a dxf format to be imported in the GIS environment.

3.3 Field surveys

In the 3D model obtained as previously described, some elements as vegetation and street furniture could not be included because the resolution of the stereo images was too low. A real model of an urban environment needs such spatial data, that the authors solved recurring to the use of a MMS, which was mounted on a vehicle or a bike (Figure 2). These data can be used for photogrammetric plotting to define the position of the objects.

Figure 2. Mobile Mapping System on car and bicycle

In this investigation, the Authors used a specific MMS (Cina et al., 2008), which is able to acquire georeferenced data with low-cost sensors (e.g. webcam, action-cam and integrated positioning system). The image acquisition system was installed both on an instrumented vehicle with sensors of better characteristics (single and dual-frequency GNSS receiver), and on a bicycle to acquire data from bikeways (Figure 2).

On the vehicle, at the ends of the bar two webcams were connected to the PC managed by the operator inside of the car (Figure 2). Two different action-cams with GPS were also employed to obtain the driving trajectory in addition to video of the path. Finally, the GPS receiver u-blox was placed in a central position and then connected to a PC. The acquisitions were carried out along the central axis of the main carriageways in the two driving directions. One contour cam and two GoPro cams were mounted on the bike to obtain stereoscopic images of the urban environment to be implemented in the 3D model. The bike path presents in the case study was surveyed in both directions.

As well known (McGlone et al., 2004), each optical device has radial and tangential distortion due to optical lens, which lead to have an image that is not a central projection. In order to use these images for photogrammetry, it is fundamental to estimate the radial and tangential distortion parameters. In this case, this step was realized using a specific MATLAB toolbox called "Camera Calibrator" and a checkerboard pattern, (Aicardi et al, 2014).

3.4 Road space construction

The couple of images collected from the MMS system was used into the modern procedures based on Structure from Motion (SfM) algorithms. This procedure allows the automatic extraction of a point cloud starting from a sequences of images with a high percentage of overlapping. In this investigation, the high quantity of images allowed the use of the SfM technology to extract a new 3D model to be integrated in the initial model, even considering street furniture, vegetation, fences, and parking lanes.

Some element were included using a specific software named "Viewer" written by the author in Fortran, which allows to correlate the frames to GPS track and allows to identify and locate in the correct spatial position visible objects within images (Ajmar et al., 2011).

To generate and complete the 3D model, the 2D model previously generated was imported in the application ArcScene, with entities such as buildings that were extruded to obtain the 3D model of the urban environment. This model was used to generate the DSM. In a first stage, a fixed height was assumed for the extruded objects. Therefore, the surveys carried out with the MMS were used to correct the odds of possible obstructions on the road infrastructure. Furthermore, the parking lane placed all along the roadways have been considered as occupied by cars, thus considering the worst driving conditions with sight obstructions placed close to the driving path. Cars were extruded at a height of 1.60 m above the ground.

The applications and the software chosen to perform the sight analysis require the conversion of data format MultiPatch as integration of the DTM. The MultiPatch is a type of geometry made of planar 3D rings and triangles, used in combination to model objects that occupy discrete area or volume in the 3D space. MultiPatches may represent geometric objects like spheres and cubes, or objects like buildings and trees.

3.5 DSM generation

The sight analysis needs an accurate and updated DSM. In this case, starting from the available regional model with 5 m of resolution, the models have been updated by DSM generated with SfM algorithms (Hartley and Zisserman, 2003) using aerial stereopair images. Moreover, a local 3D models has been considered, which has been defined using the georeferenced images collected with the low-cost devices, and the models has been created also with SfM procedure.

From the preliminary analysis, it was clear that such DSM was not sufficiently detailed to allow the identification of possible sight obstructions. Hence, a new DSM (Figure 3) with a grid of 10 cm was built converting the 3D objects from the MultiPatch format. ArcGIS® allows to perform such operation through the toolbox called "Raster Calculator", in which the difference between the original DTM and the model obtained from the conversion of MultiPatch in raster was set.

3.6 Integration techniques and updating of the GIS model

In order to consider the real scenario and any possible changing, an "updating" of the GIS model has to be considered. Firstly, it was investigated the possibility to integrate the model with a DSM including buildings (facades, walls, and fences) generated by aerial stereo pair images.

After the DSM generation, it was resampled to a resolution equal to 10 cm, which is coherent with the original DSM. The union of the model was carried out using the ARCGIS® toolbox called "Mosaic" (Figure 4). In fact, when working with different DSMs, it is important to have continuity at ground level in the border of the final model. This tool allows to generate a new model from two or more raster settings, considering the overlap between the areas and a threshold tolerance. During the tests, it was decided to perform the overlap through an operator of type "maximum" (the output data generated by the overlapping of the areas is given by the maximum value among the overlapping cells), while the tolerance of mosaicking was set equal to 0.

3.7 Point clouds generated from images captured by smartphone

Sometimes, it could be possible that some small or partially hidden entities that constitute obstructions in the sight analyses, are not detectable with the procedures described previously. A solution to update the existing DSM with the 3D model of the object is offered by the use of SfM approach, but using images captured by smartphone (Figure 5). In this case, the acquisition of the images requires a maximum care, considering that each point must appears in at least three images. The entire object must also be acquired from different angles and with different smartphone-object distances.

The product of these operations is a 3D surface that can be imported on GIS software and converted to raster files for integration with the existing DSM. To get a georeferenced model, it is possible to define coordinates (e.g. using topography) of some special points (marker) on the surface of the object. During the alignment of the images, various software allows to collimate manually the various markers and assign the reference coordinates.

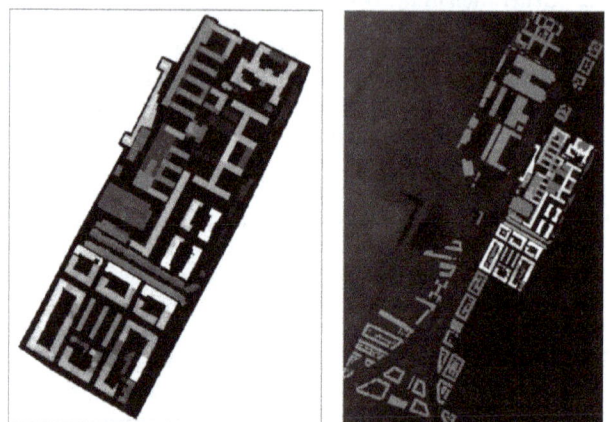

Figure 4. Example of integration DSM (grid = 20 cm); its integration in the DSM of the road infrastructure

Figure 5. 3D model formation with images captured by smartphone

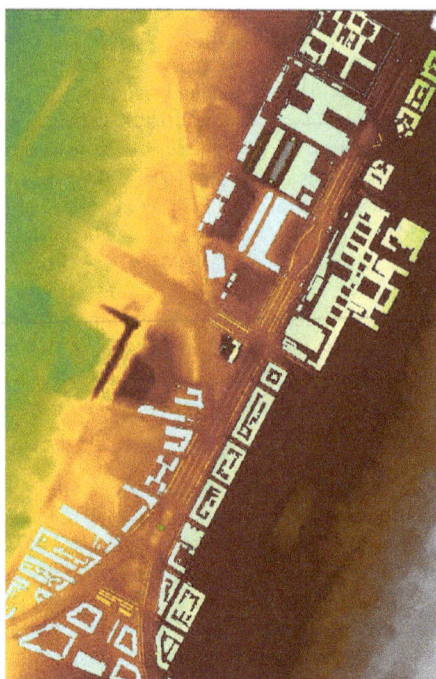

Figure 3. DSM of the test site

4. MODEL ANALYSIS

The DSM built according to the procedure described in Paragraph 3 (Figure 6) was explored considering the positions of the driver's point of view according to the road standards (Ministero delle Infrastrutture e dei Trasporti, 2001) to derive the ASD considering the presence of obstructions that limit the visibility along the road.

To pursue such objective, in the GIS environment a specific toolbox that allows the use in sequence of several geoprocessing tools was formulated and implemented with the ArcGIS® ModelBuilder . The main advantage in using the ModelBuilder is that the operations can be automated. In fact, the process is first saved and then run whenever it being necessary, or change it by modifying the input parameters to produce new results. As a result, the ModelBuilder allows setting models with a complex workflow replicable through a sequences of geoprocessing tools, where the output of the first becomes the input of the following one. The instruction can be dragged on the main screen where it is asked to indicate the data input and the output. The toolbox can be used both for 2- or 3D analyses.

ArcGIS® has several applications dedicated to the spatial analysis; between them, "Line of Sight" has been specifically develop to support the analysis of visibility. The first instruction to be used in the "Line of Sight" application is named "Construct Sight Line" that connect the point of observation to the target with a line. Working in a GIS environment, these data are organized in a georeferenced database and in order to create a unique connection between an observation point and the corresponding target identified as ObjectID, these database are related precisely through field which identify the entities themselves.

The sight lines obtained become the input of the second tool, which analyses and evaluates any spatial intersections with the DSM. The products of this second process are the lines of sight (Figure 7), which in the interface of the software assume a green color when they do not intersect the DSM, and red as long as they intersect a sight obstruction as part of the DSM. The coordinates of the point of obstruction constitute the second output, to which the 3D coordinates are evaluated with the command *Add XY Coordinates*.

Figure 7 shows the procedure adopted in the estimation of the ASD along the portion of the model included in Figure 8. The driving trajectory was assumed in the centre of the lane. The points of observation (P point in Figure 7) was then placed at 1.1 m above the road surface; a step of 50 m along the driving path was assumed to investigate the ASD. Figure 8 shows the position on the model of the eight points considered in the sight analysis.

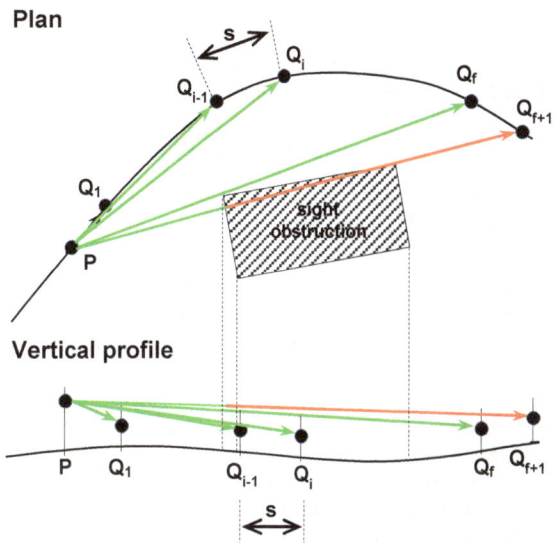

Figure 7. Representation of the algorithm used for the estimation of ASD

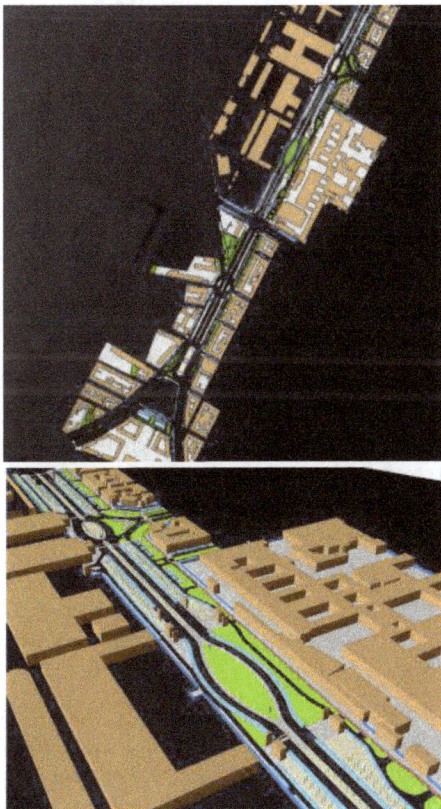

Figure 6. 2D and 3D GIS model of the test site

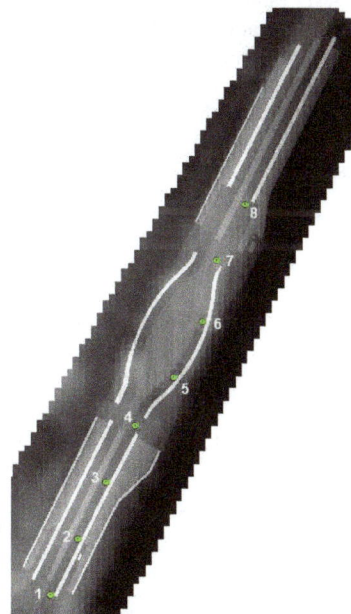

Figure 8. DSM model of the case study with eight points investigated for the ASD estimation

The target point was placed along the driving path at a height of 0.10 m above the road surface. A step of 1 m (s in Figure 7) was considered to evaluate the position of the more distant target point. A dedicated algorithm was created to generate the position of points Q of Figure 7. When the first target obstructed is identified along the driving path (Q_{f+1} in Figure 7), the algorithm selects the point Q_f as the more distant target visible from point P. Finally, all the data were collected and exported to an Excel file.

Figure 9 reports the synthesis of the results of ASD evaluation. Results shows the effects of sight limitations due to road elements (curbs, road furniture) and most of all the vehicles parked along the roadside. Minimum values of ASD where observed in points 4 and 6 which are placed closed to the shortest horizontal curves along the driving path (respectively equal to 31 and 45 m). According to the Italian standards (Ministero delle Infrastrutture e Trasporti, 2001), at a design speed of 60 km/h, which is appropriate for the specific case study, the stopping sight distance (SSD) on wet pavement is equal to 70 m, therefore ASD < SSD. To solve the sight limitation, the dismissal of the parking lane along the section indicated in Figure 9 is here proposed.

5. CONCLUSIONS

Sight analysis aims to estimate the space visible to the driver (available sight distance, ASD) which needs to be detected in advance to avoid limitations due to the presence of possible obstruction along the route. The existence of an ASD greater than the distance necessary to perform specific manoeuvres is a pre-requisite to create safe driving condition in a harmonious road environment.

The procedure described in the paper leads to the creation of a 3D model that was explored and investigated in the GIS environment. Such model was able to give back the necessary information about ASD, which can be further used to improve the safety of existing infrastructure. When the comparison of ASD with the sight distances that need to be satisfied is negative, the sight obstruction can be removed or replaced to increase the ASD and satisfy the safety needs. In this investigation, the integration of different data formats required the use of several software, which certainly affects the DSM quality with a consequent reduction in accuracy. This is the reason why in the future the authors intend to develop an automatic procedure to implement the 3D model to be used in the ASD analysis.

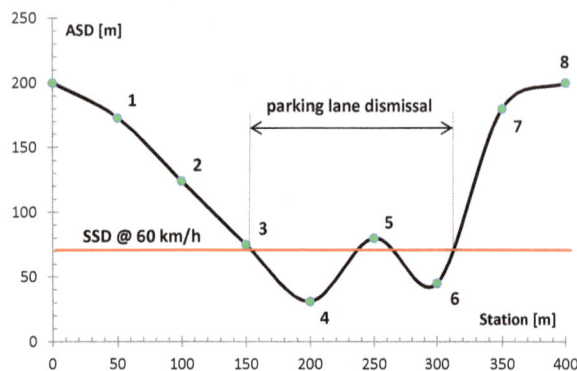

Figure 9. ASD evaluated for the case study, and comparison with SSD @ 60 km/h

Part of future investigation will be dedicated to a greater use of data acquired by mass-market instruments, like smartphone, webcam, action-cam and low-cost GNSS receiver. To take into account the fact that the urban scenario develops continuously, the idea is to create a dynamic model in which dedicated algorithms will allow the extraction of information from the recognition of geometric samples within the image and to the encoding of signals or symbols, in particular street horizontal markings and vertical signals. Finally, future research activities will be addressed to the evaluation of the ASD in correspondence to intersections, which are always affected by higher accident frequencies than longitudinal road sections.

ACKNOWLEDGEMENTS

The authors thank the "ICT- Poli di Innovazione Regione Piemonte" and FINPIEMONTE who funded the Pro-VISION Project. The other project's partner, Beanet s.r.l. and Synarea s.r.l., which have cooperated with the authors, are greatly acknowledged.

REFERENCES

Aicardi I., Lingua A., Piras M., 2014. Evaluation of Mass Market Devices for the Documentation of the Cultural Heritage. *Inter. Archives of Photogrammetry, Remote Sensing and Spatial Information Sciences*, XL-5, pp. 17-22.

Ajmar A., Balbo S., Boccardo P., Giulio Tonolo F., Piras M., Princic, 2011. A Low-Cost Mobile Mapping System (LCMMS) for Field Data Acquisition: a Potential Use to Validate Aerial/Satellite Building Damage Assessment. *International Journal of Digital Earth*, 1(21), pp. 1-21

American Association of State Highway and Transportation Officials, 2011. *A Policy on Geometric Design of Highways and Streets*. 4th edition, ISBN: 1-56051-156-7.

Automobil Club Italia, Istituto Nazionale di Statistica, 2014. *Road Accidents in Italy in the Year 2013*. http://www.aci.it/fileadmin/documenti/studi_e_ricerche/dati_sta tistiche/incidenti/Road_accidents_in_Italy_2013.pdf (June 10, 2015).

Bassani, M., Lingua, A., Piras, M., De Agostino, M., Marinelli, G. and G. Petrini, 2012. Alignment Data Collection of Highways using Mobile Mapping and Image Analysis Techniques. Transportation Research Board of the National Academies, 91st Annual Meeting, No. 12-0312, Washington D.C.

Baz, I., Kersten, T., Büyüksalih, G., & Jacobsen, K., 2008. Documentation of Istanbul Historic Peninsula by Static and Mobile Terrestrial Laser Scanning. *The International Archives of the Photogrammetry, Remote Sensing and Spatial Information Sciences*, Beijing, Vol. XXXVII. Part B 5.

Castro, M., Anta, J.A., Iglesias, L., Sánchez, J.A., 2014. GIS-Based System for Sight Distance Analysis of Highways. *Journal of Computing in Civil Engineering*, 28(3), pp. 04014005.

Cina, A., Lingua, A., Piras, M., 2008. *Low-Cost Mobile Mapping Systems: an Italian experience. IEEE/ION Position Location and Navigation Symposium*. Monterey, California, May 5-8, pp. 1033-1045.

Environmental Systems Research Institute, 2010. ArcObjects library reference, http://edndoc.esri.com/arcobjects/9.2/ComponentHelp/esriGeo Database/ISurface_GetLineOfSight.htm (accessed July 6th, 2015).

Hartley, R., Zisserman, A., 2003. *Multiple View Geometry in Computer Vision.* Cambridge University Press. ISBN 0-521-54051-8.

Karara, H.M., 1989. *Non-topographic Photogrammetry.* American Society for Photogrammetry and Remote Sensing, 2nd edition, ISBN 0-944426-10-7.

McGlone, J.C., Mikhail, E., Bethel, J., 2004. *Manual of Photogrammetry*, American Society for Photogrammetry and Remote Sensing, 5th edition, Bethesda, MD.

Ministero delle Infrastrutture e dei Trasporti, 2001. *Norme Funzionali e Geometriche per la Costruzione delle Strade.* Decreto Ministeriale 6792, Roma, Italia.

25

INTEGRATED ESTIMATION OF SEISMIC PHYSICAL VULNERABILITY OF TEHRAN USING RULE BASED GRANULAR COMPUTING

H. Sheikhian[a], M.R. Delavar[b] and A. Stein[c]

[a] MSc. Student, GIS Dept., School of Surveying and Geospatial Eng., College of Eng., University of Tehran, Tehran, sheikhain@ut.ac.ir
[b] Center of Excellence in Geomatic Eng. in Disaster Management, School of Surveying and Geospatial Eng., College of Eng., University of Tehran, Tehran, Iran, mdelavar@ut.ac.ir
[c] Department of Earth Observation Science, University of Twente, The Netherlands, a.stein@utwente.nl

Commission IV, WG IV/7

KEY WORDS: Granular Computing (GrC) algorithm; Geospatial Information System (GIS); Earthquake Physical Vulnerability Assessment; Multi-Criteria Decision Making

ABSTRACT:

Tehran, the capital of Iran, is surrounded by the North Tehran fault, the Mosha fault and the Rey fault. This exposes the city to possibly huge earthquakes followed by dramatic human loss and physical damage, in particular as it contains a large number of non-standard constructions and aged buildings. Estimation of the likely consequences of an earthquake facilitates mitigation of these losses. Mitigation of the earthquake fatalities may be achieved by promoting awareness of earthquake vulnerability and implementation of seismic vulnerability reduction measures. In this research, granular computing using generality and absolute support for rule extraction is applied. It uses coverage and entropy for rule prioritization. These rules are combined to form a granule tree that shows the order and relation of the extracted rules. In this way the seismic physical vulnerability is assessed, integrating the effects of the three major known faults. Effective parameters considered in the physical seismic vulnerability assessment are slope, seismic intensity, height and age of the buildings. Experts were asked to predict seismic vulnerability for 100 randomly selected samples among more than 3000 statistical units in Tehran. The integrated experts' point of views serve as input into granular computing. Non-redundant covering rules preserve the consistency in the model, which resulted in 84% accuracy in the seismic vulnerability assessment based on the validation of the predicted test data against expected vulnerability degree. The study concluded that granular computing is a useful method to assess the effects of earthquakes in an earthquake prone area.

1. INTRODUCTION

Earthquakes are among the most hazardous natural disasters. They are unpredictable in time, location and intensity. They seriously affect the population, building constructions and infrastructure, especially in urban areas. They often occur close to geological faults and plate boundaries. The city of Tehran is located on several faults that have had a long period of inactivity. The faults thus contain a high risk for releasing a large amount of seismic energy, thus exposing the city to a catastrophic earthquake followed by destruction of thousands of buildings, its infrastructure, a number of fatalities and leaving many injured inhabitants. The city has suffered huge earthquakes in cycles of approximately every 150 years. Since there have not been any large earthquakes (greater than 6 at the scale of Richter) in Tehran in the past 185 years, seismologists expect a large earthquake to happen in Tehran soon (JICA, 2000). This confirms the need to estimate the expected damage and the associated loss caused by an earthquake in order to effectively mitigate its consequences.

Several institutions have carried out risk assessment in Tehran, in particular the International Institute of Earthquake Engineering and Seismology (Zare et al., 1999, Zaré and Memarian, 2003, Boustan and Shafiee, 2011) that classified potential Tehran earthquake damage from a geotechnical point of view and the Japan International Cooperation Agency (JICA, 2000) that produced seismic micro-zoning maps for the city.

Defining earthquake physical vulnerability as a multi-criteria decision making depends upon various parameters including building properties such as the material, the number of floors and earthquake characteristics such as intensity, surface topography attributes like slope and expert's judgments. All of these contain large uncertainties (Aghataher et al., 2005, Silavi et al., 2006, Amiri et al., 2008, Samadi Alinia and Delavar, 2011, Jahanpeyma et al., 2007, Khamespanah et al., 2013a,b, Panahi et al., 2013, Moradi et al., 2014a,b).

Abundant research efforts were carried out to address the earthquake modelling problem. Examples are coseismic displacement modelling (Yaseen et al., 2013a,b), hybrid models (Kappos et al., 1998) and spatio-temporal models (van Lieshout and Stein, 2012).

In this context, several researchers have focused on utilizing multi-criteria evaluation methods to define the seismic vulnerability of buildings in Tehran and handle the associated uncertainty aspects. For instance, Aghataher et al. (2005) implemented a fuzzy logic and analytical hierarchical process (AHP) approach to obtain weights of vulnerability factors to perform human loss probabilities in Tehran. Silavi et al. (2006) considered an AHP improved with intuitionistic fuzzy to obtain pessimistic and optimistic maps to assess human and physical seismic vulnerability assessment in the city. Amiri et al. (2008) used dominance-based rough sets to approximate the partition of a set of predefined and preference-ordered of the vulnerability grades. In this framework, Tehran metropolitan areas have been sorted with respect to their vulnerability degrees by the means of decision rules in the form of "IF–THEN" statements including both exact and non-exact rules. Majority voting in spatial group multi-criteria decision making supported by density induced ordered weighting average operator was applied to the problem by Moradi et al. (2014a).

In this study we turn towards Granular Computing (GrC) to extract compatible and accurate rules from a training data set to classify the whole study area (Samadi Alinia and Delavar,2011). GrC uses granules of information to find appropriate solutions (Zadeh, 1998,Yao, 2001, 2004). This model was implemented in the basic form by Samadi Alinia and Delavar (2011) and in an integrated form with Dempster-Shafer theory (Shafer,1992) by Khamespanah et al. (2013b) to assess Tehran seismic vulnerability.

Here, a new model for rule extraction process by GrC is proposed. It uses generality and absolute support for rule extraction and coverage as well as entropy to determine quality of the rules to be used in forming the granular tree. In the past, attention was focused on determining Tehran seismic vulnerability in the case of activation of the North Tehran fault (Samdi Alinia and Delavar, 2011, Khamespanah et al., 2013a), or activation of the three faults separately (Moradi et al., 2014a). In this paper, seismic vulnerability is assessed against aggregated activation of the North Tehran fault, the Mosha fault and the Rey fault simultaneously. In this way, the maximum seismic vulnerability imposed by the three faults is obtained and used to determine the physical seismic vulnerability of Tehran as the worst case scenario of the simultaneous activation of the three faults.

Geospatial information system (GIS) has been used as a spatial modelling and fusion framework, where GrC has been applied. Tehran urban statistical units have been considered as objects and six physical vulnerability criteria are taken into account as attributes of the objects forming an information table.
The rest of the paper is organized as follows. The theory of GrC and the customized model used in this paper is presented in Section 2. Data characteristics and obtained results are presented in Section 3. Section 4 finally presents the discussion and conclusions of the paper.

2. GRANULAR COMPUTING ALGORITHM

GrC is the science of processing data in different granularity levels (Bargiela and Pedrycz, 2003, Pawlak, 1982, Hobbs, 1985, Zadeh and Kacprzyk, 1999, Nguyen et al., 2001, Miao and Fan, 2002, Keet, 2008, Yao, 2008). In order to do so, information is divided into subsets, which are called granules of information (Yao, 2001, Lin, 2003, Yao, 2008).

The basic idea of information processing in GrC presents the information table, which is a finite set of objects commonly named the universe described by a finite set of describing attributes presented by Equation (1) (Pawlak, 1982):

$$S = (U, A_t, L, \{V_a \mid a \in A_t\}, \{F_a \mid a \in A_t\}) \quad (1)$$

where U is a finite non-empty set of objects, A_t is a finite non-empty set of attributes, L is a language defined by using attributes in A_t, V_a is a non-empty set of values of $a \in A_t$, $I_a: U \rightarrow V_a$ is an information function mapping an object from U to exactly one possible value of attribute a in V_a (Pawlak, 1982). To classify a data set by means of GrC, a set of rules is extracted. In a number of studies of machine learning and data mining, an IF–THEN statement paraphrases a rule, "If an object satisfies Φ, then the object satisfies Ψ.". In this way, a rule can be expressed in the form of $\Phi \Rightarrow \Psi$, where Φ and Ψ are intensions of the two concepts (Gupta et al., 1979, Pawlak, 1982). The interpretation suggests a

cause and effect relationship between Φ and Ψ (Yao, 2001). GrC applies several measures for a single-granule properties, a relationship between two granules, and a relationship between a granule and a set of granules.

2.1 Generality

The generality of concept Φ displays the relative size of constructive granule of the concept Φ, as defined in Equation (2). It confirms that a larger granule will result in the greater generality index (Pawlak, 1982):

$$G(\phi) = \frac{|m(\phi)|}{|U|} \quad (2)$$

where $|m(\Phi)|$ is the size of granule that constructs the concept Φ and $|U|$ is the size of constructive granule of the whole universe.

2.2 Absolute Support

For the two given concepts Φ and Ψ, the absolute support (AS) or confidence that Φ provides to the Ψ, is defined by Equation (3) displaying the conditional probability of a situation that a randomly selected object satisfying Ψ, also satisfies Φ (Yao, 2001):

$$AS(\phi \rightarrow \psi) = \frac{|m(\phi \wedge \psi)|}{|m(\phi)|} \quad (3)$$

where $|m(\Phi \wedge \Psi)|$ is the size of granule which supports both concepts Φ and Ψ. The quantity AS is between 0 and 1 and expresses the degree to which Φ implies Ψ (Yao, 2001).

2.3 Coverage

The coverage of concept Φ provided by concept Ψ is defined by Equation (4) (Yao, 2001):

$$CV(\Phi \rightarrow \Psi) = \frac{|m(\Phi \wedge \Psi)|}{|m(\Psi)|} \quad (4)$$

where $|m(\Psi)|$ is the size of constructive granule of concept Ψ, and $|m(\Phi \wedge \Psi)|$ is the size of granule constructing both concepts Φ and Ψ. This quantity displays the conditional probability of a randomly selected object to satisfy Φ, when satisfies Ψ and shows the coverage of Ψ upon Φ (Yao, 2001, 2008).

2.4 Conditional Entropy

For formulas Φ a family of formulas of $\Psi = \{\Psi_1, \Psi_2, \ldots, \Psi_n\}$ is considered that induces a partition $\pi(\Psi) = \{m(\Psi_1), \ldots, m(\Psi_n)\}$ of the universe. The conditional entropy $H(\Psi|\Phi)$ that reveals the uncertainty of formulas Φ based on formulas Ψ, is defined by Equation (5) (Yao, 2008):

$$H(\Psi|\Phi) = -\sum_{i=1}^{n} p(\Psi i|\Phi) \, log \, (p(\Psi i|\Phi)) \quad (5)$$

where: $p(\Psi i \mid \Phi) = \frac{|m(\Phi \wedge \Psi i)|}{|m(\Phi)|}$.

2.5 Mining association rules

In this research, generality and absolute support are used as the effective criteria for extracting confident rules. Entropy and coverage are then used to prioritize the extracted rules to build the granule tree. The procedure for extracting association rules is illustrated in Figure 1. This procedure comprises of extracting rules and constructing granules from training data set, until algorithm extracts the best rule set for the predictions.

2.6 Mining exception rules

According to Yao (2001), a major drawback of the association rules extracted by the original GrC is the possibility of existing rules relating some concepts in the data that may not exist in reality, or rules that are not extractable by applying this model, making association rules incomplete to classify the dataset appropriately.

For example, based on the association rule extraction principles, for the two concepts Φ and Ψ, the rule $\Phi \rightarrow \Psi$ may have a high absolute support, whereas it has not been extracted as an association rule. However, if Ψ is a concept with high generality, considering the absolute support formula, it could be concluded that in reality Φ supports Ψ negatively and association does not exist (Yao, 2001, Khamespanah et al., 2013a). Existing of suitable rules for classifying data that may not have a high generality but not extracted by the association rules is also possible (Yao, 2001).

Exception rules can be extracted for a rule like $\Phi \rightarrow \Psi$ if the formula Φ' is found and added to the initial rule and a converse result to initial rule is obtained for instance, $\Phi' {}^{\wedge} \Phi \rightarrow \neg \Psi$; in which it has high absolute support, no matter how low the generality is (Yao, 2001).

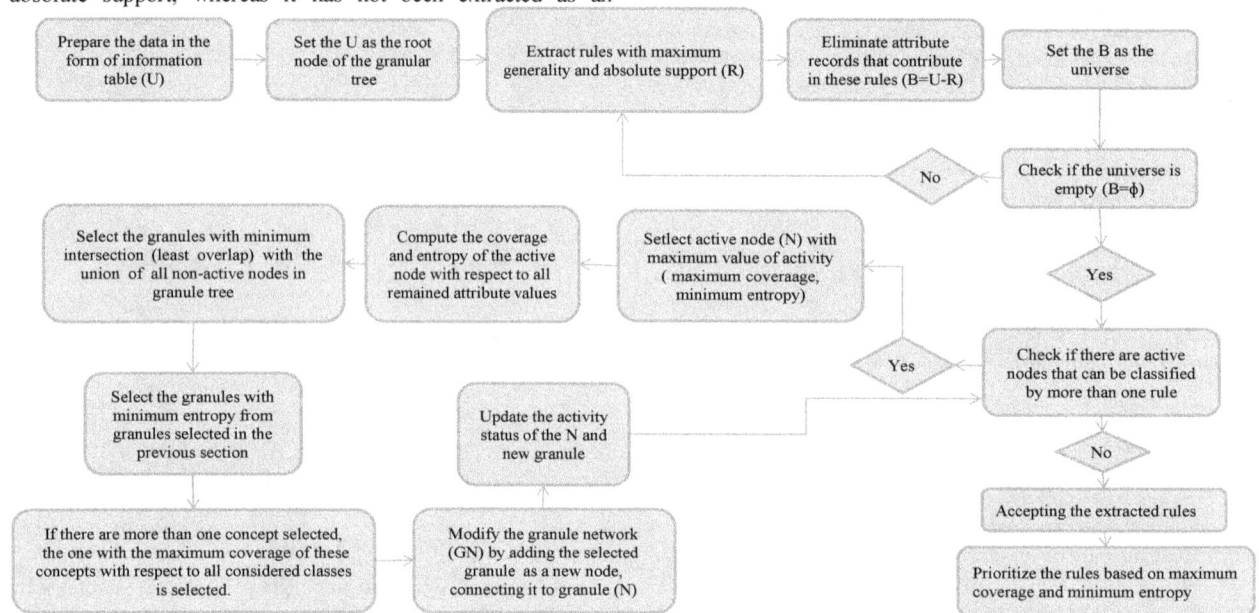

Figure 1: Schema of obtaining classification rules by the GrC algorithm

3. EXPERIMENTAL RESULTS

In this section, the procedure of data preparation, model implementation and the obtained results are discussed.

3.1 Data preparation

The physical vulnerability of Tehran against earthquake is based on the activation of the three major faults, i.e. the North Tehran fault, the North and South Rey faults and the Mosha fault as presented in Figure 2.

The North Tehran Fault is 90 Km long located at the southernmost piedmont of Central Alborz. It has an E–W to ENE–WSW strike, a dip of less than 75º, and a thrust mechanism (Berberian and Yeats, 1999). It proved the major active fault threatening directly the city due to several historical earthquakes recorded. The shape of this fault does not have a distinct scarp (Berberian and Yeats, 1999). Some authors suspect that events in the past, such as the 855 A.D and 856 A.D earthquakes, could be associated to this fault (Berberian and Yeats, 1999, 2001). The Mosha Fault is located at the northeast side of Tehran with a length of 150 Km. It has experienced several earthquakes with magnitude greater than 6.5 in the past (Berberian and Yeats, 1999). The South and North Ray Faults are located south of Tehran (JICA, 2000). The North Ray Fault has a length of 16.5 Km, in the W–E direction and a dip towards the north. The South

Ray Fault is 18 km long with ENE–WSW direction (Berberian and Yeats, 2001).

Census data of 2000 were used because of data availability, although the most recent Tehran census data was carried out in 2010. This data set contains 3175 statistical units in the Tehran metropolitan area.

The average slope of the land, the intensity of earthquake in MMI and building parameters are considered as effective seismic parameters (Aghataher et al., 2005, Silavi et al., 2006, Amiri et al., 2008, Samadi Alinia and Delavar, 2011, Khamespanah et al., 2013). Material and the number of floors of building are taken as the most important building parameters in assessing the seismic vulnerability. The percentage of materially weak-constructed and less-than-or-equal-to-four-floors buildings and the percentage of weakly founded buildings of more than four floors in any given urban statistical unit are taken as the two major parameters in building seismic vulnerability (Aghataher et al., 2005, Silavi et al., 2006, Amiri et al., 2008, Samadi Alinia and Delavar, 2011, Khamespanah et al., 2013a,b).

Since Iranian regulations for building designs have been approved in 1966, buildings constructed before this date are considered as non-standard constructions. Moreover, fortification regulations against earthquakes were applied for the first time in 1988. In this regard, the percentage of buildings constructed before 1966 and the percentage of buildings constructed between 1966 and 1988 are considered as effective parameters (Khamespanah et al., 2013a).

To determine physical seismic vulnerability for each statistical urban sample unit, experts were asked to rank the degree of vulnerability for 100 randomly selected units, using numbers from one to five, corresponding to the classes of 'very low vulnerability', 'low vulnerability', 'intermediate vulnerability', 'high vulnerability' and 'very high vulnerability', respectively. The effective parameters were divided into four intervals of equal length and output vulnerabilities were divided into five classes. Views for 15 selected samples are shown in Table 1.

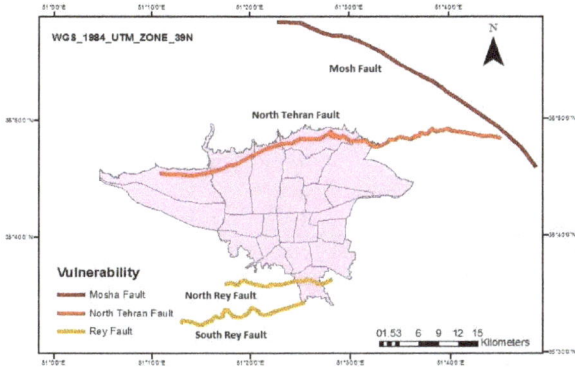

Figure 2: Position of North Tehran fault, Mosha fault and Rey faults

Sample number	Slope	MMI	Buil_less4	Bef-66	Bet-66-88	Buil_more4	Expert remark
132	1	2	1	1	4	1	2
256	3	1	2	1	2	2	1
33	2	1	1	1	3	1	2
456	1	2	2	1	4	1	2
2335	1	2	2	1	2	1	2
6	1	2	1	1	4	2	2
745	1	2	1	1	2	1	1
678	4	2	1	1	1	1	1
1129	1	1	1	1	4	1	3
103	3	1	3	1	4	1	5
11	4	2	2	1	3	2	2
2799	1	2	1	1	4	1	2
1342	1	2	4	1	4	3	5
144	3	2	1	1	2	1	1
335	1	2	4	1	4	1	3

Table 1: Classified vulnerability information for 15 out of 100 randomly selected building with Slop: Slope, MMI: MMI, Build_less4: Percentage of weak buildings having less than or equal to 4 floors, Build_more4: Percentage of buildings having more than 4 floors, Bef-66: Percentage of buildings built before 1966 and Bet-66-88: Percentage of buildings built between 1966 and 1988 (Samadi Alinia and Delavar, 2011).

3.2 Extracting rules

The rules satisfying maximum generality and absolute support parameters were selected as the effective rules. These rules were assessed by maximum coverage and minimum entropy to ensure that rules of the highest quality for the classification of seismic physical vulnerability of Tehran were extracted. The extracted rules employing the granular tree to determine the seismic vulnerability of Tehran are illustrated in Figure 3. In this tree, extracted rules are placed in prioritized order from left to right, showing confidence accounted for each rule based on minimum entropy and maximum coverage. Attributes are abbreviated as delineated in Table 1. This tree is used to assign seismic physical vulnerability class to the Tehran urban statistical units.

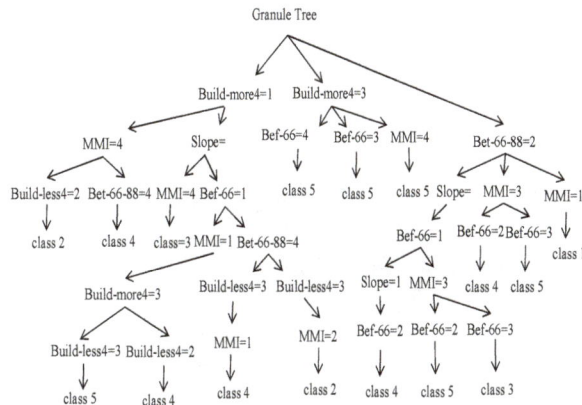

Figure 3. The granule tree of the extracted rules from training data set for classifying statistical units of Tehran with respect to their seismic physical vulnerability

3.3 Applying granular tree

The extracted granular tree was used to classify Tehran statistical units into the seismic vulnerability classes. The seismic physical vulnerability maps resulted for Mosha fault, Rey fault and North Tehran fault using GrC are illustrated in Figures 4, 5 and 6, respectively. These maps present the degree of vulnerability imposed to each statistical urban unit by the considered faults. It can be interpreted that activation of the North Tehran fault will have the highest destructive impact among the three faults, whereas activation of the Mosha fault will have the least destructive impact on Tehran, due to its distance to the city compared to that of the other two faults.

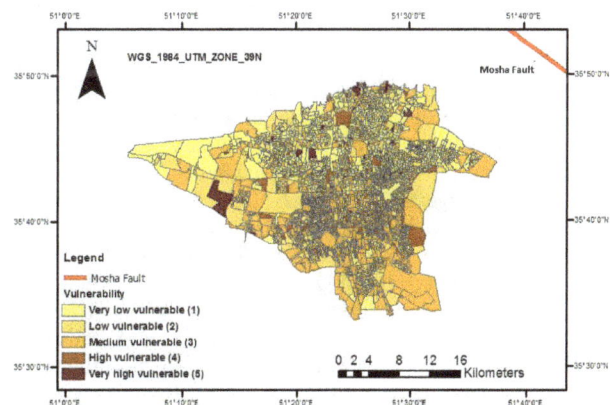

Figure 4. Tehran seismic physical vulnerability map against Mosha fault activation using GrC

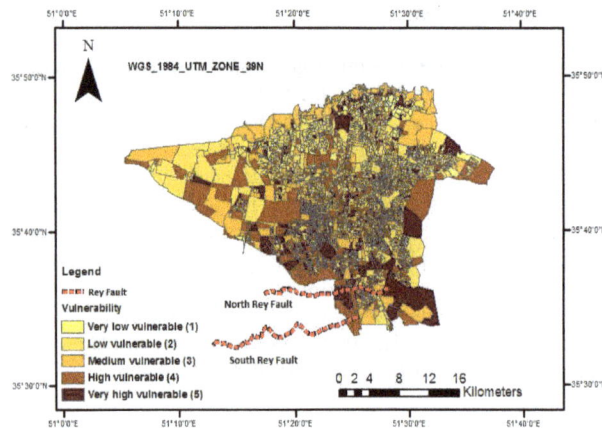

Figure 5. Tehran seismic physical vulnerability map against Rey fault activation using GrC

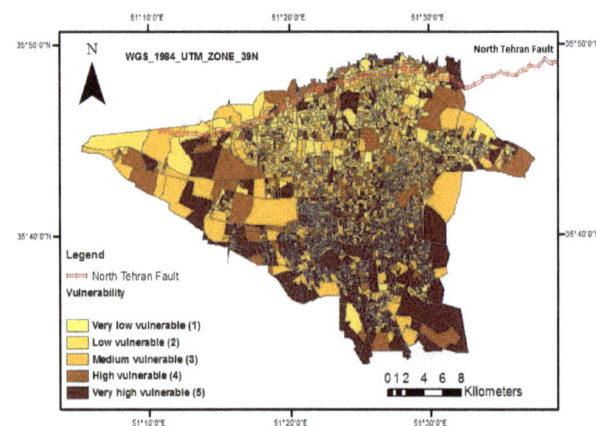

Figure 6. Tehran seismic physical vulnerability map against North Tehran fault activation using GrC

3.4 Aggregated value of Tehran seismic vulnerability

Next, we considered the aggregated value for Tehran's seismic physical vulnerability. We aimed to identify the effects at the areas with the highest seismic vulnerability. For all statistical urban units of Tehran, vulnerabilities from the North Tehran fault, the Mosha fault and the Rey fault have been determined using the extracted granular tree. For each unit, the highest resulted seismic physical vulnerability is considered as the aggregated measure of vulnerability. Aggregated vulnerability therefore shows the worst case scenario that may happen for a particular statistical unit in Tehran.

Table 2 demonstrates the aggregated seismic vulnerability value for the 15 statistical urban sample units presented in Table 1, indicating the imposed vulnerability from the three major faults and the maximum value that considered to be the worst case for that sample. Figure 7 illustrates Tehran physical seismic vulnerability map considering the worst case scenario using GrC, which shows the maximum possible vulnerability for the statistical units. In addition, Figure 8 demonstrates percentage of Tehran statistical units allocated to each vulnerability degree for different scenarios considered. According to Figure 8, activation of the Mosha fault and the Rey fault, will result in more than 70% of the units falling into the medium and low vulnerability classes. Activation of the North Tehran fault, however, will result in more than 50% of the units to have a high or very high degree of vulnerability. In the worst case model, therefore, more than 90% of the statistical units occur in the high and very high vulnerability classes.

Statistical unit number	Physical seismic vulnerability			Maximum Value
	North Tehran fault	Mosha fault	Rey fault	
132	3	2	1	3
256	3	1	2	3
33	2	1	1	2
456	2	3	5	5
2335	5	4	4	5
6	2	2	2	2
745	1	2	1	2
678	4	2	1	4
1129	3	2	1	3
103	5	4	4	5
11	3	3	2	3
2799	2	2	1	2
1342	5	4	3	5
144	3	4	3	4
335	3	1	1	3

Table 2. Selected samples of Tehran statistical units and their aggregated seismic physical vulnerability

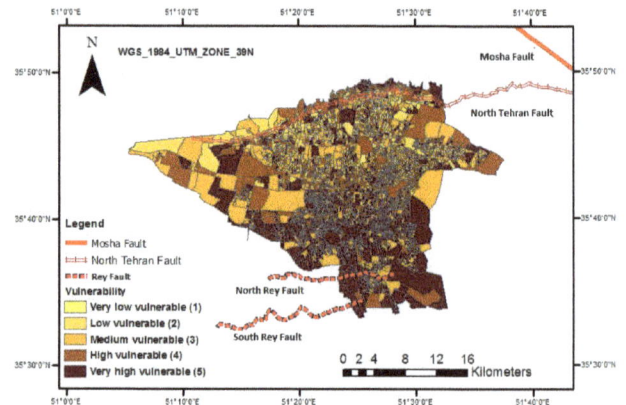

Figure 7. Tehran seismic physical vulnerability map considering the worst case scenario using GrC

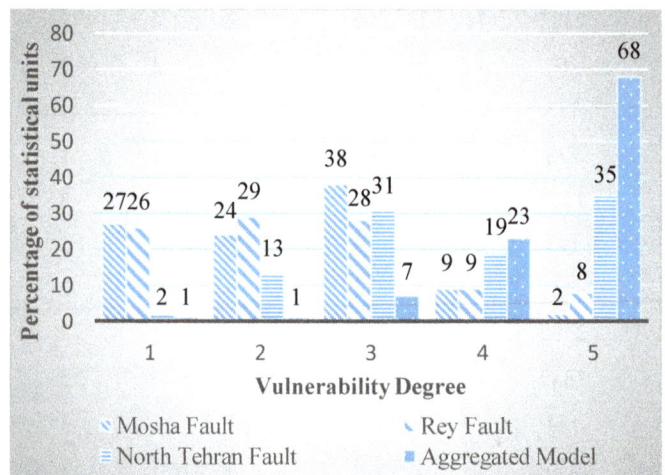

Figure 8. Comparison of Tehran seismic physical vulnerability imposed by the four scenarios of the fault activation

3.5 Validation of the results

Since there is no large earthquake happened in Tehran within about last 150 years of a magnitude above 6 at the scale of Richter, the only way to validate the model output is to assess the accuracy of the classification based on the data used to extract the classification rules. The model is therefore assessed to realize the number of the units correctly assigned to the degree of vulnerability as defined by the experts. The overall classification accuracy of the seismic physical vulnerability is calculated using $\frac{k}{k+n}$, where k is the statistical urban sample unit correctly classified and n is the number of incorrectly classified units, based on the available data set and experts' remarks (Yao, 2005, Pedrycz et al., 2008). The final accuracy for the North Tehran fault is estimated as 84% which confirms the reliability of the algorithm in comparison with previous GrC algorithms implemented by Samadi Alinia and Delavar (2011) an Khamespanah et al. (2013a), which resulted in 72% and 60% accuracy for the North Tehran fault, respectively.

4. DISCUSSION

The seismic physical vulnerability of Tehran has been assessed by historical proofs (Berberian and Yeats, 2001) where precise prediction of time, intensity and location of possible earthquakes is restricted by technological and scientific advances. Thus, it is imperative to produce seismic physical vulnerability maps to enable planners in order to develop risk reduction plans, which is the aim of this paper. In this regard, a new classification model based on the GrC algorithm was applied. Moreover, an aggregated seismic vulnerability value estimated the impact of the three major threatening faults around Tehran.

Seismic vulnerability maps showed that Tehran is highly vulnerable against earthquakes arising from the three major faults around the city. The effect of the Mosha fault was the least among the three faults, because the fault is located farthest from Tehran. Activation of the Rey fault, however, will impose a medium destructive impact on the city. Finally, the North Tehran fault is the most threatening fault because there are more statistical urban units associated to it at a high degree of vulnerability. The worst case scenario happens when these three faults are simultaneously activated, because earthquakes are not always related to a specific fault and it is possible for the three major Tehran faults to become active in response to a regional stress (Mucciarelli et al., 2001, Ashtari et al., 2005). The model developed in this study verified that for all of the three major faults of Tehran, most of the statistical urban units that have a high vulnerability are located in the south of Tehran. This corresponds with the actual building conditions in Tehran, because most of the aged buildings and non-standard constructions exist in the southern part of the city.

The new GrC algorithm implemented in this paper uses generality and absolute support for rule selection and prioritizes the extracted rules by entropy and coverage. Since there is no real earthquake happened in Tehran in the past 185 years, the only way to validate the results is to compare the acquired results with the experts' judgments. This algorithm led to 84% accuracy in classification of Tehran statistical units into seismic physical vulnerability classes, which exceeds previous attempts on Tehran seismic vulnerability classification by GrC algorithm implemented by Samadi Alinia and Delavar (2011) and Khamespanah et al. (2013a), resulting in 72% and 60% accuracy, respectively.

5. CONCLUSION

Natural disasters have always been imposed catastrophic casualties and devastations to the human society, and even today, precise prediction of the disaster is not adequately addressed by scientific and technological advancements. However, prognostication of the natural disasters seems insufficient to shield the peoples and their holdings from disaster consequences. Improving resistance of the urban environments by comprehensive planning can be regarded as the solution to the problem. Earthquake, is one of the most calamitous disasters endangering human kind all over the world, mostly because of its abrupt nature. Tehran, capital of Iran, is a highly populated city with numerous non-standard construction and aged buildings, surrounded with several known and unknown faults exposing the city to possibly huge earthquakes. In order to enable the urban managers to develop seismic damage reduction plans, this paper aimed at evaluating the seismic physical susceptibility of Tehran in the form of vulnerability classification maps.

This paper proposed a new granular computing algorithm which uses generality and absolute support for rule extraction, and utilizes coverage and entropy for rule ordering, to obtain the seismic physical vulnerability considering the effects of the three major known faults of Tehran, namely North Tehran fault, Mosha fault and Rey fault. Slope, seismic intensity in term of MMI, height and age of the buildings were considered to be the effective criteria in the classification procedure, accompanied with expert knowledge.

The new granular algorithm led to a higher accuracy in seismic vulnerability classification than precedent algorithms applied to the problem. Moreover, an aggregated model of seismic vulnerability is implemented to investigate the worst possible situation imaginable for Tehran.

6. REFERENCES

Aghataher, R., M. Delavar and N. Kamalian. 2005. Weighing of contributing factors in vulnerability of cities against earthquakes. Map Asia Conference Jakarta, Indonesia, Aug. 13 2005, pp. 22-25.

Amiri, A., M.R Delavar, S. Zahrai and M. Malek. 2008. Earthquake Risk Assessment in Tehran Using Dominance-Based Rough Set Approach. Proc. the ISPRS Workshop on Geoinformation and Decision Support Systems, Tehran, Iran, Jan. 14 2008. pp. 13-26.

Ashtari, M., D. Hatzfeld and N. Kamalian. 2005. Microseismicity in the region of Tehran. *Tectonophysics* 395(3), pp 193-208.

Bargiela, A. and W. Pedrycz. 2003. Granular Computing: an Introduction. Kluwer Academic Publishers, Boston. 452p.

Berberian, M. and R. S. Yeats. 1999. Patterns of historical earthquake rupture in the Iranian Plateau. *Bulletin of the Seismological Society of America* 89(1), pp 120-139.

Berberian, M. and R. S. Yeats. 2001. Contribution of archaeological data to studies of earthquake history in the Iranian Plateau. *Journal of Structural Geology* 23(2), pp 563-584.

Boustan, E. and A. Shafiee. 2011. Fuzzy-Probabilistic seismic hazard assessment of Tehran region. *Journal of the Earth 6(20)*, pp 17-29.

Gupta, M. M., R. K. Ragade and R. R. Yager. 1979. *Advances in Fuzzy Set Theory and Applications*, North-Holland Publishing Company. 770p.

Hobbs, J. R. 1985. Granularity. Proc. the Ninth International Joint Conference on Artificial Intelligence, University of British Columbia, Vancouver, Canada, Aug 6-7 1985, pp. 432-435.

Jahanpeyma, M.H., Delavar, M.R., Malek, M.R., Kamalian, N., 2007, Analytical evaluation of propagation of uncertainty in assessment of seismic vulnerability of Tehran using geospatial information system , Proc. the 5th International Symposium on Spatial Data Quality, June 13-15 2007, Enschede, The Netherlands, pp. 25-32.

JICA. 2000. The study on seismic microzoning of the Greater Tehran Area in the Islamic Republic of Iran. *Pacific Consultants International Report, OYO Cooperation, Japan* 01, 390p.

Kappos, A., K. Stylianidis and K. Pitilakis. 1998. Development of seismic risk scenarios based on a hybrid method of vulnerability assessment. *Natural Hazards* 17(2), pp. 177-192.

Keet, C. M. 2008. A Formal Theory of Granularity, PhD Thesis, KRDB Research Centre, Faculty of Computer Science, Free University of Bozen-Bolzano, Italy. 298p.

Khamespanah, F., M. Delavar and M. Zare. 2013a. Uncertainty management in seismic vulnerability assessment using granular computing based on covering of universe. *ISPRS-International Archives of the Photogrammetry, Remote Sensing and Spatial Information Sciences* 1(1), pp. 121-126.

Khamespanah, F., M. R. Delavar, H. S. Alinia and M. Zare. 2013b. Granular Computing and Dempster–Shafer Integration in Seismic Vulnerability Assessment. *Intelligent Systems for Crisis Management*. S. Zlatanova, Peters, R., Dilo, A., Scholten, H., Springer, pp. 147-158.

Lin, T. Y. 2003. Granular computing. *Rough Sets, Fuzzy Sets, Data Mining, and Granular Computing*, Springer, pp 16-24.

Miao, D.Q. and S.D. Fan. 2002. The Calculation of Knowledge Granulation and its Application. *Systems Engineering-theory & Practice* 1, pp. 7-14.

Moradi, M., M. Delavar, B. Moshiri and F. Khamespanaha. 2014a. A novel approach to support majority voting in spatial group MCDM using density induced OWA operator for seismic vulnerability assessment. *ISPRS-International Archives of the Photogrammetry, Remote Sensing and Spatial Information Sciences* 1, pp. 209-214.

Moradi, M., M. R. Delavar and B. Moshiri. 2014b. A GIS-based multi-criteria decision-making approach for seismic vulnerability assessment using quantifier-guided OWA operator: a case study of Tehran, Iran. *Annals of GIS 22(3)*, pp 1-14.

Mucciarelli, M., P. Contri, G. Monachesi, G. Calvano and M. Gallipoli. 2001. An empirical method to assess the seismic vulnerability of existing buildings using the HVSR technique. *Pure and Applied Geophysics* 158(12), pp. 2635-2647.

Nguyen, S. H., A. Skowron and J. Stepaniuk. 2001. Granular computing: A rough set approach. *Computational Intelligence* 17(3), pp. 514-544.

Panahi, M., F. Rezaie and S. Meshkani. 2013. Seismic vulnerability assessment of school buildings in Tehran city based on AHP and GIS. *Natural Hazards and Earth System Sciences Discussions* 1(5), pp. 4511-4538.

Pawlak, Z. 1982. Rough sets. *International Journal of Computer & Information Sciences* 11(5), pp 341-356.

Pedrycz, W., S. Bassis and D. Malchiodi. 2008. *The Puzzle of Granular Computing*, Springer.

Samadi Alinia, H. and M.R. Delavar. 2011. Tehran's seismic vulnerability classification using granular computing approach. *Applied Geomatics* 3(4), pp. 229-240.

Shafer, G. 1992. The Dempster-Shafer theory. *Encyclopedia of Artificial Intelligence*, pp. 330-331.

Silavi, T., M. Delavar, M. Malek, N. Kamalian and K. Karimizand. 2006. An integrated strategy for GIS-based fuzzy improved earthquake vulnerability assessment. ISPRS International Symposium on "Geo-information for Disaster Management (Gi4DM), Goa, India, pp , Sep. 25-26 2006. 16-26.

van Lieshout, M. and A. Stein. 2012. Earthquake modelling at the country level using aggregated spatio-temporal point processes. *Mathematical Geosciences* 44(3), pp. 309-326.

Yao, Y. 2004. A partition model of granular computing. *Transactions on Rough Sets I*, Springer, pp 232-253.

Yao, Y. 2005. Perspectives of granular computing. IEEE International Conference on Granular Computing, Beijing, China, 25-27 July 2005, pp. 85-90.

Yao, Y. 2008. A Unified Framework of Granular Computing. *Handbook of Granular Computing*. W. Pedrycz, A. Skowron and V. Kreinovich (Eds), John Wiley & Sons, pp. 401-410.

Yao, Y. Y. 2001. On Modeling data mining with granular computing. 25th Annual IEEE International Conference on Computer Software and Applications, Chicago, US, Oct. 8-12 2001, pp. 638-643.

Yaseen, M., N. A. Hamm, V. Tolpekin and A. Stein. 2013a. Anisotropic kriging to derive missing coseismic displacement values obtained from synthetic aperture radar images. *Journal of Applied Remote Sensing* 7(1), pp. 1-18.

Yaseen, M., N. A. Hamm, T. Woldai, V. Tolpekin and A. Stein. 2013b. Local interpolation of coseismic displacements measured by InSAR. *International Journal of Applied Earth Observation and Geoinformation* 23, pp. 1-17.

Zadeh, L. A. 1998. Some reflections on soft computing, granular computing and their roles in the conception, design and utilization of information/intelligent systems. *Soft Computing-A Fusion of Foundations, Methodologies and Applications* 2(1), pp. 23-25.

Zadeh, L. A. and J. Kacprzyk. 1999. *Computing with words in Information/Intelligent systems 1: Foundations*, Springer. 518p.

Zare, M., P.-Y. Bard and M. Ghafory-Ashtiany. 1999. Site characterizations for the Iranian strong motion network. *Soil Dynamics and Earthquake Engineering* 18(2), pp. 101-123.

Zaré, M. and H. Memarian. 2003. Macroseismic intensity and attenuation laws: A study on the intensities of the Iranian earthquakes of 1975–2000. Fourth International Conference of Earthquake Engineering and Seismology, Tehran, Iran, April 4-6 2003, pp. 12-14.

26

A GRAPH BASED MODEL FOR THE DETECTION OF TIDAL CHANNELS USING MARKED POINT PROCESSES

A. Schmidt [a], F. Rottensteiner [a], U. Soergel [b], C. Heipke [a]

[a] Institute of Photogrammetry and GeoInformation, Leibniz Universität Hannover, Germany -
(alena.schmidt, rottensteiner, heipke)@ipi.uni-hannover.de
[b] Institute of Geodesy, Chair of Remote Sensing and Image Analysis, Technische Universität Darmstadt, Germany -
soergel@geod.tu-darmstadt.de

KEY WORDS: Marked point processes, RJMCMC, graph model, digital terrain models, coast

ABSTRACT:

In this paper we propose a new method for the automatic extraction of tidal channels in digital terrain models (DTM) using a sampling approach based on marked point processes. In our model, the tidal channel system is represented by an undirected, acyclic graph. The graph is iteratively generated and fitted to the data using stochastic optimization based on a Reversible Jump Markov Chain Monte Carlo (RJMCMC) sampler and simulated annealing. The nodes of the graph represent junction points of the channel system and the edges straight line segments with a certain width in between. In each sampling step, the current configuration of nodes and edges is modified. The changes are accepted or rejected depending on the probability density function for the configuration which evaluates the conformity of the current status with a pre-defined model for tidal channels. In this model we favour high DTM gradient magnitudes at the edge borders and penalize a graph configuration consisting of non-connected components, overlapping segments and edges with atypical intersection angles. We present the method of our graph based model and show results for lidar data, which serve of a proof of concept of our approach.

1. INTRODUCTION

In general, strategies for the automatic detection of objects from images can be divided into top-down and bottom-up approaches. For the latter, basic image processing methods such as segmentation are employed and the results are assigned to scene objects afterwards. In contrast, top-down approaches integrate knowledge about the objects and search for matches with the input data. Stochastic methods such as *marked point processes* (Daley and Vere-Jones, 2003), which achieve good results in various disciplines, belong to the top-down approaches. Marked point processes do not work locally, but rather enable to find a global optimum of the configuration of objects. Model knowledge can be integrated in different ways. To this end Lafarge et al. (2010) developed a flexible approach for the detection of different kinds of objects in images. For that purpose, the authors set up a library composed of simple geometric patterns such as rectangles, lines and circles, which are defined by their lengths, orientations or radii. The geometric patterns are randomly fitted to the images and their conformity with the input data is evaluated based on the mean and standard deviations of the gray values inside and outside the proposed objects as well as the overlapping area between them. The authors achieve good results for different fields of application such as the extraction of line networks, buildings or tree crowns. Tournaire et al. (2007) used marked point processes for the extraction of dashed lines representing road markings from very high resolution aerial images. Here, the objects are modelled by rectangles whose conformity with the input data is evaluated based on the gray values of the pixels inside the rectangles and in their local neighbourhood. Both are modelled to be homogenous while pixels inside the rectangles have to have bright pixels. The authors penalize configurations in which the distance between rectangles differs from the expected one

depending on the road type. Road networks are extracted in the approach of Lacoste et al. (2005). This is done by taking into account different types of relations between segments and evaluating their connectivity and orientation. The input data are considered by calculating the homogeneity of the gray values of the road and the gray level variation between the road and the background. A similar approach was developed by Sun et al. (2007) for the automatic extraction of vessels on angiograms. The vessels are modelled as line segments; their configuration is optimized by penalizing segments which are not connected at both ends. The conformity with the input data is determined based on a *vesselness enhancement measure* which is calculated from the two eigenvalues of the Hessian of the image function. Perrin et al. (2005) used marked point processes for the detection of tree crowns from remotely sensed images with infrared information in order to derive tree parameters such as the diameter or the density. The trees are modelled by ellipses whose overlapping area is penalized depending on the way they overlap. A further term is introduced in order to favour regular alignments in the configuration. Furthermore, extreme closeness of objects is penalized by a constant parameter.

We aim to take advantage of the benefits of marked point processes (the global optimization and their flexibility due to the fact that the approach is a stochastic one) in order to detect tidal channels in a digital terrain model (DTM) derived from lidar data. Tidal channels are a special type of rivers furrowing the mudflat areas in coastal zones. Because they have large impact on the sedimentation and the ecology of the flat coastal waters, a detailed understanding of their characteristics is essential. Moreover, they significantly change position over time due to the influence of the tides and storm events. Extracting them automatically from remote sensing data is a challenging task.

In our previous work, we developed a two-step approach for that purpose (Schmidt et al., 2014). First, we fitted rectangles to the data using marked point processes. High DTM gradient magnitudes on the rectangle borders and non-overlapping objects were introduced to model the tidal channels. In a second step, we determined the principal axes of the rectangles and their intersection points. Based on this result a graph was constructed in which nodes represented junction points or end points, respectively, and edges straight line segments located in between. While overall the results were promising, they revealed that the topology of the tidal channel network was not always correctly described. Furthermore, the connectivity of all channels was not guaranteed. That is why we intended to integrate this characteristic directly in the sampling process.

In this paper, we adopt the method of Chai et al. (2013) and modelled tidal channels by a graph and optimize its configuration. In the graph nodes represent junction points and edges correspond to straight line segments with a certain width. We integrate the evaluation criteria of our previous work into this approach. Thus, we still favour high DTM gradient magnitudes at the segment borders of each junction point and penalize the overlaps between neighbouring objects. Moreover, we evaluate the graph and penalize a configuration in which several components are not connected as well as intersection angles between adjacent edges which differ from those typically found in tidal channel networks.

We first describe the mathematical foundation of the stochastic optimization using marked point processes (Section 2). Then, we present the proposed graph based model for our application (Section 3). In Section 4, we show some experiments and results on test data of the German Wadden Sea. Finally, conclusions and perspectives for future work are presented.

2. THEORETICAL BACKGROUND

2.1 Marked point process

Point processes belong to the group of stochastic processes, a collection of random variables. They are used for the description of phenomena in which the random variable can be observed only at particular points of time (for random variables with temporal reference) or at particular locations (for random variables with spatial reference). Then, a point process can be defined as sequence of random points of time t_1, t_2, ..., t_n at which a phenomenon occurs, or, respectively, a sequence of random locations l_1, l_2, ..., l_n where a phenomenon can be observed. For a formal definition of point processes the reader is referred to (Daley and Vere-Jones, 2003).

In object extraction, point processes are used to find the most probable configuration of objects in a scene given the data. An object is described by its location l_i. In a marked point process, a mark m_i, a multidimensional random variable, is added to each point describing an object of a certain type. If we characterise an object $u_i = (l_i, m_i)$ by its location and mark, a marked point process can be thought of as a stochastic model of configurations of an unknown number n of such objects in a bounded region S (here: a digital image, thus the points l_i exist in R^2).

There are different types of point processes. The *Poisson point process* ranks among the basic ones, it assumes a complete randomness of the objects, and the number of objects n follows a discrete Poisson distribution with parameter λ, also called intensity parameter, which corresponds to the expected value

for the number of objects. In the *Poisson point process*, the objects are independent and uniformly distributed in S, given n. In practice, a complete randomness of the object distribution is often not applicable. Instead, more complex models are postulated. However, they are described with respect to a reference point process, which is usually defined as the *Poisson point process*. In our model, we define the probability density function h of the object configuration by a Gibbs energy $U(.)$ with $h \propto \exp -U(.)$. As in Markov Random Fields the Gibbs energy can be modelled by the sum of a data energy $U_d(.)$ and a prior energy $U_p(.)$ as used for instance by Mallet et al. (2009):

$$U(.) = \beta U_d(.) + (1-\beta)U_p(.). \tag{1}$$

The relative influence of both energy terms is modelled by $\beta \in [0,1]$. The data energy $U_d(.)$ measures the consistency of the object configuration with the input data. The energy $U_p(.)$ introduces prior knowledge about the object layout; our models for these two energy terms are described in Section 3. The optimal configuration $\hat{u} = \{u_1, ..., u_n\}$ of objects can be determined by minimizing the Gibbs energy $U(.)$, i.e. $\hat{u} = \arg \min U(.)$. Since the number of possible object configurations increases exponentially with the size of the object space, the global minimum can only be approximated. This is done by coupling a RJMCMC sampler and a simulated annealing relaxation.

2.2 Reversible Jump Markov Chain Monte Carlo

Markov Chain Monte Carlo (MCMC) methods belong to the group of sampling approaches. The special feature of the method is that the samples are not drawn independently. On the contrary, each sample X_t, $t \in \mathbb{N}$, has a probability distribution that depends on the previous sample X_{t-1}. Thus, the sequence of samples forms a Markov chain which is simulated in the space of possible configurations. If the number of objects constituting the optimal configuration $\hat{u}$ were known or constant, MCMC sampling (Metropolis et al., 1953, Hastings, 1970) can be used for its determination. Reversible jump Markov Chain Monte Carlo (RJMCMC) is an extension of MCMC that can deal with an unknown number of objects and a change of the parameter dimension between two sampling steps (Green, 1995). In each iteration t, the sampler proposes a change of the current configuration from a set of pre-defined types of changes according to a density function. Each of the types is also associated with a density function Q_i called *kernel*. Each kernel Q_i is reversible, i.e. each change can be reversed applying another type of kernel. The type of changes and the kernels in our model are described in Section 3. A kernel Q_i is chosen randomly according to a proposition probability p_{Q_i} which may depend on the kernel type. The configuration X_t is changed according to the kernel Q_i, which results in a new configuration X_{t+1}. Subsequently, the Green ratio R (Green, 1995) is calculated:

$$R = \frac{Q_i(X_{t+1} \to X_t)}{Q_i(X_t \to X_{t+1})} \exp - \left(\frac{U(X_{t+1}) - U(X_t)}{T_t} \right). \tag{2}$$

In (2), T_t is the temperature of simulated annealing in iteration t (cf. Section 2.3). $\frac{Q_i(X_{t+1} \to X_t)}{Q_i(X_t \to X_{t+1})}$ is the ratio of the probabilities for the change of the configuration from X_{t+1} to X_t and vice versa. (cf. Section 3.2.5). The acceptance rate α of the new configuration X_{t+1} is computed from R using

$$\alpha = \min(1, R). \tag{3}$$

Following Metropolis et al. (1953) and Hastings (1970) the new configuration X_{t+1} is accepted with the probability α and rejected with the probability 1-α. The four steps of (1) choosing a proposition kernel Q_i, (2) building the new configuration X_{t+1}, (3) computing the acceptance rate α, and (4) accepting or rejecting the new configuration are repeated until a convergence criterion is achieved.

2.3 Simulated Annealing

In order to find the optimum of the energy, the RJMCMC sampler is coupled with simulated annealing. For that reason, the parameter T_t referred to as *temperature* (Kirkpatrick et al., 1983) is introduced in equation 2. The sequence of temperatures T_t tends to zero as $t \rightarrow \infty$. Theoretically, convergence to the global optimum is guaranteed for all initial configurations X_0 if T_t is reduced ("cooled off") using a logarithmic scheme. In practice, a geometrical cooling scheme is generally introduced instead. It is faster and usually gives a good solution (Salamon et al., 2002; Tournaire et al., 2010).

3. METHODOLOGY

We fit objects to the data using a RJMCMC sampler and simulated annealing as described in Section 2. Our model is based on our previous work (Schmidt et al., 2014) where we adapted the approach of (Tournaire et al., 2010) to our application of tidal channel detection in a DTM, but differs in the object model we use. Instead of rectangles, we now take into account the network structure of our object of interest (see Section 3.1) and optimized the configuration of an undirected graph. Following Chai et al. (2013), nodes in the graph represent junction points with outgoing segments varying in number and representing the tidal channels in the DTM. A segment becomes an edge in the graph if it links two nodes (Section 3.1). In the sampling process, nodes are iteratively added to or removed from the graph in order to find the best object configuration. Moreover, we allow for a modification of the node parameters. The changes of the configuration are described in Section 3.2. In each iteration we evaluate the graph configuration based on a global energy function which is minimized during the sampling process. Here, we favour high gradients at the segment borders and penalize the overlapping area of segments, segments which are not connected to the tidal channel system on both ends and angles between edges which differ from those typically found in tidal channel networks (Section 3.3).

3.1 Object model

We characterize an object $u_i = (l_i, m_i)$ by its location $l_i = (x,y)$ and its parameters m_i. In our approach, the objects are junction points similar to the approach of Chai et al. (2013). Each junction point possesses n outgoing segments s_j ($j = 1,..., n$), each corresponding to a part of a tidal channel in the DTM (Fig. 1). The minimum number of segments is $n = 1$ (the junction point becomes an end-point in the network); it can rise to $n = k$ whereby k is the predefined maximum number of tidal channels converging on a junction point. We specify the segments by their directions (the counterclockwise angles β_j relative to the positive y-axis) and their widths w_j. In total, the junction point parameters are $m_i = (\beta_1, ..., \beta_n, w_1, ..., w_n)$.

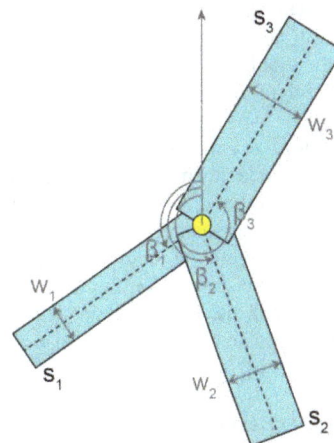

Figure 1. The object in our configuration is a junction point with n outgoing segments (here: $n = 3$). Each segment s_j is characterized by its width w_j and its direction β_j (the counterclockwise angle relative to the positive y-axis).

3.2 Changes of the configuration

We integrate four different types of changes in our approach. On the one hand, an object can be added to or removed from the current configuration, which is accomplished by the birth and death kernels, Q_B and Q_D, respectively. On the other hand, the parameters of an object of the current configuration can be modified. Here, we define the translation of a junction point which is equivalent to a change of its position (translation kernel Q_T) and the modification of one of its segment parameters (modification kernel Q_M). The type of changes and the associated kernels are described in more detail in the following section.

3.2.1 Birth event: For the birth event, the position and the number of segments of a new junction point are generated. It should be noted that all of the junction point parameters (position, number of segments, their directions and widths) may be generated based on learned probability density functions. The junction point is added to the graph as a node. Then, neighbouring junction points are determined in a local neighbourhood and the feasibility of adding an edge between them and the new node is checked. For that purpose, we integrate three criteria. First, we do not allow the segment corresponding to this edge to intersect with another edge in the graph. In this way, we take into account that tidal channels do not intersect without a junction point at the point of intersection. Second, we do not allow cycles in the graph, bearing in mind that water flows only in one direction (namely downhill). By building up an acyclic graph, we accept that we cannot detect two channels circling round an island. However, this particular case rarely occurs in our application. Third, we introduce a threshold for the angle between two edges in order to avoid too much overlap of segments in a node. If a neighbouring node does not violate any of these criteria, we add an edge connecting it with the new node (Fig. 2a). The neighbouring nodes are considered in the order of their distance to the node. Finally, we modify the neighbouring node involved and add the new segment or replace an existing segment based on an angle criterion.

3.2.2   Death event: A junction point is randomly chosen and the corresponding node as well as its outgoing edges are removed from the graph. The neighbouring nodes retain a segment in the direction of the removed node, however (Fig. 2b).

3.2.3   Translation event: A displacement vector for a (randomly chosen) node in the graph is randomly generated within a local neighbourhood. For outgoing edges we check for an intersection of the corresponding segment with the existing segments in the configuration and the angle to the other segments (similar to the criteria for the birth kernel). If these criteria are not violated, the junction point is translated and its position and directions are modified. We also update affected neighbouring nodes (Fig. 2c).

3.2.4   Modification event: We randomly choose a junction point and one of its segments and change its width (within predefined thresholds for the minimum and maximum width of a segment). If this segment corresponds to an edge, the neighbouring node is modified by updating the width for the corresponding segment (Fig. 2d)[1].

3.2.5   Kernels: In equation (2), the Green ratio is calculated based on the kernel ratio which takes into account the probability for the change of the configuration from X_{t+1} to X_t and vice versa. Following Chai et al. (2013), we model the kernel ratio for the birth and the death events by

$$\frac{Q_{BD}(X_{t+1} \rightarrow X_t)}{Q_{DB}(X_t \rightarrow X_{t+1})} = \frac{p_D}{p_B} \frac{\lambda}{n} \tag{4}$$

$$\frac{Q_{DB}(X_{t+1} \rightarrow X_t)}{Q_{BD}(X_t \rightarrow X_{t+1})} = \frac{p_B}{p_D} \frac{n}{\lambda}. \tag{5}$$

In (4) and (5) p_D and p_B correspond to the probability for choosing a birth or death event, respectively. λ is the intensity value of the *Poisson point process* which serves as reference process in our model (cf. Section 2.1) and n is the current number of nodes in the graph.

For the translation and the modification event, we set the kernel ratio to 1:

$$\frac{Q_T(X_{t+1} \rightarrow X_t)}{Q_T(X_t \rightarrow X_{t+1})} = \frac{Q_M(X_{t+1} \rightarrow X_t)}{Q_M(X_t \rightarrow X_{t+1})} = 1. \tag{6}$$

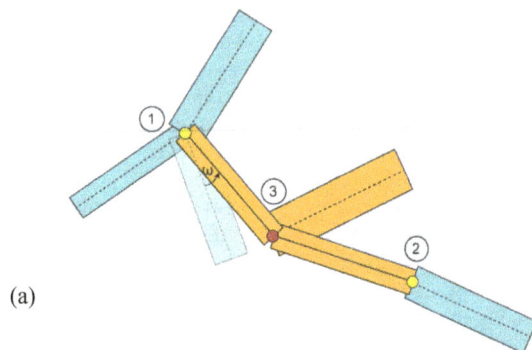

(a)

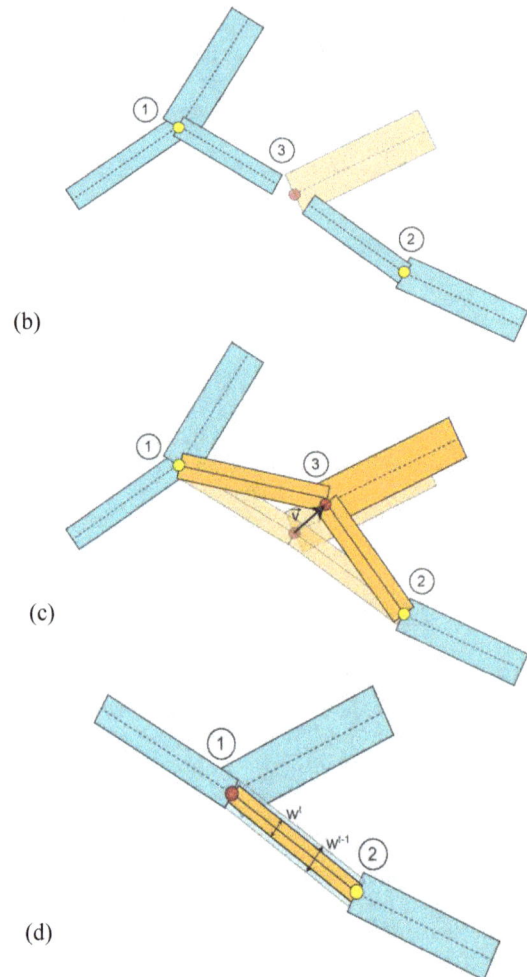

(b)

(c)

(d)

Figure 2: Four types of changes are defined for the sampling process. Nodes and segments which are chosen for a change are depicted in red and orange. The configuration before the change is depicted slightly transparent. The medial axes of the segments are illustrated by a solid line if they belong to an edge in the graph. **(a) Birth kernel:** Node 3 is added to the graph. Neighbouring nodes are connected by edges if they fulfil some criteria. Segments of the neighbouring nodes whose angle ω to the new direction is below a predefined threshold are removed (transparent segment of node 1). **(b) Death kernel:** Node 3 is removed from the configuration. Nodes 1 und 2 retain a segment in the direction of the removed node. **(c) Translation kernel:** The position of node 1 is changed by the displacement vector $\vec{v}$. Edges to neighbouring nodes are retained if they do not lead to crossing edges. **(d) Modification kernel:** The parameters of one of the outgoing segments of node 1 are changed. Because the segment belongs to an edge, the parameters of node 2 are changed, too.

3.3   Energy model

Each configuration is evaluated based on the Gibbs energy which is minimized during the sampling process. In our model, the Gibbs energy consists of two terms: the data energy and the prior energy. We set a maximum value for the length of those segments which do not correspond to an edge and, thus, do not have a neighbouring node.

---

[1] Additional modifications are part of further research.

3.3.1   Data Energy: The data energy $U_d(X_t)$ (equation 1) checks the consistency of the object configuration with the input data. Tidal channels are characterized by locally smaller heights and have high DTM gradient magnitudes on each bank and flat gradient magnitudes between these banks in the areas that may or may not be covered by water (depending on the tides). We adapt and slightly modify the data term of our previous model (Schmidt et al., 2014) where we fitted rectangle to the data. Here, we determine the gradients of the segments borders by

$$U_d(X_t) = \sum_{s_j \in X_t} \left(c - \sum_{m=1}^{2} \frac{1}{n} \sum_{k_j=1}^{n_j} \nabla^{\perp}_{\text{DTM}_{k_j}} \right). \qquad (7)$$

In (7), $\nabla^{\perp}_{\text{DTM}_k}$ is the component of the gradient of the DTM at pixel k in direction of the normal vector of edge m of the segment s_j. The gradients of the n pixels k all have equal weights. The normal vector is defined to point from the interior of the segment to the outside. We take into account only the two borders of the segment corresponding to the channel banks. A constant weight c is introduced in order to ensure a minimum value for the sum of the gradients.

3.3.2   Prior energy: Prior knowledge is integrated into the model in order to favour certain object configurations. We define the prior energy by three terms. First, we favour an object configuration in which the segments do not overlap. In this way, the accumulation of objects in regions with high data energy can be avoided. Second, we aim to end up in a configuration with only one graph and, thus, rate the connectivity of the nodes in the graph. Third, we favour those angles between edges which typically occur within tidal channel networks. In total, the prior energy is modelled by

$$U_p(X_t) = U_o(X_t) + U_s(X_t) + U_a(X_t). \qquad (8)$$

The energy U_o is composed of the sum of the overlap area a of all combinations of segments s_i and s_j

$$U_o(X_t) = \sum_{i \neq j} a(s_i, s_j). \qquad (8)$$

Because segments inevitably overlap near a junction point, we do not calculate the overlapping area of two segments belonging to the same node (Fig. 3).
In order to evaluate the connectivity of the graph, we penalize all segments which do not link two nodes by

$$U_s(X_t) = n_s \cdot f_s \qquad (9)$$

where n_s is the number of segments in the graph which are connected to only one segment and f_s is a constant penalizing factor.
Finally, favouring certain angle configuration between two edges is achieved by

$$U_a(X_t) = \sum_{i=1}^{n_a} |(\omega_e - \omega_i)| \cdot f_a. \qquad (10)$$

In (10) n_a is the number of nodes which are linked by more than one edge. For each of these nodes the angle ω_i between two edges is subtracted from the expected angle ω_e for typical tidal channel configuration which may be learned from the data or integrated as prior knowledge. f_a is a constant penalizing factor.

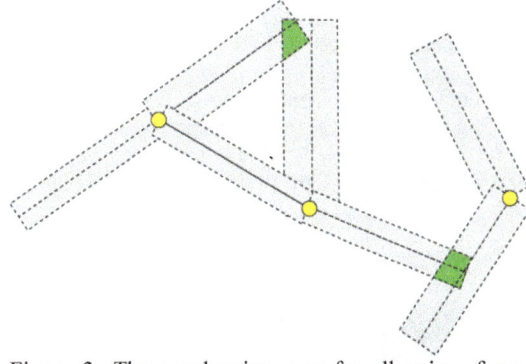

Figure 3: The overlapping area for all pairs of segments is calculated if both segments do not belong to the same node. Here, the green areas correspond to our prior energy term.

4.   EXPERIMENTS

4.1   Test data

We evaluate our method on lidar data from the German Wadden Sea. The test site is located in the south of the East Frisian island Norderney and covers an area of 440 x 150 m². The flight campaign took place in spring 2012 using a Riegl LMS-Q 560 sensor. The heights of the raw lidar point cloud (average point density 5.9 points / m²) are interpolated to generate a DTM of 1 m grid size.

4.2   Parameter settings

In our experiments, we give more weight to the prior energy and, thus, set the parameter β in equation 1 to $\beta = 0.3$. The proposal probabilities of the kernels are set to $p_{QM} = p_{QT} = 0.45$ and $p_{QB} = p_{QD} = 0.05$, whereas the probabilities for choosing a birth or death event in equation 4 and 5 are $p_D = p_B = 0.5$. In order to speed up the computations, we set a threshold for the heights in our data and propose a new junction point in the birth event only at pixels whose height is below this threshold. The intensity parameter λ of the reference *Poisson point process* is set to 50. For the prior knowledge about tidal channels we restricted the maximum width of a segment to 15 m (which corresponds to the maximum width of channels in our scene) and set the maximum number of segments for one node to $k = 3$. The expected angles ω_e for adjacent edges are set to $\omega_e = 180°$ or $\omega_e = 135°$ for junction points with two segments which are the most common combination in our data. We calculate the difference for both of the values and take the smaller one in our prior term. For junction points with three segments we set the expected angles between the segments to 180°, 135° and 45°. The penalizing factors in equation 9 and 10 are set to $f_s = 100$ and $f_a = 5$; the parameter c in equation 7 is set to $c = 200$; and the temperature T in equation 2 to $T = 100$.

4.3   Results

The results for our test data are shown in Figure 4. Junction points are added to or removed from the configuration or their parameters are changed. In this way the graph is iteratively built. After 30 million of iterations the energy converges to a minimum (Fig. 6). Then, the main tidal channel in the middle of the scene is nearly completely covered by segments. Most of the nodes are linked. Only in the parts marked by circles (Figure 4, bottom) the graph is not connected. Another good result is that segments only connected to one node are completely eliminated during the sampling process (apart from one on the left part of the scene which exceeds the image boundary). However, the smaller channels are not detected. This might be explained by the parameter settings which require large values for the gradient magnitudes at the segment borders. If we reduce c from 200 to 100, some of the smaller tidal channels are detected in the upper part of the scene (Fig 5). However, due to favouring a fully connected graph, they are wrongly connected in the marked area.

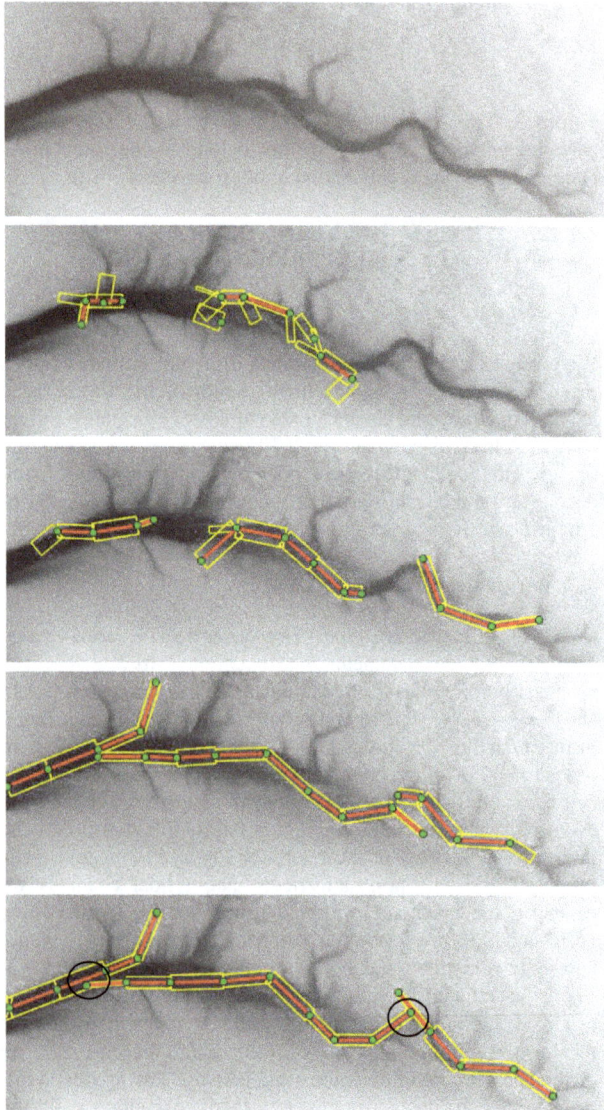

Figure 4: DTM with tidal channels (top) and the results after $1 \cdot 10^6$, $10 \cdot 10^6$, $20 \cdot 10^6$ and $30 \cdot 10^6$ iterations. The optimized segments are depicted in yellow, the nodes and edges of the graph are shown in green and red.

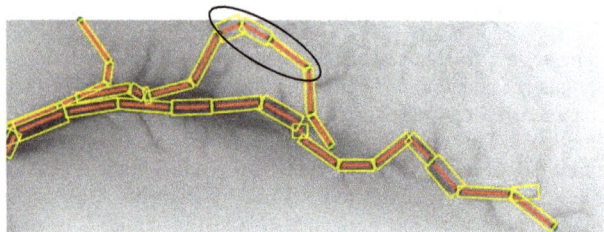

Figure 5: If we reduce the weight for the data energy, also smaller tidal channels are detected. However, they are wrongly connected.

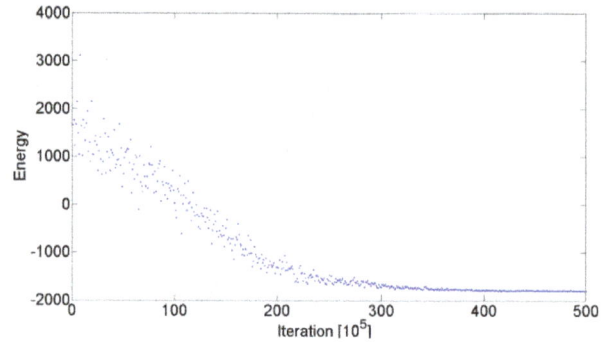

Figure 6: The energy decreases during the optimization process and converges after 30 million iterations.

5. CONCLUSION AND OUTLOOK

In this paper, we present a stochastic approach based on marked point processes for the automatic extraction of tidal channel networks. We model the tidal channels as an undirected, acyclic graph which is iteratively built during the optimization process. The approach is evaluated on a DTM derived from lidar data. The most relevant the tidal channels are detected, apart from some smaller channels. The results show a proof of concept, but the method needs to be refined in a number of ways. For instance, we intend to consider the flow direction of the water between adjacent objects and, thus, use a directed graph. We also plan to learn those parameters from training data which are set based on prior knowledge about typical tidal channel structures so far. More complex data are needed for a more comprehensive test. The detection of river networks from a DTM or line-networks in images are further interesting applications for our method.

ACKNOWLEDGEMENTS

We would like to thank the Ministry of Environment, Energy, and Climate Protection as well as the Ministry of Science and Culture of Lower Saxony for the financing of this work within the project WIMO (Scientific monitoring concepts for the model region German Bight). We would also like to thank the State Department for Waterway, Coastal and Nature Conservation (NLWKN) of Lower Saxony for providing the lidar data.

REFERENCES

Ashley, G. M., Zeff, M. L., 1988. Tidal channel classification for a low-mesotidal salt marsh. *Marine Geology*, 82, pp. 17–32.

Chai, D., Foerstner, W., and Lafarge, F., 2013. Recovering line-networks in images by junction-point processes. *IEEE Conference on Computer Vision and Pattern Recognition (CVPR)*, Portland, pp. 1894 - 1901.

Daley, D. J., Vere-Jones, D., 2003. An Introduction to the Theory of Point Processes: Elementary Theory and Methods, Springer, New York, Second Edition.

Green, P., 1995. Reversible Jump Markov Chains Monte Carlo computation and Bayesian model determination. *Biometrika*, Vol. 82(4), pp. 711–732.

Hastings, W., 1970. Monte Carlo sampling using Markov chains and their applications. *Biometrika*, Vol. 57(1), pp. 97–109.

Kirkpatrick, S., Gelatt, C. D., Vecchi, M. P., 1983. Optimization by Simulated Annealing. *Science* 220, pp. 671-680.

Lacoste, C., Descombes, X., Zerubia, J., 2005. Point processes for unsupervised line network extraction in remote sensing. *IEEE Trans. on Pattern Analysis and Machine Intelligence*, Vol. 27, No. 2, pp. 1568–1579.

Lafarge F., Gimel'farb G., Descombes X., 2010. Geometric Feature Extraction by a Multi-Marked Point Process. *IEEE Trans. on Pattern Analysis and Machine Intelligence*, Vol. 32(9), pp. 1597-1609.

Mallet, C., Lafarge, F., Roux, M., Soergel, U., Bretar, F. Heipke, C., 2010. A marked point process for modeling lidar waveforms, *IEEE Transactions on Image Processing*, 19 (12), pp. 3204–3221.

Metropolis, M., Rosenbluth, A., Teller, A., and Teller, E., 1953. Equation of state calculations by fast computing machines, *Journal of Chemical Physics*, Vol. 21, pp. 1087–1092.

Perrin, G., Descombes, X., Zerubia, J., 2005. Adaptive simulated annealing for energy minimization problem in a marked point process application. *Proc. Energy Minimization Methods in Computer Vision and Pattern Recognition*, St Augustine, United States.

Perrin, G., Descombes, X., Zerubia, J., 2005. Adaptive simulated annealing for energy minimization problem in a marked point process application. *Energy Minimization Methods in Computer Vision and Pattern Recognition*. Springer, Berlin-Heidelberg, Germany, Vol. 3757, pp. 3-17.

Salamon, P., Sibani, P., Frost, R., 2002. Facts, Conjectures and Improvements for Simulated Annealing. Society for Industrial and Applied Mathematics, Philadelphia, USA.

Sun, K., Sang, N., Zhang, T. (2007). Marked point process for vascular tree extraction on angiogram. *IEEE Conference on Computer Vision and Pattern Recognition (CVPR)*, pp. 467–478.

Tournaire, O., Paparoditis, N., Lafarge, F., 2007. Rectangular road marking detection with marked point processes. *International Archives of Photogrammetry, Remote Sensing and Spatial Information Sciences*, 36 (Part 3/W49A), pp. 73–78

Tournaire O., Bredif, M., Boldo D., Durupt, M., 2010. An efficient stochastic approach for building footprint extraction from digital elevation models. *ISPRS Journal of Photogrammetry and Remote Sensing*, Vol. 65(4), pp.317 -327.

27

ASSESSING MODIFIABLE AREAL UNIT PROBLEM IN THE ANALYSIS OF DEFORESTATION DRIVERS USING REMOTE SENSING AND CENSUS DATA

J.F. Mas[a]*, A. Pérez Vega[b], A. Andablo Reyes[c], M.A. Castillo Santiago[d], A. Flamenco Sandoval[a]

[a] Centro de Investigaciones en Geografía Ambiental, Universidad Nacional Autónoma de México, 58190 Morelia, Mexico - jfmas@ciga.unam.mx
[b] Universidad de Guanajuato, 4500 Guanajuato, Mexico -azu_pvega@hotmail.com
[c] Centro de Investigación en Alimentación y Desarrollo, 83304 Hermosillo, Mexico - aandablo@ciad.mx
[b] El Colegio de la Frontera Sur, San Cristobal de las Casas, Mexico - m.castillo.santiago@gmail.com

Commission II, WG II/4

KEY WORDS: Modifiable areal unit problem, Scale, Aggregation, Spatial analysis, Deforestation drivers

ABSTRACT:

In order to identify drivers of land use / land cover change (LUCC), the rate of change is often compared with environmental and socio-economic variables such as slope, soil suitability or population density. Socio-economic information is obtained from census data which are collected for individual households but are commonly presented in aggregate on the basis of geographical units as municipalities. However, a common problem, known as the modifiable areal unit problem (MAUP), is that the results of statistical analysis are not independent of the scale and the spatial configuration of the units used to aggregate the information. In this article, we evaluate how strong MAUP effects are for a study on the deforestation drivers in Mexico at municipality level. This was done by taking socio-economic variables from the 2010 Census of Mexico along with environmental variables and the rate of deforestation. As population census is given for each human settlement and environmental variables are obtained from high resolution spatial database, it was possible to aggregate the information using spatial units ("pseudo municipalities") with different sizes in order to observe the effect of scale and aggregation on the values of bivariate correlations (Pearsons r) between pairs of variables. We found that MAUP produces variations in the results, and we observed some variable pairs and some configurations of the spatial units where the effect was substantial.

1. INTRODUCTION

Land use/cover change (LUCC) is significant to a large range of aspects related to global environmental change and has received increasing attention from scientists and decision makers. Over the last decades, a broad range of studies have been carried out to monitor, evaluate and project LUCC with a particular emphasis on deforestation. Many studies of LUCC are based on remote sensing and census data using spatial analysis approaches. Multidate images are classified in order to monitor LUCC and spatial variables, expected to be the drivers of changes, are integrated in a GIS database. Then, the rate of change (e.g. rate of deforestation) is often compared with environmental and socio-economic variables such as slope, soil suitability or population density in order to identify and assess the effects of the drivers by means of a statistical index. Socio-economic information is obtained from census data which are collected for individual households but are commonly presented in aggregates on the basis of geographical units such as counties, municipalities or states. A common problem is that the results of statistical analysis are dependent of the scale and the spatial configuration of the units used to aggregate the information. According to Openshaw (1984) , this problem, known as the modifiable areal unit problem (MAUP), has two components: the scale problem and the aggregation (or zoning) problem. The scale problem is the variation in results observed when the data are aggregated into sets of increasingly larger units of analysis. The zoning problem is related to the variations in results observed when the analysis is done using the same number of alternative units. Some works indicate that the MAUP can cause variations of the correlations from -1 to +1 by judicious placement of zone boundaries (Openshaw, 1984; Openshaw and

Rao, 1995). However, Flowerdew (2011) used a large data set from the English Census and did not found large differences between correlations at different scales in the majority of the cases. In Mexico, most of census information is available at municipal level. In 2010, there were 2456 municipalities, which area ranges from a few km^2 to more than 53,000 km^2 with an average area of 796 km^2. The objective of this study is to evaluate how strong MAUP effects are on the assessment of deforestation drivers in Mexico using municipality-based data.

2. MATERIALS AND METHODS

We used socio-economic variables from the 2010 Census of Mexico from the National Institute of Statistics and Geography IN-EGI (2010) at human settlement level along with the marginalisation index calculated by the National Commission of Population CONAPO (2010) using information of housing, schooling and incomes from INEGI. We used also topographic indices (slope and elevation) obtained from the *Shuttle Radar Topography Mission* digital elevation model (http://www2.jpl.nasa.gov/srtm/) and the forest tree cover and forest loss data from the *Global Forest Change* database (http://earthenginepartners.appspot.com/science-2013-global-forest; Hansen et al., 2013). Table 1 shows the source and the resolution of the variables used in the study. All spatial and statistical analysis were carried out using the open source program R (R Core Team, 2013; Hijmans, 2015).

Study area encompasses about 111,360 km^2 located in the central part of Mexico. Based on the municipal average area, expected number of municipalities for this area is 140. To test the zoning effect of MAUP we generated Thiessen polygons around 140 random points. Each Thiessen polygon was used as an analysis unit

---

Variables	Characteristics	
	Source	Resolution
Forest loss	Forest Change	30 m
Number of inhabitants	2010 census INEGI	Settlement
Illiterate population (%)	2010 census INEGI	Settlement
Houses with dirt floor (%)	2010 census INEGI	Settlement
Marginalisation index	2005 CONAPO	Settlement
Elevation (m)	STRM DEM	90 m
Slope (degree)	STRM DEM	90 m

Table 1: Input variables characteristics

("pseudo municipality"). We computed, for each unit, the average elevation, average slope, population density, proportion of illiterate population, proportion of houses with dirt floors and the rate of deforestation, computed as the proportion of forest (tree cover $> 10\%$) which presents loss during 2000-2012. As a following step, we calculated the bivariate correlations (Pearson's r) between pairs of variables. In order to evaluate the zoning effect, this experiment was repeated 20 times in order to assess the variations of the values of correlation depending on the configuration of the units. In order to evaluate the scale effect, the number of polygons of Thiessen varied from a $1/4^{th}$, $1/2^{th}$, twice and four times the expected number of municipalities taking into account the average municipal area in Mexico. The variation of Pearson correlation values depending on zoning and scale effects was assessed by means of the coefficient of variation.

3. RESULTS

Figure 1 shows the first configuration of the 140 spatial units above the digital elevation model.

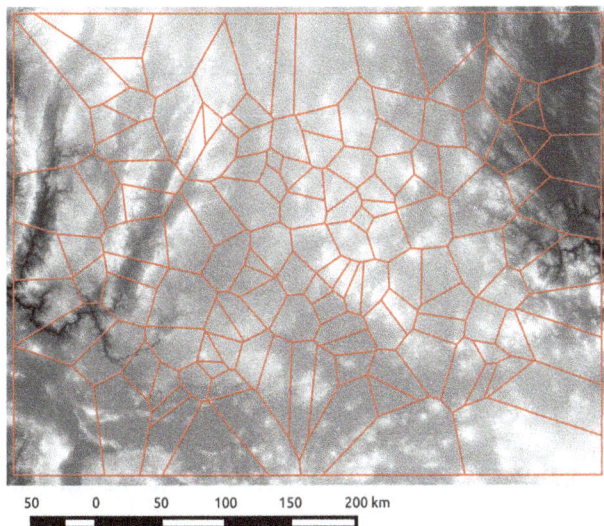

50   0   50   100   150   200 km

Figure 1: Limits (red) of the 140 random spatial units ("pseudo municipalities") above the digital elevation model (grey scale)

In table 2, which shows the variation of the correlation index depending on the zoning effect, it can be observed that the coefficient of variation ranges between 3 and 600% depending on the pair of involved variables. However, high values of the coefficient of variation correspond to weak correlation: When the coefficient of Pearson is superior to 0.5, the coefficient of variation is below 10%. It can also be observed that the minimum and maximum values of the coefficient are often different from the mean importantly. These differences mean that some specific configurations of the aggregation units can conduce to very contrasting results in the statistical analysis.

Var1	Var2	Min	Max	Mean	Stdev	CoeffVar
tdef	Pden	0.02	0.46	0.21	0.11	53
tdef	marg	-0.39	-0.00	-0.25	0.10	41
tdef	dirt	-0.37	0.05	-0.21	0.11	52
tdef	ille	-0.32	-0.05	-0.22	0.07	34
tdef	slop	-0.57	-0.34	-0.46	0.05	11
tdef	elev	-0.31	-0.09	-0.19	0.06	31
Pden	marg	-0.49	-0.28	-0.39	0.05	14
Pden	dirt	-0.27	-0.16	-0.23	0.03	14
Pden	ille	-0.43	-0.27	-0.36	0.05	13
Pden	slop	-0.23	-0.09	-0.15	0.04	28
Pden	elev	-0.19	-0.03	-0.13	0.04	34
marg	dirt	0.80	0.92	0.87	0.03	3
marg	ille	0.81	0.94	0.90	0.03	3
marg	slop	0.20	0.51	0.35	0.10	28
marg	elev	-0.13	0.21	0.06	0.09	140
dirt	ille	0.61	0.88	0.75	0.06	8
dirt	slop	0.25	0.53	0.39	0.08	21
dirt	elev	-0.35	-0.02	-0.17	0.09	53
ille	slop	0.13	0.44	0.29	0.10	33
ille	elev	-0.20	0.12	-0.01	0.08	598
slop	elev	-0.14	0.02	-0.06	0.04	74

Table 2: Minimum, maximum, mean, standard deviation and coefficient of variation of the values of the Pearson coefficient of correlation with 140 units (zoning effect)

Figures 2 and 3 are box-plots which show the variation of the Pearson coefficient values between the rate of deforestation and the slope and the index of marginalisation respectively. In the case of slope, we used the absolute value of the coefficient of correlation to make the interpretation of the graph easier. The variation of the value of correlation is due to the change in the number of units (scale effect). As Fotheringham and Wong (1991) noticed the correlation coefficient increases when the analysis is based in larger units due to a smoothing effect by averaging, so that the variation of a variable tends to decrease as the aggregation increases. In the box-plot, it can also be observed outlier values of correlation which correspond to particular configuration of the units which produce extreme values of correlation. The results obtained using 35, 70, 280 and 560 spatial units are presented in the appendix.

4. DISCUSSION AND CONCLUSION

In this study, we observed the smoothing effect related with scale. For simple statistical analyses as correlation analysis and linear regression, such variations can be theoretically expected and therefore are relatively well understood (Fotheringham and Wong, 1991; Jelinski and Wu, 1996). At the contrary, the zoning problem is more complex and much less well understood, even for simple statistical analyses. In the present study, we observed unpredictable results related with some specific configuration of the units used to compute the indices. Figure 4 shows four different spatial configurations of the same number of aggregation units ("pseudo municipalities") and can help understanding this behaviour. In the two top figures (a y b), the cluster of deforestation patches belongs to one single unit, a large one for the upper left figure (a), a small one for the upper right one (b), leading to moderate and high rate de deforestation for the corresponding unit. In the two figures at the bottom (c y d), the cluster of deforestation is distributed among three and four aggregation units leading to even rates of deforestation.

Some studies reported that correlations varied from -1 to 1 due to the MAUP effect (Openshaw and Rao, 1995). However, these correlations were obtained using highly convoluted and there-

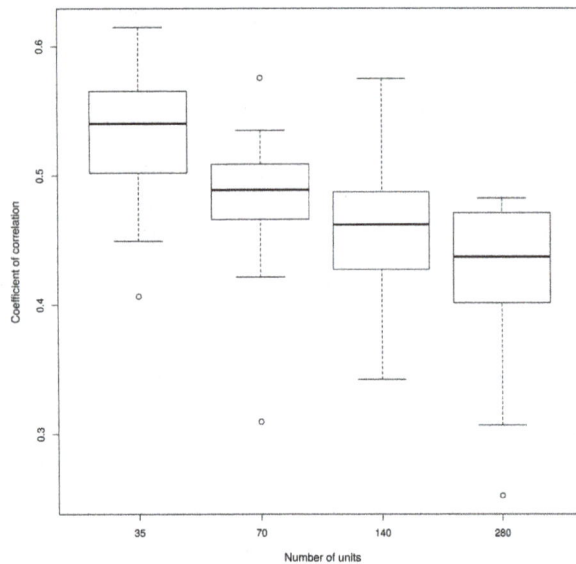

Figure 2: Variation of the Pearson coefficient between the deforestation rate and the slope due to change in the number of units

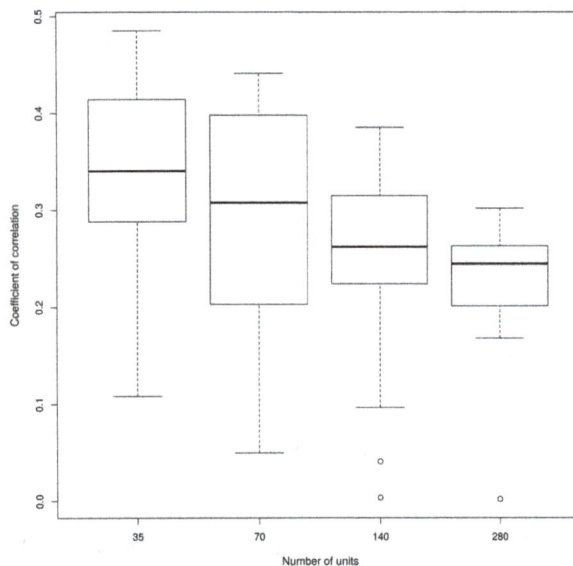

Figure 3: Variation of the Pearson coefficient between the deforestation rate and the index of marginalisation due to change in the number of units (scale effect)

fore implausible boundaries between units. In the present study, boundaries are more simple than true boundaries due to the use of Thiessen polygons. However, as the centroid of each unit is a random point, pseudo municipalities are not realistic. For instance, some units can encompass little or, in some cases, no population at all. In future research, we will examine the effect a such unrealistic feature on the design of units choosing randomly existing settlements with a minimum population as municipality seat (administrative center) and centroid of a spatial unit of analysis.

We found that, in most of the cases, MAUP does not make large difference to the results as reported in some previous studies. However, we observed some variable pairs and some specific

Figure 4: Four configurations of aggregation units ("pseudo municipalities") with the same number of units (zoning effect)

configurations where the effect was substantial. In future research, we will assess the effect of MAUP in global and local models of regression and evaluate the potential solutions reviewed by Dark and Bram (2007).

ACKNOWLEDGEMENTS

This study was supported by the *Consejo Nacional de Ciencia y Tecnología* (CONACYT) and the *Instituto Nacional de Estadística y Geografía* (INEGI) through the project *Análisis espacio-temporal de la vulnerabilidad del paisaje utilizando percepción remota y métodos espaciales: un estudio interdisciplinario y multiescalar en cuatro regiones del país.*

References

CONAPO, 2010. Índices de Marginación. Technical report, Consejo Nacional de Población, México D.F.

Dark, S. J. and Bram, D., 2007. The modifiable areal unit problem (MAUP) in physical geography. Progress in Physical Geography 31(5), pp. 471–479.

Flowerdew, R., 2011. How serious is the Modifiable Areal Unit Problem for analysis of English census data? Population Trends 145(145), pp. 106–118.

Fotheringham, A. S. and Wong, D. W. S., 1991. The modifiable areal unit problem in statistical analysis. Environment and Planning A 23, pp. 1025–1044.

Hijmans, R. J., 2015. raster: Geographic data analysis and modeling.

INEGI, 2010. Censo de Población y Vivienda 2010. Technical report, Aguascalientes.

Jelinski, D. E. and Wu, J., 1996. The modifiable areal unit problem and implications for landscape ecology. Landscape Ecology 11(3), pp. 129–140.

Openshaw, S., 1984. The modifiable areal unit problem. Geo-Books, Norwich, England.

Openshaw, S. and Rao, L., 1995. Algorithms for reengineering 1991 Census geography. Environment and Planning A 27(3), pp. 425–446.

R Core Team, 2013. R: A Language and Environment for Statistical Computing. R Foundation for Statistical Computing, Vienna, Austria.

APPENDIX

Var1	Var2	Min	Max	Mean	Stdev	CoeffVar
tdef	Pden	0.01	0.50	0.24	0.13	53
tdef	marg	-0.49	-0.11	-0.34	0.10	29
tdef	dirt	-0.54	-0.18	-0.40	0.08	20
tdef	ille	-0.42	0.02	-0.26	0.11	41
tdef	slop	-0.61	-0.41	-0.53	0.05	10
tdef	elev	-0.30	0.03	-0.14	0.10	68
Pden	marg	-0.64	-0.37	-0.52	0.07	14
Pden	dirt	-0.44	-0.24	-0.34	0.06	18
Pden	ille	-0.62	-0.35	-0.48	0.07	14
Pden	slop	-0.37	0.02	-0.17	0.10	60
Pden	elev	-0.29	-0.03	-0.17	0.08	44
marg	dirt	0.73	0.93	0.88	0.05	6
marg	ille	0.89	0.97	0.94	0.02	2
marg	slop	-0.05	0.52	0.25	0.15	62
marg	elev	-0.21	0.31	0.03	0.13	424
dirt	ille	0.55	0.91	0.82	0.08	10
dirt	slop	0.16	0.58	0.35	0.14	39
dirt	elev	-0.38	0.10	-0.15	0.12	81
ille	slop	-0.06	0.48	0.21	0.15	73
ille	elev	-0.33	0.22	-0.06	0.13	197

Table 3: Minimum, maximum, mean, standard deviation and coefficient of variation of the values of the Pearson coefficient of correlation with 35 units (zoning effect)

Var1	Var2	Min	Max	Mean	Stdev	CoeffVar
tdef	Pden	0.00	0.59	0.23	0.14	58
tdef	marg	-0.44	0.06	-0.27	0.14	52
tdef	dirt	-0.47	0.20	-0.25	0.19	74
tdef	ille	-0.39	0.09	-0.22	0.13	60
tdef	slop	-0.58	-0.31	-0.48	0.05	11
tdef	elev	-0.49	-0.07	-0.22	0.11	52
Pden	marg	-0.57	-0.26	-0.46	0.09	20
Pden	dirt	-0.35	-0.15	-0.28	0.06	23
Pden	ille	-0.50	-0.26	-0.42	0.07	18
Pden	slop	-0.31	0.03	-0.14	0.08	52
Pden	elev	-0.26	-0.04	-0.16	0.06	39
marg	dirt	0.81	0.94	0.88	0.04	5
marg	ille	0.87	0.96	0.93	0.02	3
marg	slop	-0.04	0.50	0.28	0.15	53
marg	elev	-0.16	0.28	0.08	0.12	153
dirt	ille	0.68	0.91	0.79	0.07	8
dirt	slop	0.05	0.56	0.36	0.14	40
dirt	elev	-0.46	0.22	-0.13	0.15	115
ille	slop	-0.08	0.44	0.24	0.14	58
ille	elev	-0.18	0.21	0.01	0.12	1605

Table 4: Minimum, maximum, mean, standard deviation and coefficient of variation of the values of the Pearson coefficient of correlation with 70 units (zoning effect)

Var1	Var2	Min	Max	Mean	Stdev	CoeffVar
tdef	Pden	0.06	0.32	0.18	0.08	43
tdef	marg	-0.30	0.00	-0.23	0.07	30
tdef	dirt	-0.26	0.08	-0.17	0.08	46
tdef	ille	-0.28	-0.07	-0.20	0.05	26
tdef	slop	-0.48	-0.25	-0.42	0.06	15
tdef	elev	-0.30	-0.09	-0.19	0.04	22
Pden	marg	-0.39	-0.23	-0.29	0.04	13
Pden	dirt	-0.23	-0.12	-0.17	0.02	15
Pden	ille	-0.35	-0.21	-0.27	0.03	12
Pden	slop	-0.20	-0.07	-0.14	0.03	25
Pden	elev	-0.14	-0.06	-0.11	0.02	18
marg	dirt	0.84	0.91	0.87	0.02	2
marg	ille	0.82	0.90	0.88	0.02	2
marg	slop	0.30	0.49	0.41	0.06	14
marg	elev	0.03	0.18	0.11	0.04	41
dirt	ille	0.68	0.80	0.74	0.03	4
dirt	slop	0.34	0.53	0.43	0.06	13
dirt	elev	-0.17	0.00	-0.09	0.05	58
ille	slop	0.25	0.41	0.33	0.05	16
ille	elev	-0.04	0.09	0.02	0.04	244

Table 5: Minimum, maximum, mean, standard deviation and coefficient of variation of the values of the Pearson coefficient of correlation with 280 units (zoning effect)

Var1	Var2	Min	Max	Mean	Stdev	CoeffVar
tdef	Pden	-0.00	0.20	0.10	0.06	61
tdef	marg	-0.27	-0.01	-0.18	0.08	42
tdef	dirt	-0.20	0.10	-0.11	0.09	79
tdef	ille	-0.23	-0.03	-0.16	0.06	37
tdef	slop	-0.44	-0.08	-0.35	0.11	31
tdef	elev	-0.22	-0.12	-0.17	0.03	17
Pden	marg	-0.28	-0.16	-0.23	0.03	13
Pden	dirt	-0.16	-0.09	-0.12	0.02	13
Pden	ille	-0.26	-0.15	-0.21	0.03	13
Pden	slop	-0.16	-0.06	-0.10	0.03	26
Pden	elev	-0.13	-0.05	-0.09	0.02	22
marg	dirt	0.81	0.87	0.84	0.01	2
marg	ille	0.76	0.87	0.83	0.02	3
marg	slop	0.33	0.50	0.41	0.04	10
marg	elev	0.07	0.23	0.15	0.04	30
dirt	ille	0.57	0.73	0.67	0.05	7
dirt	slop	0.29	0.50	0.41	0.05	12
dirt	elev	-0.12	0.04	-0.05	0.04	89
ille	slop	0.17	0.38	0.30	0.05	17
ille	elev	-0.00	0.14	0.06	0.04	63

Table 6: Minimum, maximum, mean, standard deviation and coefficient of variation of the values of the Pearson coefficient of correlation with 560 units (zoning effect)

DEVELOPMENT AND TESTING OF GEO-PROCESSING MODELS FOR THE AUTOMATIC GENERATION OF REMEDIATION PLAN AND NAVIGATION DATA TO USE IN INDUSTRIAL DISASTER REMEDIATION

G. Lucas [a, b, *], Cs. Lénárt [c], J. Solymosi [d],

[a] Research Institute of Remote Sensing and Rural Development, Károly Róbert College, Gyöngyös, Hungary –
gregory.luc4s@gmail.com
[b] Doctoral School of Military Engineering, National University of Public Service, Budapest, Hungary
[c] Research Institute of Remote Sensing and Rural Development, lenart.dr@gmail.com
[d] National University of Public Service, Budapest, Hungary - jozsef.solymosi@somos.hu

Commission IV, WG IV/7

KEY WORDS: Remediation, Clean-up, Industrial Disaster, Automatic Planning, Geo-processing Models, GIS, Guidance, Navigation, Python, ArcPy.

ABSTRACT:

This paper introduces research done on the automatic preparation of remediation plans and navigation data for the precise guidance of heavy machinery in clean-up work after an industrial disaster. The input test data consists of a pollution extent shapefile derived from the processing of hyperspectral aerial survey data from the Kolontár red mud disaster. Three algorithms were developed and the respective scripts were written in Python. The first model aims at drawing a parcel clean-up plan. The model tests four different parcel orientations (0, 90, 45 and 135 degree) and keeps the plan where clean-up parcels are less numerous considering it is an optimal spatial configuration. The second model drifts the clean-up parcel of a work plan both vertically and horizontally following a grid pattern with sampling distance of a fifth of a parcel width and keep the most optimal drifted version; here also with the belief to reduce the final number of parcel features. The last model aims at drawing a navigation line in the middle of each clean-up parcel. The models work efficiently and achieve automatic optimized plan generation (parcels and navigation lines). Applying the first model we demonstrated that depending on the size and geometry of the features of the contaminated area layer, the number of clean-up parcels generated by the model varies in a range of 4% to 38% from plan to plan. Such a significant variation with the resulting feature numbers shows that the optimal orientation identification can result in saving work, time and money in remediation. The various tests demonstrated that the model gains efficiency when 1/ the individual features of contaminated area present a significant orientation with their geometry (features are long), 2/ the size of pollution extent features becomes closer to the size of the parcels (scale effect). The second model shows only 1% difference with the variation of feature number; so this last is less interesting for planning optimization applications. Last model rather simply fulfils the task it was designed for by drawing navigation lines.

1. INTRODUCTION

On October 4th, 2010 Hungary faced the worst environmental disaster in its history when the embankment of a toxic waste reservoir failed and released a mixture of 600,000 to 700,000 m³ of red mud and water. Lower parts of the settlements of Kolontár, Devecser, and Somlóvásárhely were flooded. Ten people died and another 120 people were injured. The red mud flooded 4 km² of the surrounding area.

The idea motivating this research work came after considering the clean-up work done on the impacted area of Kolontár (to the north of Balaton). Whereas digital maps figuring the contour of the contaminated areas and the pollution thickness were available[1] (Burai et al., 2011), the excavation work was performed in a traditional way, without the support of positioning and navigation technologies. So accurate and detailed information produced in the early stage of the remediation process was not efficiently exploited.

In a broader context, our research work aims at developing methodologies and tools to assure a continuum with the geographic information exploitation/support through a precise remediation process. The GI gathered during the disaster assessment phase should be adapted and used in the planning phase; this would provide plans and navigation data for the clean-up phase. Additionally technologies integration (remote sensing (detection), GIS (planning), positioning and navigation (clean-up)) should also be researched.

Our bibliographic research demonstrated that the use of geoinformation technologies in remediation is mainly done during the early stage of remediation for the detection and mapping of the pollution. Aerial survey (Burai et al., 2011) and soil sampling are used for data acquisition. GIS, geo-statistical analysis and 3D modelling (Guyard, 2013; Mathieu et al., 2009; Hellawell et al., 2001; Franco et al., 2004; Webster et al.) are then employed for visualizing the pollution extents, estimating the volume to process (project dimensioning and costs) and planning /monitoring the remediation work (at site level). In contrast, the use of geographic information technologies during the clean-up stage seems quite limited. In the case of in-situ remediation, injection and recovery wells can be precisely positioned with GPS based on planning optimized with geo statistic calculations. In the case of ex-situ[2] remediation,

---

[1] from aerial survey and remote sensing processing methods

* Corresponding author

[2] ex-situ remediation is opposed to on-site remediation and requires the excavation of soil and its transportation out of the site.

literature does not mention the use of navigation and positioning technologies for the excavation work done by heavy machineries. As positioning technologies are routinely employed in civil engineering for the guidance of heavy equipment for precise and efficient work it seems the shortcomings in the case of ex-situ remediation lays in the capacity to generate adequate remediation plans and in the lack of adapted GIS tools, models, methods and practice [1]. In response, this work develops models in order to be able to produce a plan containing "clean-up parcels" and derived navigation data.

"Clean-up parcel" is a central concept and the geographic feature of interest in this work. Clean-up parcel in the real world corresponds with the surface covered by a dozer shovel until it gets filled to capacity (in other words the dozer's maximum work footprint). In the GIS model a clean-up parcel consist of a rectangular feature in a polygon feature class. Its width is equal with the dozer's blade width. Its length (length $_{Max}$) is derived from the bulldozer characteristics and the thickness of pollution to collect (1).

$$\text{Volume\_blade}_{Max} = \text{width}_{dozer} \times \text{length}_{Max} \times \text{thickness} \quad (1)$$

The area of interest (contaminated area) is presented in figure 1. It is a polygon shapefile which was created from classified hyperspectral imagery (Burai et al., 2011). The area covers 4 km^2, is 16 km long in longitude and 5 km long in latitude. Because the catastrophe was a flood, the polygon features of the contaminated area have an orientation that generally follows the direction of the flood.

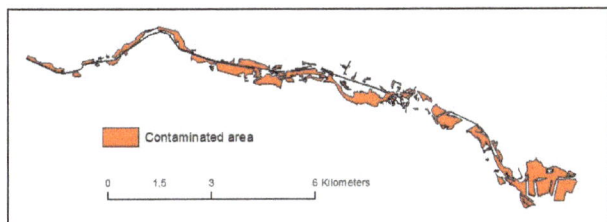

Figure 1. Overview of the source dataset "contaminated_area"

This paper focuses exclusively on precisely describing the conception of the algorithms, architecture and how geo-processing is done rather than providing line per line calculation details and scripts. The latter can be requested from the author using the email details provided.

Readers should notice that this exploratory work is relevant for ex-situ remediation (remediation where excavation is done) on extended areas where industrial disaster took place (red mud, nuclear, chemical, etc.). In such cases heavy machinery is used and it makes sense to try to plan their moves precisely in order to save effort, time and money, in a similar way as precision agriculture or civil engineering do.

The research firstly develops models through the design of algorithms and their transcription in Python scripts. Secondly the models are tested with a test dataset derived from the red mud disaster impact assessment. The first test control if geo-processing is done without errors. The second test control if the model shows efficiency in its tasks consisting of optimizing the clean-up parcel plan (i.e. reducing the number of parcels). Last, the efficiency of the model is assessed in regard to time efficiency. Based on the results of the tests diverse proposals are

formulated for the development of the final version of the models.

2. CLEAN-UP PARCELS MODEL DEVELOPMENT

2.1 Description of the objectives

This model generates a polygon feature class, containing rectangular features with a unique shape that represents the clean-up parcels. The parcel's width is inherited from the bulldozer's blade width. The parcel's length is derived from the blade capacity. Dividing the contaminated area into clean-up parcels should be done automatically. The parcels should properly cover the whole contaminated area. The pattern designed should be optimal, meaning that technically it ensures the proper removal of pollution and economically it ensures the highest efficiency.

Considering those requirements it appears that rectangular grid pattern model is optimal.

2.2 Algorithm's raw architecture

The feature class will be similar to a grid with rectangular polygons. The algorithm could be divided into two parts:

- the first part makes the calculations in order to point out to locations organised in a grid pattern with the appropriate orientation.
- the second part calculates the parcel corners' coordinates and draws rectangular polygons.

Iterations (done with loops implementing repeat/while commands) will succeed ranges calculations deriving from the geographic extent of "Contaminated_area". This calculation can be separated in a function.

As the process will be automatized, it could be useful to test different plans with different orientations of the parcels. An algorithm with four different orientations (0°, 90°, 45° and 135°) was drafted. The best result was selected can be done by counting the number of features in each feature class created and selecting the one with the fewest parcels.

2.3 Data requirement (input)

- A polygon feature class where features' geometry represents the polluted areas.
- Width (in meter), length (in meter), orientation (in degree).

2.4 Algorithm architecture

Procedure createRectangleAtPoint(x, y, length, width, orientation, layer)

This procedure draws one rectangle according to the coordinates of a corner starting point, the orientation, the width and length of the rectangle. The vertices of the rectangle are attributed in the clockwise direction (figure 2).

Function extent(fc)

This function extracts the geographical extent (xmax, xmin, ymax, ymin) from a reference layer; i.e. the contaminated area layer. It is used later in the calculation of the maximal limit for the iteration in the loops building the grids. This function already existed and we have simply re-used it. [2]

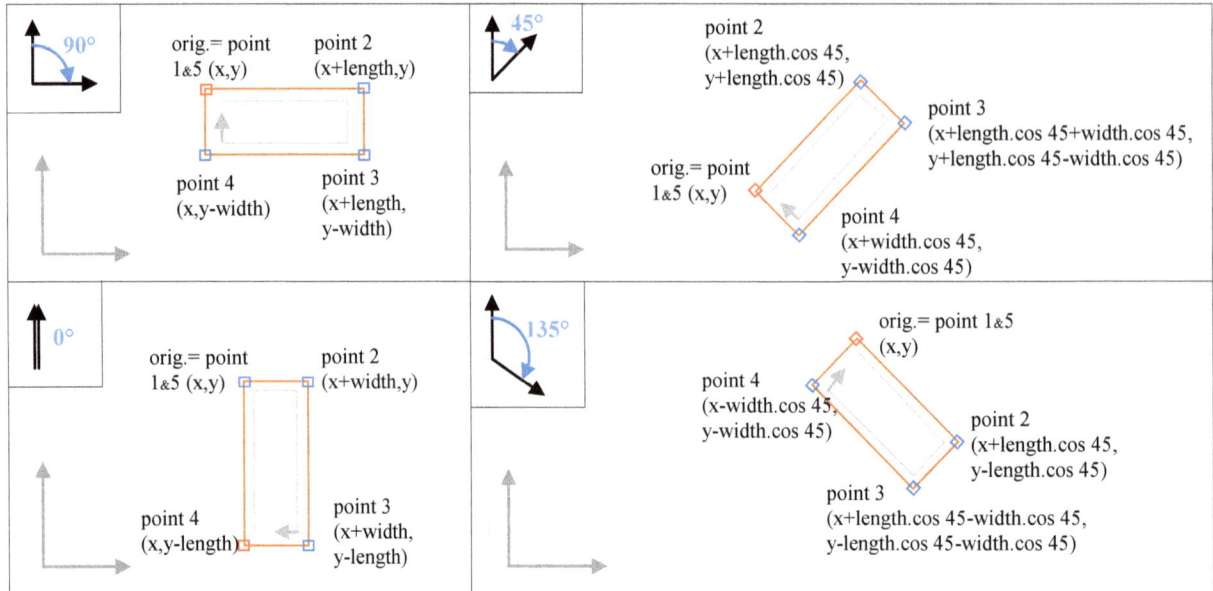

Figure 2. Details of the coordinate calculations with vertices of the parcel and drawing method in createRectangleAtPoint procedure with the 4 orientations cases

Procedure Make_Grid(length, width, layer_name, grid_orientation)

This procedure primarily draws a grid pattern taking into consideration the orientation, the width and length provided as parameters. For each point of the grid the procedure calls the createRectangleAtPoint procedure which draws a rectangle. With the 0° and 90° orientations the procedure loops top-down with the lines and left-right inside line. With the 135° and 45° patterns the procedure proceeds in two steps (step 1 is presented in blue colour, step 2 in green on Figure 3.). Step 1: loop creates features in diagonal starting from the top left corner moving towards the bottom right corner and a second loop control jumping one line down under the start of previous line using a backup of previous line start coordinates. Then in a second step, it moves diagonally going down (loop 2) but the second loop's implementation positions the next line on top of the previous one so that the grid can cover the second half of the area (above the step one). As many features are created out of the area of interest, a clean-up is necessary at the end. Selection is done on the features that intersect the "polluted_area" layer. They are copied in a new layer and all temporary layers are deleted at the end.

2.5. Script body

The algorithm uses the Procedure_Make_Grid and Procedure_CreateRectangleAtPoint in order to create four feature classes with 0°, 90°, 45° and 135° orientations. Finally a "get count" method is used to retrieve the number of features from each feature class. The feature class containing the smallest number of features is selected and saved; the other feature classes are deleted from the map document.

3. OFFSET EFFECT TESTING: MODEL DEVELOPMENT

3.1 Description of the objectives

The model should move the features of clean-up parcel altogether following a grid pattern (so both in vertical and

horizontal direction). The grid is oriented in the same way as the clean-up parcel feature class and the sampling distance of the grid is equal with a fifth of the parcel width. Each time the feature class is drifted the model counts how many features are located in the area of interest. The "get count" result with the smallest number of features shows the best offset to be applied.

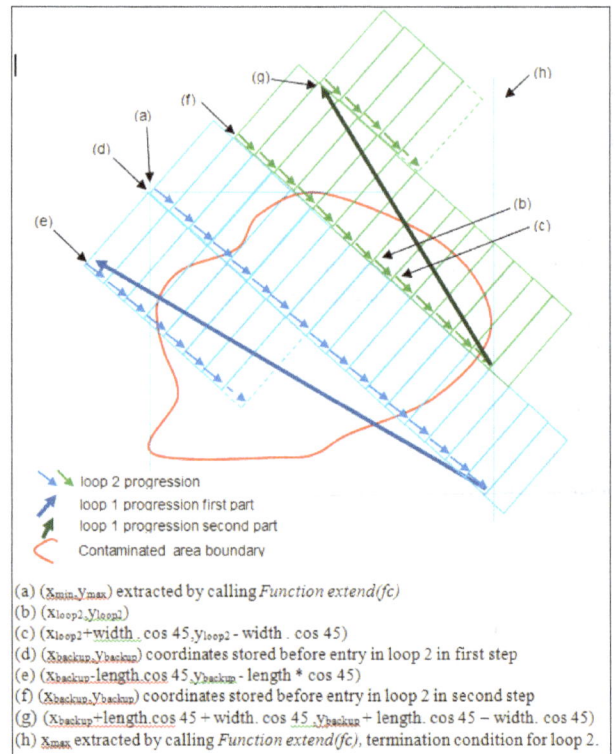

loop 2 progression
loop 1 progression first part
loop 1 progression second part
Contaminated area boundary

(a) (x_{min}, y_{max}) extracted by calling *Function extend(fc)*
(b) (x_{loop2}, y_{loop1})
(c) $(x_{loop2} + width \cdot \cos 45, y_{loop2} - width \cdot \cos 45)$
(d) (x_{backup}, y_{backup}) coordinates stored before entry in loop 2 in first step
(e) $(x_{backup} - length \cdot \cos 45, y_{backup} - length * \cos 45)$
(f) (x_{backup}, y_{backup}) coordinates stored before entry in loop 2 in second step
(g) $(x_{backup} + length \cdot \cos 45 + width \cdot \cos 45, y_{backup} + length \cdot \cos 45 - width \cdot \cos 45)$
(h) x_{max} extracted by calling *Function extend(fc)*, termination condition for loop 2.

Figure 3. Conceptual representation of clean-up parcel model in the 45° case

3.2 Data requirement (input)

- the original area of interest is required to perform a selection based on intersection.
- a new clean-up parcel feature class is necessary. It is similar to the one generated in model 1 with optimal orientation but the reference area of interest differs (extended).
- The extended reference area of interest is the original area of interest extended with a buffer zone of the parcel width. If this precaution is not implemented, the clean-up parcel extent is too limited and for example an empty area appears on the left when x receives a positive drift.

3.3 Algorithm's raw architecture

1. The model should generate a new area of interest with a buffer of "length" size around the original area of interest.
2. Clean-up parcel feature class should be recreated based on the new target area (this is done in order not to have an empty area when the features will be shifted (maximal shift will be equal to parcel length)).
3. Calculate the shift values based on parcel width, length, orientation and store them in a three dimensional list.
3. All the features of this new clean-up feature class are shifted applying the offset values stored in the matrix (x,y). The grid x and y range are fixed at one-fifth of the parcel width.
4. Each time the feature class is shifted, a selection of the features intersecting with the original target area is done and the result of "getcount" is stored in a two dimensional list.
5. Unselect all features
6. Inverted shift is applied to set the feature back in place.
7. Next shift is applied, etc.
8. When all the shifting x,y values are passed, a search in the list value returns the smallest getcount.
9. From the minimal getcount, to retrieve the optimal x,y shift values.
10. Apply a final shift with the optimal x,y shift values.

3.4 Algorithm architecture

Function_calculate_drift_matrix(length, width, orientation)
This function returns a three dimensional matrix containing the shift coordinates corresponding to each point of the grid. The grid is oriented according to the parameter "orientation". The step of the grid is width/5 both with "rows" and "columns". For example if parcels are 30mx3m at 90°, there are 50 columns and 5 rows in the grid and the step is 3/5m. This function has four parts for the four different orientations.

Funtion_shift_features(in_features, x_shift=None, y_shift=None)
This function uses the arcpy.da module's UpdateCursor. By modifying the SHAPE@XY token, it modifies the centroid of the feature and shifts the rest of the feature to match. This function was available online and usable without changes, so it was simply copied [3].

Main procedure
The procedure deals with the creation of the extended area of interest; appeals the two functions described above and deals with the searches in the list "getcount".

4. NAVIGATION LINES MODEL DEVELOPMENT

4.1 Description of the objectives

The model should create a polyline feature class with navigation lines. The navigation lines should:
- be located in the middle of parcels,
- follow their length
Input: "clean-up parcel" shape file
Output: "Navigation_lines" shape file

4.2 Algorithm's structure

Function_ExtractVerticesCoordinateFromFeature(input_feature_class)
This function extracts the vertices' coordinates from a polygon feature class geometries and returns a two dimensional list storing the coordinates. SearchCursor method is employed on each row of the feature class. The result is appended to the list. Most of the script derives from the example of Reading polyline or polygon geometries of ESRI resources help [4].

Function_CalculateMiddlePoints(list_corners)
This function receives the coordinates of the four corners of a rectangle and returns the values of the coordinates of the two points located in the middle of the shortest sides.

Procedure_WriteaLine(point_1, point_2, layer)
This procedure writes a polyline feature between two given points (coming from function_CalculateMiddlePoints) in the given layer. SearchCursor method is applied to enter new geometry.

Procedure_DrawNavigationLines(Ouput_Navigation_Lines, Source_feature_class)
This procedure makes use of the functions and procedures above to draw a new polyline feature class with the navigation lines. Createfeatureclass_management method is used to create the output feature class.

The algorithms were successfully converted into scripts in Python language and models tested first with a subset of the pollution thickness layer derived from the processing of hyperspectral aerial survey data of Kolontar red mud disaster. (Burai et al., 2011)

5. RESULTS AND DISCUSSION

5.1 Clean-up parcel model

During its development the script was tested on a small feature extracted from the "Contaminated_area" shapefile.
Figure 4 shows an example of result with the four intermediary feature classes generated by the clean-up parcels model with 0°, 90°, 45° and 135° orientation, 3 meter width and 30 meter length on a sample of the contaminated area.

After correcting mistakes in the script the geo-processing model was applied to the whole "contaminated_area" shapefile. It resulted in very long geo-processing (more than 3 days to generate 0° and only a part of 45° orientation clean-up parcels layers).

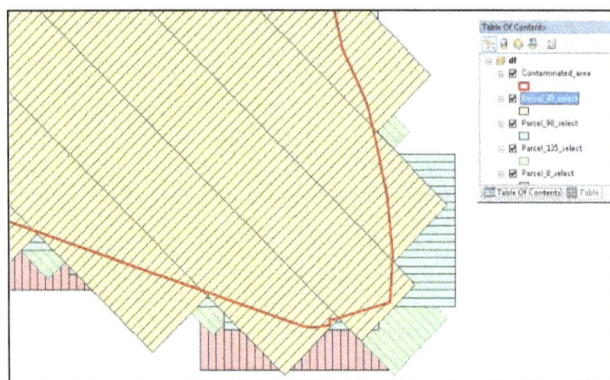

Figure 4. Intermediary results of clean-up parcel model with 0°, 90°, 45° and 135° orientation plan overlay

Processing was voluntarily stopped before geo-processing was completed. This long calculation was caused:

1/ by the extent and geometry of the target area (containing a lot of empty space where it was useless to have the geo-processing run),

2/ by the huge number of parcels to generate (around 60,000); a direct effect of the extent of contaminated_area,

3/ by the procedure_Make_grid which is not efficient with geo-processing (a lot of unnecessary geo-processing is done during iteration outside of the area of interest).

To cope with these various problems the second test was run on the same data but split into 8 zones (11 shapefiles as zone 7 was split in four).

The number of features generated per zone with the four orientations is summarized in the table 1. Smallest values are highlighted in green and highest in red background.

Orientation	0	45	135	90
Zone_1	13048	13690	13151	13062
Zone_2	13133	13869	12514	13112
Zone_3	25442	25522	24358	23416
Zone_4	1887	1938	1695	1614
Zone_5	5033	4431	5089	4486
Zone_6	2489	2795	2276	2370
Zone_7a	19	61	52	52
Zone_7b	147	165	205	203
Zone_7c	112	127	151	138
Zone_7d	22	65	61	64
Zone_8	297	457	359	387

Table 1. Number of features with the different orientations within the 8 zones.

At first we can observe a significant difference in the number of features obtained after geo-processing with different orientations. It appears the orientation of the parcel pattern is an important parameter to consider in optimizing planning.

Table 2 provides statistics per zone. First we calculated a classic measure of deviation of $\sigma/\bar{x}$ (standard deviation divided by mean). As the number of entities vary significantly per sample (zone) and in order to have values of the same order it was necessary to divide deviation by mean. The deviation varies from 2% to 20%. The second value provided in the table is more relevant in our opinion because it better expresses the important difference between the extremes and better pulls out the efficiency of the algorithm (subtraction of the maximum feature number with the minimum feature number divided by the maximum feature number and expressed in percent). This value can be interpreted as the ability of the algorithm to "reduce" the number of parcels by x%. The feature number reduction ranges from 4% to 38%.

Zone	σ	$\bar{x}$	$\sigma/\bar{x}$	(Max-Min)/Max (in %)
1	305	13238	2%	5%
2	555	13157	4%	10%
3	998	24685	4%	8%
4	154	1784	9%	17%
5	349	4760	7%	13%
6	226	2483	9%	19%
7	81	411	20%	38%
8	66	375	18%	35%

Table 2. Statistics with the 8 zones.

As a second conclusion the variations in the results can be very high (up to 38 %). This is definitely significant information for the planning strategy. Last, such a difference should be investigated and explained.

Figure 5 shows the geometry and size of the 8 zones in order to be able to cross the statistical results from table 2 with spatial information. The following observation can be formulated: the smallest zones show bigger variances (reported to the mean) than the biggest zones.

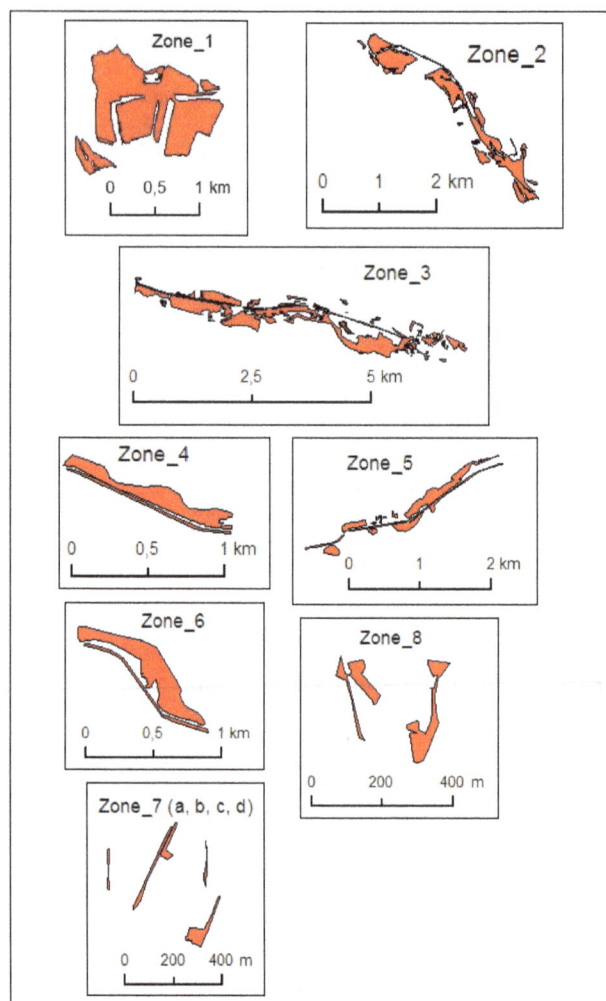

Figure 5. Overview of the 8 zones

Hypothesis 1: the reduced number of features is the cause of the bigger variance. Orientation matches more efficiently with smaller number of features because much of them are oriented in the same way. On the contrary when there are more features, their orientation varies more and the efficiency of the model decreases.

Hypothesis 2: the cause for important variance is a scale effect because the model efficiency works with a border effect. On smaller areas the features could be smaller; the ratio boundary/area is more in favour of the boundary compared to massive area and orientation becomes much more important.

After additional tests we could conclude that both hypotheses seem valid. When comparing the results between zone 4 that has two oriented features in the same direction and zone 7 a, b, c, d with small and long features; feature number decrease by 38 % with zone 7 whereas the feature number is only decreased by 17% for zone 4.

In terms of practice with the preparation of "Contaminated_area"; in order to optimize the geo-processing, the user should pay attention to three things:
1/ to prepare zones as small as possible in order to reduce empty areas (time consideration).
2/ to the extent possible have features with the same orientation inside one zone. If necessary, a zone should be split into several parts in order to ensure the features' general orientation is as similar as possible (example is 7 a, b, c, d).
3/ to split feature if their geometry is complex. The result should be the creation of sub-features with simpler and oriented geometries.

In order to validate the presumptions mentioned above, the method was implemented on Zone 1 (figure 6) (where the algorithm showed the lowest efficiency) which was divided following the above recommendations. New results are summarized in table 3.

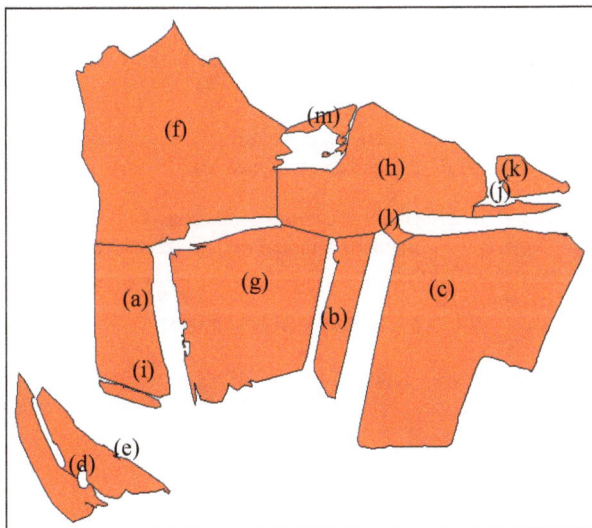

Figure 6: splitting of zone 1 into several parts

An additional reduction of 3.7% could be reached by applying an appropriate cut with zone 1 compared to the previous result.

Orientation	0	45	90	135
Zone_1a	887	1065	913	956
Zone_1b	507	605	588	593
Zone_1c	3076	3240	3089	3166
Zone_1d	459	527	483	453
Zone_1e	572	756	554	534
Zone_1f	3183	3449	3197	3216
Zone_1g	2037	2224	2052	2113
Zone_1h	1760	1836	1691	1770
Zone_1i	123	171	92	102
Zone_1j	166	170	102	161
Zone_1k	224	303	210	237
Zone_1l	71	84	61	73
Zone_1m	164	170	148	184
Sum min	12562		Sum max	14043

Table 3. Counting of the number of features with the different orientation and the different sub-zones.

Further developments

Regarding the reduction of geo-processing time, a test will be added inside the scripts implementing iteration. Before calling createRectangleAtPoint procedure an "IF" condition will be applied to check if the corner point (x,y) of the rectangle to be drawn falls into the area of interest (extended with a buffer zone of the parcel length). If x,y falls out no action will be taken, if it falls in then the rectangle will be written.

The orientation clearly appeared as a key parameter to control in order to optimize the remediation plan design. In our approach (which was exploratory) we decided to limit the number of orientation to 4 (with 0°, 45°, 90° and 135°). In order to increase the efficiency of the model the optimal orientation could be identified with 1° accuracy. This means the algorithm should be improved to take the following actions:
1/ isolate each individual polygon
2/ calculate polygon's orientation (with 1° accuracy)
3/ apply a modified version of the clean-up parcel algorithm in order to design a clean-up plan with x° orientation for the feature considered.
With such an implementation, the optimized clean-up parcels are designed directly and it is no longer necessary to run the same script (clean-up parcel) four times with the four different orientations. So it would solve two issues: reducing the time processing and improving algorithm efficiency while reducing the number of parcel.
Dividing the contamination-area into subparts with homogenous orientation and limited geographic extents is the task of the user.

5.2 Shift testing results

Model 2 testing showed very limited results with the reduction of feature number. Only 1% percent difference with the number of features could be modelled. After further considerations, it seems that due to the irregular shape of the area of interest and the scale ratio, a shift is useless because on average as parcels disappear on one border others appear on the opposite border. If the AOI is regular (rectangle for example) and the scale ratio much smaller (AOI area compared to parcel area), this tool could achieve significant results. In our case -a large scale industrial disaster with relatively large irregular areas- the tool shows limited efficiency; consequently we decided not to go further with the development.

5.3 Draw navigation line model

Figure 7 and figure 8 below show the result of the Navigation lines model.

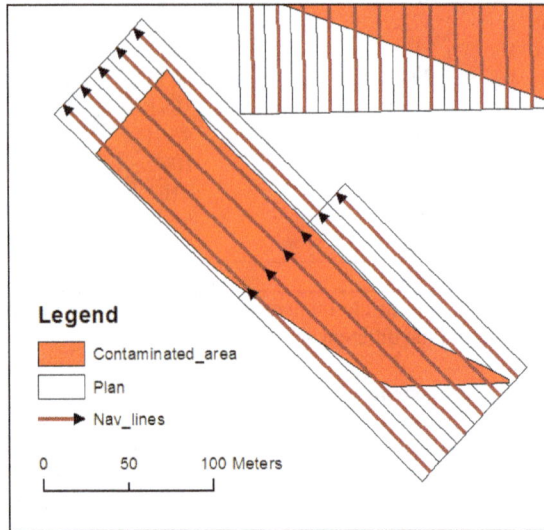

Fig.7. Navigation lines feature class generated by the navigation line model overlaying the clean-up parcels feature class

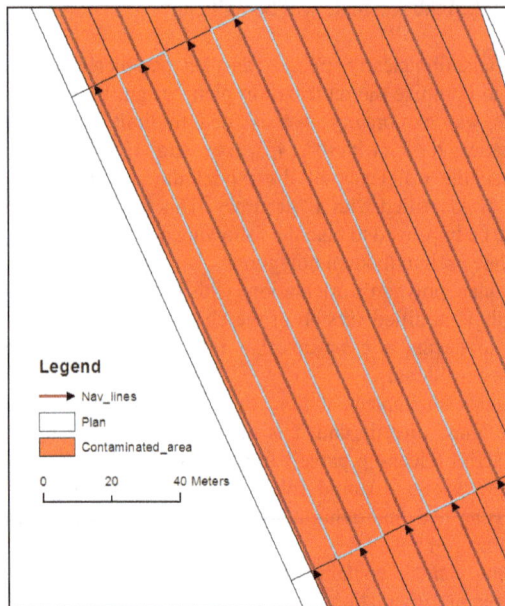

Fig.8. Zoom in the navigation lines

The algorithm orients all the geometries of the lines in the same direction by default. Some additional algorithms development could be foreseen if it turns out that navigation requirement would need a pre-planned navigation direction and computation of the optimal order of visit. As the operation in the field constantly changes, this kind of development is not considered a priority at the moment.

REFERENCES

Burai P., Smailbegovic A., Lénárt Cs., Berke J., Tomor T., Bíró T., 2011. Preliminary analysis of red mud spill based on aerial imagery. *Acta Geographica Debrecina, Landscape & Environment Series*, Volume 5, I: pp. 47-57.

Franco, C., Delgado, J., Soares, A., 2004. Impact Analysis and Sampling Design in The Pollution monitoring Process of The Aznalcollar Accident using Geostatistical Methods. In: *The International Archives of the Photogrammetry, Remote Sensing and Spatial Information Sciences*, Istanbul, Turkey, Vol. XXXV, Part B7, pp.373-378.

Guyard, C., 2013. Dépollution des sols et nappes : un marché sous pression. L'eau, l'industrie, les nuisances, n°359, pp.23-40. http://www.ianesco.fr/web/revue%20de%20presse/eau_industrie_nuisances_d%C3%A9pollution_sols_et_nappes_f%C3%A9v_2013_n%C2%B0359.pdf (accessed: 01 Jul. 2015)

Hellawell, E.E, Kemp, A.C, Nancarrow, D.J, 2001. A GIS raster technique to optimize contaminated soil removal. *Engineering Geology*, Vol. 60, Issues 1–4, pp. 107-116

Mathieu J.B., Garcia, V., Garcia, M., Rabaute, A., 2009. SoilRemediation: a plugin and workflow in Gocad for managing environmental data and modeling contaminated sites. Gocad Meeting, Nancy June 2-5, 2009.

Lindsay, J., Simon, T., Graettinger, G., 2001. Application of USEPA FIELDS GIS technology to support Remediation of Petroleum Contaminated Soils on the Pribilof Islands, Alaska. 2nd Biennial Coastal GeoTools Conference, Charleston, SC, January 8-11. http://proceedings.esri.com/library/userconf/proc00/professional/papers/PAP789/p789.htm (accessed: 01 Jul. 2015)

Webster, I., Ciccolini, L. Solving contaminated site problems cost effectively: plan, use geographical systems (GIS) and execute.http://www.projectnavigator.com/downloads/webster_ciccolini_solving_contaminated_sites_04.09.02.pdf (accessed: 01 jul. 2015)

[1] Global remediation technologies. MAPPING & MODELING http://grtusa.com/services/mapping-modeling/ (accessed: 01 jul. 2015)

[2] ArcGIS Help 10.1 / Extent (Arcpy) http://resources.arcgis.com/en/help/main/10.1/index.html#//018z00000072000000 (accessed: 01 Jul. 2015)

[3] ArcPy Café / Shifting features https://arcpy.wordpress.com/2012/11/15/shifting-features/ (accessed: 01 Jul. 2015)

[4] ArcGIS Help 10.1 / Reading geometries http://resources.arcgis.com/en/help/main/10.1/index.html#//002z0000001t000000 (accessed: 01 Jul. 2015)

ACKNOWLEDGEMENTS

This research was supported by the TÁMOP programme to the Károly Robert College (TÁMOP-4.2.2.D-15/1/ KONV-2015-0010).

29

CHANGE DETECTION BASED ON PERSISTENT SCATTERER INTERFEROMETRY

C.H. Yang*, U. Soergel

Institute of Geodesy, Technische Universität Darmstadt, Germany - (yang, soergel)@geod.tu-darmstadt.de

KEY WORDS: Persistent Scatterer Interferometry (PSI), Change Detection, Urban Monitoring

ABSTRACT:

Persistent Scatterer Interferometry (PSI) is a technique to extract subtle surface deformation from sets of scatterers identified in time-series of SAR images which feature temporally stable and strong radar signal (i.e., Persistent Scatterers, PS). Because of the preferred rectangular and regular structure of man-made objects, PSI works particularly well for monitoring of settlements. Usually, in PSI it is assumed that except for surface motion the scene is steady. In case this is not given, corresponding PS candidates are discarded during PSI processing. On the other hand, pixel-based change detection relying on local comparison of multi-temporal images typically highlights scene modifications of larger size rather than detail level.

In this paper, we propose a method to combine these two types of change detection approaches. First, we introduce a local change-index based on PSI, which basically looks for PS candidates that remain stable over a certain period of time, but then break down suddenly. In addition, for the remaining PS candidates we apply common PSI processing which yields attributes like velocity in line-of-sight. In order to consider context, we apply now spatial filtering according to the derived attributes and morphology to exclude outliers and extract connect components of similar regions at the same time. We demonstrate our approach for test site Berlin, Germany, where, firstly, deformation-velocities on man-made structures are estimated and, secondly, some construction-sites are correctly recognized.

1. INTRODUCTION

Synthetic Aperture Radar (SAR) is a remote sensing technique providing radar images. Due to the active sensor principle and signal wavelength in centimeter scale, SAR is capable for night vision and independent from weather conditions, respectively. In addition, SAR may cover vast areas in short time with considerable fine spatial resolution (e.g., satellite TerraSAR-X, TSX, acquires in extreme case of Stripmap mode stripes of 30 *km* by 1500 *km* along orbit with geometric resolution of about 3 *m*). Therefore, in particular the techniques based on spaceborne SAR is widely used for tasks like monitoring and change detection. In this paper, we combine two established change detection methods based on SAR, which are complementary with respect to the kind of changes looked at, the exploited features, and the typical size of extracted results.

The first technique is called Persistent Scatterer Interferometry (PSI) (Ferretti et al., 2000; Ferretti et al., 2001; Ferretti et al., 2011; Hooper et al., 2004; Kampes, 2006), which aims at surface deformation by processing of time-series of SAR images. For this purpose the image stack is scanned for SAR resolution cells of temporally stable and strong radar signal that are called Persistent Scatterers (PS). PSI is an opportunistic geodetic approach, this means, on the one hand, we take advantage of PS no matter what object they cause or wherever they are located, but, on the other hand, the PS density can vary strongly and voids of considerable size frequently occur. Due to the preferred rectangular and regular structure of man-made objects, PSI works particularly well for urban areas. Subsequent PSI processing is conducted in a network of the identified PS set. After filtering of nuisances like atmospheric phase delay and thermal noise, by spatial-temporal double-differences of the signal phase, the relative motion in line-of-sight of the SAR

sensors is obtained for each PS with respect to a reference PS. In order to avoid under-sampling, the relative PS-motion of two consecutive SAR images must neither temporally nor spatially exceed a 2π phase cycle (in case of TSX 1.5 *cm*). PSI has proven to be useful to monitor surface deformation in cities in the order of some millimeter per year (Crosetto et al., 2008; Gernhardt and Bamler, 2012; Perissin and Ferretti, 2007). Such deformation may be triggered by physical processes of various kinds leading to different motion behaviour such as linear long-term subsidence (Dixon et al., 2006; Liu et al., 2011; Osmanoğlu et al., 2011) or even sinusoidal pattern due to seasonal expansion of steel construction (Colesanti et al., 2003; Gernhardt et al., 2010; Monserrat et al., 2011) Even though a PS network is processed, standard PSI can be regarded as a local method because apart from post-processing like spatial low-pass filtering (e.g., according to a correlation length derived from some geophysical model of the underlying deformation process) the PS are essentially treated individually. And there is another limitation: Usually, in PSI processing it is assumed that except for surface motion the scene is steady: For instance, a building populated by certain substructures which produce some PS remains unchanged over the image stack. In case this is not true and such structures substantially alter or even vanish, corresponding PS candidates are discarded during PSI processing.

However, areas affected by bigger changes can be efficiently detected by the second common technique, we turn to now: pixel-based change detection applied to pairs or time-series of multi-temporal images. In the most basic approaches the grey values are directly compared, for instance, by image differencing, image rationing, or regression analysis (Hussian et al., 2013). No matter whether the amplitude, the coherence, or the differential phase of SAR images is used for analysis, noise might lead to undesired false alarms. Since noise is modelled to be spatially uncorrelated, such effect is mitigated by averaging preferably over connected image components. However, this inevitably comes along with loss of detail.

In this paper, we propose a method to combine these two types of change detection approaches. In Section 2, first we briefly outline PSI processing focussing on steps which are crucial for our further workflow. Based on this we introduce a local change-index derived from PSI, which basically looks for PS candidates that remain stable over a certain period of time, but then break down suddenly. In addition, for the remaining PS candidates we apply common PSI processing which yields features like velocity in line-of-sight. In order to consider context, we apply now filtering according to the derived attributes and morphology to exclude outliers and extract connect components of similar regions at the same time (Section 3). In short, the proposed method can determine two types of change instead of combining the results of PSI and pixel-based methods. In addition, we are able to combine attributes of PS-points (e.g. geographic coordinates) and extended features like regular grids of PS at façades in further applications. In Section 4, the approach is demonstrated for the example of the inner city of Berlin, Capital of Germany, where in the last decade many new buildings have been erected whereas others have been torn down.

2. CHANGE DETECTION BASED ON PERSISTENT SCATTERER INTERFEROMETRY

PSI requires a set of N multi-temporal SAR images taken from same orbit covering a study area. A master image is selected to pair with the other slave images to form $N-1$ interferograms. The optimal master image is chosen based on small-baseline constraint (Berardino et al., 2002; Lanari et al., 2004) to reduce the phase noise (caused by geometrical and temporal decorrelations) in the interferograms. The phases of pixels in interferograms (indicated by p) are expressed as

$$\phi^p(x) = \phi_r^p(x) + \phi_a^p(x) + \phi_n^p(x) \tag{1}$$

where x indicates pixel, $\phi_r^p(x)$ is the phase term related to the line-of-sight (LOS) distance between SAR sensor and target, $\phi_a^p(x)$ models the atmospheric phase component, and $\phi_n^p(x)$ is due to noise.

To determine the LOS-deformation-velocities (LOS will be omitted for simplicity) and the heights of PS-points, $\phi_r^p(x)$ can be decomposed into

$$\phi_r^p(x) = C_v^p(x)v(x) + C_h^p(x)h(x) \tag{2}$$

where $v(x)$ and $h(x)$ are the deformation-velocity and the height with respect to some point of reference, and $C_v^p(x)$ and $C_h^p(x)$ are the coefficients of $v(x)$ and $h(x)$, respectively. Hypothesize $v(x)$ is linear, $C_v^p(x)$ is expressed as

$$C_v^p(x) = \frac{4\pi}{\lambda} B_T^p \tag{3}$$

where λ is the wavelength of SAR signal, and B_T^p is the temporal baseline of interferogram. $C_h^p(x)$ is expressed as

$$C_h^p(x) = \frac{4\pi}{\lambda} \cdot \frac{B_\perp^p(x)}{r(x)\sin\theta(x)} \tag{4}$$

where $B_\perp^p(x)$ is the normal baseline of the interferogram, $r(x)$ is the LOS distance between SAR sensor and target, and $\theta(x)$ is the local incidence angle based on reference ellipsoid. To reduce the unknown range in estimating $h(x)$, the topography-height $h_{topo}(x)$ is subtracted from $h(x)$ to obtain residual-height $\Delta h(x)$ (In this paper, $h_{topo}(x)$ was provided by ASTER Global Digital Elevation Model, ASTER GDEM, in this paper). In short, $C_v^p(x)$ and $C_h^p(x)$ can be calculated from system parameters of the SAR sensor. Subsequently, $\phi^p(x)$ is the phase observation for solving the unknown $v(x)$ and $\Delta h(x)$ in the following Periodogram.

The estimated $\hat{v}(x)$ and $\Delta\hat{h}(x)$ are obtained by the Periodogram often called temporal coherence (Ferretti et al., 2001)

$$\arg \max_{v(x),\Delta h(x)} \left\{ |r| \Big|_{[0\sim1]\in R} = \left| \frac{1}{N-1} \sum_{p=1}^{N-1} e^{j(\phi^p(x) - C_v^p v(x) - C_h^p \Delta h(x))} \right| \right\} \tag{5}$$

where high ensemble temporal coherence $|r|$ indicates high accuracy of estimation, j is $\sqrt{-1}$, and the searching ranges and iteration-increments for $v(x)$ and $\Delta h(x)$ depend on a priori information. In addition, temporal coherence also reflects the phase stability of a pixel through the whole multi-temporal SAR image stack. Consequently, PS-points are selected if their temporal coherences are higher than a threshold (0.75 was used in this paper as empirical value). The choice of the threshold is a trade-off between accuracy of estimated unknowns and amount of PS-points. In urban areas, this threshold can be set higher because dense PS-points are expected anyway.

To increase the accuracy of $\hat{v}(x)$ and $\Delta\hat{h}(x)$, a second estimation step is implemented by Periodogram (5) after the estimated $\hat{\phi}_a^p(x)$ is removed from $\phi^p(x)$. Based on (1) and (2), $\phi_a^p(x)$ can be expressed as

$$\phi_a^p(x) = \phi_{res}^p(x) - \phi_n^p(x) \tag{6}$$

where the residual phase $\phi_{res}^p(x)$ is modelled as

$$\phi_{res}^p(x) = \phi^p(x) - C_v^p \hat{v}(x) - C_h^p \Delta\hat{h}(x). \tag{7}$$

Term $\phi_a^p(x)$ is assumed to be temporally uncorrelated and spatially correlated (Ferretti et al., 2000; Ferretti et al., 2001; Hooper et al., 2004). Thus, $\hat{\phi}_a^p(x)$ is obtained from $\phi_{res}^p(x)$

by using a spatiotemporal filter which consists of low-pass filter in space and high-pass one in time. In summary, the accuracy in second estimation can be improved because the atmospheric interferences have been diminished in the phase observations.

2.1 Change Index

In order to carry some PS-point, the underlying structure must not undergo any changes other than LOS-motion for the entire period of time spanned by PSI processing; else the PS candidate is most probably sorted out. This means, any "big change" (BC), for instance, demolition or construction of a certain building, occurring in the considered lapse of time almost inevitably comes along with loss of PS for the entire site of change. However, for a certain shorter period of time there might be PS found from the first SAR image to the event of demolition or from the end of construction to the last image, respectively. In order to identify such events, we run two PSI processing steps to be compared later. Firstly, in the I-PSI case we process the entire SAR image set (I- indicates "inclusive of big change"). PS candidates which undergo any big change are assumed to be discarded due to low temporal coherences $|r^I(x)|$. Secondly, the F-PSI ("free of big change") case consists of a certain temporal sub set of SAR images without any big change. The temporal coherences $|r^F(x)|$ of points accordingly should not be affected by appearing or vanishing objects. In short, $|r^I(x)|$ and $|r^F(x)|$ should be close if no big change occurred; otherwise $|r^I(x)|$ is expected to be smaller than $|r^F(x)|$.

To retrieve BC-points, we introduce a change-index

$$\mathrm{CI}(x) \underset{[-1 \sim +1 \in R]}{=} \left| r^F(x) \right| - \left| r^I(x) \right| \qquad (8)$$

to quantify the possibility of big changes. Absolute values of CI close to 1 indicate strong hints to such modifications. Accordingly, BC-points are selected if their change-indices exceed a threshold. As a result, these BC-points are marked as big changes in a deformation-velocity image (I-PSI case) to form a new-style of change detection image (CD-image).

The city of Berlin still undergoes frequent activities in terms of renewal and construction of infrastructure, business districts, and residential buildings. In this paper, an area of the town is investigated where several buildings have been torn down recently (Figure 1). All PS-points in the F-PSI case are selected as BC-points if their change-indices exceed 0.3. At the time being this threshold was set manually. It may depend on many factors, for example, the number of SAR images and the a priori probability of big changes. The correlation between change-index and big changes will be explored in depth in future work.

3. OUTLIER-FILTERS

There are several causes of outliers such as: noise in phase observations, misfit of linear deformation-velocity hypothesis, inaccuracy of the topography-height model, inappropriate sampling of the Periodogram, under- or overestimated atmospheric phase item, and suboptimal setting of thresholds applied for extracting PS- and BC-points. Three types of outlier are described below as well as their corresponding removal

strategy. All filters are designed based on sliding-window operation.

Isolated outliers: We deal with objects like buildings, industrial plants and infrastructure of a certain minimal size. Therefore, we remove PS- or BC-point which are isolated inside a 5 x 5 pixel sliding window.

Inconsistent outliers: One type of point will emerge as an outlier when it is surrounded mostly by heterogeneous points as connected components are expected to comprise single object. For example, in practice, a PS-point is unlikely to last when a construction event (full of BC-points) covers it. The PS- or BC-points are removed if their amount within a 3 x 3 window is less 3.

PS-points with peculiar velocity estimates: A PS-point whose deformation-velocity is implausibly large or is singular compared with the other neighboring points is recognized as outlier. Two filters are developed here. Firstly, a PS-point is removed as outlier if its deformation-velocity exceeds the tolerance range: -2 to +2 (mm/year). Secondly, the average of and standard deviation of deformation-velocities are calculated within a 3 x 3 window. Then the center PS-point is deleted if the difference between its deformation-velocity and the average exceeds 0.5 mm/year or 3-multiple of the standard deviation.

4. RESULTS AND DISCUSSION

4.1 Materials and Study Area

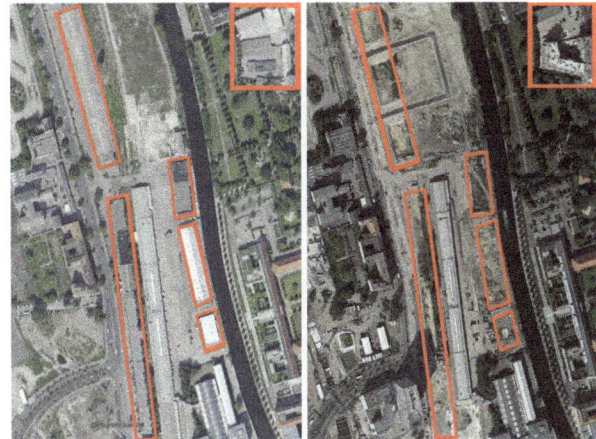

(a) 2010.09.12   (b) 2014.09.05

Figure 1. Study area (0.3 km^2) near the north of Berlin Central Station, Berlin. (a) Google-Earth image taken on 2010.09.12; (b) Google-Earth image, 2014.09.05. Red rectangles: construction-sites (the original buildings have been demolished but reconstruction if any is still in preparation stage).

In this experiment, we generated a CD-image presenting the construction-sites and deformation-velocities of steady buildings over Berlin. A cut-out from this image is used as the study area (Figure 1) in this paper. Here, the type of modification is restricted to demolished buildings. The construction-sites (Figure 1) are confirmed by using Ground truth in form of six historic images available in Google Earth. They have been taken in the acquisition period of the forty TSX-images (from 2010.10.27 to 2014.09.04). These images are used in the I-PSI and F-PSI cases. The ground resolution is resampled into 1 m.

In order to compare with pixel-based approaches, ratio and coherence-difference images are calculated (Figures 2(a, b)). They are generated based on (Rignot and van Zyl, 1993) and (Liu et al., 2001), respectively. Although the construction-sites can be at least partly recognized by naked eye, ubiquitous image noise renders any automatic pixel-based change detection to be very hard or even impossible.

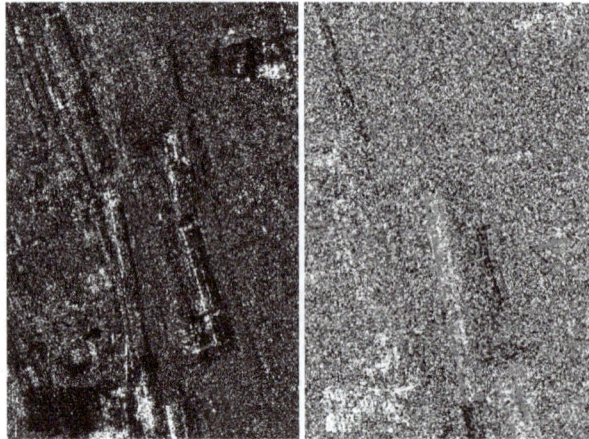

(a) Ratio image   (b) Coherence-difference image

Figure 2. Pixel-based change images. (a) Ratio image (ratio of first TSX-image (2010.10.27) to last one (2014.09.04)). (b) Coherence-difference image (subtract first coherence image from second one). First coherence image is generated from two TSX-images taken on 2010.10.27 and 2013.08.15; Second coherence image, 2013.08.15 and 2014.09.04. In both images white or black pixels indicate changes (i.e., large difference).

4.2   Results of PSI Processing

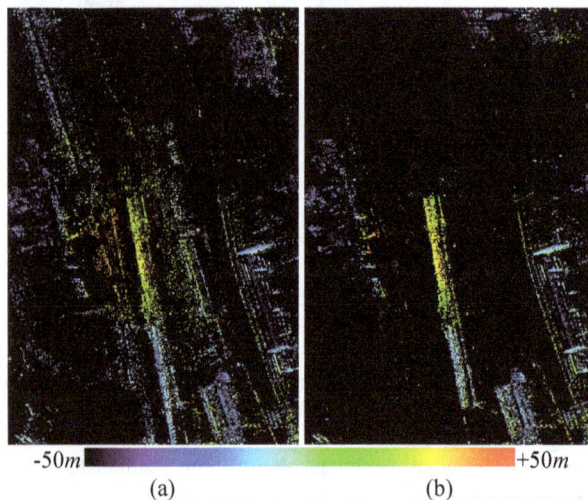

-50m ███████████████████ +50m
     (a)   (b)

Figure 3. (a) 24937 PS-points in F-PSI case; (b) 14652, I-PSI case. The height of the PS relative to the ASTER GDEM is coded in false color.

The I-PSI stack consists of the entire set of forty TSX-images in which the construction-sites (Figure 1) were captured. All demolished buildings have been torn down after the sixteenth TSX-image (2012.02.12) was acquired. Thus, the F-PSI case comprised these first sixteen TSX-images only to avoid degrading temporal coherences of points due to ongoing construction activities.

The PS-points (Figure 3) in the F-PSI and I-PSI cases show good agreement with the shape of the buildings. We observed

that 10285 PS-points on the construction-sites in the F-PSI case (Figure 3(a)) disappear in the I-PSI case (Figure 3(b)). Due to thresholding according to the temporal coherence value, the noise prone regions visible in Figure 2(a, b) are mostly suppressed now. This enables to continue the analysis by a sequence of low-level image processing methods which are described in more detail below.

4.3   Change-index Image

The change-index image (Figure 4) is generated by subtracting the temporal coherence image in I-PSI case from the one in F-PSI case. Generally, the change-indices of points on the construction-sites (Figure 1) exceed 0.3. Thus, such points tend to show up in yellow or red in the false-color change-index image. On the contrary, the points on the steady buildings have change-indices near to 0 indicated by light-green. Otherwise, the remaining points appear as noise-like pattern if they are not located on buildings. In short, we observe that from the change-index image construction-sites can be discriminated from endured buildings. Their positions and outlines are clearer visible compared to the ones in the ratio and coherence-difference images (Figures 2(a, b)).

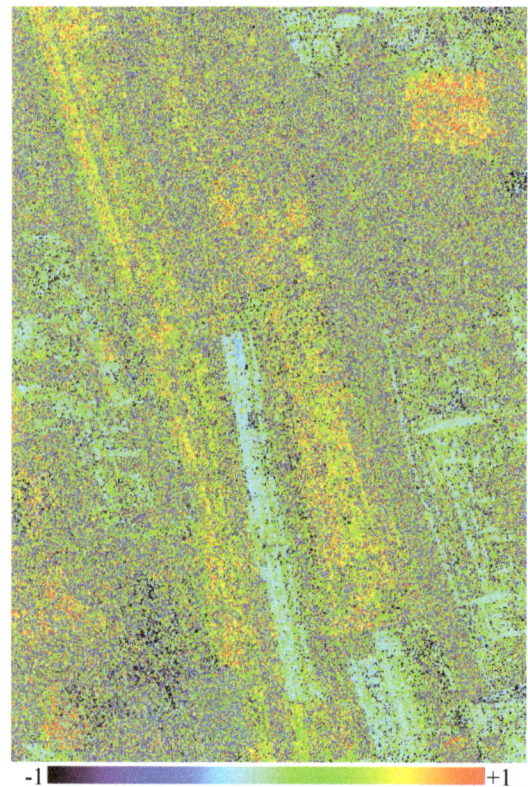

-1 ███████████████████ +1

Figure 4. Change-index image. Color table: change-index.

The 5318 BC-points (Figure 5) are extracted from the PS-points in F-PSI case (Figure 3(a)) if their change-indices are equal to or above 0.3. We observe that the number of these BC-points (5318) is less than the one of the disappeared PS-points (10285, Figure 3). In our opinion this mainly is due to the following reason. Instead of relying on a global hard threshold we use a relative measure which is more appropriate to adapt to local conditions. For example, a certain façade may be populated by small sub-structures that provide only weak PS-points. In such case we deal with stable signal of however moderate temporal coherence only. Hard thresholding might result in entire loss of such structure. This is avoided by the index which provides a relative measure. Nevertheless, the threshold of 0.3 ensures that we yield only significant BC-points.

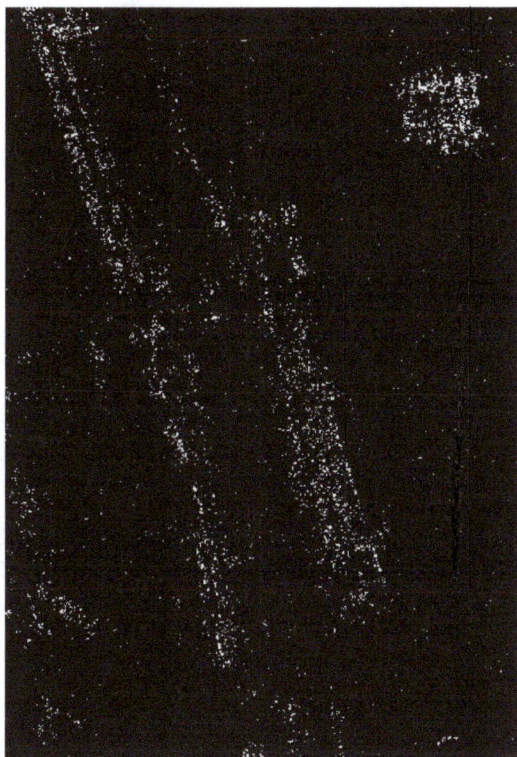

Figure 5. 5318 BC-points extracted from change-index image (Figure 4).

Overall, the positions of most BC-points (Figure 5) correspond to the construction-sites (Figure 1). However, we also recognize the noise-like BC-points (outliers) which need to be removed by the outlier-filters in the following steps.

4.4 New-style Change-detection Image

The new-style CD-image (Figure 6) is generated by adding the BC-points (Figure 5) to the deformation-velocity image (I-PSI case). This CD-image presents both the construction-sites (indicated by BC-points) and the deformation-velocities of steady buildings (indicated by PS-points). The positions of most BC-points are consistent with the construction-sites (Figure 1). In contrast to BC-points, PS-points are located on the steady buildings. Their deformation-velocities are quite uniform, which indicates plausible results. Overall, the distribution of BC- and PS-points is consistent with Ground Truth (Figure 1). However, we observe three types of outliers which have to be filtered in the subsequent steps:

- The PS- and BC-points outside the buildings.
- The PS-points on the construction-sites.
- The BC-points on the steady buildings.

The filtered CD-image (Figure 7) is generated by applying the outlier-filters described in Section 4 to the initial CD-image (Figure 6). The 2921 outliers have been removed from the 12003 BC- and PS-points, whereas the construction-sites and the steady buildings are maintained.

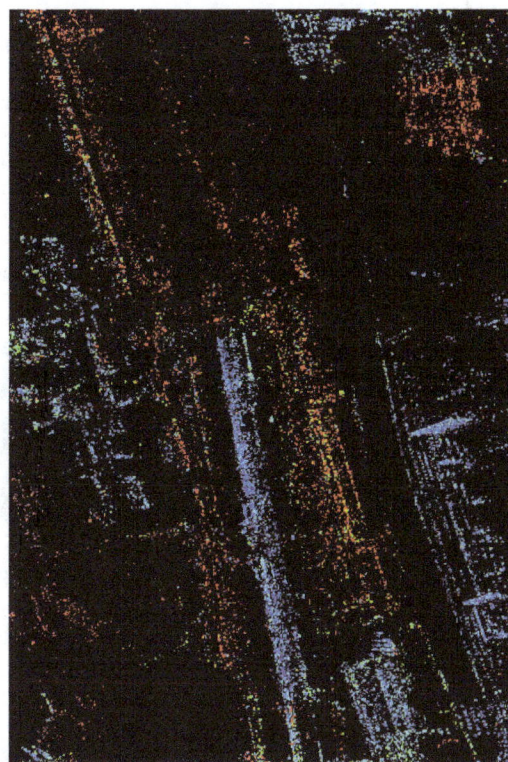

-5(*mm/year*)   +5(*mm/year*)

Figure 6. CD-image (before outlier-filterings). Color table, deformation-velocity of PS-point. Positive value in color table means deformation moves away from satellite; Negative value: towards. Red points: BC-points.

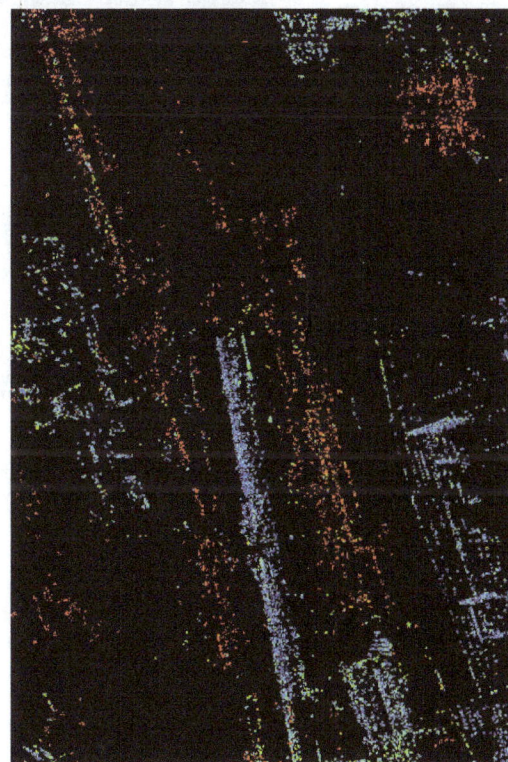

-5(*mm/year*)   +5(*mm/year*)

Figure 7. CD-image (after outlier-filterings). Color table, deformation-velocity of PS-point. Positive value in color table means deformation moves away from satellite; Negative value: towards. Red points: BC-points.

Three examples of outlier removal are discussed now using the details of the filtered CD-image (Figure 7). In Figure 8 a cut-out is depicted that shows an area covered by deciduous trees and lawn. Usually, in such environment hardly any PS-point occurs. The few and scattered PS- and BC-point candidates, left after PSI processing, have been successfully removed in the filtered image. Figure 9 shows an example where false PS-points on the construction-site have been removed. Finally in Figure 10 the BC-points on the steady building have been removed. Since few outliers still remain, developing more sophisticated filtering methods is required in future work.

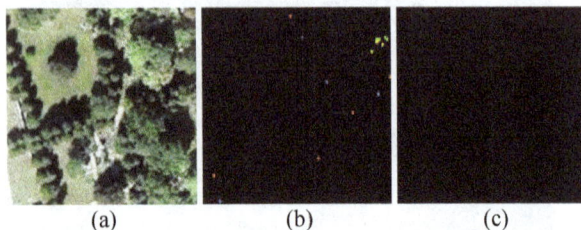

(a)   (b)   (c)

Figure 8. PS- and BC-points in vegetated area have been removed in filtered CD-image by outlier-filters. (a) Google-Earth image. (b) Cut-out of CD-image (Figure 6). (c) Cut-out of filtered CD-image (Figure 7). Non-red colors: PS-points, red: BC-points.

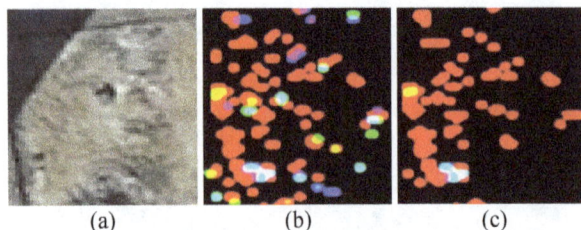

(a)   (b)   (c)

Figure 9. PS-points on construction-sites have been removed in filtered CD-image by outlier-filters. (a) Construction-site (Google-Earth image). (b) Cut-out of CD-image (Figure 6). (c) Cut-out of filtered CD-image (Figure 7). Non-red colors: PS-points, red: BC-points.

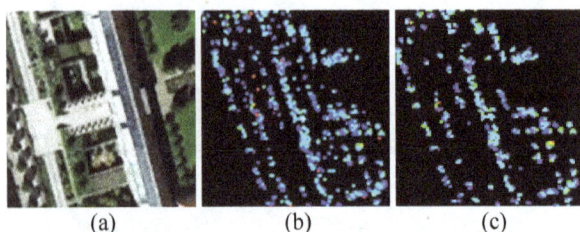

(a)   (b)   (c)

Figure 10. BC-points on steady buildings have been removed in filtered CD-image by outlier-filters. (a) Steady buildings (Google-Earth image). (b) Cut-out of CD-image (Figure 6). (c) Cut-out of filtered CD-image (Figure 7). Non-red colors: PS-points, red: BC-points.

In the lower left of Figure 7 we see some hints to changes highlighted in red not discussed so far. Comparing with Figure 1 we notice that some new buildings have been erected in this area. However, it is yet unknown when exactly these constructions took place, but it seems that those changes occurred at a later stage than those discussed so far. Also an image-by-image check of the entire SAR data stack did not provide more insight because of severe disturbance by noise even for the large building at the very bottom of the scene. Nevertheless, the change detection result reflects those changes at least to some extent.

4.5 Examples of Different Applications over Berlin

Here we demonstrate four examples to show the ability of the proposed method to monitor urban area. All of them are located in the inner city area of Berlin.

In the first example (Figure 11), the PS- and BC-points can be correctly recognized on steady buildings and construction-sites, respectively. The deformation-velocities on each building are homogeneous and therefore plausible. The sites with construction activities give rise to areas densely populated by BC-points.

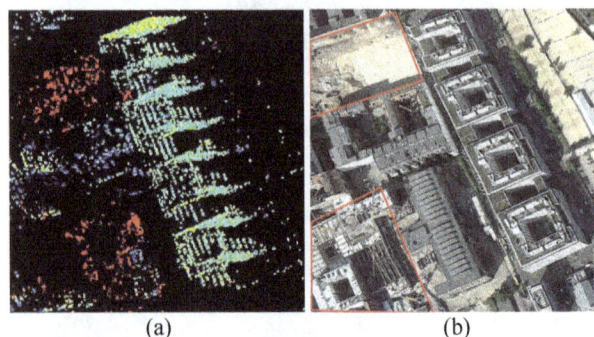

(a)   (b)

Figure 11. Example of monitoring buildings. (a) Cut-out of filtered CD-image covering Berlin. Color table, shown in Figure 7. Non-red colors: PS-points, red: BC-points. (b) Cut-out of Google-Earth image (2014.09.05). Red polygons: construction-sites.

In Figure 12 the area around Berlin Central Station is presented. The big change of roof-structure of the building at the bottom right is detected in addition to the other construction-sites. Moreover, the outline of waterways can also be recognized. This finding implies that big change along waterway (e.g. flood or extension) can be detected by the proposed method. A further example (Figure 13) is related to a sports and recreation area including a stadium (Friedrich-Ludwig-Jahn-Sportpark). Fortunately, there is no undesired motion of the stadium itself; however we can clearly see big changes due to construction at the right wing of the hall in the upper part.

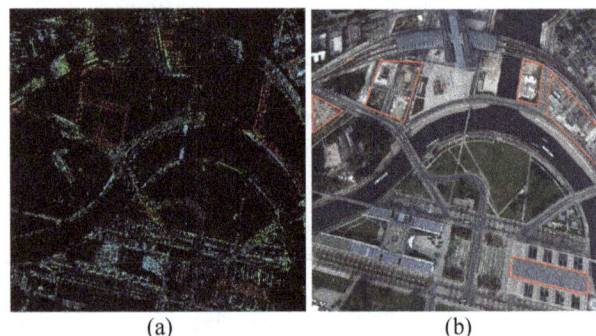

(a)   (b)

Figure 12. Example of monitoring area around Berlin Central Station. (a) Cut-out of filtered CD-image covering Berlin. Color table, shown in Figure 7. Non-red colors: PS-points, red: BC-points. (b) Cut-out of Google-Earth image (2012.05.20). Red polygons: construction-sites.

Finally, an interesting case is shown in Figure 14 where two round areas are highlighted. The left one indicated by the red circle coincides with so called Marx-Engels-Forum, a former central square dedicated to communist leaders, which currently undergoes modification leading to BC-points. The right one (blue circle) shows a monument called Neptunbrunnen located on an open place, where PSI indicates some LOS-motion of unknown origin.

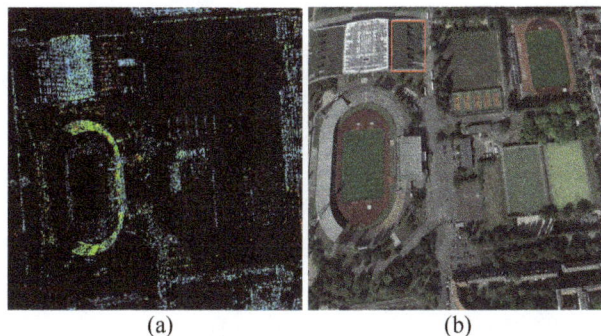

(a)   (b)

Figure 13. Example of monitoring a stadium (Friedrich-Ludwig-Jahn-Sportpark). (a) Cut-out of filtered CD-image covering Berlin. Color table, shown in Figure 7. Non-red colors: PS-points, red: BC-points. (b) Cut-out of Google-Earth image (2014.09.05). Red rectangle: construction-site.

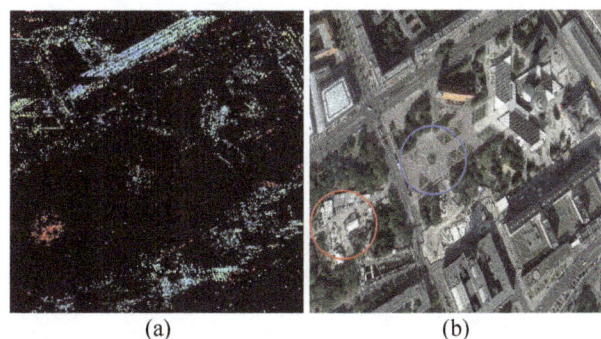

(a)   (b)

Figure 14. Example of monitoring round areas. (a) Cut-out of filtered CD-image covering Berlin. Color table, shown in Figure 7. Non-red colors: PS-points, red: BC-points. (b) Cut-out of Google-Earth image (2014.09.05). Red circle: construction-site of Marx-Engels-Forum, blue circle: Neptunbrunnen.

5. CONCLUSIONS AND FUTURE WORK

We proposed a method to combine PSI and pixel-based change detection from a given stack of suitable SAR images. In this manner we are able to detect both changes at object level (e.g., emergence or vanishing of buildings) and surface deformation. By combining these two complementing processes we can get a richer picture of ongoing changes in urban areas.

The findings presented in this paper represent outcomes of our first experiments. Of course, there is still much room for improvement; for instance, we have set a few thresholds manually. In future work, these thresholds shall be set according to analysis of the given data and a priori knowledge about the scene. At the time being, we have selected the period of time assuming no big changes occurred (F-PSI stack) globally and manually for our test area. In order to work towards an operational approach, this has to be done in a smarter manner: Firstly, we need a local approach since construction activities in different quarters are usually not correlated. Secondly, we want to find the optimal temporal partition in a data driven manner. This means we have to run several PSI frameworks spanning different time lapse for F-PSI and to find the best set-up for a certain part of the scene.

REFERENCES

Berardino, P., Fornaro, G., Lanari, R. and Sansosti, E., 2002. A new algorithm for surface deformation monitoring based on small baseline differential SAR interferogram. *IEEE Trans. Geosci. and Remote Sens.*, 40(11), pp. 2375-2382.

Colesanti, C., Ferretti, A., Novali, F., Prati, C. and Rocca, F., 2003. SAR monitoring of progressive and seasonal ground deformation using the permanent scatterers technique. *IEEE Trans. Geosci. and Remote Sens.*, 41(7), pp. 1685-1700.

Crosetto, M., Biescae, E., Duro, J., Closa, J. and Arnaud, A., 2008. Quality assessment of advanced interferometric products based on time series of ERS and Envisat SAR data. *Photogramm. Eng. Remote Sens.*, 74(4), pp. 443-450.

Dixon, T. H., Amelung, F., Ferretti, A., Novali, F., Rocca, F., Dokka, R., Sella, G., Kim, S. W., Wdowinski, S. and Whitman, D., 2006. Subsidence and flooding in New Orleans. *Nature*, 441, pp. 587-588.

Ferretti, A., Prati, C. and Rocca, F., 2000. Nonlinear subsidence rate estimation using permanent scatterers in differential SAR interferometry. *IEEE Trans. Geosci. and Remote Sens.*, 38(5), pp. 2202-2212.

Ferretti, A., Prati, C. and Rocca, F., 2001. Permanent scatterers in SAR interferometry. *IEEE Trans. Geosci. and Remote Sens.*, 39(1), pp. 8-20.

Ferretti, A., Fumagalli, A., Novali, A., Prati, C., Rocca, F. and Rucci, A., 2011. A new algorithm for processing interferometric data-stacks: SqueeSAR. *IEEE Trans. Geosci. and Remote Sens.*, 49(9), pp. 3460-3470.

Gernhardt, S. and Bamler, R., 2012. Deformation monitoring of single buildings using meter-resolution SAR data in PSI. *ISPRS J. Photogramm. Remote Sens.*, 73, pp. 68-79.

Gernhardt, S., Adam, N., Eineder, M. and Bamler, R., 2010. Potential of very high resolution SAR for persistent scatterer Interferometry in urban areas. *Ann. GIS*, 16(2), pp. 103-111.

Hooper, A., Zebker, H., Segall, P. and Kampes, B., 2004. A new method for measuring deformation on volcanoes and other natural terrains using InSAR persistent scatterers. *Geophys. Res. Lett.*, 31(23), pp. 1-5.

Hussain, M., Chen, D., Cheng, A., Wei, H. and Stanley, D., 2013. Change detection from remotely sensed images: from pixel-based to object-based approaches. *ISPRS J. Photogramm. Remote Sens.*, 80, pp. 91-106.

Kampes, B. M., 2006. *Radar Interferometry: Persistent Scatterer Technique.* Springer.

Lanari, R., Mora, O., Manunta, M., Mallorquí, J. J., Berardino, P. and Sansosti, E., 2004. A small baseline approach for investigating deformations on full resolution differential SAR interferograms. *IEEE Trans. Geosci. and Remote Sens.*, 42(7), pp. 1377-1386.

Liu, J., Black, A., Lee, H., Hanaizumi, H. and Moore, J. McM, 2001. Land surface change detection in a desert area in Algeria using multi-temporal ERS SAR coherence images. *Int. J. Remote Sens.*, 22(13), pp. 2463-2477.

Liu, G., Jia, H., Zhang, R., Zhang, H., Jia, H., Yu, B. and Sang, M., 2011. Exploration of subsidence estimation by persistent scatterer InSAR on time series of high resolution TerraSAR-X images. *IEEE J. Sel. Topics Appl. Earth Observ. in Remote Sens*, 4(1), pp. 159-170.

Mittermayer, J., Wollstadt, S., Prats-Iraola, P. and Scheiber, R., 2014. The TerraSAR-X staring spotlight mode concept. *IEEE Trans. Geosci. and Remote Sens.*, 52(6), pp. 3695-3706.

Monserrat, O., Crosetto, M., Cuevas, M. and Crippa, B., 2011. The thermal expansion component of persistent scatterer Interferometry observations. *IEEE Geosci. Remote Sens. Lett.*, 8(5), pp. 864-868.

Osmanoğlu, B., Dixon, T. H., Wdowinski, S., Cabral-Cano, E. and Jiang, T., 2011. Mexico city subsidence observed with persistent scatterer InSAR. *Int. J. Appl. Earth Obs. Geoinf.*, 13, pp. 1-12.

Perissin, D. and Ferretti, A., 2007. Urban target recognition by means of repeated spaceborne SAR images. *IEEE Trans. Geosci. and Remote Sens.*, 45(12), pp. 4043-4058.

Rignot, E. J. M. and van Zyl, J. J., 1993. Change detection techniques for ERS-1 SAR data. *IEEE Trans. Geosci. and Remote Sens.*, 31(4), pp. 896-906.

30

GENERALISATION AND DATA QUALITY

N. Regnauld

1Spatial, Tennyson House, Cambridge Business Park, Cambridge CB4 0WZ, UK, nicolas.regnauld@1spatial.com

Commission II, WG II/4

KEY WORDS: Data Quality, Quality Criteria, Automatic Generalisation, Validation, Fitness for Purpose

ABSTRACT:
The quality of spatial data has a massive impact on its usability. It is therefore critical to both the producer of the data and its users. In this paper we discuss the close links between data quality and the generalisation process. The quality of the source data has an effect on how it can be generalised, and the generalisation process has an effect on the quality of the output data. Data quality therefore needs to be kept under control. We explain how this can be done before, during and after the generalisation process, using three of 1Spatial's software products: 1Validate for assessing the conformance of a dataset against a set of rules, 1Integrate for automatically fixing the data when non-conformances have been detected and 1Generalise for controlling the quality during the generalisation process. These tools are very effective at managing data that need to conform to a set of quality rules, the main remaining challenge is to be able to define a set of quality rules that reflects the fitness of a dataset for a particular purpose.

1. WHO NEEDS SPATIAL DATA QUALITY?

The quality of spatial data is important to both the data producer and the user of the data. Someone using bad quality data will encounter difficulties in using it, introducing delays and costs if the data needs fixing before it can be used. Alternatively the user can alter the process it is using to cope with the shortfalls in the quality of the data it uses. This too has a cost. Sometimes the bad quality data is not spotted straight away, but leads to bad materials being produced from it, and potentially bad decisions made as a result. For the data producer, bad quality data means unhappy customers, who come back to them with complaints. This means additional and usually unplanned work required to fix it. Data quality is therefore a major concern to both the data producer and the data user. The ISO 19157:2013 (Geographic information - data quality) standard has been defined to facilitate the description and evaluation of quality of spatial data. The quality criteria defined by this standard are limited to checking that the data is geometrically and topologically clean, geometrically and semantically accurate, and does not contain omissions or commissions. This is however not sufficient to describe the quality of generalised data aimed for mapping, as these need to be readable at a given scale. For example, nothing in the ISO 19157:2013 checks that features are large enough to be seen on the target map or that the cartographic symbol of a road won't obliterate a building alongside it.

In this paper we discuss both the impact of data quality on generalisation and the impact of generalisation on data quality. We describe how 1Spatial products are dealing with data quality, to assess it, improve it and maintain it during processes such as generalisation. This leads us to a discussion on quality criteria, and the open challenge of finding a set of quality criteria that can reflect the fitness for purpose of a set of data.

2. DATA QUALITY AND GENERALISATION

Generalisation is a process that transforms data and is affected by data quality in two ways; Firstly it relies on the quality of the source data to operate properly: the source data must be suitable for the generalisation process. Secondly it needs to deliver generalised data fit for a purpose. The quality of the generalised data is an evaluation of how closely the data fulfils the specified requirements.

The simplest way to ensure that a dataset is suitable for a generalisation process is to design the generalisation process to cope with the state of the data, improving or enriching the data along the way when needed. For example, if the source data contains a river network which is not properly connected (no links for sections covered my manmade features, as often happens in urban environment), it will be difficult for a generalisation process to identify which river sections are part of the main channel. This could result in sections of the main channel being removed because they are small and seem isolated. One way to overcome this is to add a data enrichment process that will automatically look for breaks in the network and automatically deduce the missing links, sometimes using additional data sources. In a similar way, the easiest way to make sure the generalised data is fit for purpose, is to design a specific process. This is the approach which was followed for all the existing systems performing automatic generalisation in production today, like the production of OS VectorMap District at Ordnance Survey GB (Regnauld et al. 2013), the production of the 1:25000 maps at IGN France (Maugeais et al. 2013), the production of 1:50000 maps at the Dutch Kadaster (van Altena et al. 2013), or the production of 1:50000 and 1:100000 Maps in some German Länders (Urbanke and Wiedemann 2014). These bespoke systems are costly to develop and difficult to maintain or to reuse. The process needs to be heavily updated if the source data changes or if the required characteristics of the output changes.

Many studies to overcome this problem have been conducted, with the aim to build a system that can take formal descriptions of the requirements as input, and automatically adapt the process to achieve it. Research has been conducted in many relevant directions: how to formalise requirements and how to capture them. Assuming these requirements are available in a machine readable way, how can they be used to automatically adapt the generalisation process? Constraint based systems (Burghardt et al 2007), optimisation techniques (Sester 2005), multi agent model (Ruas and Duchêne 2011) among others have been proposed. A comprehensive review on the research available to date on these subjects can be found in (Burghardt et al 2014). As noted by the editors of the book in their concluding section, on-demand mapping remains one of the main challenges faced by the community of researchers on the subject.

2.1 Quality management during generalisation: 1Generalise approach

At 1Spatial, we also believe that we are not yet ready to produce a system that can automatically adapt to the source data and a formal description of the expected result. We definitely see it as an exciting area of expansion in the future though. We do have an Agent-based optimisation technology available in 1Generalise that can be used, but we are lacking the formal description of expected generalised results to use it effectively. 1Generalise provides predefined generalisation processes based on reusable components, working in both nationwide creation and change only update modes. The whole system is designed to facilitate the quick reuse of parts of existing systems (simple or complex) to build new ones. This also provides the building blocks to go further in the future and add to these components, the metadata that will allow more intelligent mechanisms to use them when they are relevant, progressing towards a system to build maps on demand.

In its current version, 1Generalise encapsulates a full generalisation process in a Flowline. Typically a Flowline will be built to derive a specific type of map, for example a general purpose topographic 1:25k map. The default Flowlines coming with 1Generalise are generic, they are data provider independent. They can then be extended or modified by a Flowline designer to meet their specific requirements. These Flowlines are made of a sequence of Subflows. Each Subflow is a sequence of Steps. Both Subflows and Steps are reusable components. A Subflow typically contains the steps required to generalise a theme. For example, a Subflow to generalise vegetation could be made of three steps: 1) amalgamate touching polygons, 2) remove small holes and 3) simplify the outline. Each of the steps encapsulates the logic that decides where the generalisation algorithms should be applied and validates the results. In this example, the amalgamation step would identify adjoining vegetation features and perform the amalgamation of their geometries. This step could also contain a rule that prevents the amalgamation of the two adjoining vegetation features if a fence runs along their boundary. The third step of this Subflow, in charge of simplifying the outline, could have a rule that checks that the result does not intersect other features and if so, either apply some corrective action or reject the result and simplify only parts of the features which are not close to the conflict area. The steps are editable using a rule language, so that the logic used to trigger algorithms and validate their results can easily be changed by the Flowline designer, without having to write code in a programming language like C or Java.

In this way, the steps can encapsulate pre and post conditions that allow the Flowline designer to integrate quality checks, and possibly corrective actions at any stage in the generalisation process. The actual generalisation algorithms, usually coded in C or Java, are triggered by these rules, so the Flowline designer has a high level of control over them without needing to understand how they are implemented.

2.2 Quality management pre and post generalisation: 1Validate and 1Integrate approaches

While managing the quality during generalisation is a good idea, it is not always possible or practical. Usually, during the generalisation, after a specific generalisation algorithm, we add quality checks to trap known potential side effects of the algorithm. If we were to track all possible problems after each application of each algorithm, the performance would quickly become unacceptable. Similarly, checking the conformance to standard for the input data for each algorithm would result in the same hit on performances. Not performing systematic full checks opens the door to problems though, as a slightly odd result from one algorithm ends up as input to the next which can then produce unexpected results. So it makes sense to have an initial pre-process that checks that the source data is of suitable quality, and a post-process that checks that the result meets the requirements. These are completed by targeted local checks performed within generalisation steps.

Both initial and final quality checks can be handled by 1Validate, 1Spatial's dedicated software to check the conformance of a dataset against a set of rules. These rules can be authored by the user, using a rule editor and a number of functions performing geometric and topologic checks. The screenshot in Figure 1 shows a 1Validate rule that checks that two distinct buildings do not intersect unless they simply touch (one special case of intersection where the interior of the geometries do not meet). The rules used in 1Generalise use the same syntax.

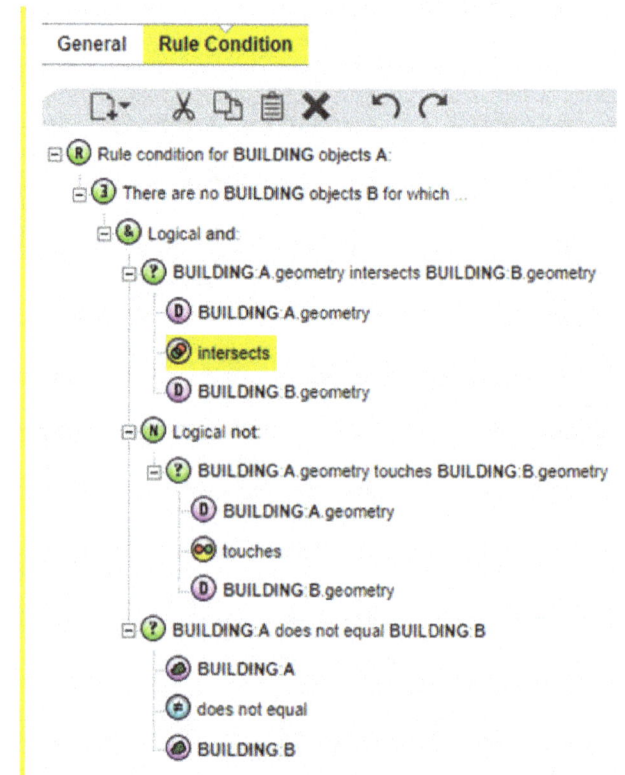

Figure 1: 1Validate rule editor

When validation failures are found in the source dataset, it must be corrected. This can either be done by manual intervention, using an editor, or sometimes automatically. For fixing issues automatically, 1Integrate can be used. 1Integrate is an extension of 1Validate that lets the user write actions, using the same but extended rule language as in 1Validate. The same rules can be written in 1Integrate, and they can be mapped to corrective actions. So for each object in the database that does not pass the rule, the mapped action will be triggered on it. These are particularly useful to fix topological issues, for example when network edges don't exactly meet, creating little gaps or overshoots. These are often not identifiable visually, unless using a level of zoom which will make the analysis of the whole dataset unpractical. While these issues are difficult to see, they will disrupt processes that perform network analysis tasks. For example, a generalisation process focusing on pruning a path network will usually preserve paths which are part of long routes. Bad connectivity could make the algorithm remove sections apparently disconnected to the rest while they are in fact important links in the network. 1Integrate can fix such issues in two steps. First a rule identifies all edges that end very close but not on another edge. For each instance of this, it triggers an action that moves the culprit edge end on the closest position on the nearby edge.

For validation failures on the generalised data, correction is rarely possible without the source data as reference. Usually, if the evaluation of the generalisation is not satisfactory, then either the generalisation process needs to be altered to avoid creating the issue, or a manual editing process can be added to correct the issue. Which way is the most cost effective depends on the frequency with which the problem occurs, the complexity of fixing the generalisation process to prevent it, and the cost of fixing it manually for the life time of the system.

3. QUALITY CRITERIA

While we have tools to express quality rules and to evaluate them against a set of data, identifying what quality criteria are required to ensure the input or generalised data are fit for purpose remains a challenge.

3.1 For input data

The ISO 19157:2013 standard categorises the quality criteria into five main families: Completeness, Topological consistency, Positional accuracy, Temporal consistency, Thematic accuracy. It also proposes methods to measure these. The main difficulty is to aggregate these measures in a way that can help decide if a set of data is fit for a particular use. While the criteria related to clean geometries (no double points, well formed polygons, no self intersection, etc.) and topology (no overshoots or undershoots at junctions, no overlaps, no double points; well formed polygons, no self intersection, etc.) can fairly easily be checked and sometimes automatically fixed, the others are more difficult to check, and would require a reference dataset. For example, looking at a single dataset, it is often impossible to detect that a feature is missing or misclassified. Exceptions exist, when semantic inconsistencies can be detected, for example, when a section of river has been misclassified as a road. The combination of the fact that a section of road is isolated from the rest of the road network and the fact that this section also connects two dead end river nodes could lead to the conclusion that this section of road has been misclassified and should be reclassified as a river. Writing such rules is extremely time consuming though, as combinations of potential misclassifications and the context in which they occur are almost endless. Such rules are worth writing once we have identified that a particular type of misclassification occurs frequently. Some simpler types of checks can still be done though, like checking that all the features have a valid height attribute if one is expected.

3.2 For generalised data

While the above criteria are also relevant to generalised data, interpreting the results can be difficult. By its nature, generalisation will reduce the content, so completeness will be affected. Formulas like the ones proposed by (Töpfer and Pillewizer 1966) can be used as a guide to how much features should be preserved. This is by no means a universal law, the thematic focus of the map can have a strong influence on the ratio of features that should be preserved. Positional accuracy is also affected by generalisation, as map features sometimes need to be exaggerated and displaced to make them readable at a smaller scale. So accuracy must be interpreted according to the scale of the map.. However, these criteria designed for reference data are not enough for evaluating the quality of generalised data. The main concern is to make sure that the characteristics of the data meet the requirements of their expected use. In particular, for generalised data aimed at mapping, the readability of the map also needs to be evaluated (Burghardt et al 2008).

For generalised data, the same set of criteria is therefore relevant but needs to be extended by a number of additional ones, which vary based on the expected use of the product. [João 1998] provides a complete review on the effect of generalisation, which is a good source of information to define additional criteria.

For a map product, criteria must be defined to reflect the readability of the map (minimum size, width of features, minimum distances between features, maximum density of features). This may not be sufficient though, a readable map might have been overly generalised and not contain the information required for its intended use. This requires additional criteria to be defined, they could relate to the positional accuracy of the features, criteria for their selection (based on individual characteristic or density measure over a given theme), the level of detail required (geometric and semantic). The difficulty is to find a set of criteria that reflects a specific targeted use and can be interpreted. Generalisation is an abstraction of the reality that produces a dataset which is a trade-off between the preservation of the reality and the readability of the result. What trade-off is adequate for what usage?

4. CONCLUSION

1Spatial has many experts in spatial data management and data quality. Through the 1Spatial Management Suite, 1Spatial proposes a set of software to capture spatial data, maintain a spatial database, integrate data from several sources, generalise data and publish map. In all these steps, validation plays a key role to ensure that the quality of the data is always under control.

Evaluating the quality of generalised data is an area where a lot more could be done. We are able to let our users define the criteria that they want and check them, but we would like to propose a set of predefined standard criteria that collectively provide a good insight on the quality of the generalised dataset. This could then be tweaked and extended to satisfy the specific requirements of each customer. Such standard criteria could then be integrated with the generalisation process itself, to progress towards the goal of building a system capable of performing on demand mapping. This is an active field of research. A workshop was organised in March 2015 by the International Cartographic Association commission on Generalisation and Multiple Representations to study the use of ontologies to formalise the knowledge required to support on demand mapping [Mackaness et al. 2015]. We did participate in this workshop and are most interested in pursuing this collaboration. One of its outcomes could be the definition of standard quality criteria and user requirements for generalisation.

REFERENCES

Burghardt, D., Duchêne, C., Mackaness, W. (Editors), 2014. *Abstracting Geographic Information in a Data : Methodologies and Applications of Map Generalisation.* Springer, Lecture Notes in Geoinformation and Cartography, Publications of the International Cartographic Association (ICA).

Burghardt, D., Schmid, S., Stoter, J. 2007. Investigations on cartographic constraint formalization. *In Proceedings of the 11th ICA workshop on generalisation and multiple representation.* Moscow, Russia, 2007.

Burghardt, D., Schmid, S., Duchêne, C., Stoter, J., Baella, B, Regnauld, N., Touya G, 2008. Methodologies for the evaluation of generalised data derived with commercial available generalisation systems. *In Proceedings of the 12th ICA workshop on generalisation and multiple representation.* Montpellier, France, 2008.

João, E.M., 1998. Causes and consequences of map generalization. London: Taylor & Francis

Mackaness, W., Gould, N., Bechhofer, S., Burghardt, D., Duchene, C., Stevens, R., Touya, G., 2015. Thematic workshop on building an ontology of generalisation for on-demand mapping. http://generalisation.icaci.org/images/files/workshop/ThemWor kshop/ThematicOntologyOnDemand_Paris2015.pdf

Maugeais, E., Lecordix, F., Halbecq, X., Braun A., 2011. Dérivation cartographique multi échelles de la BDTopo de l'IGN France: mise en œuvre du processus de production de la Nouvelle carte de base. *In Proceedings of the 25th International Cartographic Conference*, Paris, France, July 2011.

Regnauld, N., Lessware, S., Wesson, C., Martin, P., 2013. Deriving products from a multi-resolution database using automated generalisation at Ordnance Survey. *In Proceedings of the 26th Internationall Cartographic Conference.* Dresden, Germany, August 2013.

Ruas, A., Duchêne, C., 2011. A Prototype Generalisation System Based on the Multi-Agent System Paradigm. *In Generalisation of Geographic Information : Cartographic Modelling and Applications*, Mackaness, Ruas and Sarjakoski eds, pp. 269-284.

Urbanke, S., Wiedemann, A. , 2014. AdV-Project « ATKIS : Generalisation » - Map Production of DTK50 and DTK100 at LGL in Baden-Wurttemberg. *In Abstracting Geographic Information in a Data Rich World.* Burghardt, Duchêne and Mackaness Eds, Springer, pp. 369-373.

Sester, M., 2005. Optimization approaches for generalization and data abstraction. *International Journal of Geographic Information Science*, 19 (8-9), pp. 871-897.

Töpfer, F., Pillewizer, W., 1966. The principles of selection: a means of cartographic generalization. *The Cartographic Journal*, 3(1), pp. 10-16.

Van Altena, V., Nijhuis, R., Post, M., Bruns, B, Stoter, J., 2013. Automated generalisation in production at Kadaster NL. In *Proceedings of the 26th International Cartographic Conference*, Dresden, Germany, August 2013.

31

IMAGE BASED RECOGNITION OF DYNAMIC TRAFFIC SITUATIONS BY EVALUATING THE EXTERIOR SURROUNDING AND INTERIOR SPACE OF VEHICLES

A. Hanel[a], H. Klöden[b], L. Hoegner[a], U. Stilla[a]

[a]Photogrammetry & Remote Sensing, Technische Universitaet Muenchen, Germany - (alexander.hanel, ludwig.hoegner, stilla)@tum.de
[b]BMW Research & Technology, Muenchen, Germany - horst.kloeden@bmw.de

KEY WORDS: vehicle camera system, crowd sourced data, image analysis, machine learning, object detection, illumination recognition, traffic situation recognition

ABSTRACT:

Today, cameras mounted in vehicles are used to observe the driver as well as the objects around a vehicle. In this article, an outline of a concept for image based recognition of dynamic traffic situations is shown. A dynamic traffic situation will be described by road users and their intentions. Images will be taken by a vehicle fleet and aggregated on a server. On these images, new strategies for machine learning will be applied iteratively when new data has arrived on the server. The results of the learning process will be models describing the traffic situation and will be transmitted back to the recording vehicles. The recognition will be performed as a standalone function in the vehicles and will use the received models. It can be expected, that this method can make the detection and classification of objects around the vehicles more reliable. In addition, the prediction of their actions for the next seconds should be possible. As one example how this concept is used, a method to recognize the illumination situation of a traffic scene is described. This allows to handle different appearances of objects depending on the illumination of the scene. Different illumination classes will be defined to distinguish different illumination situations. Intensity based features are extracted from the images and used by a classifier to assign an image to an illumination class. This method is being tested for a real data set of daytime and nighttime images. It can be shown, that the illumination class can be classified correctly for more than 80% of the images.

1. OBJECT DETECTION AS BASIS FOR VEHICLE ADVANCED DRIVER ASSISTANCE SYSTEMS

Today, recognizing the position of objects around a vehicle is an important task for advanced driver assistance systems. Because of their high vulnerability, the focus of such systems should be put especially on pedestrians being in the instantly following drive way of a vehicle. Therefore, the estimation of the pedestrian's position with a high certainty is a valuable contribution to avoid accidents and injuries. In order to recognize pedestrians, modern vehicles are equipped with camera systems in many cases (Figure 1). On images of these cameras, object detectors can be applied in order to detect pedestrians in front of a vehicle.

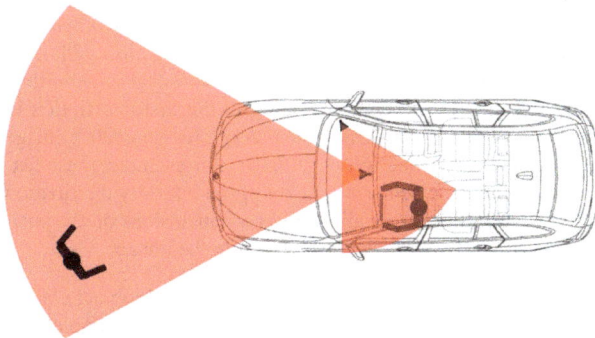

Figure 1: Position and orientation (black triangles) and field of view (red) of a exterior and interior looking vehicle camera used for recording of the scene in front of a vehicle and the driver, respectively. Images can be used to detect pedestrians (black top-down shape).

Before the detector can be used, an underlying model and the corresponding parameters must be determined. These values can be learned in a training step by evaluating a large set of training samples. These samples must contain pedestrians in all kinds of appearances, which should be detectable. To handle changes in the appearance of objects over the time, the detector must be learned incrementally. The necessary repeated recording and labeling of training data can easily take a few hundred hours and is therefore a key problem of incremental learning. It is another advantage of a series of recordings, that it can also be used to learn a motion model to predict the actions of pedestrians for the next seconds. Knowing the probable actions of pedestrians can be a contribution to avoid accidents.

In many cases, separate detector models are trained for different appearances within an object class. A high detection reliability for an individual object can only be achieved, if the detector uses a model trained on the same appearance as the individual object has. An important influence on the appearance of objects has the illumination of a scene (Figure 2). If the illumination goes down, this will lead to a decrease of the image contrast, which makes it more difficult or even impossible to recognize texture information from the image. Together with a lower signal-to-noise ratio, the reliability of gray value based and texture based features for detectors will be lowered. To be able to use the object detector with a suitable parameter set, it is necessary to recognize the typical illumination situation before.

To avoid accidents, it is advisable, not only to detect the objects around a vehicle, also the driver of the recording vehicle should be considered. If a driver is aware of the current traffic situation, he will for example be able to do an emergency braking, if another car from the side road misses a red traffic light. If the driver is not aware of the situation, he will probably not recognize the other car and therefore take the risk of a collision. To consider the driver in an advanced driver assistance system, it is useful to get information about the grade of his attention. Therefore, a vehicle camera looking at the driver can be taken.

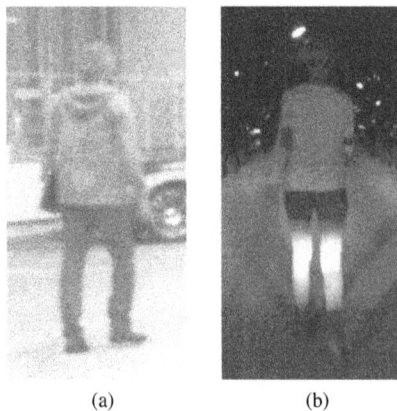

(a)   (b)

Figure 2: Images of a vehicle camera system with original resolution, but increased brightness. Different appearance of pedestrians: a) during a cloudy day, b) in the night.

Depending on the grade of attention, a driver assistance system can take different measures like showing an optical warning message to increase the situation awareness of the driver or do an automated braking to avoid a collision.

2. STATE OF THE ART

Recently, some projects handling crowd sourced data have started. Crowd sourced data can be described as big set of data obtained from a high number of sources. This strategy can be used to collect traffic data using several vehicles as sources (Jirka et al., 2013). For instance, the vehicle velocity can be obtained from sensors in vehicles. This data is acquired by many vehicles, sent with radio networks over the internet and collected on a central server (SmartRoad project, Hu et al. 2013a). In the SmartRoad project, vehicle status data acquired together with GPS data from 35 cars driving 6,500 km in total was used to detect and identify traffic lights and stop signs (Hu et al., 2013b). A similar approach with more than 200,000 traces from a vehicle fleet is used by Ruhhammer et al. (2014) to extract multiple intersection parameters like the number of lanes or the probability of turning maneuvers. In contrast, Wang et al. (2013) use the term "social sensing" for applications, where observations are collected by a group of sources, e.g. individuals and their mobile phones. They address the problem of possible unreliability of social sensing. As solution, they propose to introduce physical constraints, which for example allow to understand, which variables can be observed by a source at a certain location. Madan et al. (2010) show the high capability of social sensing systems by using only location information from mobile phones carried by individuals to derive information about their health status. With this data basis, they were able to analyse, that illness of a person leads to a change in behavior and movements during a day.

Machine learning is often used to evaluate a large amount of data. Supervised learning techniques base on the availability of ground truth information for all training samples. This ground truth information is not available for every learning task. Active learning is a strategy for supervised learning if ground truth information is only available for a small number of training samples or the effort for getting ground truth information is very high (Settles, 2010; Sivaranam & Trivedi, 2010). The basic idea of active learning is to obtain the highest possible amount of information for a classifier model by specific selection of only a number of training samples, for which ground truth data must be obtained. The selection is controlled by a selection function, which combines information about the distribution of all training data samples and the current classifier model. From this information, the expected improvement of the classification results is estimated and the ground truth information of the samples with the most influence on it requested by the classifier. The selection and training process is done iteratively. Wuttke et al. (2015) present a method, how the usefulness for each unlabeled training sample can be rated. They create a hypothesis based on analyses of the following three components. First, information about the structure of the unlabeled data is extracted. Furthermore the change of the hypothesis between before and after including ground truth information for new samples is considered. As the last part, the results of the prediction step of a classifier is compared with the ground truth information.

Machine learning techniques can also be used for object detection. A wide variety of standard methods for machine learning on features extracted from images exists already. In many cases, a combination of a descriptor describing the image information and a classifier separating these descriptions into different object classes is used. HOG-SVM is such a descriptor-classifier model used to detect pedestrians (Dalal and Triggs, 2005). This descriptor calculates for each image block-wise information about the orientation of dominant gray scale edges. The SVM classifier with a linear kernel allows to determine a hyper plane separating two classes in a passable time also for a high number of training samples. Dalal & Triggs (2005) use a model, which describes objects as one part. Felzenszwalb et al. (2009) extend the HOG-SVM descriptor by introducing part-based models, which can handle deformations within an object.

Besides the HOG features, in the literature there are several measures describing the quality of images, for example blur, noise or compression (Avcibas et al., 2002). For example, quality measures exist based on low-level image features like the Minkowski measure (De Ridder, 1992), which can be obtained from the pixelwise difference of image intensities. Furthermore, in Nill (1992) more complex methods based on the transformation of the image information into the frequency domain can be found. Another approach described by Saghiri et al. (1989) is the modeling of the human visual system (HVS), in order to derive quality measures from it.

Image features are for example histogram based features, as described by Ross (2010). They can be calculated from statistical measures (e.g. mean value, standard deviation) of the gray value histogram of an image. With these features, it is possible to characterize on the one hand the histogram itself and on the other hand the image. The combination of intensity information with their corresponding geometrical distribution in an image is described by central image moments, according to the first description in Hu (1962) and an analysis of their usage for geometric image transformations (Huang & Leng, 2010). For example, these features determine the center of gravity of the intensity distribution in an image and therefore allow to draw conclusions on the position and strength of light sources shown in the image.

A common classification technique to distinguish object classes represented by features is the Bayes classifier, whose implementation is described for instance by Fukunaga (1990). This classifier can be categorized into the group of supervised classifiers. The Bayes classifier determines for each data sample the probability of belonging to each object class. Finally, the sample will be associated to the object class with the highest probability. According to Sokolova & Lapalme (2009) and Congalton (1991), the overall accuracy (OA), the user's accuracy (UA) and the producer's accuracy (PA) are common reliability measures for the evaluation of the classification results.

3. CONCEPT FOR IMAGE BASED RECOGNITION OF DYNAMIC TRAFFIC SITUATIONS

In this section, the concept for image based recognition of dynamic traffic situations is shown. The functionality will be embedded in a client-server system (process flow see Figure 3).

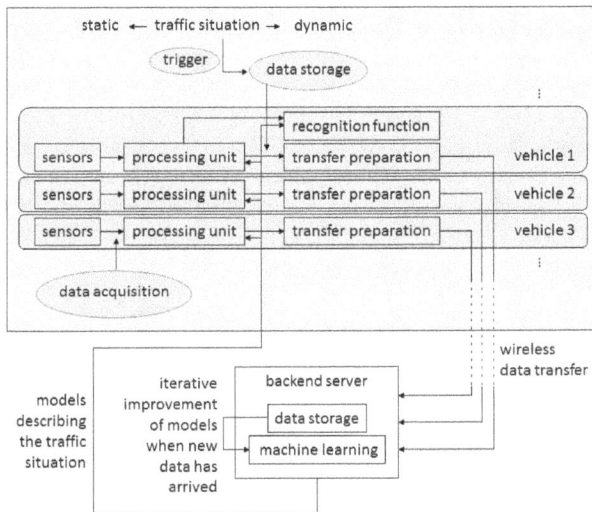

Figure 3: Process flow of the client-server system for image based recognition of dynamic traffic situations.

The key objective of this work is to develop new strategies to use unsupervised or semi-supervised machine learning techniques (e.g. active learning or transfer learning) based on image data in order to recognize dynamic traffic situations. In sum, several vehicles (clients) will acquire data about the traffic situation they are driving in and send it to a server. On the server, the data will be aggregated continuously and used iteratively for machine learning. The resulting machine learning models describing the traffic situation will be sent back to the vehicles and used there for the recognition of the current traffic situation.

3.1 Traffic situations and sensor setup

The recognition of traffic situations is based on images of cameras, which are mounted in vehicles (Figure 1). One camera in each vehicle will look at the driver and the other one will look at the street scene in front of the vehicle. The area aside or behind a recording vehicle will not be taken into account at the moment. In addition, internal measuring instruments (GPS receiver, IMU) of the vehicles will be used to acquire vehicle status data. This sensor setup will be integrated in a vehicle fleet, whereby all vehicles will get the same devices.

In this paper, a traffic situation consists of a recording vehicle and its driver as well as of the objects surrounding this vehicle. A traffic situation can be described as static or dynamic. The dynamic description contains all objects of the traffic situation and their current movements and their intentions. The static description can be seen as a single snapshot of the dynamic traffic situation. Only objects being around and present at the same time as the recording vehicle are relevant for the description of the traffic situation. The space, which is considered for other objects around the recording vehicle depends on the possibility, that these objects may influence the recording vehicle and has to be restricted

due to the range of the measuring instruments. Objects can also be dynamic (e.g. vehicles, pedestrians) or static (e.g. light poles, trees).

3.2 Data acquisition and triggering for storage

The first part of the processing chain is the acquisition of images and vehicle status data. This will be done by the vehicle fleet driving around for a certain time and capturing data about the traffic situations they are in. In other machine learning projects, this is done by only a single vehicle or a small number of vehicles due to the high cost and time efforts. For example, in a project described by Dollar et al. (2009), a single driver acquired data for approximately 10 hours containing only around 2,300 pedestrians (cf. data set used for training in chapter 4). By using a fleet of vehicles, a much larger amount of data can be obtained and used for machine learning.

After acquisition, the images as well as the vehicle status data will be transmitted to a processing unit in the vehicle. The acquisition and transmission of data will be done continuously with a frequency depending on the instruments (e.g. exposure time of a camera) and on the underlying physical principles (e.g. position from GPS is not needed as often as accelerations from IMU). The data collected in the processing unit will be used as input for the transfer preparation (see subsection 3.3).

The transmission to the transfer preparation unit is controlled by a trigger. Triggering will be started in road spots, which are interesting for a certain function of an advanced driver assistance systems. For example, intersections can be interesting road spots if the behavior of road users at intersections should be learned. Interesting road spots and features to recognize them must be defined prior to the data acquisition. The features should ensure a low false detection rate and therefore avoid manual intervention. For the example described above, triggering will start when an algorithm detects traffic lights or stop signs from images. Then the acquired data will be transmitted to the transfer preparation unit and the following processing tasks are started. Triggering will be stopped after a few seconds depending on the type of road spot. As of now, the acquired data will be deleted until triggering is started again.

3.3 Transfer preparation

The objective of this step is to reduce the amount of data, before it is transmitted (subsection 3.4) to and stored on the server. This step is necessary to be able to send the data wireless from the vehicles to the server, because restrictions on the data rates of the wireless transmission techniques must be met. Especially, this is important for memory consuming image data. As a further advantage, the reduction of data allows to send more different types of data and to increase the sending frequency of data samples. In this step, the number of data samples will not be reduced, because this would have an negative influence on the reliability of the machine learning.

The reduction can be performed by a bundle of methods and is done in each vehicle separately. First, already during the acquisition step, triggering is activated only for a few seconds at relevant road spots. Therefore, most data will never be prepared for transfer. In this step, the dimensionality of the data samples will be reduced. This will be done using data compression algorithms like compressive sensing (e.g. described in Baraniuk (2007)). This method can transform the representation of an image, so that only a fraction of the original memory space is needed, but the information loss is kept minimal. A further reduction of the dimensionality will be performed by the extraction of features, which describe relevant parts of the data. If extraction is performed, only the features will be stored, but not the acquired raw data. The feature extraction can be applied especially for memory consuming image data and is less relevant for memory saving numerical data.

3.4 Data transfer

After the preparation of the data in each vehicle of the fleet, it will be transferred wireless from the vehicle via the internet to the server and stored there. For transfer, either WiFi networks or mobile radio networks are used (Figure 4). The decision which technique is used is based on the availability of hot spots and on the vehicle velocity. The key problem of wireless data transfer is, that a data packet has to be sent again, if the connection to the hot spot gets lost while sending (e.g. because the vehicle is moving). The influence of this problem becomes bigger if the vehicle velocity increases and may prevent, that as much data is transferred as it is acquired.

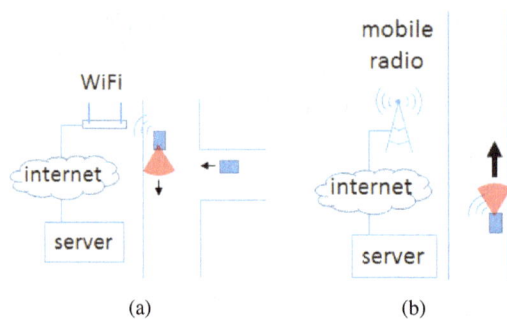

Figure 4: Techniques for wireless data transfer between vehicles and a server. The black arrows indicate the driving direction; the red circular segments indicate the field of view of a vehicle camera: a) WiFi, used for communication at low vehicle speed, e.g. at intersections, b) mobile radio, used at higher speeds or if WiFi is not available.

Transfer via WiFi (Figure 4a) uses public WiFi networks, which are offered in some cities. This method is preferred over the other one, because of higher data rates. A vehicle will search for WiFi networks, if the velocity is lower than a threshold around 10 km/h. This threshold ensures, that the connection does not get lost too fast due to the small covered area of a single WiFi hot spot, if the vehicle is moving fast. When having found a hot spot, the connection to it will be established and the data packets will be sent. The data transfer will end and the connection will terminate, when the velocity threshold is exceeded.

Transfer via mobile radio networks (Figure 4b) will only be used if no public WiFi network is available or the vehicle velocity is higher than the threshold. Due to the larger covered area of cellular hot spots, it is possible to transmit data with mobile radio networks for a longer time period even if the vehicle is moving.

3.5 Machine learning on the server

When data from the vehicles is available on the server, features used to model traffic situations have to be extracted from it. As an advantage compared to the processing unit in the vehicles, a higher computational power is available on the server, which allows to extract more complex features. Additionally, the feature calculation in the "post-processing" on the server can take more time than the processing of continuously acquired data allows.

As an example for low-level features used for object detection from images, HOG features (see section 2) describing gray value edges in images can be extracted. Also a sequence of images can be used to derive features. For example, information about the relative distance between the camera and an object can be derived from consecutive images. The position data of the vehicle obtained from the vehicle status data can integrate scale information. As an example for high-level features, the attention of the driver can be described by calculating the frequency of his eye blinks or the frequency how often he turns his head from images. These values and their changes allow to draw conclusions on the grade of attention.

These features are the input, on which new strategies for machine learning techniques will be applied. The first training step will take place, when a certain amount of data has arrived on the server. Up-to-date learning approaches like active learning or transfer learning will be used. With active learning, ground truth information has to be determined only for a small subset of all training samples (samples used for a pedestrian detector see Figure 8). In the first iteration, ground truth (example: does the sample contain a pedestrian or not?) will be provided for each training sample. In the following iterations, only the samples requested by the algorithm will be labeled. This reduces high labeling effort, which can take easily 40 times as much time as needed for data recording (Dollar et al., 2009). Transfer learning should be used to cover a wider variety of object classes.

The machine learning algorithms on the server will learn a model of the driver of the recording vehicle as well as models of the objects surrounding this vehicle. The mutual dependency of these two kinds of models will be considered. Especially, geometric and semantic links between the models will be taken into account for the learning and analysed subsequently. By using image sequences, the behavior of the driver and the objects will also be modeled. The resulting models describing the driver and the surrounding objects will be transmitted back to each vehicle of the fleet. There, they will be integrated into advanced driver assistance systems and used for the recognition function for dynamic traffic situations. Underlying models for the detection and classification of objects will also be sent to the vehicles and integrated in the existing detection and classification systems.

During and after the learning, the acquisition, preparation and transfer will continue. The learning algorithms will be applied again, when a certain amount of new data has arrived on the server. The models resulting from the iterative learning will be sent to the vehicles again. To make the strategy more efficient, the learning will be done incrementally with only the new data and update the already learned models.

3.6 Recognition of the illumination situation for object detection

To be able to handle the influence of the scene illumination on object detection, it is important to recognize the illumination in advance. In this subsection, a method to recognize the illumination situation of a traffic scene from images is described (process flow see Figure 5). Therefore, the client-server architecture of the proposed concept is used. Features which extract information about the illumination of the scene are derived from images on the central server. After learning of a model, each image can be assigned to an illumination class. This part of the proposed concept is already finished and evaluated in more detail.

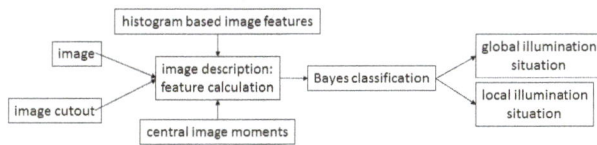

Figure 5: Flow chart for recognizing the illumination situation. For an image or image cutout, an image description is obtained. A classification based on this description leads to the illumination class for the global (whole scene) or local (only parts of the scene) illumination situation of the scene, which is shown in the image.

3.6.1 Definition of the term "illumination situation"

In this subsection, the term "illumination situation" describes the illumination of a scene, which consists of the combination of several light sources with different positions, orientations and power (example see Figure 6). The illumination of a scene can be influenced by natural and artificial light sources. Natural sources like the sun or the moon illuminate the scene widely. In the following text, the term "global illumination situation" is used, if a light source has influence on the illumination of the whole scene. The global illumination situation can be distinguished into daylight and nighttime. Artificial light sources can be for instance street lights or vehicle head lights, which illuminate a local area of the scene. The "local illumination situation" is defined as the combination of global and local light sources. In general, the illumination of a scene captured in an image is not uniform. Especially, this is important for images of night drives, which are taken in a scene with an illumination consisting of different light sources like vehicle head lights and street lights. If looking at different spots in the image, they will have different light exposures in the image depending on the viewing angle and the orientation of the camera and the current illumination situation in the scene.

Figure 6: Example image (gray scale), taken at night in a residential area. The illumination of the scene in the image is mainly influenced by the vehicle head lights. For better presentability, the contrast of the image was increased.

The wide variety of possible illumination situations is divided in a few illumination classes, in order to use different parameter sets for object detectors depending on the illumination. Illumination situations with low illumination can be separated in descending order into the three classes *night: urban area* (head light of recording and other vehicles and street lights), *night: residential area* (head light of recording vehicle and street lights) and *night: country road* (only head lights of recording vehicle). Illumination situations with bright illumination are summarized in the class *daylight*, which contains sunny, shadowed as well as clouded areas of the scene in one class.

3.6.2 Image description and classification

As image description, histogram based features as well as central image moments

(section 2) are used. Both feature groups consist of a high number of single features. The value for each single feature can be calculated for a certain image or a cutout of the image. The composition of all feature values for a certain image or image cutout forms the feature vector, which is used as the image description. To obtain information about the global illumination, the composition of the feature vector is done for the feature values of a whole image. To get information about the local illumination, this composition is done for an image cutout. The cutout is represented by a search window, which is moved in a sliding window approach with a grid shape over the whole image. The position, shape and size of this window can be influenced by the search window of an object detector. For instance for a pedestrian detector, a vertically aligned rectangle suits very well.

Finally, the feature vector can be evaluated with a classifier algorithm. It assigns an image or a cutout to an illumination class. For this purpose, the Bayes classifier (section 2) will be used. If the feature vector of the whole image is being evaluated, the illumination class for the global illumination situation is determined. If on the other hand only a local image cutout is being evaluated, the illumination class for the local illumination situation is determined.

In a previous step, the classifier must be trained on the server with a data set (see chapter 4) of images and image cutouts of all illumination classes. This step is made separately for whole images and for image cutouts. To use the data samples for training, the feature vectors must be calculated and the true class (ground truth) must be known. After this, it is possible to assign images or image cutouts without known ground truth to a certain illumination class using the classifier model. This assignment will be done standalone in vehicles for images acquired by the cameras.

By combining the global and local illumination class of a cutout, profound information about the illumination of the part of the scene shown in the cutout can be given. Thereby it is possible to link the global illumination of a scene with the local variations of the illumination within the scene.

3.6.3 Improvement of object detections

If the global and local illumination situation is known for a certain image cutout, it is possible to choose a suitable detector model for an object detector (example see Figure 7). An illumination specific detector model will give the object detector additional a-priori knowledge about the appearance of the objects and therefore improve the reliability of the detector. A look-up table can be used to link the illumination information and the detector model. This table stores for each combination of global and local illumination situation the information in a 2d matrix, which descriptor and classifier with which parameters fits best for object detection.

Figure 7: Vehicle camera image. Pedestrian detector already applied on the image. Bounding rectangles (multiple detections on different scales) of detected pedestrian in red, ground truth data in green.

4. EXPERIMENTS

For the recognition of the illumination situation (subsection 3.6), 32,000 images are recorded by the exterior vehicle camera. These images are used as data set for training and testing. Therefore, cutouts with a fixed size of 320x160 px are taken from these images. Positive cutouts (Figure 8a) are obtained using ground truth information and contain the picture of a single pedestrian with a certain background in each direction around the person. Negative cutouts (Figure 8b) are sampled randomly and contain the picture of arbitrary objects in the scene, but no pedestrians. Altogether, in all illumination classes around 18,000 positive and 60,000 negative image cutouts are used. Such an domination of negative samples is typical for object detectors, as shown by Dalal and Triggs (2005). Approximately 38% of all images are from the illumination class *night: residential area*, 29% from *night: urban area*, around 19% from *night: country road* and 14% from the *daylight* class. This imbalance is caused by the higher number of pedestrians in urban areas and therefore a higher importance of such areas for pedestrian detection. For all this data the correct class membership (ground truth) is known. For the whole images as well as for the two types of cutouts, the features are calculated and the feature vectors obtained separately. The total data set is divided randomly in a 80% part used for training and a 20% part used for validation of the detector.

(a)   (b)

Figure 8: Image cutouts used to train the pedestrian detector: a) positive cutout showing a pedestrian, b) negative cutout showing no pedestrians.

Having the feature vector, the proposed Bayes classifier will be trained with the training data. Finally, with the learned model, the validation data samples can be assigned to one of the illumination classes. The proposed classifier reliability measures can be calculated by comparing the classifier results with the ground truth information.

5. RESULTS AND DISCUSSION

In this part, the expected results of the recognition of the traffic situation and the recognition of the illumination situation are described.

5.1 Expected results of the recognition of the traffic situation

It can be expected, that the quality of the recognition of dynamic traffic situations will be improved with the proposed concept. For comparison, a recognition system based only on numeric data

will be considered (e.g. described in Ruhhammer et al. (2014)). In more detail, precise information about the current traffic situation should be obtained by the system in the vehicle. Furthermore, the detection and classification rate of objects will raise probably. The advantage of using image data compared to using only position data (e.g. vehicle trajectory from GPS) will be shown and evaluated. The completeness and the geometrical accuracy of the detected objects can be evaluated by comparing with manually labeled image data used as ground truth. The quality of the prediction of the intention of the driver and the intention of other traffic users should be analysed in respect with the results of the estimation of the attention of the driver. The evaluation of the prediction for few seconds can be made with the consecutively recorded image data. Furthermore, the different reduction methods should be compared considering the trade-off between the grade of reduction and loss of information. For evaluation, real use cases in the topic of vehicle safety will be applied. The decision between dangerous and non-dangerous traffic situations for a vehicle by using the recognition for dynamic traffic situations should be taken as example.

5.2 Results of recognizing the illumination situation

In the following part, the results of the illumination classification are shown. The method is being evaluated for positive cutouts (Table 1), negative cutouts (Table 2) and whole images (Table 3) with the proposed classifier reliability measures (see section 2).

Class	UA	PA	OA
Night: Residential Area	69.63%	86.21%	
Night: Urban Area	94.51%	67.47%	82.63%
Night: Country Road	66.81%	95.77%	
Daylight	98.31%	96.13%	

Table 1: Quality measures overall accuracy (OA), user's accuracy (UA) and producer's accuracy (PA) for evaluation of the illumination recognition for positive image cutouts.

Class	UA	PA	OA
Night: Residential Area	77.68%	77.87%	
Night: Urban Area	83.44%	85.21%	86.12%
Night: Country Road	58.62%	48.30%	
Daylight	97.34%	95.42%	

Table 2: Quality measures overall accuracy (OA), user's accuracy (UA) and producer's accuracy (PA) for evaluation of the illumination recognition for negative image cutouts.

Class	UA	PA	OA
Night: Residential Area	89.42%	99.33%	
Night: Urban Area	99.46%	88.45%	96.01%
Night: Country Road	98.19%	100%	
Daylight	100%	99.78%	

Table 3: Quality measures overall accuracy (OA), user's accuracy (UA) and producer's accuracy (PA) for evaluation of the illumination recognition for whole images.

The results show, that the proposed method is able to do the illumination recognition for whole images nearly without errors. The overall accuracy has a value of 96%, which means, that approximately only every 20th image will be assigned to a wrong class. For image cutouts, no matter whether positive or negative, slightly lower values of the overall accuracy can be obtained. By comparing among each other, the classification of the negative cutouts has a little bit better results as the classification of positive cutouts. This difference can probably be drawn back to the strongly varying appearance of pedestrians in the image fore-ground in the positive cutouts. Especially, pedestrians can have bright or dark cloths. On the other hand, the negative cutouts showing an arbitrary part of a scene contain more often objects from the background of the scene, like parked cars or the street surface. Such objects have especially in night images a lower variety in the appearance.

Confusions between the classes occur mainly between the different night classes, what can be seen at the lower values for UA and PA compared to the *daylight* class. In this context, it is re-markable, that the class *night: urban area* has a far lower value for PA compared to the other classes for positive cutouts. In-stead, the other two night classes have a notable lower value for the UA. This difference in the values for the UA and PA shows, that samples of the class *night: urban area* are assigned to one of the other two night classes sometimes. For instance, a pedestrian shown in a cutout belonging in fact to the class *night: urban area* is illuminated directly and strongly by the head light of a vehicle, but only hardly by street lights. Then the illumination classifier might assign this cutout with a high probability to the class *night: country road* or *night: residential area*. The reason may be, that for these two classes pedestrians are typically illuminated only by the vehicle head light, but not by other light sources. This conclu-sion can be turned around for negative cutouts. This means, that negative cutouts belonging in fact to the class *night: urban area* are assigned to this class correctly in the vast majority of cases, which can be seen from the high value for the PA. Instead, for the other two night classes, this value decreased a lot.

Compared to the cutouts, for whole images wrong classifications do not occur for any of the three night classes in a notable number, what can be seen from the permanently high values of UA and PA. All kinds of images and cutouts have it in common, that the *daylight* class can be assigned correctly with a higher reliability than the night classes.

For a pedestrian detector based on the HOG-SVM model, using an unique parameter set for all illumination classes gives an av-erage precision of around 28% for the pedestrian detection. Sep-arate models for each illumination class give values from 65% up to 85% for the night classes. Only the average precision of the *daylight* class remains with around 23% at a low level. This means, that the quality of the pedestrian detection can be im-proved by considering the illumination situation for the selection of the detector parameters. A more detailed discussion of the re-sults of the pedestrian detector is provided in a preceding master's thesis (Hanel, 2015).

6. CONCLUSION

In this article, a concept to recognize dynamic traffic situations from images is described. The concept proposed to mount cam-eras in vehicles to use them to observe the driver as well as the objects around the vehicle. The images will be acquired by a vehicle fleet and aggregated on a server. On the server, differ-ent types of low-level and high-level features will be extracted from images and used as input for learning. New strategies for machine learning will be used to learn models describing traffic situations. After back-transfer of the models to the vehicles of the fleet, they will be used there for standalone recognition of traffic situations. It can be expected, that the models allow to separate different situations reliably. The prediction of actions of road users for the next seconds should be possible. Further, the mod-els will probably make the detection and classification of objects around the vehicles more reliable. One of the learned models is

used to recognize the illumination situation of a scene from im-ages. Therefore, low level features calculated from an image or an image cutout are used by a classifier to assign the image to an illumination class. With this information, it is possible to use specific parameters for an object detector depending on the rec-ognized illumination situation. The results show, that only sim-ple image features based on the image intensities are necessary to recognize the correct illumination situation. No transforma-tions with a high computational effort or features using reference images are needed. The quality of the illumination recognition is very high for whole images as well as for cutouts of images. The results of the illumination recognition lead to a higher over-all accuracy and reliability of a pedestrian detector compared to leaving out the recognition of the illumination situation. It can be shown, that the greatest negative influence on the reliability of the illumination classification is caused by confusions between night time classes.

Future work on this part of the concept can be done on finding features, which allow a more robust classification. Additionally, more illumination classes should be identified and distinguished from the current ones in order to increase the overall accuracy. In this article, for daylight scenes only one illumination class was taken. It can be assumed, that for example frontlighting (low po-sition of the sun in the evening) or drives through tunnels can cause special illumination conditions, which should be captured by additional illumination classes. Also different weather con-ditions (e.g. fog, rain, snowfall) can lead to special illumination conditions, which should be evaluated.

REFERENCES

Avcibas, I., Sankur, B. and Sayood, K., 2002. Statistical evalu-ation of image quality measures. *Journal of Electronic Imaging*, 2002. Vol. 11(2), pp. 206-223.

Baraniuk, R., 2007. Compressive Sensing [Lecture Notes]. *IEEE Signal Processing Magazine*, Vol. 24(4), pp. 118-121.

Congalton, R., 1997. A review of assessing the accuracy of classi-fication of remotely sensed data. *Remote Sensing of Environment*, Vol. 37(1), pp. 35-46.

Dalal, N. and Triggs, B., 2005. Histograms of oriented gradients for human detection. In: *Computer Vision and Pattern Recogni-tion, 2005. CVPR 2005. IEEE Computer Society Conference on.* Vol. 1, pp. 886-893.

De Ridder, H., 1992. Minkowsky Metrics as a Combination Rule for Digital Image Coding Impairments. In: *Human Vision, Visual Processing, and Digital Display III, 1992*, pp. 17-27.

Dollar, P., Wojek, C., Schiele, B. and Perona, P., 2009. Pedes-trian detection: A benchmark. In: *Computer Vision and Pattern Recognition, 2009. CVPR 2009. IEEE Computer Society Confer-ence on*, pp. 304-311.

Felzenszwalb, P.F., Girshick, R.B., McAllester, D., Ramanan, D., 2010. Object Detection with Discriminatively Trained Part-Based Models. In: *Pattern Analysis and Machine Intelligence, IEEE Transactions on*, Vol. 32(9), pp. 1627-1645.

Fukunaga, K., 1990. Introduction to Statistical Pattern Recogni-tion. Academic Press Processional Inc, Waltham, Massachusetts, USA.

Hanel, A., 2014. Bestimmen der Beleuchtungsverhältnisse und Optimieren der Fußgängerdetektion aus Bildern eines

Fahrzeugkamerasystems. Master's thesis. Technische Universität München, Faculty of Civil, Geo and Environmental Engineering, Photogrammetry and Remote Sensing.

Hu, M.-K., 1962. Visual pattern recognition by moment invariants. In: *Information Theory, IRE Transactions on*. Vol. 8(2), pp. 179-187.

Hu, S., Liu, H., Su, L., Wang, H. and Abdelzaher T.F., 2013a. SmartRoad: A Mobile Phone Based Crowd-Sourced Road Sensing System. Technical Report, University of Illinois at Urbana-Champaign.

Hu, S., Liu, H., Su, L., Wang, H. and Abdelzaher T.F., 2013b. SmartRoad: A Crowd-Sourced Traffic Regulator Detection and Identification System. Technical Report, University of Illinois at Urbana-Champaign.

Huang, Z., Leng, J. 2010. Analysis of Hu's Moment Invariants on Image Scaling and Rotation. In: *Computer Engineering and Technology, 2010 2nd International Conference on*. Vol. 7, pp. 476-480.

Jirka, S., Remke, A., Bröring, A., 2013. enviroCar - Crowd Sourced Traffic and Environment Data for Sustainable Mobility. Technical Report, 52°North Initiative for Geospatial Open Source Software GmbH.

Madan, A., Cebrian, M., Lazer, D., Pentland, A., 2010.Social Sensing for Epidemiological Behavior Change. In: *Proceedings of the 12th ACM International Conference on Ubiquitous Computing*, pp. 291-300.

Nill, N., 1992. Objective image quality measure derived from digital image power spectra. *Optical Engineering*, Vol. 31(4), pp. 813-825.

Ross, J., 2010. Introductory Statistics. Elsevier Science, Amsterdam, Netherlands.

Ruhhammer, C., Hirsenkorn, N., Klanner, F. and Stiller, C., 2014. Crowdsourced intersection parameters: A generic approach for extraction and confidence estimation. In: *Intelligent Vehicles Symposium Proceedings, 2014 IEEE*, pp. 581-587.

Saghiri, J., Cheatham, P. and Habibi, A., 1989. Image quality measure based on a human visual system model. *Optical Engineering*, Vol. 28(7), pp. 813-818.

Settles, B., 2010. Active Learning Literature Survey. Technical Report, University of Wisconsin-Madison.

Sivaraman, S., Trivedi, M.M., 2010. A General Active-Learning Framework for On-Road Vehicle Recognition and Tracking. In: *Intelligent Transportation Systems, IEEE Transactions on*, Vol. 11(2), pp. 267-276.

Sokolova, M. and Lapalme, G., 2010. A Systematic Analysis of Performance Measures for Classification Tasks. *Information Processing & Management*, Vol. 45(4), pp. 427-437.

Wang, D., Abdelzaher, T., Kaplan, L., Ganti, R., Hu, S., Liu, H., 2013. Exploitation of Physical Constraints for Reliable Social Sensing. In: *Real-Time Systems Symposium (RTSS), 2013 IEEE 34th*, pp. 212-223.

Wuttke, S., Middelmann, W. and Stilla, U., 2015. Concept for a compound analysis in active learning for remote sensing. In: *PIA15+HRIGI15 – Joint ISPRS conference*, Vol. XL-3/W2, pp. 273-279.

32

SIMPLE APPROACHES TO IMPROVE THE AUTOMATIC INVENTORY OF ZEBRA CROSSING FROM MLS DATA

P. Arias [a], B. Riveiro [b], M. Soilán [a], L. Díaz-Vilariño [a], J. Martínez-Sánchez [a]

[a] Applied Geotechnologies Group, Dept. Natural Resources and Environmental Engineering, University of Vigo, Campus Lagoas-Marcosende, CP 36310 Vigo, Spain (parias, msoilan, lucia, joaquin.martinez)@uvigo.es
[b] Applied Geotechnologies Group, Dept. Materials Engineering, Applied Mechanics and Construction, University of Vigo, Campus Lagoas-Marcosende, CP 36310 Vigo, Spain – belenriveiro@uvigo.es

Commission III, WG III/4

KEY WORDS: Point clouds, LiDAR, Automatic Processing, Traffic signs, Road Inventory, 3D Modelling.

ABSTRACT:

The city management is increasingly supported by information technologies, leading to paradigms such as smart cities, where decision-makers, companies and citizens are continuously interconnected. 3D modelling turns of great relevance when the city has to be managed making use of geospatial databases or Geographic Information Systems. On the other hand, laser scanning technology has experienced a significant growth in the last years, and particularly, terrestrial mobile laser scanning platforms are being more and more used with inventory purposes in both cities and road environments. Consequently, large datasets are available to produce the geometric basis for the city model; however, this data is not directly exploitable by management systems constraining the implementation of the technology for such applications.

This paper presents a new algorithm for the automatic detection of zebra crossing. The algorithm is divided in three main steps: road segmentation (based on a PCA analysis of the points contained in each cycle of collected by a mobile laser system), rasterization (conversion of the point cloud to a raster image coloured as a function of intensity data), and zebra crossing detection (using the Hough Transform and logical constrains for line classification). After evaluating different datasets collected in three cities located in Northwest Spain (comprising 25 strips with 30 visible zebra crossings) a completeness of 83% was achieved.

1. INTRODUCTION

Cities are increasingly large and complex areas that require integrated technologies for an effective management and to ensure productivity, continued economic growth and environmental sustainability. The city management is increasingly supported by information technologies, leading to paradigms such as smart cities, where decision-makers, companies and citizens are continuously interconnected. 3D modelling turns of great relevance when the city has to be managed making use of geospatial databases or Geographic Information Systems. In this sense, the new concepts of 3D modelling not accounting only for geometry but also for semantic and topologic data became of essential importance. Standard schemas such as cityGML contribute to define how 3D models need to be structured in order to be used under an interoperable perspective. On the other hand, laser scanning technology has experienced a significant growth in the last years, and particularly, terrestrial mobile laser scanning platforms are being more and more used with inventory purposes in both cities and road environments. Consequently, large datasets are available to produce the geometric basis for the city model, however, this data is not directly exploitable by management systems constraining the implementation of the technology for such applications.

In the last times, an intense activity on automating data processing has been reported by the literature. Serna and Marcotegui (2014) published an interesting review of methods that focused on urban environments, including several approaches for tackling the detection, segmentation, and classification of urban objects. Intense work in the detection an classification of road markings has been published in the last years (Hervieu et al., 2015; Kumar et al., 2014; Guan et al.,

2014). Different approaches such as the creation of images (Zhu et al., 2010), voxelization (Douillard et al., 2011), mathematical morphology (Hernández and Marcotegui, 2009) may contribute to the detection of urban objects from point cloud data. Classification requires more complex approaches after extracting appropriate descriptors for objects. An example is the usage of decision trees with support vector machine (SVM) algorithms as proposed by (Owechko et al., 2010; Golovinskiy et al., 2009).

This work presents different approaches to automatically detect and classify urban objects such as traffic signs and markings from large LiDAR datasets. Particularly, it will be shown how reducing data to 2D space can contribute for an efficient and robust segmentation of the 3D data. Also, using rasterization approaches permits the application of well-validated image processing algorithms (such as mathematical morphology, the Generalized Hough Transform, clustering algorithms, etc.) that contribute to a more robust segmentation of point clouds.

In the case of road marking, an algorithm was developed that consists of several subsequent processes starting with a road segmentation by performing a curvature analysis for each laser cycle, then intensity images are created in order to detect zebra crossing using the Generalized Hough Transform and logical constrains. To optimize the results, image processing algorithms are applied including binarization, median filtering, and mathematical morphology operations. Once the road marking is detected its position and orientation are calculated with inventorying purposes using Geographic Information Systems.

Real scenarios are scanned using the Mobile Laser Scanners (on-the-fly mode) and used to test the methods and algorithms presented in the paper. The initial results are promising, with efficiency ratios over 83%. The accuracy of the results, as well as the acceptable computation time, recommend extending the approaches developed in this paper for traffic signs to other objects of the city. Additionally, by appropriately structuring the segmentation results using the cityGML schema eases the implementation of the MLS technology for the extensive inventory and 3D modelling of cities and their components.

2. METHODOLOGY

The main steps of the algorithm for automatic data extraction of zebra passes position and orientation are shown in figure 1. Next, the main processes of the algorithm are described which were implemented using Matlab software.

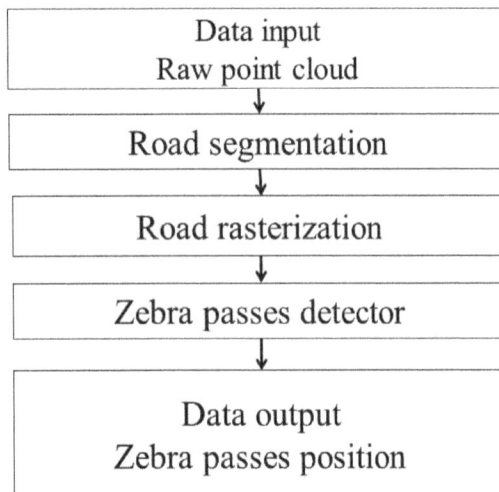

Figure 1. General workflow of the proposed algorithm for zebra crossing detection.

2.1 Road Segmentation

The principal component analysis (PCA) is a mathematical procedure that seeks to reduce the dimensionality of a set of variables to a new set of variables (principal components) that are linear combinations of the initial variables and are also uncorrelated to each other. In this work, a PCA is used in order to detect geometric changes into the profile defined by the points contained in each scanner line (each cross-section generated by the scanner when rotating 360°) [13, 15].

Since the data of each cross-section is represented into 2D space, analysis is performed using the altitude of each point (Z coordinate) and deflection angle. These two variables allow to easily detecting peaks denoting the transversal limits of the road when PCA analysis is performed into the local neighbourhood (10 points) of each of the points contained in each scanner cross-section. Figure 2 shows the value of the second eigenvalue in the PCA analysis (using 2 variables: z coordinate and deflection angle) with respect to deflection angle. As can be seen, in the range of angles than correspond to road the second eigenvalue is almost zero, and peaks appear at those points corresponding to the borders of the road.

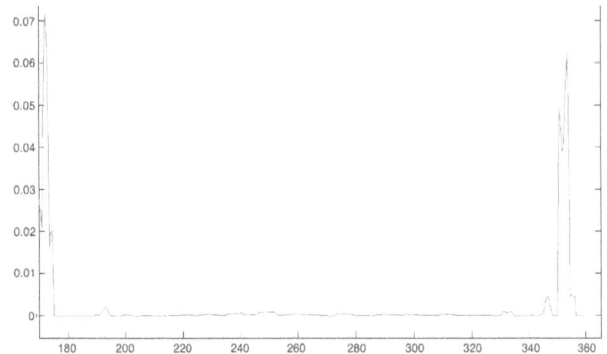

Figure 2. Segmentation of road points using a peak detection based on the second eigenvalue during PCA analysis of a cross section of the road.

Figure 3a presents an example of point cloud to be segmented. After applying the filter to each individual cross section of the road, an irregular segmentation may happen in those cross sections where other objects alter the profile of the street to the segmentation does not happen in a coherent manner as shown in figure 3b. To improve the segmentation avoiding the effect caused by cars or other objects in the street, the angle of segmentation was averaged having into account the precedent cross-sections. Figure 3c shows the sample point cloud after correcting the angle of partition.

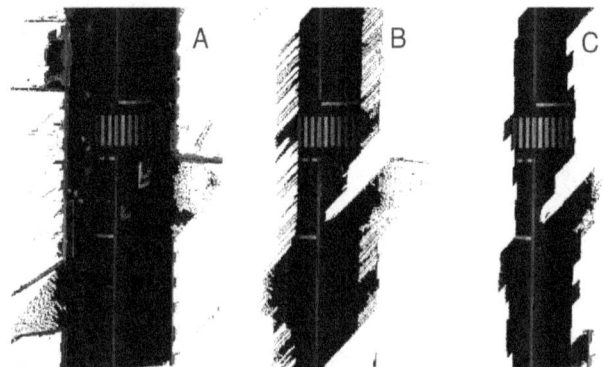

Figure 3. Segmentation of road from the original point cloud: a) original point cloud; b) noise segmentation evaluating each laser cross-section; c) improved segmentation averaging the angles of segmentation with the angles of precedent cross-sections.

Once the road is segmented from the rest of the point cloud this down sampled point cloud is partitioned into strips of the same length in order to normalize the data processing. A length of 18 m was established for each strip, which is computed from the navigation data provided by the Mobile Laser Scanner used to collect the data.

An overlap of 50% between adjacent strips is programmed to avoid the loss of relevant information about the urban parts (i.e. zebra crossings). Figure 4 shows an example of road cutting in strips. For example, the zebra crossing is completely recorded in strip 7. However, if only strips 6 and 8 would be acquired, the data will not provide the complete information from the zebra crossing so the road marking would be hardly detected.

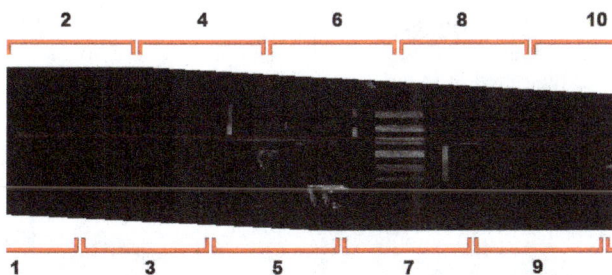

Figure 4. Example of road segmentation and strips with overlap of 50 %.

2.2 Point cloud rasterization

Point cloud data show geometric and radiometric information that can be easily converted in an image by applying image processing algorithms. The radiometric information of each coordinate comes from the reflective intensity of the laser light and depends on the distance between laser emitter and target surface, reflection angle, and target reflectivity. The painted road, as occurs in a zebra crossing, gives high reflectivity in comparison to their surroundings. This characteristic will be used to transform the point cloud data from the road to 2D imaging data.

The first step in road rasterization consists of the projection of the point cloud data on the plane that best fits to the road strip. Next, a nearest neighbour algorithm is used to assign the intensity data to a regular matrix created on the projection plane. The algorithm calculates the Euclidean distance between each node of the matrix and the neighbourhood points. Once the closest point is defined, the value of intensity is assigned to the node. Distance between nodes was selected to be 5 cm to have an enough spatial resolution in the images. Finally, the intensity from the images (12 bits: 0 – 4095) is normalized between 0 and 1. Figure 5 shows an example of the rasterized data.

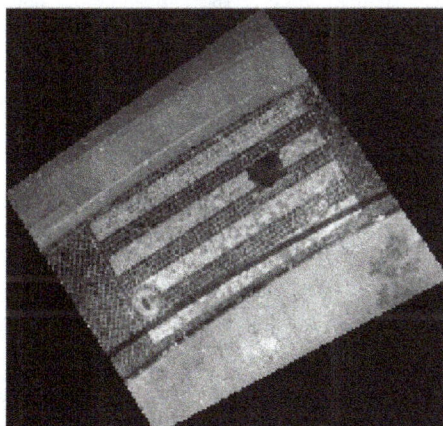

Figure 5. Example of the raster image of a zebra crossing from a point cloud.

2.3 Image preparation

Image processing techniques are used in this step of the algorithm in order to ease the detection of the road marks, particularly zebra crossing. Road marking detection is based on the higher reflectivity of the element and in the characteristic parallel lines of the edges between the painting area and the pavement.

The first step to detect reflective paintings consists of the noise reduction applying a median filter (figure 6b). Then threshold is applied to the previously filtered image. The thesholding is calculated using the Otsu method (Otsu, 1979), which chooses the threshold that minimize the intraclass variance of the black and white pixels (figure 6c).

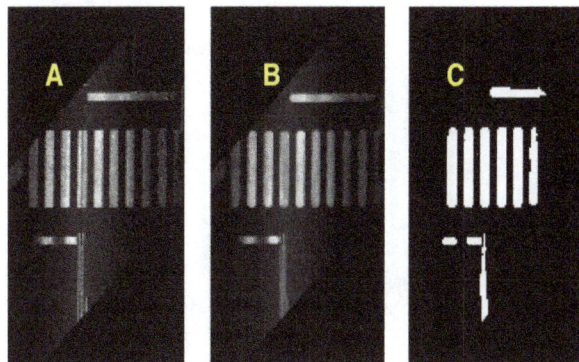

Figure 6. Image processing to detect reflective painted marks. A intensity image; b) filtered image using median filter; c) binary image using the Otsu method.

To improve the edges of the markings Mathematical morphology is used. Particularly a closing operation (dilation followed by and erosion) of the image using a 5 x 5 square structure is performed (that correspond to a square element side length of 25cm on ground). The purpose of the closing is to fill the gaps between the pixels in the border of white marks. This morphological operation is critical to achieve robust edge detection.

The second step consists of the edge detection of the zebra crossing. Edge detection is performed using the Canny method, which finds edges by looking for local maxima of the gradient of the image. The gradient is calculated using the derivative Gaussian filter. As it can be observed in Figure 7 (left), edges appear very thin. To improve the edge thickness a dilation operation (Figure 7 – right) is performed with a square structure element (5x5).

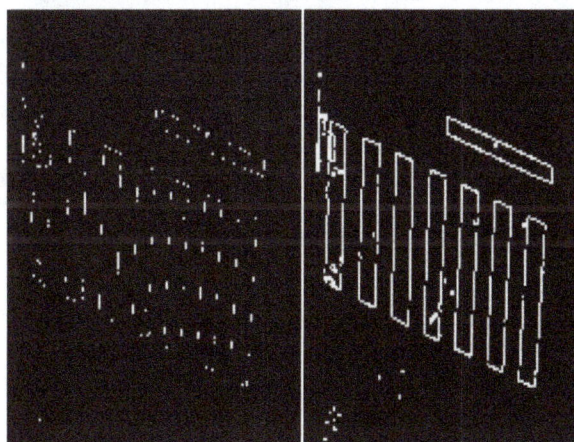

Figure 7. Edge detection of the zebra crossing (left). Image after a dilation (right).

2.4 Line classification using Hough Transform

The last step refers to the detection of lines in a binary image using the Hough Transform. The result of detection is shown in Figure 8 where the lines detected are depicted in green. Some lines are not detected for the constrains defined in the Hough Transform (principally lines were incomplete and their length did not satisfied the minimum length expected for a zebra crossing mark).

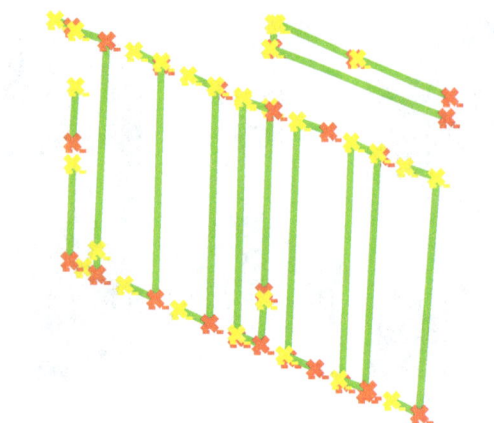

Figure 8. Line detection using the Hough Transform

To define detection of lines as a zebra crossing the authors established logical criteria. Lines represent a zebra crossing when they can be represented by a set of parallel lines with similar angle to the direction of the vehicle and with similar length. The authors establish a limit of 6 lines to avoid false positives. This threshold corresponds to a zebra pass with a minimum of 3 painting areas. The process of checking the number of lines from the Hough Transform detector counting is automatically performed.

The location of the zebra crossing is obtained from averaging the points defining the lines correctly used to identify the road mark from the HT detector. As the pixel coordinates from image come from the point cloud data, the 3D coordinates are directly obtained and the data can be transferred to Geographic Information Systems, for example to perform the inventory of a road. The direction and size of the zebra crossing is computed from the angle defined by lines and their length, respectively. Further details of the method can be seen in Riveiro et al., 2015.

3. CASE STUDIES

3.1 Study sites

Three mobile LiDAR datasets were used for this work. The mobile LiDAR survey was done without interrupting the traffic conditions using the Optech Lynx mobile mapper of the University of Vigo. In all cases the dilution of precision of the global navigation satellite system was kept below 2.5 to obtain accurate data. The acquisition frequency of the inertial measurement unit of the Optech was configured to 200 Hz to acquire an accurate trajectory data. LiDAR sensors were set up at 500 kHz (laser measurement rate) and 200 Hz (scan frequency) to provide a dense point cloud, with the maximum possible resolution.

LiDAR datasets were all collected in the Northwest region of Spain. The first dataset was collected in the Augas Férreas

Square in the city of Lugo; the second area was collected in Samil Avenue in the city of Vigo; and finally, the third dataset was registered in Progreso Street in the city of Ourense. Augas Ferreas data comprises 11 strips with a total of 34.352.512 points, Samil Avenue comprises 12 strips with 35.026.765 points, and Progreso Street has 2 strips with 6.181.733 points. An example view of the Progreso Street is shown in figure 9.

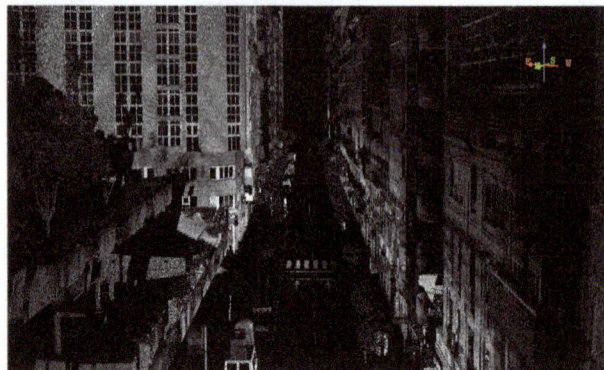

Figure 9. A 3D view of the point cloud collected in the Progreso Street (Ourense, Spain).

3.2 Instrumentation.

The mobile LiDAR used for surveying was the Optech Lynx Mobile Mapper (Figure 10). The systems consist of a navigation system with global navigation and inertial units, two LiDAR sensors and four digital cameras. All the systems are geometrically boresighted and time stamped to give an accurate point cloud from the environment after a mobile surveying. The maximum measurement range (according technical specifications from the manufacturer) is 200 m. Range precision is 8 mm and absolute accuracy 5 cm [14]. The laser measurement rate is programmable between 75 and 500 kHz. The system can detect up to 4 returns from each laser pulse. The scan frequency is also programmable between 80 and 200 Hz. The scanner field of view is 360°. It requires a power of 12 VDX and 30 A. Operating temperature ranges between -10° C and 40°C.

Figure 10. Optech Lynx mobile mapper. Draw (right) and survey van (left).

4. RESULTS

Table 1 presents the results obtained from the 3 different study sites. A total of 30 zebra crossing were evaluated with a completeness of 83 %. The results were checked comparing the output of the algorithm with the manual detection performed by a human operator.

Site	Strip	Points	X (m)	Y (m)	Completeness
Progreso Street	P1	3796143	593371	4688086	3/3
	P2	2385590	593424	4687855	3/3
Samil Avenue	S1	1756687	518695	4674416	1/1
	S2	2359182	518841	4674519	1/1
	S3	2467759	519038	4674518	1/1
	S4	5290571	-	-	0/0
	S5	2929814	519210	4674388	1/1
	S6	5193961	519445	4674352	1/2
	S7	3197622	-	-	0/1
	S8	2629765	-	-	0/0
	S9	1316111	-	-	0/1
	S10	1610806	520138	4673855	1/1
	S11	1188348	-	-	0/0
	S12	5086139	520391	4673226	1/1
Augas Ferreas Square	A1	4243692	618488	4761271	1/1
	A2	5049528	618549	4761179	1/1
	A3	2294035	618644	4761173	1/1
	A4	1279948	618343	4761314	1/1
	A5	794258	618591	4761444	1/1
	A6	1490190	618445	4761509	1/1
	A7	3777648	618345	4761643	1/1
	A8	5000282	618288	4761578	2/2
	A9	3983655	618417	4761590	1/2
	A10	2046422	618332	4761442	1/1
	A11	4393954	618235	4761379	1/2

Table 1. Results found for the three study sites.

Figure 11 shows an example of a correctly detected zebra crossing. Some zebra crossing were not detected, although false positives are not presented in the results.

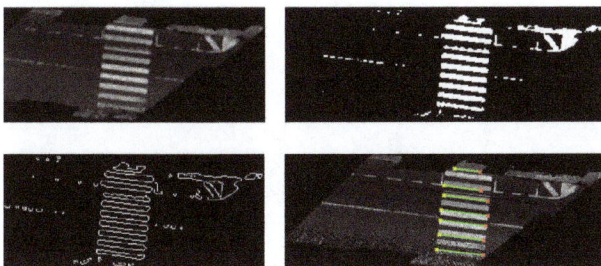

Figure 11. Correct detection of a zebra crossing in Samil Avenue dataset.

Those existing zebra crossing that were not detected by the algorithm, were deeply analysed in order to understand the causes of failure in the algorithm. For example, the algorithm does not detect another zebra crossing that can be visually located in the same strip of Samil Avenue dataset. The road contains occlusions due to vehicles above the zebra crossing in figure 12 (top – left image). As can be seen this would provoke a malfunction of the algorithm because those lines that are detected using the Hough transform would not satisfy the conditions of parallelism nor length established during the development of the algorithm.

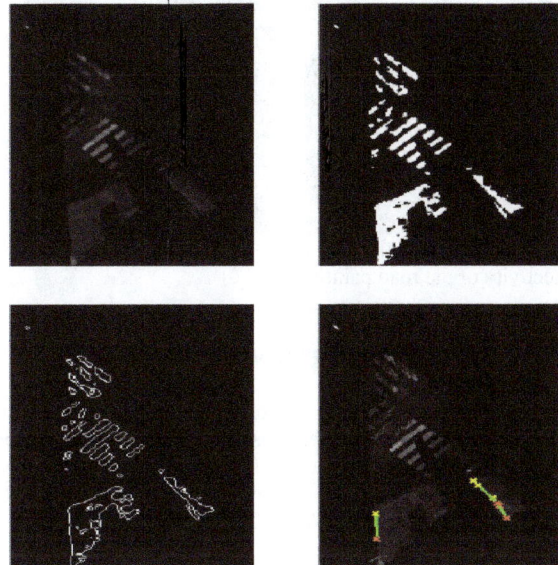

Figure 12. Malfunction of the algorithm due to occlusions of the point cloud.

In Figure 13 binarization from the image (top – right) appears slightly deficient, probably by the low reflectivity of the paint of the zebra crossing. This fact diminishes the contrast between the asphalt and the road markings, and decreases the quality of the binarization. It also affects to the edge detection operation (bottom – left) and consequently it also affects to the line detection using the Hough transform (bottom – right).

5. CONCLUSIONS

An algorithm for the automatic detection of zebra crossing is developed with the aim of automate the detection and inventory of zebra crossings. The algorithm is divided in three main steps: road segmentation (based on a PCA analysis of the points contained in each cross-section of collected by a mobile laser system), rasterization (conversion of the point cloud to a raster image coloured as a function of intensity data), and zebra crossing detection (using the Hough Transform and logical constrains for line classification).

A completeness of 83% was found when testing the algorithm over 25 strips with 30 zebra crossing. Non-detected zebra crossing were analysed and in all cases the reasons of failure come from two main reasons. The first comes from the loss of paining of the road marking due to time provoking a low contrast between the road and the zebra crossing. The second reason is motivated by occlusions caused by other vehicles in the point cloud produced by other urban objects during the survey. All the test were done using Matlab software.

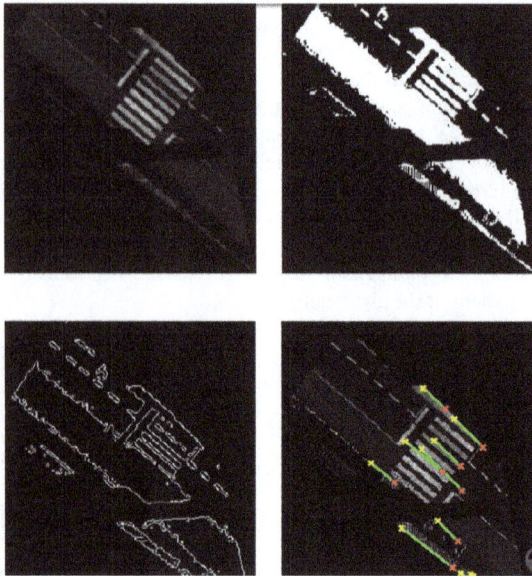

Figure 13. Malfunction of the algorithm due to a deficient reflectivity of the road paintings.

ACKNOWLEDGEMENTS (OPTIONAL)

This work has been partially supported by the Spanish Ministry of Economy and Competitiveness through the project "HERMES-S3D: Healthy and Efficient Routes in Massive Open-Data based Smart Cities (Ref.: TIN2013-46801-C4-4-R), by Xunta de Galicia (Grant No. CN2012/269). Human resources program FPI is acknowledged (Grant No. BES-2014-067736 and FPU AP2010-2969).

REFERENCES

Douillard, B., Underwood, J., Kuntz, N., Vlaskine, V., Quadros, A., Morton, P., Frenkel, A., 2011. On the segmentation of 3D LIDAR point clouds. In: *IEEE International Conference on Robotics and Automation*, ICRA'11, pp. 2798–2805.

Hernández, J., Marcotegui, B., 2009. Point cloud segmentation towards urban ground modeling. In: *The 5th GRSS/ISPRS Joint Urban Remote Sensing Event (URBAN2009)*. Shangai, China, pp. 1–5.

Hervieu, A., Soheilian, B., & Brédif, M., 2015. Road Marking Extraction Using a MODEL&DATA-DRIVEN Rj-Mcmc. ISPRS Annals of Photogrammetry, Remote Sensing and Spatial Information Sciences, 1, 47-54.

Holgado-Barco A, González-Aguilera D, Arias-Sánchez P, and Martínez-Sánchez J., 2014. An automated approach to vertical road characterization using mobile LiDAR systems: Longitudinal profiles and cross-sections. *ISPRS Journal of Photogrammetry and Remote Sensing* 96: 28 – 37.

Kumar, P., McElhinney, C. P., Lewis, P., & McCarthy, T., 2014. Automated road markings extraction from mobile laser scanning data. International Journal of Applied Earth Observation and Geoinformation, 32, 125-137.

Golovinskiy, A., Kim, V.G., Funkhouser, T., 2009. Shape-based recognition of 3D point clouds in urban environments. *In: 12th IEEE International Conference on Computer Vision*, pp. 2154–2161.

Guan, H., Li, J., Yu, Y., Wang, C., Chapman, M., & Yang, B., 2014. Using mobile laser scanning data for automated extraction of road markings. ISPRS Journal of Photogrammetry and Remote Sensing, 87, 93-107.

Owechko, Y., Medasani, S., Korah, T., 2010. Automatic recognition of diverse 3-D objects and analysis of large urban scenes using ground and aerial LiDAR sensors. *In: Conference on Lasers and Electro-Optics (CLEO) and Quantum Electronics and Laser Science Conference (QELS)*, pp. 16–21.

Otsu, N., 1975. A threshold selection method from gray-level histograms. Automatica, 11(285-296), 23-27.

Riveiro, B., González-Jorge, H., Martínez-Sánchez, J., Díaz-Vilariño, L., Arias, P., 2015. Automatic detection of zebra crossings from mobile LiDAR data. *Optics & Laser Technology*, vol. 70 pp 63-70.

Serna, A. and Marcotegui, B., 2014. Detection, segmentation and classification of 3D urban objects using mathematical morphology and supervised learning. *ISPRS Journal of Photogrammetry and Remote Sensing*, vol 93, pp. 243–255.

Zhu, X., Zhao, H., Liu, Y., Zhao, Y., Zha, H., 2010. Segmentation and classification of range image from an intelligent vehicle in urban environment. In: *IEEE/RSJ International Conference on Intelligent Robots and Systems (IROS2010)*, pp. 1457–1462.

33

DEVELOPMENT OF A CARTOGRAPHIC STRATEGY AND GEOSPATIAL SERVICES FOR DISASTER EARLY WARNING AND MITIGATION IN THE ECOWAS SUBREGION

L. A. Gueye [a], M. S. Keita [b, *], J. O. Akinyede [c], O. Kufoniyi [d], G. Erin [a]

[a] ECOWAS Commission 101 Yakubu Gowon Crescent, Asokoro District, Abuja, Nigeria. (latgueye@yahoo.com, gbengaerin@yahoo.com);
[b] Regional Centre for Training in Aerospace Surveys (RECTAS), OAU Campus, Ile Ife, Osun State, Nigeria (keita@rectas.org);
[c] Department of Remote Sensing and Geosciences Information System, Federal University of Technology, Akure, Ondo State, Nigeria (jakinyede@yahoo.com);
[d] Department of Geography, Obafemi Awolowo University, Ile-Ife, Osun State, Nigeria (jidekufoniyi@yahoo.com).

Commission IV, WG IV/7

KEY WORDS: Cartographic Strategy, Geospatial Services, Disaster, Early Warning, Mitigation

ABSTRACT:

The West Africa Sub-region has been crisis and disaster ridden in recent times with enormous challenges for disaster mitigation. The crisis/disasters range from conflicts fuelled by political upheaval to epidemics that take their tolls on the population of some countries in the sub-region. The crisis and disaster events have overwhelming magnitudes and are highly dynamic, requiring a well-articulated plan for immediate response in order to mitigate their effects. A study carried out by the Early Warning Directorate (EWD) of the Economic Commission of West African States (ECOWAS) highlighted the risks and vulnerabilities of the region despite the considerable progress made in development and peace consolidation in some parts of the region. The study identified apparent institutional and infrastructural deficiencies, such as the lack of up-to-date geospatial data and information, and inadequate platforms for data gathering and data sharing among the relevant national agencies, which have made much of the region particularly vulnerable to the emerging threats. It is against the foregoing that the development of a Cartographic Strategy and Geospatial Services for EWD and the ECOWAS is being proposed. In addition to the resolution of the crucial need of reliable geospatial data capacity of member states, this initiative will spearhead the realisation of a Geospatial Data Infrastructure for ECOWAS Commission, through the appropriate policy formulation and implementation. Through the proper implementation of the Cartographic Strategy and Geospatial Services, ECOWAS will have the capacity to provide geospatial analysis and mapping support focusing on areas related to conflict prevention and resolution, regional planning for food security, early warning of viral diseases and epidemics, disaster preparedness, mitigation and response, infrastructural development and refugee resettlement, and a host of other vital projects/programmes for promoting ECOWAS regional integration agenda. This paper discusses the outcome of the preliminary studies and activities carried out by an Expert Group commissioned by the EWD to develop a Cartographic Strategy and propose a framework for its implementation. These include the assessment of the status of mapping and Geographic Information System uptake in member states, the formulation of policy and the realisation of a work plan for its successful implementation in the region.

1. INTRODUCTION

One of the sustainable development challenges of any nation is its capacity to manage its environment and disasters. Both natural and manmade disasters can have devastating effects on people, properties and economies (*Akinyede*, *2005*). Today, the frequency and magnitude of natural hazards and disasters threatening large populations living in diverse environments have shown a dramatic rise. The impact of disasters on the global environment has become increasingly severe over the last decades. The reported number of disasters has risen dramatically, as well as the number of people affected and the cost to global economy. About 95% of the deaths due to disasters occur in developing countries (*Kufoniyi, 2007*).

West African nations have experienced many serious disaster challenges including drought, flood, coastal erosion, wild fires, epidemics, conflicts, etc. The sahelian drought of the 1980s affected West African countries and led to the loss of thousands of people, livestock and livelihoods. The increasing desertification and loss of prime agricultural lands occasioned by climate change and increased desiccation of the sudano-sahelian region led to landuse conflicts, communal and ethnic clashes and loss of lives (*Adeniyi, 2009*). The occurrence and severity of droughts and floods in

the sub-region are happening with increasing frequency, accompanied by various diseases. Communities are more and more vulnerable to the unfamiliar hazards and cannot cope with the shocks leading to a constant rise in the number of people needing humanitarian assistance. Moreover, the number of emergencies in Africa per year has almost tripled since the mid-1980s to about 25 million people in 2005 (IFRC, 2008).

To face these challenges, efforts are being deployed in the countries through national entities which include Emergency Management Agencies, Civil Protection Agencies and Early Warning Departments. Nevertheless, despite the contributions of those agencies, tackling disaster challenges still remains a difficult task for decision makers as demonstrated by the recurrent calamities, crisis and diseases in many countries of the region in recent times. To overcome the inconsistencies, regional bodies like the Economic Community of West African States (ECOWAS) have embarked on initiatives using geospatial technology to find solutions to the problems.

The advances in geospatial technology, specifically earth observation satellite and Geographical Information System (GIS), have increased their potential to meet challenges in scientific and social fields of studies. In disaster management,

---

* Corresponding author

space-based technologies and more efficient GIS tools are used to monitor the environmental changes at certain intervals and to carry out risk analysis. Indeed, geospatial technologies have contributed to providing significant solutions in all phases of disaster management areas (early warning, mitigation, preparedness, emergency response, relief and rehabilitation) in both developed and developing countries (*Stevens, 2004*).

However, access to relevant geospatial data needed for planning, development and management of resources has remained limited in Africa, particularly in West African countries. In addition, accurate and up-to-date base maps which provide accurate locational information as basic requirement for emergency response are not always readily available. Although national capabilities in the use of the technologies are increasing at a significant rate, yet there is a definite need to support the technology transfer, while proposing methodological approaches that are appropriate to the specific needs of the countries in the planning and decision-making on disaster management.

The current initiative of the ECOWAS Early Warning Directorate aims at assisting the ECOWAS and its Member states to improve the capabilities to manage the disaster events (pre-disaster, response and post-disaster) by using geospatial technology. The proposed Cartographic Strategy and Geospatial Services will surely provide appropriate solutions for the prevention of disaster events and emergency responses.

The main purpose of this paper is to do a systematic analysis of the geospatial capability of ECOWAS Member States in overcoming the disaster challenges affecting the region and present an overview of the Cartographic Strategy and Geospatial Services initiative of the Early Warning Directorate aimed at mainstreaming geospatial data and services. The expected results, in terms of contribution to the improvement of the ability to deal with emergencies and mitigation of major disasters, are also highlighted.

2. DISASTER CHALLENGES IN WEST AFRICA

The West African sub-region is comprised of 15 countries: Benin, Burkina Faso, Cabo Verde, Cote d'Ivoire, the Gambia, Ghana, Guinea, Guinea-Bissau, Liberia, Mali, Niger, Nigeria, Senegal, Sierra Leone and Togo.

Recurrent natural and human-induced disasters, such as drought, trans-boundary animal and human diseases, plant pests and diseases, as well as socio-economic and political crises have been more frequent in the sub-region of West Africa. The consequences include significantly increasing food security, child malnutrition and vulnerability of populations to shocks. The sub-region's vulnerability to disaster is already exacerbated by a number of factors, such as: poverty, low education levels and a lack of access to basic services, political instability, conflicts, poor governance, weak economic dependent on international markets, high population growth and a trend towards unplanned urbanisation and exodus from rural areas. Indeed, as of 2010, all the countries in the sub-region, with the exception of Cabo Verde , were among the countries with the lowest level of human development (FAO, 2010).

These crises in West Africa are aggravated by the impact of climate change: over the last fifty (50) years. The sub-region has undergone changes in rainfall patterns that have translated not only into serious, extensive droughts, especially in 2005, but also irregular and violent storms causing destructive floods damaging public infrastructures, homes, crops and livestock.

Such increasingly complex and diverse disasters and crises, the majority of which are transnational, require urgent commitment and intervention of regional bodies like the Economic Community of West African States (ECOWAS) taking advantage of its regional integration agenda.

3. ECOWAS' CONTRIBUTION THROUGH EARLY WARNING IN WEST AFRICA

3.1 Why Early Warning Systems?

Early warning Systems are essential in order to assess, to monitor risks and to warn of a potential crisis. In West African countries, there were local and national initiatives which often tended to prioritise the analysis of food supplies (food security) rather than other disaster challenges. This has prevented them from being able to identify vulnerabilities and formulate adequate preparation and response plans.

Since the preparedness and emergency intervention capacities and capabilities of local populations and national governments to deal with the extensive and frequent crises were weak, it became necessary to establish more efficient systems to strengthen the national capacities and coordinate the activities at regional level.

Therefore, the limited capacity of the national systems to find solutions to the crises and disaster challenges has necessitated ECOWAS Commission to initiate the establishment of a sub-regional peace and security observation system, pursuant to the protocol relating to the mechanism for conflict prevention, management, resolution, peace-keeping and security.

3.2 ECOWAS Early Warning / Observation Monitoring Centre

The ECOWAS Early Warning System (EWS) was established under Chapter IV of the 1999 Protocol Relating to the Mechanism for Conflict Prevention, Management, Resolution, Peacekeeping and Security. The Early Warning Directorate's activities include collecting open sourced information largely based on 66 pre-determined indicators of regional peace and security, analysing and submitting timely reports with recommendations to the Office of the President of ECOWAS Commission through the Office of the Commissioner, Political Affairs, Peace and Security. The EWD performs these activities through the ECOWARN Field Reporter, a web-based field monitoring, interactive analysis, and visualization tool customized for the ECOWAS region. It functions by a network of 77 field monitors, collecting early warning data and feeding it to the ECOWARN Situation Room in Abuja, the headquarters of the system.

Recent developments in the system seeks to operationalise the National Early Warning and Response Mechanisms (NEWRMs) in Member States, a decentralized early warning and response system that involves all key stakeholders at the local, national and regional levels.

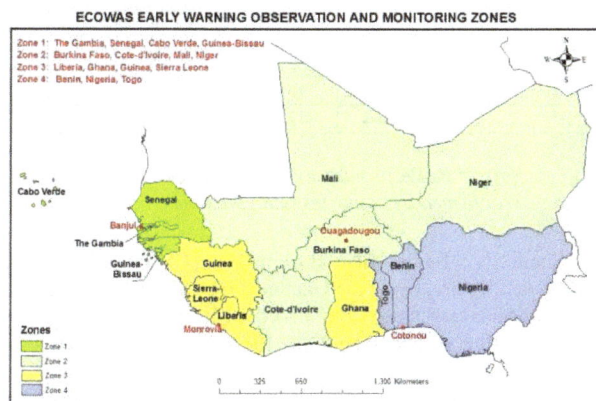

Figure 1. - ECOWAS Early Warning Observation zones by country

The Directorate works in partnership with representatives of ECOWAS Member States and Civil Society Organisations like West African Network for Peacebuilding (WANEP), West African Civil Society Forum (WACSOF) and research Institutes. It also collaborates with other Regional Economic Communities (RECs) and the African Union (AU) in the establishment of the Continental Early Warning System (CEWS).

3.3 Best practice in risk prevention and disaster mitigation

Experience has shown that the difference in the level of vulnerability of hazards in the developed and developing countries is not really in the number and magnitude of occurrence of hazards, but in the level of disaster management - disaster preparedness (risk reduction), mitigation, and emergency responses. This is made possible through the use of geographic data i.e. geospatial data (*Adeniyi, 2009*). The actual questions to respond are **how prepared** are we for disaster management? **How much do we know** about our **environment**? and **how much data** and **information** do we have that can help us respond to crisis situation if it occurs?

Today, geospatial technology finds responses to those questions and many other pertinent questions. It has been identified as the best applied technology for risk prevention and disaster mitigation. Geospatial technology comprises many other technologies which include Remote Sensing, Geographic Information System (GIS) and Global Positioning System (GPS). While Remote Sensing remains the most comprehensive and economically cost effective tool for obtaining synoptic geospatial data required for disaster monitoring, GIS plays a major role in the acquisition and management of related to the spatial characteristics and dimensions of disasters and hazards. For high precision information on disasters location on the Earth surface, the GPS technology provides accurate field point data required

for a GIS. The adoption of Spatial Data Infrastructure (SDI) as a veritable tool in mitigating the impact of disaster has also become very obvious and widely acknowledged (Kufoniyi *et al*, 2013).

Many countries in West Africa sub-region are already familiar with these technologies and they have the expertise and technical know-how required for disaster management using geospatial technology. Nigeria, for example, is a member of the Disaster Monitoring Constellation (DMC), which, via collaboration, is able to provide data to and access data from other partners in the event of any disaster. (*Akinyede, 2005*).

One of the best solutions to explore for efficient disaster management is therefore the development of a plan of action at the EWD that would coordinate the efforts being deployed by national agencies/institutions, and organise a co-operative network making use of geospatial technology. The adoption of such best practice will greatly encourage the governments of Member States, while servicing the vulnerable communities.

4. USE OF GEOSPATIAL TECHNOLOGY IN DISASTER MANAGEMENT IN WEST AFRICA

4.1 Status of mapping in West Africa

Following the setting up of a GIS unit within the ECOWAS Early Warning Department, a meeting of the Directors General of the National Mapping Organisations of the 15 Member States was held in Dakar, Senegal in February, 2010. The goal of that meeting was to review the mapping status of the region and discuss a global cartographic strategy for ECOWAS. The meeting was attended by all the 15 directors, the representatives of development partners (French Cooperation) and other experts. The different presentations highlighted the status of mapping of each country, the availability of digital data, geodetic networks, equipment and National Mapping Strategy/Policy.

As follow-up actions, the inventory of geospatial data and policies was undertaken, using a questionnaire in the first instance and a visit to each of the member states to validate the questionnaires collected.

The **criteria** used in assessing the "*geospatial performance*" of Member States are presented in table1.

Table 1. : Scoring sheet of geospatial performance criteria

S/N	Criteria	Code	Score			
			Very Good	Good	Fair	Poor
1.	Status of National Mapping	C1	10	8	6	2
2.	Number of agencies visited	C2	8 (>3)	6 (=3)	3 (=2)	2 (=1)
3.	Availability of Geosptial Data	C3	10	8	5	2
4.	Synergy between partners	C4	10	9	6	2
5.	Evidence of SDI initiative	C5	10	9	7	1
6.	Level of commitment	C6	10	8	6	1

The resulting "**geospatial performance**" of each Member State is presented in tables 2a.and 2b

Table 2a. : Geospatial performance of Member States (1)

Criteria	Country's code							
	BN	BF	CV	CI	GA	GH	GU	GB
C1	7	8	6	8	6	6	5	5
C2	8	8	2	4	4	5	6	4
C3	7	8	6	8	6	6	6	5
C4	5	6	2	4	2	5	4	6
C5	2	7	7	2	2	2	2	2
C6	6	8	7	6	6	6	6	6
P*	35	45	30	32	26	30	29	28

From the above findings of the investigation, the status of mapping of some of the countries has been found as very low. That demonstrated the urgent need to undertake the updating of existing maps produced in the 60s, still in analogue form in most cases. It was further noted that the deficiency observed in the mapping industry in the region has contributed to a large extent to the widening gap between the countries and the developed world. That has seriously influenced the successful implementation of many projects including disaster management programmes.

4.2 Approach of establishing a Cartographic Strategy and Geospatial Services for ECOWAS

In order to achieve its intended effect of positively affecting sustainable development, geospatial information must be readily available and easily accessible to the users who need it for various decision-making processes in economic planning and development. This informed the decision to set up a standing committee of Experts to work with the Commission to spearhead the rapid development of a Cartographic Strategy and Geospatial Services for ECOWAS.

Table 2b. : Geospatial performance of Member States (2)

Criteria	Country's code						
	LB	ML	NIG	NG	SN	SL	TO
C1	5	7	7	8	7	5	6
C2	6	8	7	8	4	7	4
C3	5	7	7	8	7	5	6
C4	4	7	6	8	9	4	5
C5	2	7	7	9	8	2	2
C6	2	8	8	8	8	2	6
P*	24	44	42	49	43	25	29

Note: * P (Performance) = C1+C2+C3+C4+C5+C6 *(confirmed by countries and validated by ECOWAS)*

Where BN = Benin, BF = Burkina Faso, CV = Cabo Verde, CI = Cote d'Ivoire, GA = the Gambia, GH = Ghana, GU = Guinea, GB = Guinea-Bissau, LB = Liberia, ML = Mali, NIG = Niger, NG = Nigeria, SN = Senegal, SL = Sierra Leone and TO = Togo.

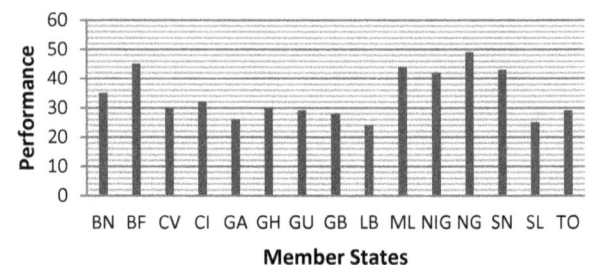

Figure 2. - Geospatial performance of ECOWAS Member States

During the first stakeholders meeting held in Dakar in February, 2010, the main activities recommended by the participants include the following:

-   Identification and collection of geospatial data available in ECOWAS' member states and the Multinational Geospatial Co-production Program (MGCP) project through the French Co-operation;
-   Definition and adoption of a Cartographic Strategy and Geospatial Policy for ECOWAS;
-   Building and strengthening capacity and capability of Mapping agencies and other stakeholders of member States;
-   Development of an ECOWAS Geospatial Data Infrastructure (EGDI).

Sequel to the meeting held in Dakar, different workshops and seminars were held in Ouagadougou (2011), Banjul (2012), Abidjan (2013) and Cotonou (2014) to discuss the development and implementation of the ECOWAS Cartographic Strategy and Geospatial Data Infrastructure initiative. A plan of action was designed and adopted, based on activities to be executed in collaboration with all the stakeholders, in synergy with experts and resource persons under the coordination of ECOWAS Early Warning Directorate.

4.2.1 Geospatial data and technology availability in ECOWAS Member States

The mission undertaken to do the inventory of geospatial data and technology available in Member States permitted to identify the data available, the data gaps and the problems of accessibility. The available data include topographic and thematic maps, geodetic networks, databases and other statistical data. The inventory also identified the types of data and the difficulties of data sharing and dissemination between stakeholders.

The inventory shows that the required geospatial technology platform including Remote Sensing technology (satellite and photography), Geographic Information Systems, Digital Cartography, ground surveying, is not fully present in all the Member States. Most countries are yet to set up a digital platform, thus the existing geospatial data are only in analogue form in many countries. The lack of internet connectivity makes in most cases the access to and dissemination of data very difficult.

However, through the Japan International Cooperation Agency (JICA) and the European Union funding, some mapping projects (large scale mapping and map revision) are currently going on in some countries (Burkina, Benin, Mali, Niger and Senegal). Furthermore, Nigeria has almost completed the analogue to digital conversion of the country's 1:50000 topographic map series and is currently undertaking the 1:25000 topographic mapping of the country.

4.2.2 Cartographic Strategy and Geospatial Data Infrastructure Policy

The need to acquire reliable and up-to-date geospatial data arose when the GIS Unit was put in place at ECOWAS Early Warning Directorate in 2010. For the production of situation reports, there was need to integrate data acquired from different sources. That brought up the necessity to harmonise the geospatial data production system in all the countries. Thus, the development of a Cartographic Strategy for the sub-region.

The *Multinational Geospatial Co-production Program (MGCP)* mapping project, working on the elaboration of topographic maps of West Africa at 1:50,000 scale, using satellite imageries was initiated by the French Cooperation at ECOWAS. However, the Head of National Mapping Agencies of ECOWAS Member States, during the meeting in Dakar in January 2010, strongly recommended that each country be responsible for the production of its own national maps, using at least regional standards to be developed as guidelines. It was finally decided that each country be encouraged to implement its National Spatial Data Infrastructure, while a Regional Geospatial Data Infrastructure should be developed under the coordination of EWD.

The draft policy of the ECOWAS Mapping Strategy and Geospatial Data Infrastructure was then developed in Ouagadougou in 2011, amended by Working Group of Experts and resource persons in Banjul in 2012 and validated in Abidjan in 2013. An action plan was then finalised with concrete terms of reference for working groups comprising representatives of Member States and Experts resource persons at the Expert Group meetings in Cotonou and Banjul in 2014 and 2015 respectively.

4.2.3 Capacity building and Institutional strengthening

The action plan will start will a full awareness building and buy-in of Ministers, Political Authorities and decision makers at ECOWAS Commission and in Member States. One of the important components of the strategy will be the increasing of the current available capacities of the region through training and capacity building programmes, institutional strengthening and funding.

The capacity building exercises will target the available human resources, who are very important for the interventions during disasters, as well as experts and resource persons who will advise on relevant equipment and infrastructure to be acquired. Other selected expert will be in charge of all institutional arrangements for putting in place conducive working environment at both national and regional levels.

4.2.4 Development and Implementation of ECOWAS Geospatial Data Policy

The development and maintenance of a Regional Geospatial Data Infrastructure at ECOWAS will consist of building of synergies among institutions and Member States and enabling the sharing of data and expertise for preparation of contingency plans and interventions in case of emergencies.

The proposed ECOWAS Geospatial Data Infrastructure will:

- ✓ Ensure orderliness and common standard in the development and processes for executing mapping projects in the geographic space of ECOWAS.
- ✓ Facilitate the identification, production and sharing of fundamental datasets paramount to the regional development and integration.
- ✓ Facilitate and increase the rate of mapping activities within the member states.
- ✓ Provide a framework for standardizing the acquisition, processing, analysis, storage and dissemination of various geospatial datasets for socio-economic and environmental data needed to sustain and enhance regional integration and development.
- ✓ Eliminate duplication in the acquisition and maintenance of geospatial datasets in the region.
- ✓ Encourage regional use of EGDI-endorsed standards in order to create and maintain data at a high level of quality, consistency and harmony with the ISO standards.

✓ Promote the awareness of Geographic Information and its applications.
✓ Ensure adequate funding to maintain the EGDI vision and sustain the spirit of cooperation and collaboration amongst the member states.
✓ Promote research, training, education and capacity building related to geospatial data production, management, dissemination and usage.

The implementation of EGDI will be done under the **coordination** of EWD with the contribution of Experts and resource persons. It will be done in accordance with the Plan of Action and terms of reference defining the different activities to be undertaken by the following six (6) working groups:

a. Geospatial dataset
b. Nomenclature & Codification
c. Standards
d. Coordination, Communication & Partnership
e. Clearinghouse & Metadata
f. Capacity building & Sustainability

5.   EXPECTED OUTCOMES AND BENEFITS

For ECOWAS Geospatial Data Infrastructure to meet the stated objectives and purposes, it has to be well organised in such a way that it can provide reliable and up-to-date geospatial data and products for disaster mitigation, preparedness, emergency responses and reconstruction. It will be able to provide a Crisis Information Management System for the Early Warning Directorate, as well as relevant databases for the activities of ECOWAS Commission.

The benefits of the geospatial data and products to be generated will cover many areas of endeavour. The EGDI will be in assistance of planning and implementation of projects and programmes, as well as decision making processes in ECOWAS Commission and Member States. In the area of disaster management, conflict prevention and resolution, peace keeping and security, the following will be fulfilled:

a. Preparation of action plans for assignment of resources in response to potential crisis situations;
b. Monitoring the execution of operations, projects and programmes through the correlation and integration of information from governments, military services and other civil sources;
c. Evaluation of the potential and actual conflict situations, military force capabilities, planning for resources, use of map products, etc...;
d. Generation and updating of maps and databases that are important for disaster management, conflict prevention and peace keeping activities;
e. Preparation of reports and briefings for ECOWAS Early Warning Directorate and authorities of other departments and also for use in regional activities on disaster management.

6.   CONCLUSION

An action plan for the establishment of a geospatial data infrastructure for ECOWAS sub-region has been presented in this paper. The infrastructure will provide effective tool for disaster mitigation, emergency response and relief. The infrastructure will provide timely and essential geospatial data and information to generate resources for preparing for and responding to various forms of disaster.

Geospatial data are often used for the preparation of contingency plans. The popular adage has it that "prevention is better than cure". While we may not necessarily be able to prevent some natural hazards, we can act to prevent their disastrous effects on vulnerable people and places. (Adeniyi, 2009). Given the heavy and enduring costs of disaster management, there is no disputing the fact that making efforts to prevent them from breaking out in the first place is better than waiting until it is too late (Souare, 2006). The earlier a disaster, a disease, a dispute or a conflict can be identified, diagnosed and its causes properly addressed, the less likely it is that the situation will deteriorate into crisis, epidemic or violence. Normally, the calamities and conflicts do not occur suddenly without warning indicators. Thus, in order to guarantee the sustainability of the prevention measures, governments and regional and international organisations must ensure that there are proper mechanisms for conflict early warning in place, based on information gathering and informal and formal fact-finding. EWD is in a position to play that role in the sub-region.

Therefore, the development of a Cartographic Strategy and Geospatial Services is not going to improve only the performance of EWD, but will be useful for the successful implementation of projects and programmes for ECOWAS Commission, as well as the execution of disaster management activities in Member States.

REFERENCES

D. Stevens (2004): Space Technology and Disaster Management- A Plan-of-action for Africa

O. Akinyede (2005): Aspects of Space Science and Technology most relevant to Africa's development: Short term and Long Term., paper presented at the 1st African Leadership Conference on Space Science and Technology for Sustainable Development, November, 2005.

A.L. Gueye, A.O. Akingbade, O. Kufoniyi and P.B. Borisade (2004): The Use of GIS in Crisis Management: A Case Study of the Crisis Information Management System for the West African Sub-region".

Kufoniyi O., O. A. Ogundele and D. O. Baloye (2013), "Spatial Data Infrastructure (SDI): a plausible solution to improve disaster management processes in Nigeria", *Journal of Sustainable Development in Africa, Vol. 15(3), 2013,* Clarion University of Pennsylvania, Clarion, Pennsylvania, p130-147.

I.K. Souare (2006): Conflict Prevention and Early Warning Mechanisms in West Africa: A Critical Assessment of Progress. - African Security Review 16.3 / Institute for Security Studies.

P.O. Adeniyi (2009): Geospatial Information and Disaster Management - Geoinformation Technology & Development: A compendium of Selected papers.

34

INSTALLED BASE REGISTRATION OF DECENTRALISED SOLAR PANELS WITH APPLICATIONS IN CRISIS MANAGEMENT

Rosann Aarsen[1], Milo Janssen[1], Myron Ramkisoen[1], Filip Biljecki[2], Wilko Quak[3], Edward Verbree[3]*

[1] Delft University of Technology,
Faculty of Architecture and the Built Environment, MSc Geomatics
Julianalaan 134, 2628 BL, Delft, The Netherlands – {R.M.Aarsen; M.L.Janssen-1; M.G.W.Ramkisoen}@student.tudelft.nl
[2] Delft University of Technology,
Faculty of Architecture and the Built Environment, Department of Urbanism, 3D Geoinformation
Julianalaan 134, 2628 BL, Delft, The Netherlands – F.Biljecki@tudelft.nl
[3] Delft University of Technology,
Faculty of Architecture and the Built Environment, OTB Research Institute for the Built Environment, GIS Technology
Julianalaan 134, 2628 BL, Delft, The Netherlands – {C.W.Quak; E.Verbree}@tudelft.nl

Gi4DM and ISPRS, WG IV/7

KEY WORDS: Smart Grid, Crisis Management, Photovoltaic (PV) Panels, Registration, Linked Data

ABSTRACT:
In case of a calamity in the Netherlands - e.g. a dike breach - parts of the nationwide electric network can fall out. In these occasions it would be useful if decentralised energy sources of the Smart Grid would contribute to balance out the fluctuations of the energy network. Decentralised energy sources include: solar energy, wind energy, combined heat and power, and biogas. In this manner, parts of the built environment - e.g. hospitals - that are in need of a continuous power flow, could be secured of this power. When a calamity happens, information about the Smart Grid is necessary to control the crisis and to ensure a shared view on the energy networks for both the crisis managers and network operators. The current situation of publishing, storing and sharing data of solar energy has been shown a lack of reliability about the current number, physical location, and capacity of installed decentralised photovoltaic (PV) panels in the Netherlands. This study focuses on decentralised solar energy in the form of electricity via PV panels in the Netherlands and addresses this challenge by proposing a new, reliable and up-to-date database. The study reveals the requirements for a registration of the installed base of PV panels in the Netherlands. This new database should serve as a replenishment for the current national voluntary registration, called Production Installation Register of Energy Data Services Netherland (EDSN-PIR), of installed decentralised PV panel installations in the Smart Grid, and provide important information in case of a calamity.

1. INTRODUCTION

In case of a calamity in the Netherlands - e.g. a dike breach - a failure in the electricity network may occur. Both decentralised energy suppliers as the regular central energy suppliers are connected to the entire nationwide energy network. Current decentralised energy sources can be part of a Smart Grid; an advanced electrical grid that is controlled via information technologies to accurately regulate the energy demand and supply. Information about the Smart Grid is necessary to control the crisis caused by a calamity. Important buildings – such as hospitals - can be provided with energy via the Smart Grid if the location of decentral PV panels is known.

This study (performed for the course Synthesis Project, for the MSc Geomatics at TU Delft) is a subproject of the *CERISE-SG* project (*Dutch: Combineren van Energie- en Ruimtelijke Informatie Standaarden als Enabler – Smart Grids*). Its goal is to: "create future proof and efficient information exchange between the energy sector, eGovernment and geo-world" (Cerise, 2015) to make crisis management more effective and efficient. This study will answer the following research question:

How to create a database which can be used to validate and improve the current registration of installed decentralised PV panels in the Smart Grid usable for an application in Crisis Management?

This study focuses on photovoltaic (PV) panels as a recent development of renewable decentralised energy sources, applied by private owners in the Netherlands, and the network operator Alliander. The contribution of this study is to create a database - which currently not exists in the Netherlands - to provide important information (e.g. locations of PV panels) in case of a calamity. This database can be used to validate and improve the current registration of installed decentralised PV panels in the Smart Grid, which is not up-to-date and not usable for an application in Crisis Management. Building this database covers a new methodology in locating and combining the necessary input data from multiple sources. Furthermore, the applications that may arise from the data in this database may evoke new perspectives and interest in the accurate knowledge about PV panel locations in the Netherlands.

In 2013 the distribution of produced solar energy was 0.4% in the Netherlands (CBS, 2015). Through the recent boom in PV industries and its global deployment the risk of PV panels (e.g. fire safety, correct instalment and electric circuits) will increase (Verhees, et al., 2013). These risks will be important for safety regions, crisis managers, network operators and energy suppliers. The only way to support such cases is to know exactly (10 cm accuracy) where PV panels are installed.

The following paper includes eight sections. Section 2 presents related work. Section 3 describes the methodology of finding the required data. Section 4 provides an elaboration on our

---

* Corresponding author

newly defined database serving as replenishment for the current national voluntary registration. Section 5 provides a statistical analysis on the case study. Section 6 includes the application of our database in Crisis Management. Section 7 discusses the conclusions of this study and future work.

2. RELATED WORK

GIS has a longstanding underpinning in energy applications. For instance, in estimating the energy demand of households (Kaden and Kolbe, 2014; Strzalka et al., 2011), estimating the energy efficiency of buildings (Carrión et al., 2010), and in the consequent planning of their energy-efficient retrofit (Previtali et al., 2014). In solar energy applications, there are several approaches to estimate and map the PV energy potential (Biljecki et al., 2015; Catita et al., 2014; Hofierka and Zlocha, 2012). Such information is available through different atlases (Zonatlas, 2015; Zonnekaart, 2015). However, these studies are intended to aid in planning new energy installations, rather than to identify and/or to assess existing ones.

There is no reliable information available about the current number, physical location, and capacity of installed PV panels in the Netherlands. The same holds true for the other energy sources of the Smart Grid, like: wind turbines, biogas installations, and combined heat and power installations.

An innovative and complete way to connect, share and exchange data is via the Linked Data principle. Linked Data uses the Semantic web – where web content is incorporated into semantics via URI and RDF – to generate new links between data that has not been linked before.

Since the course of a crisis is often unpredictable, data requirements and the constellation of involved stakeholders might suddenly change. This might lead to an overload of the central crisis management system as it has to integrate constantly changing datasets (Zlatanova et al., 2014).

According to Ramona Roller (2015) Linked Data is supposed to improve Crisis Management because if all stakeholders base the data sharing on the same conceptual model (i.e. using the same terms for concepts and their relationships) then fewer interoperability problems will occur;
- Additional datasets can easily be added by generating new triples ;
- The data is available to all stakeholders ;
- Implicit relationships between data concepts can be made visible and in this way dependencies will become apparent.

According to Battle & Kolas (2011) Linked Data can aid in combining spatial data with other datasets, providing more content to users, helping solve cross knowledge domain issues and helping in aiding location based services. This method combines datasets more easily, and it is an innovative way for sharing information between different stakeholders and experts. In the future, when Link Data will become more common, this way of sharing information will be promising.

3. DATA RETRIEVAL FOR DATABASE CONTENT

In this section the methodology will be introduced of finding the required data regarding the location, maximum capacity at any time (kiloWattpiek (kWp)), and amount of power (kiloWatthour (kWh)) produced by installed decentralised PV

panels in the Netherlands. In general, three data sources were used, namely the:

1. Production Installation Register (PIR)
2. Website of *klimaatmonitor* on which solar energy data is published (Existence9)
3. Aerial imagery

These three data sources were used to analyse the reliability of the existing data sources (PIR and *klimaatmonitor*) versus the reliability of retrieving this data via aerial imagery.

3.1 Production Installation Register (PIR)

The current registration of installed decentralised PV panels is set in the Production Installation Register of Energy Data Services Nederland (EDSN-PIR). The PIR is a national registry where consumers can register (voluntarily) their sustainable decentralised production installation. The registration and cancellation of production installations by the customer (or the installation company on his behalf) can be done via a number of ways, such as via the Internet (Energieleveren, 2015). Approximately 28.000 installations are currently registered in the PIR. With the PIR network operators want to get an overview of the development of PV panels, wind turbines, combined heat and power, and other renewable installations in the Netherlands.

The following information will be registered of the production installations:
(1) address-data (street, house number, postcode, living area);
(2) connecting-data (EAN code connection, net manager);
(3) installation-data (building year, type of installation, power, date in use).

It was decided to keep the registration of the production installations for customers easily accessible. Thereon the number of input and controls is minimised. The PIR is online since about the year 2011. Therefore many old, mostly smaller installations will never be included in the PIR, because they could not be registered, since the PIR was non-existent.

The PIR data is not publicly available. Only network operators and energy suppliers have full access to the PIR database. Furthermore, the consumer registration of the PIR is voluntarily and not obliged by any law. Hence, the PIR can be regarded as an incomplete database, because many active installations might not be registered in the PIR.

3.2 Solar energy data on the website of *klimaatmonitor* (Existence9)

A part of the (anonymised) PIR data can be found in databases on the web (klimaatmonitor, 2015). This data comprises on one hand sections of PIR data, on the other hand data of the PIR supplemented with nine other sources (e.g. Novem-projects). The fact that the *klimaatbank* is using other sources to rectify the amount of PV panels of the PIR emphasize that it is known that the PIR is not up-to-date and reliable.

3.3 Aerial imagery

The aerial images were obtained via the company Aerodata International Surveys. We have investigated whether aerial imagery can be used to detect PV panels by their specific physical appearance. Images with an accuracy of 10 cm can be processed by a detection algorithm, based on Template Matching, which outputs a geometry point in the centroid of a PV panel. Therefore, specific PV panel locations with XY coordinates can be made available.

The main principle of Template Matching is that a template of pixels – e.g. a square of 3 by 3 pixels, also called kernel, filter, operator or convolution matrix – with certain intensity values, 'moves' over the entire original aerial image while multiplying the pixel intensity values of the template, with the pixel values of the original image. This multiplying, 'moving' template results in a new raster file, with new intensity values (Figure 1). In general, the pixel in this new raster that appears to fit the initial template values the best– depending on which kind of template is used – will portray maximum intensity values. One can state that on this specific location with maximum values, a certain feature is detected, namely the feature that represent the moving template.

This detection methodology is applied on the case study in this paper and is developed by Karto (2015). In this algorithm, a digital image is seen as raster data: each cell of the raster (pixel) represents simply a value (RGB light intensity). Furthermore, the algorithm is based on a combination of aerial imagery and multiple other data sources. The detection is done via object recognition techniques and further enhanced using multiple templates. Note: several sources of error can be explicated regarding the working principle of such an algorithm. A part of the detected markers of PV panels in our test case (district Stevenshof in the municipality Leiden, in the Netherlands) were obtained via the detection algorithm developed by Karto.

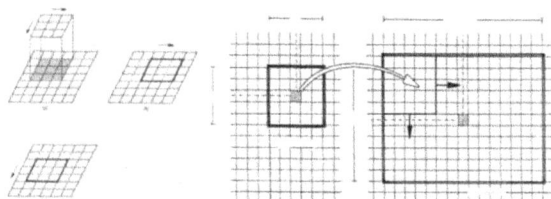

Figure 1: Principle of a moving image kernel (Lemmens, 2011)

4. DATABASE FOR INFORMATION PROVISION

This section provides an elaboration on our newly defined database, called Peer+. Before making a database (to store the data) it is necessary to create an UML diagram, to see the interrelations between all tables of the database. When the UML diagram is completed it is possible to convert this diagram to a SQL script for creating a database. First it is necessary to create a DDL schema. This DDL schema can be converted into a SQL script. For this research six tables were made (Figure 2):

Table 1: Peer+
Table 2: BAGIntermediate
Table 3: Aerial
Table 4: PIR
Table 5: NeighborhoodBoundaries
Table 6: Existence9

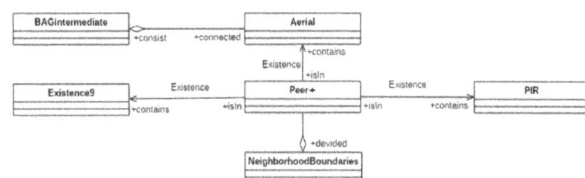

Figure 2: Database with six classes

These six tables form the database for information provision, where the location is stored of the decentralised solar panels. This Peer+ table contains combined information of the data sources (PIR, *klimaatbank* and aerial imagery), the neighbourhood boundaries (for clustering the location of PV panels to neighbourhood-, city-, province-, or national level) and BAGintermediate table for assigning the location of the PV panels – obtained by aerial imagery - to a neighbourhood.

In the Peer+ table, all information of PV panels in our test case is combined. Within this table the total amount of PV panels, households, capacity and power is shown. This data is retrieved automatically from the tables BAGIntermediate, Aerial, PIR, NeighborhoodBoundaries and Existence9 via a SQL script in pgAdmin (PostgreSQL/PostGIS). The Peer+ table also shows statistical information regarding the difference of what is not in the PIR and Existence9 tables, and what is stated in the Aerial table: e.g. the PIR table, the current national registration of PV panels can be compared with the Peer+ table to show the reliability of the PIR. This Peer+ table can be made publicly accessible and gives an up-to-date overview of all decentralised installed PV panels per neighbourhood in our test case. The Peer+ table will be published as open data at neighbourhood level, because it is privacy sensitive to publish the coordinates of every PV panel.

When the Peer+ table is filled with data from the other tables, it can be used to query this data. For example the slopes and azimuth degrees with the amount of PV panels can be queried. By having a central, reliable Peer+ database, where the accurate location of PV panels is stored in coordinates, more linking possibilities regarding other applications and stakeholders can be made. The Peer+ makes the PV panel registration data more usable.

5. STATISTICAL ANALYSIS OF THE ACQUIRED DATA

A statistical analysis is performed on the case study to compare the reliability of the existing data sources (PIR and *klimaatbank*) versus data retrieved with aerial imagery. In order to make it possible to compare the PIR table with the Existence9 table (consisting of a part of the (anonymised) PIR data) and the Aerial table, units have to be transformed. The given data from the PIR and Existence9 consist of 'Known addresses with PV panels', while the data obtained via aerial imagery consist of the 'Amount of PV panels'. A conversion of 'Amount of households' into 'Amount of PV panels' and vice versa had to be performed.

From this analysis can be concluded that one of the four neighbourhoods in our test case (Kloosterhof, Dobbewijk-North, Dobbewijk-South and Schenkwijk) had the most installed PV panels (in 2014), namely Dobbewijk-South. That is why most of the power (kWh) and the highest capacity at any time (kWp) are generated in the neighbourhood Dobbewijk-South. By using the image processing method more households with PV panels are detected in comparison to the PIR (17 more addresses were found in our test case (on a total of 74)) and *klimaatbank* (2 more addresses were found in our test case (on a total of 74)). The image processing method did not detect 2 addresses from the *klimaatbank*, but did detect all the addresses of the PIR. This shows that creating a database with the location of decentralised PV panels with aerial imagery is more accurate than the current existing sources.

By using aerial imagery of different years, the total amount of PV panels can be compared and an increase or decrease can be detected. In our test case in total 207 PV panels were detected in 2012, 422 PV panels in 2013, and 984 PV panels in 2014.

6. APPLICATION IN CRISIS MANAGEMENT

The Peer+ database is built for certain applications. A lot of parties could be interested in the exact location of PV panels, for example market- and taxation researchers. Market researchers are interested in the amount of PV panels on the roofs. The Peer+ database can help resellers and solar panel distributors to discover the solar potential of buildings. For insurance companies it would be an asset to know where PV panels are installed with the production amount, because the presence of installed PV panels may increase the risk of a fire hazard. This research focuses on one specific field of applications in crisis management: offensive indoor efforts of the fire brigade.

According to Backstrom & Dini (2011) firefighters are vulnerable to electrical and casualty hazards when mitigating a fire encompassing PV systems. In general, the fire brigade eliminates the energy power before entering the building, assuming that there are no PV panels present. PV panels deliver direct current (DC) voltage which is of a larger risk than alternating current (AC) voltage. DC voltage distributes voltage throughout the entire wiring of a building, which may become a threat. Also, when a calamity or fire has to be tackled during the evening or night time, the Fire Brigade uses bright spotlights which may influence the energy production of a PV panel installation. Even if the energy supply in a building is shut down by fire fighters, the presence of PV panels may still distribute voltage. To conclude, buildings equipped with solar power systems may introduce unfamiliar hazards that require new firefighting strategies and procedures (Grant, 2013). However, there is not much knowledge and data available to understand the risks for developing safety solutions in order to respond in a safe manner when dealing with installed PV panels on a building in case of a calamity. For more risks of the presence of PV panels during mitigating a fire one is referred to Backstrom & Dini (2011).

Fire Services the Netherlands (*Dutch: Brandweer Nederland*) has set some regulations in the form of an attention-infosheet (*Dutch: aandachtskaart*) used by firefighters in repressive situations dealing with electricity. Basically PV panels are regarded as a low-voltage system which does not obstruct the execution of their duties, albeit the regular safety measures should be taken into account. However PV panels are a risk as such, as they might obstruct access when a firefighter has to walk on the roof of the (neighbouring) premise for mitigating a fire, because they may become hot, and they may drop down from the roof. In short, the risk for firefighters caused by installed PV panels on a building is present, but the recognition of this matter is still an open issue.

7. CONCLUSIONS AND FUTURE WORK

It can be concluded that the main problem in having a reliable and up-to-date database is the cooperation between different stakeholders. A lot of data regarding PV panel registration is available, but this data is either not shared with other parties, or it is not recorded properly (for exchange and linking). In this manner, it is evident to combine multiple existing sources of data with new/other resources to get reliable data. This research

has shown that it is more accurate to retrieve the location of decentralised PV panels with aerial imagery. A notion to be aware of is the possible privacy violation, when publishing the found PV panel registration data of the concerned test area as open data, on premises level.

One of the insights which were gained during this study regarding PV panels in the Netherlands is that it is very important to set up an up-to-date and reliable database of PV panels with help of aerial imagery. During the survey in our test case it was noticed that there is a need for an algorithm (or an enhancement of the detection algorithm of Karto) which can detect for instance thin film PV panels. These Thin film PV panels are currently an error source in the detection algorithm and are hard to detect through aerial imagery. The algorithm of Karto that was adopted for this project focuses solely on standard PV panels with a frame.

By having a central, reliable Peer+ database - where the accurate location of PV panels retrieved by aerial imagery is stored in coordinates - more linking possibilities regarding other applications and stakeholders can be made. The Peer+ enables data usage of the registration of PV panels and makes various new statistical analyses possible, e.g. power estimation by PV panels aggregated on neighbourhood level. Future applications based on the Peer+ database can provide useful and up-to-date insights regarding Crisis Management.

ACKNOWLEDGEMENTS

This work is part of the CERISE Topsector Switch2SmartGrids research program (www.cerise-project.nl), which is a program that has the aim to create future proof and efficient information exchange between the energy sector, eGovernment and geo-world. The authors would like to thank in particular: Sven Briels (Karto), Leen van Doorn (Alliander), Peter Segaar (PolderPV), and Robert Voûte (CGI) for their support.

REFERENCES

Backstrom, R., & Dini, D. A., 2011. Firefighter Safety and Photovoltaic Installations Research Project, Underwriters Laboratories Inc.

Battle, R., & Kolas, D., 2011. Geosparql: enabling a geospatial semantic web. *Semantic Web Journal, 3*(4), 355-370.

Biljecki, F., Heuvelink, G. B. M., Ledoux, H., & Stoter, J., 2015. Propagation of positional error in 3D GIS to the estimation of the solar irradiation of building roofs. *International Journal of Geographical Information Science, in submission.*

Carrión, D., Lorenz, A., & Kolbe, T. H., 2010. Estimation of the energetic rehabilitation state of buildings for the city of Berlin using a 3D city model represented in CityGML. ISPRS Archives of the Photogrammetry, Remote Sensing and Spatial Information Sciences. Proceedings of the ISPRS 5th 3D GeoInfo Conference, Berlin, Germany, Vol. XXXVIII-4/W15, 31-15.

Catita, C., Redweik, P., Pereira, J., & Brito, M. C., 2014. Extending solar potential analysis in buildings to vertical facades. *Computers and Geosciences, 66*, 1–12.

Cerise, 2015. www.cerise-project.nl, (accessed on June 29th, 2015).

CBS, 2015. Aandeel hernieuwbare energie stabiel. http://www.duurzaambedrijfsleven.nl/energie/2977/cbs-aandeel-hernieuwbare-energie-stabiel, (accessed on June 29th, 2015)

Energieleveren, 2015. www.energieleveren.nl (accessed on June 29th, 2015)

Grant, Casey C., 2013. Fire Fighter Safety and Emergency Response for Solar Power Systems. In The Fire Protection Research Foundation.Quincy, USA.

Hofierka, J., & Zlocha, M., 2012. A New 3-D Solar Radiation Model for 3-D City Models. *Transactions in GIS*, *16*(5), 681–690.

Kaden, R., & Kolbe, T. H., 2014. Simulation-Based Total Energy Demand Estimation of Buildings using Semantic 3D City Models. *International Journal of 3-D Information Modeling*, *3*(2), 35–53.

Karto, 2015. www.burokarto.nl   (accessed on June 29th, 2015)

Klimaatmonitor,   2015. www.klimaatmonitor.databank.nl/report/bestanden.html, (accessed on June 29th, 2015).

Lemmens, M., 2011. Geo-information. *Geotechnologies and the Environment*. Delft, the Netherlands

PolderPV, www.polderpv.nl (accessed on June 29th, 2015)

Previtali, M., Barazzetti, L., Brumana, R., Cuca, B., Oreni, D., Roncoroni, F., & Scaioni, M., 2014. Automatic façade modelling using point cloud data for energy-efficient retrofitting. *Applied Geomatics*, *6*(2), 95–113.

Roller, R, Roes, J, Verbree, E, 2015. Benefits of Linked Data for Interoperability during Crisis Management. Submitted to Gi4DM workshop at ISPRS GEOSPATIAL WEEK 2015.

Strzalka, A., Bogdahn, J., Coors, V., & Eicker, U., 2011. 3D City modeling for urban scale heating energy demand forecasting. *HVAC&R Research*, *17*(4), 526–539.

Verhees, B., Raven, R., Veraart, F., Smith, A., & Kern, F. 2013. The development of solar PV in The Netherlands: A case of survival in unfriendly contexts. Renewable and Sustainable Energy Reviews, 19, 275-289.

Zlatanova, S. 2010. Formal modelling of processes and tasks to support use and search of geo-information in emergency response. *13th Annual International Conference and Exhibition on geospatial Information Technology and Applications,* Gurgaon, India, pp. 1-10.

Zonatlas, 2015, www.zonatlas.nl, (accessed on June 29th, 2015).

Zonnekaart, 2015, www.zonnekaart.nl, (accessed on June 29th, 2015).

35

LOCATION-BASED INFRASTRUCTURE INSPECTION FOR SABO FACILITIES

M. Nakagawa [a,*], T. Yamamoto [a], S. Tanaka [a],
Y. Noda [b], K. Hashimoto [b], M. Ito [b], M. Miyo [b]

[a] Dept. of Civil Engineering, Shibaura Institute of Technology, Tokyo, Japan - mnaka@shibaura-it.ac.jp
[b] Watanabe Engineering Co., Ltd., Fukushima, Japan - mim@watanabe-office.jp

Commission IV / WG 7

KEY WORDS: Infrastructure asset management, Mobile Device, Field-based inspection, Construction Information Modeling

ABSTRACT:

Infrastructure asset management is a framework for achieving sustainable infrastructure. Based on this framework, although we often generate a three-dimensional (3D) geometrical model as a base map in management, it is not easy to acquire details of asset attributes in 3D measurement. Therefore, we focus on field-based investigation and inspection using mobile devices, and aim at assisting investigators in infrastructure asset monitoring with location-based applications. In this paper, we propose and evaluate our location-based investigation application as follows. First, we propose an inspection flow suitable for field-based monitoring. Second, we develop a Web GIS application for field-based investigation with mobile devices. Third, we propose base map generation suitable for sabo facilities using UAV and terrestrial laser scanner. We conduct an experiment in a sediment-retarding basin consisting of dikes, bridges, and debris barriers, and explore some issues in infrastructure asset monitoring using mobile devices.

1. INTRODUCTION

Infrastructure asset management is a framework for achieving sustainable infrastructure, such as roads, bridges, railways, and water treatment facilities. A conventional flow for ground-based infrastructure inspection is shown in Figure 1.

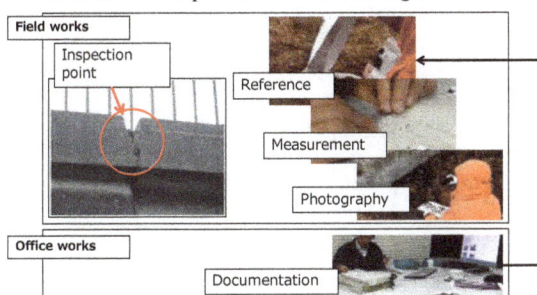

Figure 1. Ground-based infrastructure inspection

During infrastructure inspection, we generally refer to the latest inspection documents to determine an inspected position, as follows. First, the structure to be inspected is detected after the inspector's arrival in the inspection area. Next, an inspected point is detected in the structure. Then, the condition of the inspected point is recorded and compared with the latest inspection. After that, a geo-tagged photo is captured at the inspected point.

Generally, the management focuses on the low life-cycle cost in a process of construction, maintenance, rehabilitation, and replacement. Based on this framework, a 3D geometric model is often generated based on construction information modeling (CIM). Moreover, asset attributes, such as deterioration, condition, and age are acquired. To check the position of structures and structural elements and collect data related to these structures in frequent monitoring, there is a need to refer to maps, engineering drawings, databases, and technical documents (Garrett et al. 2002). Reliability, completeness, efficiency, and cost are significant indices in monitoring.

Reliability, completeness, and efficiency can be satisfied using terrestrial LiDAR, a vehicle-borne mobile mapping system (MMS), and aerial photogrammetry using an unmanned aerial vehicle (UAV), as shown in Figure 2.

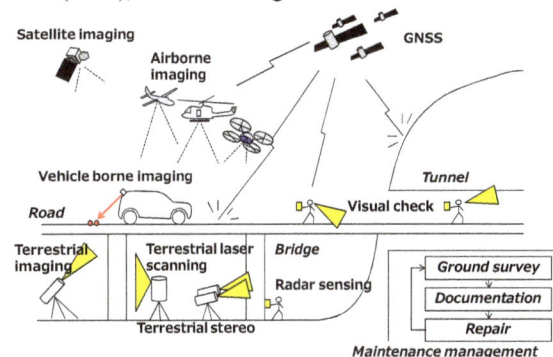

Figure 2. Infrastructure inspection

In the current state, although 3D scanners can acquire high resolution data, it is not easy to acquire details of asset attributes with 3D measurements. Therefore, we focus on ground investigation and inspection using mobile devices (Kamada et al. 2013). Field-based inspection requires some location-based applications, such as geo-tagged image acquisition (photography), database interface, and navigation (Hammad et al. 2006). Mobile devices, such as tablet PCs, smart phones, and global positioning system (GPS) cameras, have the potential to assist inspectors in infrastructure asset monitoring because of their built-in sensors and components that include cameras, assisted GPS receivers, gyro sensors, Wi-Fi, microphones, speakers, vibrators, and large storage. We aimed to assist investigators in infrastructure asset monitoring with location-based applications. The control of erosion and sediment is called Sabo. The Sabo is one of significant topics in infrastructure inspection. In this paper, we propose and evaluate our location-based investigation application for Sabo facility management.

2. METHODOLOGY

Our proposed methodology for Location-based infrastructure inspection is described in Figure 3. Our methodology consists of inspection operations with mobile devices and 3D mapping with images and point cloud to improve conventional inspection approaches.

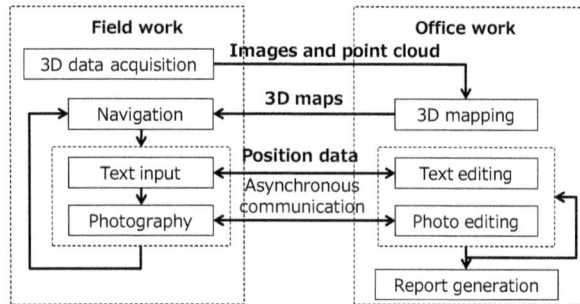

Figure 3. Proposed methodology

2.1 Mobile inspection application

The functions and performance of infrastructure inspection assistance with a mobile device, such as a tablet PC equipped with GPS, are summarized in Table 1. Category A indicates essential functions and category B indicates additional functions.

Table 1. Functions and performance of infrastructure inspection assistance with a mobile device

Category A	·Display of maps, drawings, images, movies, and technical information ·Input of characters, lines, and shapes ·Adding a postscript to technical documents
Category B	·Documentation compatible with various template sheets ·Display of various types of maps and drawings (tiff, shp, sxf, dwg, etc.) ·Navigation in facility area ·Measurement (distance and area, etc.) ·Change detection ·User intuitive operability

In addition, we propose a data model for our Web GIS-based mobile inspection application to satisfy the above-mentioned functions, as shown in Figure 4.

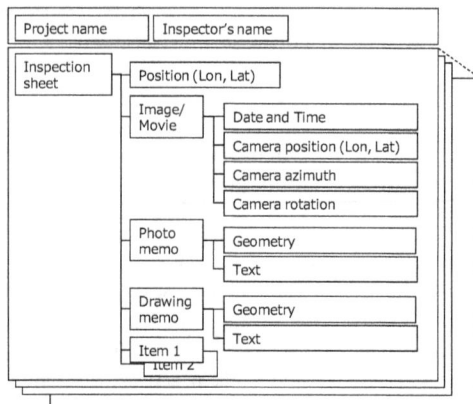

Figure 4. Data model for our Web GIS-based mobile inspection application

An inspection work is subdivided into several activities, such as geotagged image acquisition, adding a postscript to a photo, and adding a postscript to an engineering drawing. Geotagged data

generated from these works are managed with Extensible Markup Language to automate file export using an inspection template prepared by municipalities and a combination of managed data, such as maps, images, and movies, using position data as a retrieval key in inspection navigation. Acquired GPS data are mainly used for the management of location and time data. The location data included represent the position of structures, camera position data, and camera azimuth and rotation data.

2.2 Location data management

The required positioning accuracy is dynamically changed by each inspection work. For example, a closed photograph requires the same position (with approximately 1 cm accuracy) and direction (with approximately 1 degree accuracy) in the latest inspection to achieve automation of image registration for detection of any change in an infrastructure inspection (Nakagawa, Katuki, Isomatu and Kamada, 2013). On the other hand, inspection point detection requires lower positioning accuracy, from approximately 10 cm to 1 m. Moreover, in structure detection, positioning accuracy is allowed to be approximately 10 m. In addition, 100 m positioning accuracy is sufficient for an inspector's arrival in an inspection area. Thus, a definition with several steps or spatial resolutions is effective in location data management. In this research, these steps are represented as levels of details (LODs), such as LOD1: address, LOD2: structure, LOD3: inspection point, and LOD4: photography, as shown in Table 2.

Table 2. LODs in infrastructure inspection

Levels of details	Content	Required accuracy
LOD1 **Address**	Inspector's arrival in an inspection area	100m
LOD2 **Structure**	Structure detection	10m
LOD3 **Inspection**	Inspection point detection	10 cm – 1 m
LOD4 **Photo management**	Documentation - Photography - Drawing	- 1 cm - 1 degree

2.3 Base map generation

In the CIM, base maps and 3D data are required to manage processes of construction, maintenance, rehabilitation, and replacement. Online maps, such as Google Maps and OpenStreetMap, are useful for infrastructure inspection in urban areas. However, in rural areas and mountainous districts, the online maps are often insufficient for infrastructure inspection to recognize the details of natural features. Thus, base maps and 3D data should be prepared before the inspection.

In an open-sky environment, aerial photogrammetry and Structure from Motion (SfM) using UAV is more effective than ground-based scanning. On the other hand, when environments include natural obstacles, such as trees, terrestrial LiDAR is more effective than UAV and MMS.

In our research, we apply both approaches to prepare digital surface models (DSM) and digital elevation models (DEM) as base maps. Particularly, we expect that LiDAR measurement can generate colored DEM from colored point cloud acquired with a laser scanner, as shown in Figure 5. However, when we acquire point cloud around riversides which include many water surfaces, many missing points will exist because of laser reflection problems. Thus, we apply a randomized algorithm for quickly finding an approximate nearest-neighbor matches between image patches (Barnes et al. 2009), as shown in Figure 6.

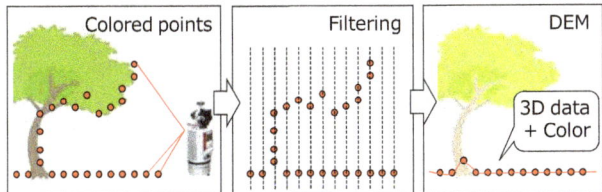

Figure 5. DEM generation from terrestrial LiDAR data

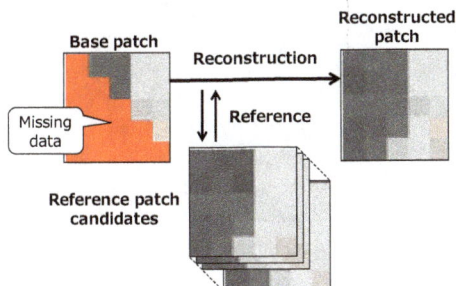

Figure 6. PatchMatch processing for ortho image generation

3. EXPERIMENT

We conducted experiments involving the daily and annual Sabo infrastructure inspection work in a sediment-retarding basin consisting of dikes, bridges, and debris barriers in Fukushima, Japan (see Figure 7).

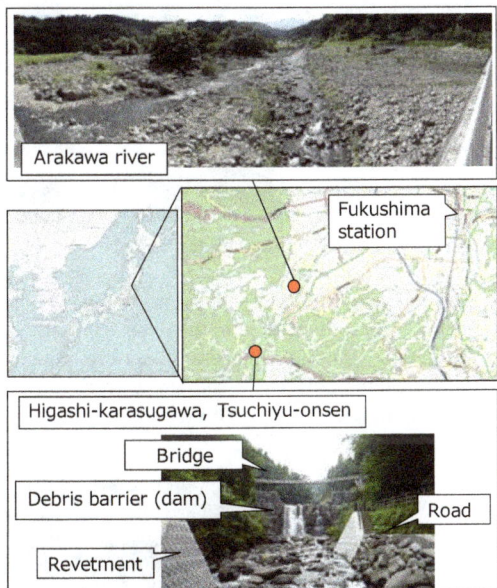

Figure 7. Study area

3.1 Attribute data acquisition

In attribute data acquisition, we record conditions of infrastructures, such as cracks, damages and displacements, based on checklists distributed by Japanese Ministry of Land, Infrastructure, Transport and Tourism (MLIT). We assigned these checklists to meta-data and main data, as shown in Figure 8. Then, we input text data and images to record the conditions

of infrastructures with some mobile devices, as shown in Figure 9.

Figure 8. Checklists in structure inspection based on MLIT's guidelines

Figure 9. Mobile devices (tablet PCs and smart phone)

In addition, omni-directional images are also acquired to record attribute data of the conditions of infrastructures. These images are used to improve the integrity in infrastructure inspection with augmented reality applications in office works.

We used two types of cameras, such as THETA m15 (RICOH) and QBiC PANORAMA (Elmo). These cameras were mounted on a monopod, as shown in Figure 10 and Figure 11. We also used a GPS logger (N-241, HOLUX) to get position data with omni-directional images.

Acquired omni-directional images were stitched to be panoramic images and movies. These images and movies are viewed with a head-mount display (Oculus Rift), as shown in Figure 12.

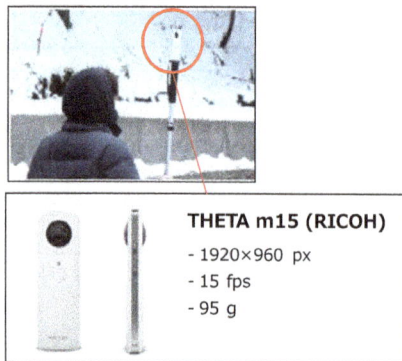

Figure 10. Panoramic video camera (THETA)

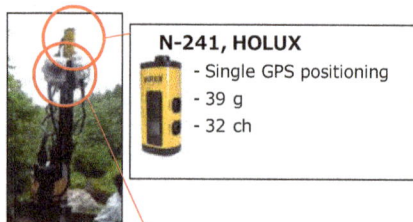

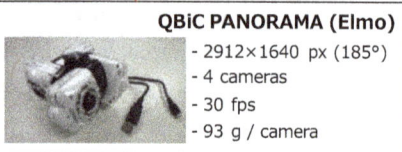

Figure 11. Panoramic video camera (QBiC)

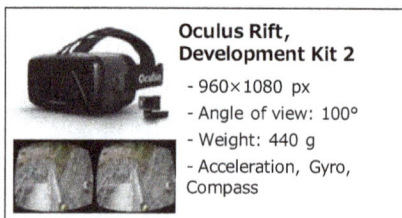

Figure 12. Panoramic video viewer

3.2 Base map generation

We used md4-1000 (Microdrones) and PEN EP-1 (Olympus) (Figure 13) to acquire aerial 1000 images from 150 m height with 90 % overlaps and 60 % side-laps for SfM processing.

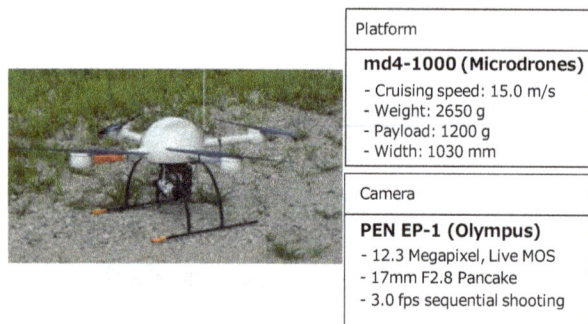

Figure 13. UAV (md4-1000 and PEN EP-1)

Moreover, we used RIEGL VZ-400 (Figure 14) in our terrestrial laser scanning. We acquired 114 million colored points (Figure 15) with from 7 points.

Figure 14. Terrestrial LiDAR (RIEGL VZ-400)

Figure 15. Point cloud acquired with terrestrial LiDAR

4. RESULTS

4.1 Attribute data acquisition

In our experiment, 213 images were acquired with mobile devices. Using geo-tag data, these images are reverse-geocoded into a map with GPS position data, as shown in Figure 16.

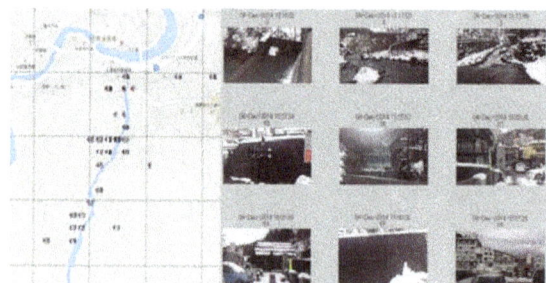

Figure 16. Geotagged images

Then, acquired images are grouped into 36 viewpoints. In a manual work, it took 3120 sec. On the other hand, it took 4 sec in our position and azimuth filtering. Therefore, we confirmed that our application drastically shorten a work time for the image retrieval, as shown in Figure 17.

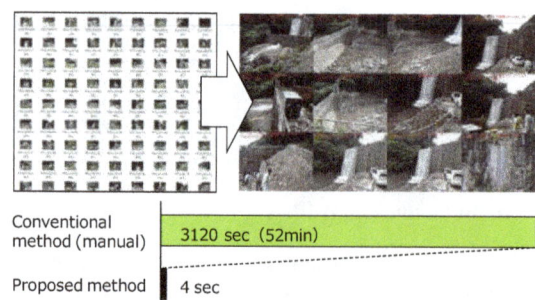

Figure 17. Position and azimuth filtering result

4.2 3D map generation

We generated a colored DSM with a ground resolution of 4 cm from the UAV. As shown in Figure 18, we have confirmed that ground surfaces, such as roads, gravels, trees, and water surfaces, were reconstructed well.

Figure 18. DSM generated with images from UAV

Moreover, we generated a colored DEM with a ground resolution of 5 cm from point cloud acquired with the terrestrial laser scanner. As shown in Figure 19, several missing areas exist in the DEM due to laser reflection problems. Thus, we applied the PatchMatch to the colored DEM from the laser scanning data. We have confirmed that missing areas were reconstructed as well as an aerial image, as shown in Figure 20.

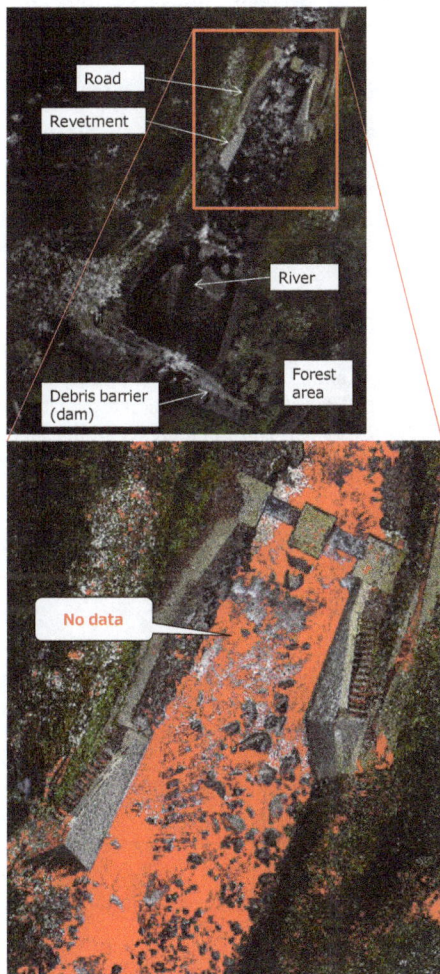

Figure 19. DEM generated point cloud

Figure 20. DEM after the PatchMatch processing

5. DISCUSSION

Positioning from LOD1 to LOD3 requires from 100 to 1 m accuracy. Thus, single GPS positioning is suitable for position data acquisition. However, LOD4 requires 1 cm accuracy with precise positioning, such as a real-time kinematic GPS (RTK-GPS). Generally, low-cost inspection restricts the use of expensive devices such as an RTK-GPS. In low-cost inspection, the performance of satellite positioning is generally improved by assisted-GPS, differential GPS and multi-GNSS positioning using GPS, GLONASS, and QZSS. Data fusion of GPS and dead reckoning also improves the performance of positioning. However, although these approaches improve availability, they have almost no effect on positioning accuracy improvement (Inaba et al. 2013). In this research, satellite positioning was assumed to have 1m accuracy, even if we could apply improvement approaches to positioning accuracy. Thus, a location data management approach using movies was applied in LOD4 (precise positioning). This approach assists inspectors to determine a position in a photography using a movie that was captured in the latest inspection and the attached approximate position data acquired with GPS.

5.1 Integrity of location data

In this research, we focused on a mobile application for infrastructure inspection, and location data management for inspection navigation. In fact, the availability of an application using location data depends on the integrity of the location data (Yabuki, 2013), and the integrity of the location data strongly affects location data browsing, and inspection recording and log browsing. Here, to improve the integrity, we focused on a combination of acquired data, such as location, photo and movie data. We classified three combinations among these data types, as follows.

a) Position (latitude and longitude), azimuth, and elevation angle

Several minutes were required for detection of the previous inspection position. The most suitable procedure for the detection of the previous inspection position in our experiment was as follows. First, we referred to longitude and latitude to detect the position. Second, we referred to the azimuth angle using magnetic sensor data taken from a tablet PC. Third, we referred to an elevation angle using gyro sensor data taken from the tablet PC.

b) Position (latitude and longitude), azimuth, elevation angle, and photograph

We determined the position for inspection with position, azimuth, and elevation angle data. Moreover, we could reconfirm the position with a photo taken in the previous inspection. However, when a difference existed between the position determined from position, azimuth, and elevation angle data and the position estimated from a geotagged photo, inspectors were unable to determine the true position for the inspection.

c) Position (latitude and longitude) and movie

We did not conduct an experiment related to video capture, which provides the shortest capture time and efficiency for navigation. However, we confirmed that an effective approach for inspection navigation was to capture a movie that shifted from a position that was far from an inspected point to one that was close. Moreover, we confirmed that a movie was a better approach than a picture for change detection in infrastructure inspection when an inspection point was in a complex environment, such as craggy places and Sabo sites.

5.2 Performance of operation using tablet PCs

We qualitatively confirmed that automation of location and time data recording is more reliable than manual paper-based recording in infrastructure inspection. On the other hand, paper-based recording offers an advantage for documentation in an outdoor location, because text input with a mobile PC is time-consuming work. Moreover, we confirmed that raindrops worsen the performance of the touch interface, even when a waterproof tablet PC is used.

Position data acquisition depends on single GPS positioning. Although our study area consisted of open-sky environments and structures, GPS positioning was insufficient for positioning in LOD3 (inspected position detection) in an area surrounded by mountains or under a bridge. On the other hand, we have confirmed that geotagged movie was effective in estimating the LOD3 and LOD4 position data. Even if position data included a positioning error caused by low dilution of precision and multipath transmission, an inspection position could be detected using movie guidance. Moreover, we could also focus on geotagged omni-directional camera data to detect an inspected position.

However, although we used Google Maps and OpenStreetMap as the base maps in our preliminary experiments, there was a difficulty in managing the frequent map updates. Therefore, we prepared DSM generated using images from UAV and DEM generated from a laser scanner as a more reliable base map. Although we prepared high resolution data, we used them as ortho images because it is not easy to view 3D data with tablet PC.

In addition, we confirmed that inspection work using a tablet PC held with both hands was dangerous on bad roads, in riverbeds, and in craggy places. Therefore, we propose to use hands-free applications using wearable devices and voice-guided applications with geofencing techniques to improve safety in inspections using a mobile device.

6. SUMMARY

In this paper, we focused on ground investigation and inspection using mobile devices. We aimed to assist investigators in infrastructure asset monitoring with location-based applications. We proposed and evaluated our location-based investigation application for facility management based on CIM. Through our experiment, we explored several issues in infrastructure asset monitoring using mobile devices. Integrity in positioning should be improved to achieve more reliable and effective inspection works. Therefore, we proposed an LOD definition for positioning data management in inspection works. Moreover, we proposed combinations of base maps and several types of data acquired with a mobile device in inspection works to improve reliability, completeness, and integrity in positioning.

ACKNOWLEDGEMENT

This work was supported by JSPS KAKENHI Grant Number 26870580. Moreover, our experiments are supported by Fukushima City and Fukushima River and National Highway Office, Tohoku Regional Development Bureau, Ministry of Land Infrastructure and Tourism.

REFERENCES

Garrett, J. H. Jr., Sunkpho, J., 2002, An Overview of the research in Mobile/Wearable Computer-Aided Engineering Systems in the Advanced Infrastructure, *VDI BERICHTE 1668*, pp.5-20.

Kamada, T., Katsuki, F., Nakagawa, M., 2013, The GPS Camera Application for the Efficiency Improvement of the Bridge Inspection, *The 13th East Asia-Pacific Conference on Structural Engineering and Construction*, 6 pp.

Hammad, A., Zhang, C., Hu, Y., Mozaffari, E., 2006, Mobile Model-Based Bridge Lifecycle Management System, *Computer-Aided Civil and Infrastructure Engineering*, Volume 21, Issue 7, pp.530-547.

Nakagawa, M., Katuki, F., Isomatu, Y., Kamada, T., 2013, Close-range stereo registration for concrete crack monitoring, *EASEC13 (The 13th East Asia-Pacific Conference on Structural Engineering and Construction)*, 8 pp, E-2-4.

Barnes, C., Shechtman, E., Finkelstein, A., Dan B Goldman, D. B., 2009, PatchMatch: A Randomized Correspondence Algorithm for Structural Image Editing, *ACM Transactions on Graphics (Proc. SIGGRAPH) 28(3)*.

Inaba, H., Nakagawa, M., 2013, Integrity Improvement In Localization : Guarantee Added Localization Methodology Using Spatio-temporal Contexts, *The 34th Asian Conference on Remote Sensing 2013*, pp.59-63.

Yabuki, N., 2013, Development and Applications of the Outdoor Augmented Reality with an Accurate Registration Technique in Construction Projects, *Proceedings of the 6th Asian Civil Engineering Conference (ACEC) and the 6th Asian Environmental Engineering Conference (AEEC)*, 13.pp.

Permissions

The contributors of this book come from diverse backgrounds, making this book a truly international effort. This book will bring forth new frontiers with its revolutionizing research information and detailed analysisof the nascent developments around the world.

We would like to thank all the contributing authors for lending their expertise to make the book truly unique. They have played a crucial role in the development of this book. Without their invaluable contributions this book wouldn't have been possible. They have made vital efforts to compile up to date information on the varied aspects of this subject to make this book a valuable addition to the collection of many professionals and students.

This book was conceptualized with the vision of imparting up-to-date information and advanced data inthis field. To ensure the same, a matchless editorial board was set up. Every individual on the board wentthrough rigorous rounds of assessment to prove their worth. After which they invested a large part of theirtime researching and compiling the most relevant data for our readers.

The editorial board has been involved in producing this book since its inception. They have spent rigoroushours researching and exploring the diverse topics which have resulted in the successful publishing of this book. They have passed on their knowledge of decades through this book. To expedite this challenging task, the publisher supported the team at every step. A small team of assistant editors was also appointed to further simplify the editing procedure and attain best results for the readers.

Apart from the editorial board, the designing team has also invested a significant amount of their time inunderstanding the subject and creating the most relevant covers. They scrutinized every image to scout for the most suitable representation of the subject and create an appropriate cover for the book.

The publishing team has been an ardent support to the editorial, designing and production team. Theirendless efforts to recruit the best for this project, has resulted in the accomplishment of this book. They are a veteran in the field of academics and their pool of knowledge is as vast as their experience in printing. Their expertise and guidance has proved useful at every step. Their uncompromising quality standards have made this book an exceptional effort. Their encouragement from time to time has been an inspiration for everyone.

The publisher and the editorial board hope that this book will prove to be a valuable piece of knowledge for researchers, students, practitioners and scholars across the globe.

List of Contributors

M. Gerke, F. Nex, P. Jende
University of Twente, Faculty of Geo-Information Science and Earth Observation (ITC), Department of Earth Observation Science, The Netherlands

J. A. Gonçalves
Faculdade Ciências - Universidade Porto, Rua Campo Alegre 4169-007 Porto, Portugal

A. Berveglieri
Univ Estadual Paulista – UNESP, Faculty of Science and Technology, Presidente Prudente, Brazil
Graduate Program in Cartographic Sciences

A. M. G. Tommaselli
Univ Estadual Paulista – UNESP, Faculty of Science and Technology, Presidente Prudente, Brazi, Department of Cartography

P. Molinaa, M. Bl´azquez, J. Sastrea, I. Colomina
GeoNumerics S.L., Parc Mediterrani de la Tecnologia, 08860 Castelldefels (Spain)

S. Baltrusch
LAiV M-V, NMCA M-V, Department for Photogrammetry, 19059 Schwerin, Germany

A. M. Manzino and C. Taglioretti
DIATI Department, Politecnico di Torino, Corso Duca degli Abruzzi 24, 10129 Torino, Italy

M.Meijer, L.A.E. Vullings, J.D. Bulens, F.I. Rip, M. Boss, G. Hazeu and M.Storm
Alterra, Wageningen University and Research Centre, Wageningen, The Netherlands

W. Ostrowski and K. Bakuła
Department of Photogrammetry, Remote Sensing and Spatial Information Systems, Faculty of Geodesy and Cartography, Warsaw University of Technology, Poland

Jean-Franc¸ois Mas and Rafael Gonz´alez
Centro de Investigaciones en Geografia Ambiental (CIGA) Universidad Nacional Aut´onoma de M´exico (UNAM) Antigua Carretera a Patzcuaro No. 8701 Col. Ex-Hacienda de San Jos´e de La Huerta C.P. 58190 Morelia Michoacan MEXICO

P. Schaer
GEOSAT SA, Route du Manège 59b, 1950 Sion, Switzerland

J. Vallet
HELIMAP SYSTEM SA, Le Grand-Chemin 73, 1066 Epalinges, Switzerland

P. Jende, Z. Hussnain, M. Peter, S. Oude Elberink, M. Gerke and G. Vosselman
University of Twente, Faculty of Geo-Information Science and Earth Observation (ITC), Department of Earth Observation Science, The Netherlands

J.F. Masa
Centro de Investigaciones en Geografía Ambiental, Universidad Nacional Autónoma de México, 58190 Morelia, Mexico

B. Soares-Filho and H. Rodrigues
Centro de Sensoriamento Remoto, Universidade Federal de Minas Gerais, Belo Horizonte 31270-900, MG, Brasil

M. Daakir
Vinci-Construction-Terrassement, 1, Rue du docteur Charcot, 91421 Morangis, France
Université Paris-Est, IGN, SRIG, LOEMI, 73 avenue de Paris, 94160 Saint-Mande, France

Y. Rabot, , F. Pichard,
Vinci-Construction-Terrassement, 1, Rue du docteur Charcot, 91421 Morangis, France

M. Pierrot-Deseilligny
Université Paris-Est, IGN, ENSG, LOEMI , 6-8 Avenue Blaise Pascal, 77455 Champs-sur-Marne, France
Université Paris-Est, IGN, SRIG, LOEMI, 73 avenue de Paris, 94160 Saint-Mande, France

P. Bosser
ENSTA Bretagne-OSM Team, 2 rue Francois Verny, 29806 Brest, France

C. Thom
Université Paris-Est, IGN, SRIG, LOEMI, 73 avenue de Paris, 94160 Saint-Mande, France

K. Jacobsen
Leibniz University Hannover, Institute of Photogrammetry and Geoinformation, Germany

M. Gerke
Faculty of Geo-Information Science and Earth Observation of the University of Twente, Netherlands

M. Khaghania and J. Skaloud
EPFL, Geodetic Engineering Laboratory TOPO, Route Cantonale,1015 Lausanne, Switzerland

S. Kerner, I. Kaufman and Y. Raizman
VisionMap, 19D Habarzel, Tel Aviv, Israel

A. Comber
School of Geography, University of Leeds, Leeds, LS2 9JT, UK

P. Mooney
Department of Computer Science, National University of Ireland Maynooth, Ireland

R.S. Purves
Department of Geography, University of Zurich, 8057 Zurich, Switzerland

D. Rocchini
Fondazione Edmund Mach, 38010 S. Michele all'Adige

A. Walz
Potsdam Institute for Climate Impact Research, 14412, Potsdam, Germany

A. Ch. Braun
Institute of Regional Science, Karlsruhe Institute of Technology (KIT) Reinhard-Baumeister-Platz 1, 76131 Karlsruhe, Germany

M. Weinmann, S. Keller and S. Hinz
Institute of Photogrammetry and Remote Sensing, Karlsruhe Institute of Technology (KIT) Englerstr. 7, 76131 Karlsruhe, Germany

R. Müller and P. Reinartz
Remote Sensing Technology Institute (IMF), German Aerospace Center (DLR) 82234 Wessling, Germany

J. F. Lichtenauera
Laan der Vrijheid 92, 2661HM Bergschenhoek, The Netherlands

B. Sirmacek
Department of Geoscience and Remote Sensing, Delft University of Technology, Stevinweg 1, 2628CN Delft, The Netherlands

Yan Wang and Xiangyun Hu
School of Remote Sensing and Information Engineering, Wuhan University, Luoyu Road 129, Wuhan 430079, China

Yuanyuan Wang
Helmholtz Young Investigators Group "SiPEO", Technische Universität München, Arcisstraße 21, 80333 Munich, Germany

Xiao Xiang Zhu
Helmholtz Young Investigators Group "SiPEO", Technische Universität München, Arcisstraße 21, 80333 Munich, Germany
Remote Sensing Technology Institute (IMF), German Aerospace Center (DLR), Oberpfaffenhofen, 82234 Weßling, Germany

G. Waldhoff, S. Eichfuss and G. Bareth
Institute of Geography, University of Cologne, Albertus-Magnus-Platz, 50923 Cologne, Germany

L.Y. Qiu and J.Y. Gu
State Key Laboratory of Information Engineering in Surveying, Mapping and Remote Sensing, Wuhan University, 430079 Wuhan, China

Q. Zhu
State-province Joint Engineering Laboratory of Spatial Information Technology for High-Speed Railway Safety, Southwest Jiaotong University, 610000, Chengdu, China
Faculty of Geosciences and Environmental Engineering, Southwest Jiaotong University, 610000, Chengdu, China

Z.Q. Du
State Key Laboratory of Information Engineering in Surveying, Mapping and Remote Sensing, Wuhan University, 430079 Wuhan, China
Collaborative Innovation Center of Geospatial Technology, 430079 Wuhan, China

M. Bassani, N. Grasso and M. Piras
Dept. of Environment, Land and Infrastructure Engineering, Politecnico di Torino, 24 corso Duca degli Abruzzi, Turin, 10024 Italy

H. Sheikhian
MSc. Student, GIS Dept., School of Surveying and Geospatial Eng., College of Eng., University of Tehran, Tehran

M.R. Delavar
Center of Excellence in Geomatic Eng. in Disaster Management, School of Surveying and Geospatial Eng., College of Eng., University of Tehran, Tehran, Iran

A. Stein
Department of Earth Observation Science, University of Twente, The Netherlands

C. Heipke, A. Schmidt and F. Rottensteiner
Institute of Photogrammetry and GeoInformation, Leibniz Universität Hannover, Germany

U. Soergel
Institute of Geodesy, Chair of Remote Sensing and Image Analysis, Technische Universität Darmstadt, Germany

A. Flamenco Sandoval and J.F. Masa
a Centro de Investigaciones en Geograf´ıa Ambiental, Universidad Nacional Aut´onoma de México, 58190 Morelia, Mexico

A. Pérez Vega
Universidad de Guanajuato, 4500 Guanajuato, Mexico

A. Andablo Reyes
Centro de Investigación en Alimentación y Desarrollo, 83304 Hermosillo, Mexico

M.A. Castillo Santiago
El Colegio de la Frontera Sur, San Cristobal de las Casas, Mexico

G. Lucas,
Research Institute of Remote Sensing and Rural Development, Károly Róbert College Gyöngyös Doctoral School of Military Engineering, National University of Public Service, Budapest, Hungary

S. Lénárt
Research Institute of Remote Sensing and Rural Development

J. Solymosi
National University of Public Service, Budapest, Hungary

C.H. Yang and U. Soergel
Institute of Geodesy, Technische Universität Darmstadt, Germany

N. Regnauld
Spatial, Tennyson House, Cambridge Business Park, Cambridge CB4 0WZ, UK

A. Hanel, L. Hoegner and U. Stilla
Photogrammetry & Remote Sensing, Technische Universitaet Muenchen, Germany

H. Klöden
BMW Research & Technology, Muenchen, Germany

P. Arias, M. Soilán, L. Díaz-Vilariño and J. Martínez-Sánchez
Applied Geotechnologies Group, Dept. Natural Resources and Environmental Engineering, University of Vigo, Campus Lagoas-Marcosende, CP 36310 Vigo, Spain

B. Riveiro
Applied Geotechnologies Group, Dept. Materials Engineering, Applied Mechanics and Construction, University of Vigo, Campus Lagoas-Marcosende, CP 36310 Vigo, Spain

L. A. Gueye and G. Erin
ECOWAS Commission 101 Yakubu Gowon Crescent, Asokoro District, Abuja, Nigeria.

M. S. Keita
Regional Centre for Training in Aerospace Surveys (RECTAS), OAU Campus, Ile Ife, Osun State, Nigeria

J. O. Akinyede
Department of Remote Sensing and Geosciences Information System, Federal University of Technology, Akure, Ondo State, Nigeria

O. Kufoniyi
Department of Geography, Obafemi Awolowo University, Ile-Ife, Osun State, Nigeria

Rosann Aarsen, Milo Janssen and Myron Ramkisoen
Delft University of Technology, Faculty of Architecture and the Built Environment, MSc Geomatics Julianalaan 134, 2628 BL, Delft, The Netherlands

Filip Biljecki
Delft University of Technology, Faculty of Architecture and the Built Environment, Department of Urbanism, 3D Geoinformation Julianalaan 134, 2628 BL, Delft, The Netherlands

Wilko Quak and Edward Verbree
Delft University of Technology, Faculty of Architecture and the Built Environment, OTB Research Institute for the Built Environment, GIS Technology Julianalaan 134, 2628 BL, Delft, The Netherlands

M. Nakagawa, T. Yamamoto and S. Tanaka
Dept. of Civil Engineering, Shibaura Institute of Technology, Tokyo, Japan

Y. Noda, K. Hashimoto, M. Ito and M. Miyo
Watanabe Engineering Co., Ltd., Fukushima, Japan

Index

A

Accuracy, 1, 3, 6, 11-14, 16-20, 22, 26-29, 35, 37-40, 42-44, 47-53, 55-59, 64, 69, 73-75, 78, 80-81, 84-85, 87-88, 94, 96-98, 100, 102, 106-108, 116, 123-124, 126-127, 131, 136, 140, 142, 147, 165, 168-169, 177, 180, 184-185, 188, 190, 200-201, 206, 208-209

Aerial Photography, 9, 87-88, 90

Agisoft Photoscan, 9-10, 42-45

Airborne Perspective, 1

Airvision Model Toolbox, 22

Amazon Deforestation, 65

Archived Aerial Phots, 10

Asift-approaches, 1

Autonomous Navigation, 81, 84

B

Basic Geodata, 25, 27, 122, 124

Brief, 3, 8, 57-58, 64, 87

C

Calibration, 3-4, 9-12, 14-16, 18-28, 30, 35-36, 43-44, 65-66, 69-80, 85, 103-107

Calibration Certificates, 9-10

Car Trajectory, 29

Communication, 20, 37, 41, 182, 198, 206

Computer Vision, 3, 8, 10, 12, 25, 35-36, 47, 64, 74, 87-88, 90, 101, 108, 117, 121, 141, 154-155, 185, 192

Consumer, 19, 26, 37-38, 41-42, 201

Corridor Mapping, 19-23

D

Data, 1-3, 7-9, 11-16, 18, 28-30, 34-35, 37-38, 40-43, 46, 48, 51, 53, 56-61, 63-64, 68-76, 78, 81, 84-87, 90-97, 99-101, 103-104, 108-115, 119, 121-140, 142, 144, 147-154, 156, 158, 160-161, 163-164, 166, 172-198, 200-210

Digital Photogrammetry, 10, 25, 64

E

Estimation, 1, 3-4, 8, 14, 35-36, 57, 60, 62, 70-71, 73-74, 85-86, 88, 90, 92, 103, 106-108, 114-116, 121, 128, 131, 135-136, 139, 142, 168-169, 173-174, 179, 184, 186, 203-204

Ets, 37, 40-41

F

Features from Accelerated Segment Test, 3

Feature Extraction, 57-59, 63, 87, 89, 118, 155, 182

Feature Matching, 57-58, 60-61, 64, 87, 89, 103

Filtering Techniques, 29

Fitness for Use, 37-38, 40-41

Fragmentation, 65, 67-68

G

Genetic Algorithm, 65-67

Geodata Post-processing, 19

Geometric Calibration, 25

Geometry, 6-8, 15-16, 18-20, 25-26, 42-47, 71-72, 75-76, 78, 90, 108-109, 115, 124, 138, 141, 160-161, 163-165, 187, 202, 206

Geonumerics' Generic Network, 22

Gnss Aerial Control, 19, 21-22

Gnss Outage, 55-56, 81, 84-86

Gps, 1, 7, 10, 13-14, 16-17, 20, 29-30, 34-35, 53, 69-74, 86, 88, 137, 160, 180-181, 184, 195, 205-210

Gps Denied Environment, 53

Gps Location, 7

I

Image Orientation, 10-14, 17-19, 26, 42-43, 57

Image Segmentation, 48-49, 51, 113

Inertial Navigation Systems (INS), 57

Integrated Sensor Orientation, 19

Integration, 1, 9, 29, 34-35, 55, 57, 81, 84-86, 113, 122, 124-125, 127-128, 130, 133, 137-138, 140, 148, 160, 193-194, 197-198

Interior Orientation (IO), 20, 22

Isprs, 1-2, 7-8, 12, 18, 23, 35-36, 47, 56, 64, 74-76, 80, 95, 100, 113, 121, 126, 147-148, 155, 173, 186, 192, 200, 203-204

K

Kaze, 1-6, 8, 58-64

Kinematic Ground Control, 19-20, 24

Kinematic Ground Control Point (KGCP), 20

L

Land Cover Database, 48

Lidar, 19, 53, 59, 74, 86, 104, 108-109, 111-113, 115-116, 121, 135, 149, 153-155, 187, 190, 192, 205-208

Long Tunnel Survey, 53

Lpis, 37-41

M

Mapkite, 19-24

Mls, 53, 56-58, 60-64, 187-188

Mm Data Products, 57

Mobile Laser Scanning (MLS), 57

Mobile Mapping, 18-19, 29, 35-36, 53, 57-59, 62-64, 135, 137, 140, 205

Motion Models, 29-30, 34-35

Multisensor, 1, 86, 126

N

National Mapping Agency (NMCA), 25

National Mapping Institutes, 9

Ndvi, 25, 28, 97, 101

O

Oblique Airborne Imagery, 42

Oblique Images, 42-47, 57

Odometry, 29, 35

Open Data, 37, 128, 133, 202-203

Orb Technique, 1

Orientation, 1-4, 9-14, 17-24, 26, 35, 42-47, 57-58, 63-64, 69-71, 74-76, 78-80, 104-105, 149, 160-165, 179-180, 183, 187-188

Oriented Fast, 3

P

Penta Camera, 75-76

Photogrammetry, 1-3, 8, 10, 13, 18, 22-23, 25, 35-36, 42-43, 46-47, 56, 64, 69-70, 74-75, 80, 87, 95-96, 101, 104, 108, 113, 121, 126, 137, 140-141, 148-149, 155, 166, 186, 192, 203, 205-206

Pix4d, 2, 7-8, 10, 42-46

Polynomials, 9, 112

Producer, 37-41, 175, 184

R

Radiometry, 14, 25-26

Ransac, 2-3, 6, 61-62, 87-90

S

Scale Invariant Feature Transform

Sift, 3

Sensor Integration, 29, 85

Solar Elevation Angle, 1

Spatial Data Quality, 37-38, 41, 148, 175

Static Ground Control Point (SGCP), 20

Stochastic Spatial Simulation, 65

Surf, 1-8, 58, 64, 90

T

Terrestrial Mobile Mapping Systems (TMMS), 19

Tie-points, 87-90, 104

Trimble Uas Master, 10

U

Ukf, 29, 34-36

Unmanned Aerial Systems (UAS), 19

Unmanned Aerial Vehicles (UAV), 1

Unmanned Aircraft (UA), 19

Unscented Kalman Filter (UKF), 29

Urban Environments, 1, 64, 103, 147, 187, 192

Urban Terrestrial Mobile Mapping (TMM), 29

V

Validation, 25-28, 37, 41, 99, 142, 146, 175-177, 184

Vehicle Dynamic Model, 81, 86

Visual Interpretation, 48-49, 51

Von Gruber Positions, 87

W

Wallis-filtering, 1, 7

www.ingramcontent.com/pod-product-compliance
Lightning Source LLC
Chambersburg PA
CBHW080630200326
41458CB00013B/4570